Machine Learning

Zhi-Hua Zhou

# Machine Learning

Springer

**Zhi-Hua Zhou**
Nanjing University
Nanjing, Jiangsu, China

*Translated by*
**Shaowu Liu**
University of Technology Sydney
Ultimo, NSW, Australia

ISBN 978-981-15-1969-7 ISBN 978-981-15-1967-3 (eBook)
https://doi.org/10.1007/978-981-15-1967-3

Translation from the language edition: *Machine Learning* by Zhi-Hua Zhou, and Shaowu Liu, © Tsinghua University Press 2016. Published by Tsinghua University Press. All Rights Reserved.


This Springer imprint is published by the registered company Springer Nature Singapore Pte Ltd.
The registered company address is: 152 Beach Road, #21-01/04 Gateway East, Singapore 189721, Singapore

# Preface

This is an introductory-level machine learning textbook. To make the content accessible to a wider readership, the author has tried to reduce the use of mathematics. However, to gain a decent understanding of machine learning, basic knowledge of probability, statistics, algebra, optimization, and logic seems unavoidable. Therefore, this book is more appropriate for advanced undergraduate or graduate students in science and engineering, as well as practitioners and researchers with equivalent background knowledge.

The book has 16 chapters that can be roughly divided into three parts. The first part includes Chapters 1–3, which introduces the basics of machine learning. The second part includes Chapters 4–10, which presents some classic and popular machine learning methods. The third part includes Chapters 11–16, which covers advanced topics. As a textbook, Chapters 1–9 and 10 can be taught in one semester at the undergraduate level, while the whole book could be used for the graduate level.

This introductory textbook aims to cover the core topics of machine learning in one semester, and hence is unable to provide detailed discussions on many important frontier research works. The author believes that, for readers new to this field, it is more important to have a broad view than drill down into the very details. Hence, in-depth discussions are left to advanced courses. However, readers who wish to explore the topics of interest are encouraged to follow the further reading section at the end of each chapter.

The book was originally published in Chinese and had a wide readership in the Chinese community. The author would like to thank Dr. Shaowu Liu for his great effort of translating the book into English and thank Springer for the publication.

**Zhi-Hua Zhou**
Nanjing, China

# Contents

# Symbols

| | |
|---|---|
| $x$ | Scalar |
| $\boldsymbol{x}$ | Vector |
| $\mathbf{x}$ | Variable set |
| $\mathbf{A}$ | Matrix |
| $\mathbf{I}$ | Identity matrix |
| $\mathcal{X}$ | Sample space or state space |
| $\mathcal{D}$ | Probability distribution |
| $D$ | Data set |
| $\mathcal{H}$ | Hypothesis space |
| $H$ | Hypothesis set |
| $\mathfrak{L}$ | Learning algorithm |
| $(\cdot, \cdot, \cdot)$ | Row vector |
| $(\cdot; \cdot; \cdot)$ | Column vector |
| $(\cdot)^{\mathrm{T}}$ | Transpose of vector or matrix |
| $\{\cdots\}$ | Set |
| $\lvert\{\cdots\}\rvert$ | Number of elements in set $\{\cdots\}$ |
| $\lVert \cdot \rVert_p$ | $\mathrm{L}_p$ norm; $\mathrm{L}_2$ norm when $p$ is absent |
| $P(\cdot), P(\cdot \mid \cdot)$ | Probability mass function, conditional probability mass function |
| $p(\cdot), p(\cdot \mid \cdot)$ | Probability density function, conditional probability density function |
| $\mathbb{E}_{\cdot \sim \mathcal{D}}[f(\cdot)]$ | Expectation of function $f(\cdot)$ with respect to $\cdot$ over distribution $\mathcal{D}$; $\mathcal{D}$ and/or $\cdot$ are omitted when context is clear |
| $\sup(\cdot)$ | Supremum |
| $\mathbb{I}(\cdot)$ | Indicator function, returns 1 if $\cdot$ is true, and 0 if $\cdot$ is false |
| $\operatorname{sign}(\cdot)$ | Sign function, returns $-1$ if $\cdot < 0$, 0 if $\cdot = 0$, and 1 if $\cdot > 0$ |

# Introduction

Table of Contents




Z.-H. Zhou, *Machine Learning*,
https://doi.org/10.1007/978-981-15-1967-3_1

## 1.1 Introduction

Following a drizzling, we take a walk on the wet street. Feeling the gentle breeze and seeing the sunset glow, we bet the weather must be nice tomorrow. Walking to a fruit stand, we pick up a green watermelon with curly root and muffled sound; while hoping the watermelon is ripe, we also expect some good academic marks this semester after all the hard work on studies. We wish readers to share the same confidence in their studies, but to begin with, let us take an informal discussion on what is *machine learning*.

Taking a closer look at the scenario described above, we notice that it involves many experience-based predictions. For example, why would we expect beautiful weather tomorrow after observing the gentle breeze and sunset glow? We expect this beautiful weather because, from our experience, the weather on the following day is often beautiful when we experience such a scene in the present day. Also, why do we pick the watermelon with green color, curly root, and muffled sound? It is because we have eaten and enjoyed many watermelons, and those satisfying the above criteria are usually ripe. Similarly, our learning experience tells us that hard work leads to good academic marks. We are confident in our predictions because we learned from experience and made experience-based decisions.

While humans learn from experience, can computers do the same? The answer is "yes", and machine learning is what we need. Machine learning is the technique that improves system performance by learning from experience via computational methods. In computer systems, experience exists in the form of data, and the main task of machine learning is to develop *learning algorithms* that build *models* from data. By feeding the learning algorithm with experience data, we obtain a model that can make predictions (e.g., the watermelon is ripe) on new observations (e.g., an uncut watermelon). If we consider computer science as the subject of algorithms, then machine learning is the subject of *learning algorithms*.

Mitchell (1997) provides a more formal definition: "A computer program is said to learn from experience $E$ for some class of tasks $T$ and performance measure $P$, if its performance at tasks in $T$, as measured by $P$, improves with experience $E$."

In this book, we use "model" as a general term for the outcome learned from data. In some other literature, the term "model" may refer to the global outcome (e.g., a decision tree), while the term "pattern" refers to the local outcome (e.g., a single rule).

E.g., Hand et al. (2001).

## 1.2 Terminology

To conduct machine learning, we must have data first. Suppose we have collected a set of watermelon records, for example, (color = dark; root = curly; sound = muffled), (color =

green; root = curly; sound = dull), (color = light; root = straight; sound = crisp), . . . , where each pair of parentheses encloses one record and "=" means "takes value".

Collectively, the records form a *data set*, where each record contains the description of an event or object, e.g., a watermelon. A record, also called an *instance* or a *sample*, describes some attributes of the event or object, e.g., the color, root, and sound of a watermelon. These descriptions are often called *attributes* or *features*, and their values, such as green and dark, are called *attribute values*. The space spanned by attributes is called an *attribute space*, *sample space*, or *input space*. For example, if we consider color, root, and sound as three axes, then they span a three-dimensional space describing watermelons, and we can position every watermelon in this space. Since every point in the space corresponds to a position vector, an instance is also called a *feature vector*.

The entire data set may also be seen as a "sample" sampled from the sample space, and therefore depending on the context, according to which a "sample" can refer to either an individual data instance or a data set.

More generally, let $D = \{\boldsymbol{x}_1, \boldsymbol{x}_2, \ldots, \boldsymbol{x}_m\}$ be a data set containing $m$ instances, where each instance is described by $d$ attributes. For example, we use three attributes to describe watermelons. Each instance $\boldsymbol{x}_i = (x_{i1}; x_{i2}; \ldots; x_{id}) \in \mathcal{X}$ is a vector in the $d$-dimensional sample space $\mathcal{X}$, where $d$ is called the dimensionality of the instance $\boldsymbol{x}_i$, and $x_{ij}$ is the value of the $j$th attribute of the instance $\boldsymbol{x}_i$. For example, at the beginning of this section, the second attribute of the third watermelon takes the value straight.

The process of using machine learning algorithms to build models from data is called *learning* or *training*. The data used in the training phase is called *training data*, in which each sample is a *training example*, and the set of all training examples is called a *training set*. Since a learned model corresponds to the underlying rules about the data, it is also called a *hypothesis*, and the actual underlying rules are called the *facts* or *ground-truth*. Then, the objective of machine learning is to find or approximate ground-truth. In this book, models are sometimes called *learners*, which are machine learning algorithms instantiated with data and parameters.

A training example is also called a *training instance*.

Nevertheless, the samples in our watermelon example are not sufficient for learning a model that can determine the ripeness of uncut watermelons. In order to train an effective prediction model, the *outcome* information must also be available, e.g., ripe in ((color = green; root = curly; sound = muffled), ripe). The outcome of a sample, such as ripe or unripe, is often called a *label*, and a sample with a label is called an *example*. More generally, we can write the $i$th sample as $(\boldsymbol{x}_i, y_i)$, where $y_i \in \mathcal{Y}$ is the label of the sample $\boldsymbol{x}_i$, and $\mathcal{Y}$ is the set of all labels, also called the *label space* or *output space*.

Learning algorithms often have parameters, and different parameter settings and training data lead to different learning outcomes.

If we consider the label as part of the data sample, then *example* and *sample* can be used interchangeably.

When the prediction output is discrete, such as ripe and unripe, it is called a *classification* problem; when the prediction output is continuous, such as the degree of ripeness, it is called a *regression* problem. If the prediction output has only two possible classes, then it is called a *binary classification* problem, where one class is marked as *positive* and the other is marked as *negative*. When more than two classes are present, it becomes a *multiclass classification* problem. More generally, a prediction problem is to establish a mapping $f : \mathcal{X} \mapsto \mathcal{Y}$ from the input space $\mathcal{X}$ to the output space $\mathcal{Y}$ by learning from a training set $\{(\boldsymbol{x}_1, y_1), (\boldsymbol{x}_2, y_2), \ldots, (\boldsymbol{x}_m, y_m)\}$. Conventionally, we let $\mathcal{Y} = \{-1, +1\}$ or $\{0, 1\}$ for binary classification problems, $|\mathcal{Y}| > 2$ for multiclass classification problems, and $\mathcal{Y} = \mathbb{R}$ for regression problems, where $\mathbb{R}$ is the set of real numbers.

The process of making predictions with a learned model is called *testing*, and the samples to be predicted are the *testing samples*. For example, the label $y$ of a testing sample $\boldsymbol{x}$ can be obtained via the learned model $y = f(\boldsymbol{x})$.

Also known as *testing instances*.

Other than predictions, another type of learning is *clustering*. For example, we can group watermelons into several *clusters*, where each cluster contains the watermelons that share some underlying concepts, such as `light color` versus `dark color`, or even `locally grown` versus `imported`. Clustering often provides data insights that form the basis of further analysis. However, we should note that the concepts, such as `light color` or `locally grown`, are unknown before clustering, and the samples are usually unlabeled.

Otherwise, the labels directly give the clusters; exceptions are discussed in Sect. 13.6.

Depending on whether the training data is labeled or not, we can roughly divide learning problems into two classes: *supervised learning* (e.g., classification and regression) and *unsupervised learning* (e.g., clustering).

It is worth mentioning that the objective of machine learning is to learn models that can work well on the new samples, rather than the training examples. The same objective also applies to unsupervised learning (e.g., clustering) since we wish the learned clusters work well on the samples outside of the training set. The ability to work on the new samples is called the *generalization* ability, and a well-generalized model should work well on the whole sample space. Although the training set is usually a tiny proportion of the sample space, we still hope the training set can, to some extent, reflect the characteristics of the whole sample space; otherwise, it would be hard for the learned model to work well on the new samples. We generally assume that all samples in a sample space follow a distribution $\mathcal{D}$, and all samples are independently sampled from this *distribution*, that is, *independent and identically distributed* (*i.i.d.*). Generally speaking, the more samples we have, the more information we know about the distribution $\mathcal{D}$, and consequently, the better-generalized model we can learn.

More precisely, unseen samples.

In practice, the size of a sample space is often huge. For example, given 20 variables each of which has 10 possible values, then the size of the sample space is already as large as $10^{20}$.

## 1.3 Hypothesis Space

*Induction* and *deduction* are two fundamental tools of scientific reasoning. Induction is the process from specialization to *generalization*, that is, summarizing specific observations to generalized rules. In contrast, deduction is the process from generalization to *specialization*, that is, deriving specific cases from basic principles. For example, in axiomatic systems of mathematics, the process of deriving a theorem from a set of axioms is deduction. By contrast, learning from examples is an inductive process, also known as *inductive learning*.

In a broad sense, inductive learning is almost equivalent to learning from examples. In a narrow sense, inductive learning aims to learn *concepts* from training data, and hence is also called *concept learning* or *concept formation*. The research and applications on concept learning are quite limited because it is usually too hard to learn generalized models with clear semantic meanings, whereas in real-world applications, the learned models are often *black boxes* that are difficult to interpret. Nevertheless, having a brief idea of concept learning is useful for understanding some basic concepts of machine learning.

The most fundamental form of concept learning is *Boolean concept learning*, which encodes target concepts as Boolean values 1 or 0, indicating true or false. Taking the training data in ◘ Table 1.1 as an example, suppose we want to learn the target concept of ripe, assume that the ripeness of a watermelon entirely depends on its color, root, and sound. In other words, whether a watermelon is ripe or not is determined once we know the values of those three variables. Then, the concepts to be learned could be "ripe is watermelon with color $= X$, root $= Y$, and sound $= Z$", or equivalently as the Boolean expression "ripe $\leftrightarrow$ (color $=$?) $\wedge$ (root $=$?) $\wedge$ (sound $=$?)", where the "?" marks are the values to be learned from training data.

More generally, we can use the disjunctive normal form $(A \wedge B) \vee (C \wedge D)$ when describing concepts.

**◘ Tab. 1.1** A toy watermelon data set

| ID | color | root | sound | ripe |
|---|---|---|---|---|
| 1 | green | curly | muffled | true |
| 2 | dark | curly | muffled | true |
| 3 | green | straight | crisp | false |
| 4 | dark | slightly curly | dull | false |

Readers may have recognized that the first row of ◘ Table 1.1 is already in the form of a concept for identifying ripe watermelons, that is, ripe $\leftrightarrow$ (color $=$ green) $\wedge$ (root $=$ curly) $\wedge$ (sound $=$ muffled). Though the concept is valid for this particular watermelon, it does not generalize to other unseen watermelons. If we just memorize all watermelons in the training set, then for sure, we can classify any watermelon that looks identical to those we have seen. However, what if an unseen watermelon does not look like any watermelons that we have seen before? For example, (color = light) $\wedge$ (root = curly) $\wedge$ (sound = muffled).

The approach of "memorizing" training examples is called *rote learning* (Cohen and Feigenbaum 1983). See Sect. 1.5.

We can think of machine learning as search in the hypothesis space for a hypothesis that is consistent with the training set, that is, the one that can correctly classify all watermelons in the training set. The hypothesis space, along with its size, is determined once its form is specified. In our watermelon problem, the hypothesis space is the collection of all hypotheses in the form of (color =?) $\wedge$ (root =?) $\wedge$ (sound =?) for all possible values of "?". For example, color may take the value green, dark, or light. Sometimes the value of color could be arbitrarily assigned, denoted by the wildcard value "$*$", such as ripe $\leftrightarrow$ (color = $*$) $\wedge$ (root = curly) $\wedge$ (sound = muffled), which says that ripe is watermelon with any color, curly root, and muffled sound. Besides, we should also consider some extreme cases. For example, what if there were no ripe watermelon in the universe, and such a concept did not exist at all? In such cases, we use $\varnothing$ to represent the hypothesis. In the watermelon problem, we have three possible values of color, three possible values of root, and three possible values of sound, resulting in a hypothesis space of size $4 \times 4 \times 4 + 1 = 65$. ◘ Figure 1.1 illustrates the hypothesis space of the watermelon problem.

Here, we assume that there is no noisy data in the training examples and do not consider negation $\neg A$, e.g., not green. Note that the $\varnothing$ hypothesis would not be applied as long as the training set contains at least one positive example.

There are many strategies to search in the hypothesis space, such as top-down, general-to-specific, bottom-up, and specific-to-general. After filtering out all hypotheses that are inconsistent with the training examples, we may end up with one or more hypotheses that can correctly classify all training examples.

The search strategy is flexible. For example, we can simultaneously search using the top-down and bottom-up strategies, and only remove hypotheses that are inconsistent with the positive examples.

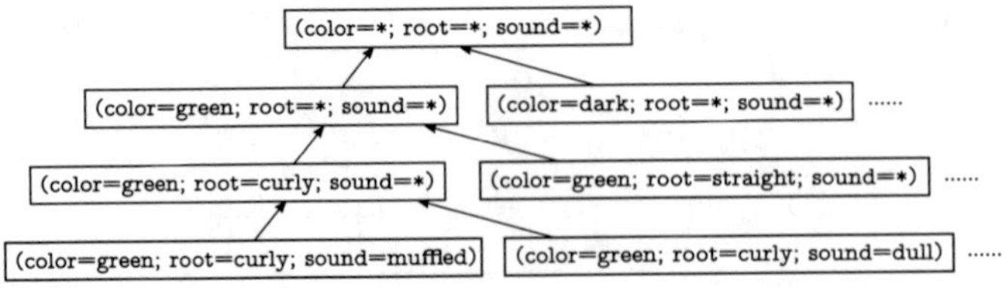


**Fig. 1.1** The hypothesis space of the watermelon problem

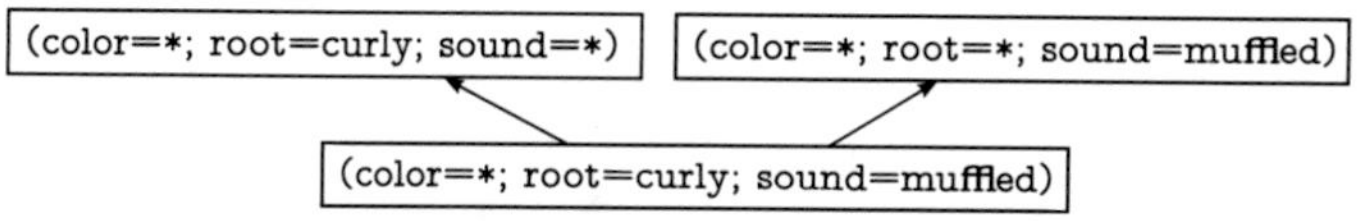


**Fig. 1.2** The version space of the watermelon problem

In practice, however, the hypothesis space is often huge, whereas the training examples are often finite. Consequently, there could be a set of hypotheses that are all consistent with the training examples, and we call such a set of hypotheses a *version space*. For example, ◘ Figure 1.2 shows the version space corresponding to the training examples in ◘ Table 1.1.

## 1.4 Inductive Bias

Since each learned model corresponds to one hypothesis in the version space, a potential problem arises here: different models may predict the new samples differently, though all hypotheses in ◘ Figure 1.2 are consistent with the training examples. For example, the hypothesis $\texttt{ripe} \leftrightarrow (\texttt{color} = *) \wedge (\texttt{root} = \texttt{curly}) \wedge (\texttt{sound} = *)$ classifies the unseen watermelon $(\texttt{color} = \texttt{green}) \wedge (\texttt{root} = \texttt{curly}) \wedge (\texttt{sound} = \texttt{dull})$ as ripe, whereas the other two hypotheses classify it as unripe. In this case, which model (or hypothesis) should we use?

In fact, if the training examples are only those as shown in ◘ Table 1.1, then none of the three hypotheses can be justified better than the others. Nevertheless, a learning algorithm must make a choice and produce a model. In such a situation, the *inductive bias* of the learning algorithm plays a decisive role. For example, if the learning algorithm prefers the model to be "as specific as possible", then it will choose $\texttt{ripe} \leftrightarrow (\texttt{color} = *) \wedge (\texttt{root} = \texttt{curly}) \wedge (\texttt{sound} = \texttt{muffled})$; on the other hand, if it prefers "as general as possible" and trusts root for some reasons, then it will choose $\texttt{ripe} \leftrightarrow (\texttt{color} = *) \wedge (\texttt{root} = \texttt{curly}) \wedge (\texttt{sound} = *)$. The bias of a learning algorithm toward a particular class of hypotheses is called the *inductive bias* or simply *bias*.

"As specific as possible" is to minimize the number of applicable situations; "as general as possible" is to maximize the number of applicable situations.

At first glance, it seems that the choice between root and sound is related to *feature selection*; however, feature selection is based on the analysis of the training set, whereas the inductive bias in our example chooses root based on the domain knowledge. See Chap. 11 for more information about feature selection.

Every effective learning algorithm must have its own inductive bias; otherwise, it will get into trouble when multiple hypotheses look the "same" on the training set, resulting in uncertain learning outcomes. Suppose there is no inductive bias, and the learned model randomly draws a hypothesis that is consistent with the training examples, then sometimes it will

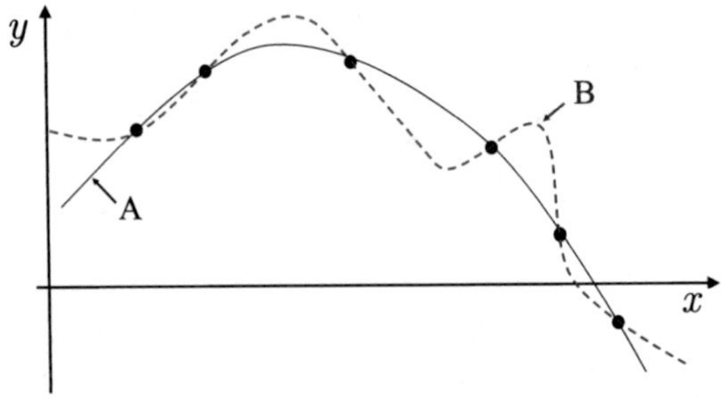


**Fig. 1.3** There are infinite curves that are consistent with the finite training set

classify the unseen watermelon (color = green) ∧ (root = curly) ∧ (sound = dull) as ripe and sometimes unripe. Such kinds of learning results are meaningless.

The regression problem in ◘ Figure 1.3 provides a more intuitive illustration of the role played by the inductive bias. In this example, each training example is shown as a point $(x, y)$, and the objective is to learn a curve passing through all training examples. Since there are infinite qualified curves for the finite training set, a learning algorithm must have its inductive bias to learn the "correct" model. For example, if the learning algorithm believes that similar samples should have similar labels (e.g., watermelons with similar attributes should have similar degrees of ripeness), then it is likely to prefer the smooth curve A over the oscillating curve B.

We can regard inductive bias as the heuristic or value philosophy of learning algorithms for search in potentially huge hypothesis spaces. It is natural to wonder that, does a general principle exist to help learning algorithms obtain the "correct" inductive bias? A fundamental and widely used principle for this question in natural science is the *Occam's razor* principle, which says that we should choose the simplest hypothesis when there is more than one hypothesis consistent with the observations. Assuming that "smoother" is "simpler", then the smooth curve A in ◘ Figure 1.3 is the preferred choice according to the Occam's razor principle. It turns out that the mathematical form of curve A, which is $y = -x^2 + 6x + 1$, is much simpler than that of curve B.

Nevertheless, Occam's razor is not the only available principle, and even if we insist on using it, there are different understandings. Indeed, applying Occam's razor is non-trivial. For example, which of these two hypotheses ripe ↔ (color = ∗) ∧ (root = curly) ∧ (sound = muffled) and ripe ↔ (color = ∗) ∧ (root = curly) ∧ (sound = ∗) is "simpler"? Answering this question is not easy and needs additional domain knowledge.

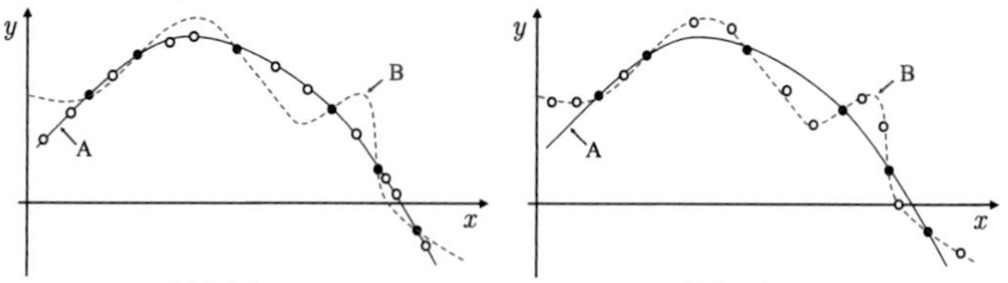


(a) A is better. (b) B is better.

**Fig. 1.4** There is no free lunch (• are training samples; ∘ are testing samples)

In fact, the inductive bias, which helps the algorithm choose a hypothesis, corresponds to a hypothesis made by the learning algorithm itself. That is, what kind of models are better than the others? In practice, whether this hypothesis matches the specific problem or not usually determines the performance of the model.

Let us revisit the example in ◘ Figure 1.3. Suppose that a learning algorithm $\mathfrak{L}_a$ learns the curve A model while another learning algorithm $\mathfrak{L}_b$ learns the curve B model. Since a smoother curve, based on our intuition, is a simpler description, we expect $\mathfrak{L}_a$ to outperform $\mathfrak{L}_b$. As expected, ◘ Figure 1.4a confirms that curve A is doing better than curve B on the testing samples, that is, curve A has better generalization ability.

Wait a moment! Though we hoped and expected $\mathfrak{L}_a$ to outperform $\mathfrak{L}_b$, is there any chance for the curve B model to be better than the curve A model like the case in ◘ Figure 1.4b? Unfortunately, this is possible. In other words, if $\mathfrak{L}_a$ outperforms $\mathfrak{L}_b$ in some situations, then $\mathfrak{L}_b$ will outperform $\mathfrak{L}_a$ in some other situations. Interestingly, this fact applies to any algorithm, even if we let $\mathfrak{L}_a$ be an advanced algorithm and let $\mathfrak{L}_b$ be just random guessing. Are you surprised? Let us have a more in-depth look.

To simplify the discussion, let both the sample space $\mathcal{X}$ and the hypothesis space $\mathcal{H}$ be discrete. Let $P(h \mid X, \mathfrak{L}_a)$ denote the probability of getting the hypothesis $h$ from the algorithm $\mathfrak{L}_a$ based on the training set $X$, and let $f$ be the ground-truth target function that we wish to learn. Then, the *out-of-sample error* of $\mathfrak{L}_a$, that is, the error on all samples except those in the training set, is

We only use basic mathematical knowledge here. However, readers who have "math phobia" can skip this part without impacting the understanding of the rest of this section, but only need to trust the unconvincing conclusion we arrived.

$$E_{\text{ote}}(\mathfrak{L}_a \mid X, f) = \sum_h \sum_{\boldsymbol{x} \in \mathcal{X} - X} P(\boldsymbol{x}) \mathbb{I}(h(\boldsymbol{x}) \neq f(\boldsymbol{x})) P(h \mid X, \mathfrak{L}_a), \tag{1.1}$$

where $\mathbb{I}(\cdot)$ is the indicator function that returns 1 for true and 0 otherwise.

In binary classification problems, the target function could be any function $\mathcal{X} \mapsto \{0, 1\}$ with a function space of $\{0, 1\}^{|\mathcal{X}|}$. Summing the errors of $f$ with respect to uniform distribution gives

If $f$ is uniformly distributed, then half of $f$ will predict $\boldsymbol{x}$ differently from $h(\boldsymbol{x})$.

$$
\begin{aligned}
\sum_f E_{\text{ote}}(\mathfrak{L}_a \mid X, f) &= \sum_f \sum_h \sum_{\boldsymbol{x} \in \mathcal{X}-X} P(\boldsymbol{x}) \mathbb{I}(h(\boldsymbol{x}) \neq f(\boldsymbol{x})) P(h \mid X, \mathfrak{L}_a) \\
&= \sum_{\boldsymbol{x} \in \mathcal{X}-X} P(\boldsymbol{x}) \sum_h P(h \mid X, \mathfrak{L}_a) \sum_f \mathbb{I}(h(\boldsymbol{x}) \neq f(\boldsymbol{x})) \\
&= \sum_{\boldsymbol{x} \in \mathcal{X}-X} P(\boldsymbol{x}) \sum_h P(h \mid X, \mathfrak{L}_a) \frac{1}{2} 2^{|\mathcal{X}|} \\
&= \frac{1}{2} 2^{|\mathcal{X}|} \sum_{\boldsymbol{x} \in \mathcal{X}-X} P(\boldsymbol{x}) \sum_h P(h \mid X, \mathfrak{L}_a) \\
&= 2^{|\mathcal{X}|-1} \sum_{\boldsymbol{x} \in \mathcal{X}-X} P(x) \cdot 1.
\end{aligned} \tag{1.2}
$$

Equation (1.2) reveals an interesting fact: the sum of errors is independent of the learning algorithm. Hence, for any two learning algorithms $\mathfrak{L}_a$ and $\mathfrak{L}_b$, we always have

$$
\sum_f E_{\text{ote}}(\mathfrak{L}_a \mid X, f) = \sum_f E_{\text{ote}}(\mathfrak{L}_b \mid X, f). \tag{1.3}
$$

In other words, no matter how smart $\mathfrak{L}_a$ or how humble $\mathfrak{L}_b$ is, their expected performance is always the same! This conclusion is known as the No Free Lunch (NFL) theorem (Wolpert 1996; Wolpert and Macready 1995).

A rigorous proof of the NFL theorem is much more complicated than our simplified discussion here.

Some readers may feel frustrated by the above conclusion since if the expected performance of all learning algorithms is comparable to random guessing, then why bother to learn machine learning at all?

Indeed, we should note that the NFL theorem relies on the critical assumption that all problems are equally likely to happen or are equally important. In practice, however, we only focus on the current problem (e.g., a specific task) and do not care whether the solution can apply to other problems or not, even if they are very similar. For example, biking is an excellent choice if we travel within our university campus, though it would be a rather inconvenient choice for traveling between cities.

The NFL theorem assumes that $f$ is uniformly distributed, which is rarely true in practice. For example, going back to our watermelon problem, we have two hypotheses: H1 : ripe $\leftrightarrow$ (color $=$ *) $\wedge$ (root $=$ curly) $\wedge$ (sound $=$ muffled) and H2 : ripe $\leftrightarrow$ (color $=$ *) $\wedge$ (root $=$ curly) $\wedge$ (sound $=$ *), which are equally good according to the NFL theorem. At first glance, this seems to be correct since H1 is better than H2 on the example (ripe, (color $=$ green) $\wedge$ (root $=$ curly) $\wedge$

(sound = muffled)) and H2 is better than H1 on the example (ripe, (color = dark) ∧ (root = straight) ∧ (sound = crisp)). However, it turns out that ripe watermelons with the attributes (root = straight) ∧ (sound = muffled) are more common, whereas ripe watermelons with the attributes (root = straight) ∧ (sound = crisp) are rare or do not exist at all.

In summary, the wisdom we learned from the NFL theorem is that debating "which learning algorithm is better" is meaningless without considering the specific task, since all learning algorithms are equally good considering all contexts. In other words, we must consider the specific learning problem when comparing different learning algorithms, and the learning algorithms that perform well on one class of problems possibly perform poorly on another class of problems. Hence, whether the inductive bias matches the specific problem or not is often the decisive factor.

## 1.5 Brief History

We now proceed with a short overview of the development of machine learning. Machine learning is an inevitable product during the progress of artificial intelligence studies. Between the 1950s and early 1970s, artificial intelligence research was in the "reasoning age" when people thought a machine could get intelligence if it can do logical reasoning. Seminal works in that period include the *Logic Theorist* program developed by A. Newell and H. Simon and later on the *General Problem Solving* program. All of these works produced highly inspiring results at that time. For example, in 1952, the *Logic Theorist* program successfully proved 38 theorems in the famous book *Principia Mathematica* written by A. F. Whitehead and B. Russell. Later on, in 1963, it proved all of the 52 theorems, and people found the Proof of the Theorem 2.85 was even more elegant than Whitehead and Russell's. The research community recognized the importance of this line of work, and hence A. Newell and H. Simon received the Turing Award in 1975.

However, as the research was advancing, people started to realize that having the ability of logical reasoning is still far away from enabling artificial intelligence. A group of researchers represented by E. A. Feigenbaum argued that machines must acquire knowledge in order to be intelligent. A new phase of development started in the mid of 1970s, where artificial intelligence research entered the so-called "knowledge age". In this period, researchers developed a large number of expert systems with numerous successful applications in a wide range of domains. E. A. Feigenbaum, who is often regarded as the father of knowledge engineering, received the Turing

So-called "knowledge is power".

In 1965, Feigenbaum and his colleagues developed the first expert system DENDRAL.

Award in 1994. Nevertheless, researchers have soon reached the "Feigenbaum's knowledge acquisition bottleneck", that is, it is difficult to extract and summarize knowledge into a form that computers can learn. Therefore, some researchers decided to explore the possibility of letting machines learn knowledge by themselves!

In fact, A. Turing had already mentioned the possibility of machine learning in his Turing test paper published in 1950. In the early 1950s, there were already some studies related to machine learning, such as the famous computer checkers program developed by A. Samuel. In the middle-late 1950s, neural-network-based connectionism learning emerged, and representative works include F. Rosenblatt's Perceptron and B. Widrow's Adaline. Between the 1960s and 1970s, the logic-representation-based symbolism learning has thrived, and representative works include the structural learning system proposed by P. Winston, the logic-based inductive learning system proposed by R. S. Michalski et al., and the concept learning system proposed by E. B. Hunt et al. Meanwhile, researchers also developed the decision-theory-based learning and reinforcement learning in this period, such as the learning machines proposed by N. J. Nilsson. This period has also witnessed some foundation works on the statistical learning theory, which becomes extremely popular 20 years later.

See Sect. 1.7 for Samuel's checkers program.

In the summer of 1980, the First International Workshop on Machine Learning (IWML) was held at Carnegie Mellon University; in the same year, the *International Journal of Policy Analysis and Information Systems* published three consecutive special issues on machine learning. In 1983, Tioga press published a book edited by R. S. Michalski, J. G. Carbonell, and T. Mitchell called *Machine Learning: An Artificial Intelligence Approach*, which summarized machine learning research in that period. In 1986, *Machine Learning*, the first journal dedicated to machine learning, was established. In 1989, *Artificial Intelligence*, the leading journal in artificial intelligence research, published a special issue on machine learning covering major research works in that period; later on, the content of this special issue appeared in the book *Machine Learning: Paradigms and Methods* edited by Carbonell (1990) and published by MIT press. To summarize, machine learning has become an independent research field in the 1980s with diverse and active research directions.

IWML is the predecessor of the International Conference on Machine Learning (ICML).

Michalski et al. (1983) divided machine learning methods into several categories: *rote learning and direct implanting of new knowledge*, *learning from instruction*, and *learning from observation and discovery*. In *The Handbook of Artificial Intelligence* (Volume III), E. A. Feigenbaum et al. divided machine learning methods into *rote learning*, *learning by being told*, *learning by analogy, and learning by induction*. Rote learning, also known as memorization-based learning, does not perform any real learning but saves all input information as it is and retrieves it when needed. Learning by being told and learning by analogy are similar to "learning from instruction" and "learning from observation and discovery" by Michalski. Since the 1980s, the most researched and applied approach is *learning from examples* (or in a broad sense, learning by induction), such as supervised learning and unsupervised learning, which form the majority content of this book. Next, let us take a quick review of the evolution history of *learning from examples*.

In the 1980s, the mainstream of learning from examples is symbolism learning, represented by decision trees and logic-based learning. Decision trees rely on information theory to simulate the tree-based decision process of humans by minimizing the information entropy. A representative work of logic-based learning is Inductive Logic Programming (ILP) which is an intersection between machine learning and logic programming. ILP employs first-order logic (i.e., predicate logic) to represent knowledge, and induces data by updating and extending the logic expressions, e.g., Prolog expressions.

See Chap. 4 for decision trees.

It is actually the predecessor of ILP.

See Chap. 15 for ILP.

The reason that symbolism learning becomes the mainstream is closely related to the history of artificial intelligence. As mentioned earlier, artificial intelligence research started from the "reasoning age" in the 1950s and moved to the "knowledge age" in the 1980s, where the "reasoning age" relied on the symbolic knowledge representation for deductive reasoning, and the "knowledge age" also relied on the symbolic knowledge representation for creating knowledge-based expert systems. Given that the symbolic knowledge representation is the underlying technology of the first two "ages", it naturally became the preferred choice in the early stage of the "learning age". In fact, decision trees, as a representative method of symbolism learning, are still one of the most commonly used machine learning techniques today. ILP has a great knowledge representation ability and can easily encode complex data relationships together with domain knowledge. Therefore, it not only can perform domain knowledge-assisted learning but also can enhance domain knowledge through the learning process. However, since the great representation ability leads to a huge hypothesis space and model complexity, ILP is impractical even for moderately sized problems, and hence became less popular since the late 1990s.

Until the middle 1990s, another mainstream technique of learning from examples is the neural-network-based connectionism learning. Though the research on connectionism learning had a big leap forward in the 1950s, it did not become the mainstream since the research community still favors symbolism learning in the early days. For example, the Turing Award winner H. Simon once said "A physical symbol system has the necessary and sufficient means for general intelligent action." Besides, connectionism learning itself also encountered some challenges at that time. For example, the Turing Award winners M. Minsky and S. Papert pointed out in 1969 that neural networks (at that time) can only handle linearly separable problems and cannot handle some simple problems such as XOR. Connectionism learning did not gain attention until J. J. Hopfield tackled the NP-hard "traveling salesman problem" using neural networks in 1983. Later on, D. E. Rumelhart et al. reinvented the renowned Backpropagating (BP) algorithm, which has a far-reaching influence. Unlike symbolism learning, which produces explicit concepts, connectionism learning produces *black boxes*, and this is clearly a weakness of connectionism learning from the knowledge acquisition perspective. Nevertheless, thanks to the effective BP algorithm, connectionism learning can still do an excellent job in many real-world tasks. Nowadays, the BP algorithm is one of the most widely used machine learning algorithms. A significant limitation of connectionism learning is "trial and error", that is, we need to tune many parameters during the learning process, and the *parameter tuning* is often manual and lacks principled guidelines. In many cases, a tiny difference in parameter tuning can lead to significant differences in the learning outcome.

See Chap. 5 for the BP algorithm.

In the middle 1990s, statistical learning, represented by Support Vector Machine (SVM) (or *kernel methods* in a broader sense), has emerged as a superstar and quickly became the mainstream. However, the foundations of statistical learning theory (Vapnik 1998) were laid more than 20 years earlier in the 1960s–1970s. For example, V. N. Vapnik proposed the concept of *support vector* in 1963, and later on, V. N. Vapnik and A. J. Chervonenkis proposed the concept of VC dimension in 1968 and the structural risk minimization principle in 1974. Though the foundations were ready, statistical learning did not become the mainstream until the middle 1990s. One reason was that although SVM was proposed in the early 1990s, its superiority was not recognized until its success in text classification in the middle 1990s. Another reason was that the limitations of connectionism learning were increasingly prominent at the time, forcing researchers to turn their attention to statistical learning techniques supported by statistical learning theory. In fact, statistical learning and connectionism learning are closely related. After the wide acceptance of SVM, kernel tricks have been used almost everywhere in machine learning, and kernel methods also become a fundamental content of machine learning.

See Chap. 6 for SVM and kernel methods.

See Exercise 6.5.

Interestingly, connectionism came back to the stage in the early twenty-first century leading a new trend called *deep learn-*

*ing*. In a narrow sense, the so-called deep learning is simply neural networks with many layers. Deep learning techniques showed superior performance in many benchmarks and contests, especially for those involving complex data types such as audio and images. In the past, a profound understanding of machine learning techniques is the key for its users to achieve good performance in applications. Nevertheless, with the extreme flexibility and complexity of deep learning models, good performance can often be achieved after hard work on parameter tuning. Therefore, deep learning is easy to be accessed by machine learning practitioners, lowering the knowledge requirement threshold for its users, though the theory behind it is yet to be developed.

See Sect. 5.6 for deep learning.

Given that deep learning works so well, why did it not become popular earlier? There are two main reasons: the data is getting massive, and so does the computation power. On the one hand, the large number of parameters in deep learning can easily lead to *overfitting* when the data is insufficient. On the other hand, given complex models and massive data, learning would not be possible without sufficient computing resources. Fortunately, we have entered the "big data era" in which both the data storage and the computing hardware have improved significantly, giving connectionism learning a new chance. Interestingly, the popularity of neural networks in the middle 1980s was also after the boom of computational power and the data access boosted by the advancement of the Intel x86 CPUs and memory devices. What is happening now to deep learning is similar to what happened to neural networks in the 1980s.

See Chap. 2 for *overfitting*.

We should note that machine learning has grown into a broad research field, and what we have discussed in this section only provides a sketch on this topic. After reading the entire book, readers will gain a more comprehensive understanding of machine learning.

## 1.6 Application Status

The past two decades have witnessed fast and significant advances in the collection, storage, transmission, and processing of data. Given that we have accumulated massive data covering all aspects of human life, effective and efficient methods for utilizing the data are urgently needed. Hence, it is not a surprise that machine learning has gained much attention since it provides solutions for unleashing the power of massive data.

Nowadays, machine learning appears in different branches of computer science, such as multimedia, graphics, network communication, software engineering, and even architecture and chipset design. Machine learning has also become one of the most critical techniques in some "applied" areas, such as computer vision and natural language processing.

Machine learning also provides support for interdisciplinary research. For example, bioinformatics is a research field that aims to employ information technology to understand biological observations and patterns, and people are excited about the human genome project and genomic medicine. In bioinformatics, the whole process, from biological observations to pattern discovery, involves data collection, data management, data analytics, and simulations. Among these tasks, data analytics is where machine learning shines, and researchers have already successfully applied various machine learning techniques to bioinformatics.

In fact, the research methodology of science is shifting from "theory + experiment" to "theory + experiment + computation", and people even use the term "data science". The importance of machine learning is increasing since computing with data is the core of both data analytics and data science. If we list the most current and eye-catching computer science technologies, then machine learning must be one of them. In 2001, scientists from NASA-JPL published an article (Mjolsness and DeCoste 2001) in the *Science* magazine pointed out that machine learning is playing an increasingly important role in supporting scientific research and is crucial for technology development. In 2003, DARPA started the PAL project, which puts machine learning to the level of national security. We see that the importance of machine learning is recognized by both NASA and DARPA, who often promote the most cutting-edge technologies in the U.S.

NASA-JPL stands for the NASA Jet Propulsion Laboratory, who developed "Spirit" and "Opportunity" robotic rovers landed on Mars.

DARPA stands for the Defense Advanced Research Projects Agency, who initiated the development of the Internet and Global Positioning System.

In 2006, Carnegie Mellon University founded the world's first school of machine learning, which is directed by Professor T. Mitchell, one of the pioneers in machine learning research. In March 2012, the Obama administration proposed the "big data research and development initiative" followed by a reinforcement project at UC Berkley led by the U.S. National Science Foundation. The initiative puts emphasis on three essential techniques: machine learning, cloud computing, and crowdsourcing. Clearly, machine learning is an essential technique in the era of big data for a simple reason: the purpose of collecting, storing, transmitting, and managing big data is to utilize big data, and it will not be possible without machine learning.

Machine learning provides data analytics, clouding computing provides data processing, and crowdsourcing provides data labeling.

When we discuss data analytics, people may think about data mining. Here, we give a brief discussion of the relationship between data mining and machine learning. Data mining research formed in the 1990s under the influence of many other research fields, where the most important ones are database, machine learning, and statistics (Zhou 2003). Data mining is about knowledge discovering from massive data, and therefore it involves the managing and analyzing of massive data. Roughly speaking, database research provides the data management facility for data mining, while machine learning and statistics provide the facility of data analytics. Research outcomes from statistics are often turned into practical learning algorithms through machine learning research, and the learning algorithms are then used by data mining. From this perspective, statistics influences data mining via machine learning, and therefore machine learning and data management are two backbones of data mining.

Data mining appeared in the statistics community a long time ago as a slightly negative term. The reason is that traditional statistics research focuses on elegant theories and ignores the practical utility. Recently, however, statisticians have started to investigate practical problems and get involved in machine learning and data mining research.

Today, machine learning strongly influences our daily life. For example, analyzing data collected from satellites and sensors using machine learning has become essential in applications such as weather forecasts, energy exploration, and environmental monitoring. In commercial applications, machine learning is now helping us to analyze sales and customer data for optimizing the inventory costs as well as designing sales strategies for targeted customer cohorts. The followings are some other aspects.

Search engines like Google Search have been changing people's lifestyles, such as searching for destination information before traveling and looking for a suitable hotel or restaurant. *Newsweek* once commented on Google Search as "*Positioning everyone a mouse-click away from the answers to the most arcane questions.*" Internet search locates the requested information by analyzing data all over the network, and the search process, from the input query to the output results, relies on machine learning technologies to establish the mapping between input and output. In fact, machine learning has made tremendous contributions to the development of Internet search that we are enjoying today, and many advanced functions, such as "search by photo", are enabled by cutting-edge machine learning techniques. All the leading technology companies, such as Google, Microsoft, Facebook, and Amazon, have their machine learning research teams or even research centers named after machine learning. The decisions made by these giant companies not only show the fast development and application of machine learning but also influence the future direction of the Internet industry.

Another notable application of machine learning is autonomous vehicles, which, hopefully, can reduce traffic collisions significantly, saving more than a million lives each year.

In the early 1990s, a neural-network-based system called ALVINN was developed to control autonomous vehicles; this project was discussed in Sect. 4.2 of the classic machine learning textbook (Mitchell 1997).

We expect autonomous vehicles to be safer since computer drivers are always experienced and are not subject to fatigue driving or drunk driving. Besides, autonomous vehicles are also valuable to the military. The U.S. has started to explore autonomous vehicles since the 1980s. The greatest difficulty is that engineers are unable to consider all possible driving situations and program them, and hence autonomous vehicles must be able to make situational decisions by themselves. If we consider the data received by sensors as input and let the control of steering, brake, and accelerator as output, then we can abstract self-driving to a machine learning problem. In March 2004, the Stanford machine learning team led by S. Thrun won the autonomous vehicle competition hosted by DARPA. Their autonomous vehicle finished 132 miles of travel in mountain and desert areas in Nevada in 6 h and 53 min. The road was rough, and driving in such conditions can be a challenging task for experienced human drivers. Later on, S. Thrun joined Google to lead its autonomous vehicle project. It is worth mentioning that research on autonomous vehicles has made some significant progress in recent years. It has attracted large amounts of research funds from electric-automobile manufacturers such as Tesla, as well as traditional automotive manufacturers, such as General Motors, Volkswagen, and BMW. In some places, we can already see autonomous vehicles on the road. In June 2011, Nevada passed a bill and became the first state of the U.S. permitting autonomous vehicles on the road, followed by Hawaii and Florida. Hopefully, autonomous vehicles, driven by machine learning technologies, will be widely adopted in the near future.

Machine learning gained its public attention for its tremendous contributions to intelligent data analytics. However, we should also be aware of another important aspect of machine learning, that is, assisting our understanding of "how humans learn". For example, when the Sparse Distributed Memory (SDM) model (Kanerva 1988) was proposed by P. Kanerva in the middle 1980s, there is no intentional imitation to the biological structure of the human brain. However, neuroscience researchers figured out that the sparse encoding mechanism in SDM widely exists in the cortex controlling vision, hearing, and olfactory, thus inspiring more neuroscience research. The motivation of conducting natural science research is fundamentally due to human curiosity about the origin of the universe, the essence of the matters, the characters of life, and the awareness of humans ourselves. Undoubtedly, "how humans learn" is an essential topic of our awareness. In some sense, machine learning is not just crucial for information science, but also has some sense of exploring the universe like natural science.

## 1.7 Further Reading

Mitchell (1997) is the first textbook dedicated to machine learning, and Duda et al. (2001), Alpaydin (2004), Flach (2012) are also excellent introductory books. Hastie et al. (2009) is a good intermediate level book, and Bishop (2006) is a great book for reference, particularly for readers who favor Bayesian learning. Shalev-Shwartz and Ben-David (2014) is suitable for readers who wish to understand more about the underlying theories. Witten et al. (2011) is an introductory book that binds with the WEKA software, which is helpful for beginners in practicing common machine learning algorithms through WEKA.

WEKA is a famous open-source machine learning package (written in JAVA) developed by researchers from Waikato University, New Zealand: ▸ http://www.cs.waikato.ac.nz/ml/weka/.

*Machine Learning—an Artificial Intelligence Approach* (Michalski et al. 1983), which collected 16 articles contributed by 20 scholars, was the most important literature in the early days of machine learning. The publication of this book has drawn much public attention, and Morgan Kaufmann Publishers published two subsequent volumes in 1986 and 1990. *The Handbook of Artificial Intelligence* series was co-edited by the Turing Award winner E. A. Feigenbaum and other scholars; its third volume (Cohen and Feigenbaum 1983) discussed machine learning and was an important literature in the early days. Dietterich (1997) provided a review and envision of the development of machine learning; this early literature is still valuable today, and some discussed ideas could regain popularity after the advancement of relevant technologies. For example, the recent trend of *transfer learning* research (Pan and Yang 2010) is in the same flavor of *learning by analogy* after significant development of statistical learning. In fact, the core idea of the popular deep learning is not much more distinguished than the neural network research back in the middle-late 1980s.

See Sect. 5.6 for deep learning.

Though machine learning research on concept learning has started in the very early days, many ideas developed at that time have a continuing influence on the entire research field. For example, decision tree learning, as a mainstream learning technique, was originated from tree-based concept formation (Hunt and Hovland 1963). The famous research work "Blocks World" (Winston 1970) connected concept learning with the generalization and specification-based search process. The argument of "learning is about searching in hypothesis space" was proposed in an early literature (Simon and Lea 1974). Later on, Mitchell (1977) proposed the concept of version space. Concept learning literature also discusses a lot of rule learning.

See Chap. 15 for rule learning.

The principle of Occam's razor suggests choosing the simplest hypothesis that matches the observations, and this is a fundamental principle widely adopted in natural science, such as physics and astronomy. For example, one of the reasons that N. Copernicus believed the heliocentric system is because it is simpler yet consistent with astronomical observations compared to Ptolemy's geocentric system. Occam's razor also has its followers in machine learning (Blumer et al. 1996). However, how to define "simple" still confuses researchers, and the utility of Occam's razor in machine learning has always been controversial (Webb 1996; Domingos 1999). We should note that Occam's razor is not the only available principle for hypothesis selection. For example, the ancient Greek philosopher Epicurus (341–270 BC) proposed the "principle of multiple explanations", which advises that all hypotheses consistent with the observations should be kept (Asmis 1984); this argument concurs with the research of ensemble learning.

See Chap. 8 for ensemble learning.

The most important international conferences in the field of machine learning include the International Conference on Machine Learning (ICML), International Conference on Neural Information Processing Systems (NeurIPS, or NIPS before 2018), and Annual Conference on Learning Theory (COLT). The most notable regional conferences include the European Conference on Machine Learning (ECML) and Asian Conference on Machine Learning (ACML). The most prestigious journals are the *Journal of Machine Learning Research* and *Machine Learning*. Machine learning papers are also regularly seen in the venues of related fields: premier AI conferences, such as the International Joint Conferences on Artificial Intelligence (IJCAI) and AAAI Conference on Artificial Intelligence (AAAI), and journals, such as *Artificial Intelligence* and *Journal of Artificial Intelligence Research*; premier data mining conferences, such as the SIGKDD Conference on Knowledge Discovery and Data Mining (KDD) and IEEE International Conference on Data Mining (ICDM), and journals, such as *ACM Transactions on Knowledge Discovery from Data* and *Data Mining and Knowledge Discovery*; premier computer vision conferences, such as Conference on Computer Vision and Pattern Recognition (CVPR), and journals, such as *IEEE Transactions on Pattern Analysis and Machine Intelligence*. Top journals in the neural networks research community also publish many machine learning papers, such as *Neural Computation* and *IEEE Transactions on Neural Networks and Learning Systems*. Besides, top journals in statistics, such as *Annals of Statistics*, also publish theoretical papers about statistical learning.

## Exercises

**1.1** What is the version space if ◘ Table 1.1 contains only the examples 1 and 4?

**1.2** Compared to single conjunction, the disjunctive normal form is more expressive. For example,

Disjunctive normal form is the disjunction of multiple conjunction clauses.

$$\begin{aligned}\texttt{ripe} \leftrightarrow &((\texttt{color} = *) \wedge (\texttt{root} = \texttt{curly}) \wedge (\texttt{sound} = *)) \\ &\vee ((\texttt{color} = \texttt{dark}) \wedge (\texttt{root} = *) \wedge (\texttt{sound} = \texttt{dull}))\end{aligned}$$

will classify both $(\texttt{color} = \texttt{green}) \wedge (\texttt{root} = \texttt{curly}) \wedge (\texttt{sound} = \texttt{crisp})$ and $(\texttt{color} = \texttt{dark}) \wedge (\texttt{root} = \texttt{straight}) \wedge (\texttt{sound} = \texttt{dull})$ as ripe. If we use at most $k$ conjunction clauses to express the hypothesis space of the watermelon classification problem in ◘ Table 1.1, then what is the total number of possible hypotheses?

Hint: pay attention to the cases of redundancy, e.g., $(A = a) \vee (A = *)$ is equivalent to $(A = *)$.

**1.3** When the data contains noise, it is possible that no hypothesis in the hypothesis space is consistent with all training examples. Design an inductive bias for selecting hypotheses in such situations.

That is, the error is unavoidable for all hypotheses.

**1.4** * The *misclassification error rate* was used when we discussed the NFL theorem in Sect. 1.4. If we change it to a different performance measurement $\ell$, then (1.1) becomes

$$E_{\text{ote}}(\mathfrak{L}_a \mid X, f) = \sum_h \sum_{\boldsymbol{x} \in \mathcal{X} - X} P(\boldsymbol{x}) \ell(h(\boldsymbol{x}, f(\boldsymbol{x})) P(h \mid X, \mathfrak{L}_a).$$

Prove that the NFL theorem is still valid.

**1.5** Describe the use of machine learning in different subcomponents of Internet search.

## Break Time

### Short Story: the naming of "machine learning"

In 1952, Arthur Samuel (1901–1990) developed a checkers program at IBM. The program possesses the learning ability to analyze past games such that it can identify "good moves" and "bad moves" given the current situation. The program improved quickly through self-learning and soon outperformed Samuel himself. In 1956, Samuel was invited by J. McCarthy ("Father of Artificial Intelligence" and the Turing Award winner in 1971) to introduce this work at the Dartmouth workshop—this workshop is widely considered as the place where Artificial Intelligence was born. Samuel invented the term "machine learning" and defined it as "field of study that gives computers the ability to learn without being explicitly programmed." In 1959, Samuel published the paper "Some studies in machine learning using the game of checkers" in *IBM Journal.* Later on, in 1961, Edward Feigenbaum ("Father of Knowledge Engineering" and the Turing Award winner in 1994) invited Samuel to provide the best game that the checkers program has ever played to be included in his renowned book *Computers and Thoughts*. Taking this opportunity, Samuel challenged the Connecticut state checkers champion (ranked 4th nationwide) with his checkers program and made a hit by winning the challenge.

This checkers program uses reinforcement learning techniques. See Chap. 16.

Samuel's checkers program significantly influenced not only the field of artificial intelligence but also the development of the entire field of computer science. Early literature in computer science thought that computers could not accomplish tasks without explicit programming. However, Samuel's program falsified this argument. Besides, the checkers program is one of the first programs ever executed on computers for non-numerical computation. The idea of logical instructions used in this program has influenced the design of the instruction set of IBM computers, which was soon adopted by other computer manufacturers.

## References

Alpaydin E (2004) Introduction to machine learning. MIT Press, Cambridge
Asmis E (1984) Epicurus' scientific method. Cornell University Press, Ithaca
Bishop CM (2006) Pattern recognition and machine learning. Springer, New York
Blumer A, Ehrenfeucht A, Haussler D, Warmuth MK (1996) Occam's razor. Inf Process Lett 24(6):377–380
Carbonell JG (ed) (1990) Machine learning: paradigms and methods. MIT Press, Cambridge
Cohen PR, Feigenbaum EA (eds) (1983) The handbook of artificial intelligence, vol 3. William Kaufmann, New York
Dietterich TG (1997) Machine learning research: four current directions. AI Mag 18(4):97–136
Domingos P (1999) The role of Occam's razor in knowledge discovery. Data Min Knowl Discov 3(4):409–425
Duda RO, Hart PE, Stork DG (2001) Pattern Classification, 2nd edn. Wiley, New York
Flach P (2012) Machine learning: the art and science of algorithms that make sense of data. Cambridge University Press, Cambridge
Hand D, Mannila H, Smyth P (2001) Principles of data mining. MIT Press, Cambridge
Hastie T, Tibshirani R, Friedman J (2009) The elements of statistical learning, 2nd edn. Springer, New York
Hunt EG, Hovland DI (1963) Programming a model of human concept formation. In: Feigenbaum E, Feldman J (eds) Computers and Thought. McGraw Hill, New York, pp 310–325
Kanerva P (1988) Sparse distributed memory. MIT Press, Cambridge
Michalski RS, Carbonell JG, Mitchell TM (eds) (1983) Machine learning: an artificial intelligence approach. Tioga, Palo Alto
Mitchell T (1997) Machine learning. McGraw Hill, New York
Mitchell TM (1977) Version spaces: a candidate elimination approach to rule learning. In: Proceedings of the 5th international joint conference on artificial intelligence (IJCAI), pp 305–310. Cambridge, MA
Mjolsness E, DeCoste D (2001) Machine learning for science: state of the art and future prospects. Science 293(5537):2051–2055
Pan SJ, Yang Q (2010) A survey of transfer learning. IEEE Trans Knowl Data Eng 22(10):1345–1359
Shalev-Shwartz S, Ben-David S (2014) Understanding machine learning. Cambridge University Press, Cambridge
Simon HA, Lea G (1974) Problem solving and rule induction: a unified view. In: Gregg LW (ed) Knowledge and cognition. Erlbaum, New York, pp 105–127
Vapnik VN (1998) Statistical learning theory. Wiley, New York
Webb GI (1996) Further experimental evidence against the utility of Occam's razor. J Artif Intell Res 43:397–417
Winston PH (1970) Learning structural descriptions from examples. Technical Report AI-TR-231, AI Lab, MIT, Cambridge, MA
Witten IH, Frank E, Hall MA (2011) Data mining: practical machine learning tools and techniques, 3rd edn. Elsevier, Burlington

Wolpert DH (1996) The lack of a priori distinctions between learning algorithms. Neural Comput 8(7):1341–1390

Wolpert DH, Macready WG (1995) No free lunch theorems for search. Technical Report SFI-TR-05-010, Sante Fe Institute, Sante Fe

Zhou Z-H (2003) Three perspectives of data mining. Artif Intell 143(1):139–146

# Model Selection and Evaluation

Table of Contents




Z.-H. Zhou, *Machine Learning*,
https://doi.org/10.1007/978-981-15-1967-3_2

## 2.1 Empirical Error and Overfitting

In general, the proportion of incorrectly classified samples to the total number of samples is called *error rate*, that is, if $a$ out of $m$ samples are misclassified, then the error rate is $E = a/m$. Accordingly, $1 - a/m$ is called *accuracy*, i.e., accuracy $= 1 -$ error rate. More generally, the difference between the output predicted by the learner and the ground-truth output is called *error*. The error calculated on the training set is called *training error* or *empirical error*, and the error calculated on the new samples is called *generalization error*. Clearly, we wish to have a learner with a small generalization error. However, since the details of the new samples are unknown during the training phase, we can only try to minimize the empirical error in practice. Quite often, we obtain learners that perform well on the training set with a small or even zero empirical error, that is, 100% accuracy. However, are they the learners we need? Unfortunately, such learners are not good in most cases.

Accuracy is often expressed as percentages: $(1 - \frac{a}{m}) \times 100\%$.

Here, the "error" refers to the expectation of errors.

Later chapters will introduce different learning algorithms for minimizing the empirical error.

The good learners we are looking for are those performing well on the new samples. Hence, good learners should learn general rules from the training examples such that the learned rules apply to all potential samples. However, when the learner learns the training examples "too well", it is likely that some peculiarities of the training examples are taken as general properties that all potential samples will have, resulting in a reduction in generalization performance. In machine learning, this phenomenon is known as *overfitting*, and the opposite is known as *underfitting*, that is, the learner failed to learn the general properties of training examples. ◘ Figure 2.1 illustrates the difference between overfitting and underfitting.

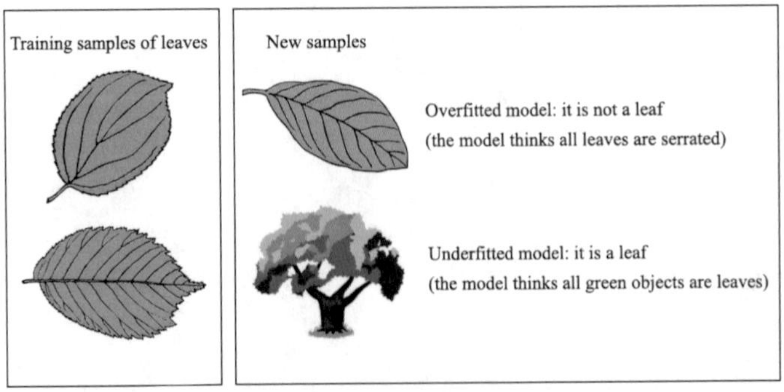


**Fig. 2.1** An intuitive analogy of overfitting and underfitting

Among many possible reasons, the overly strong learning ability is a common cause for overfitting since such learners can learn the non-general peculiarities of training examples. By contrast, underfitting is usually due to weak learning ability. In practice, underfitting is relatively easy to overcome. For example, we can do more branching in decision tree learning or adding more training epochs in neural network learning. However, as we will see later, overfitting is a fundamental difficulty in machine learning, and almost every learning algorithm has implemented some mechanisms to deal with overfitting. Nevertheless, we should realize that overfitting is unavoidable, and all we can do is to alleviate or reduce the risk of it. This argument can be briefly justified as follows. Machine learning problems are often NP-hard or even harder, but practical learning algorithms have to finish learning within polynomial time. Hence, if overfitting is avoidable, then minimizing the empirical error will lead to the optimal solution, and therefore we have a constructive proof of P=NP. In other words, overfitting is unavoidable as long as we believe in P≠NP.

In practice, there are often multiple candidate learning algorithms, and even the same learning algorithm may produce different models under different parameter settings. Then, which learning algorithm should we choose, and which parameter settings should we use? This problem is referred to as *model selection*. The ideal solution is to evaluate all candidate models and select the one with the smallest generalization error. However, as mentioned earlier, we cannot obtain the generalization error directly, while the empirical error suffers from overfitting. So, how can we evaluate and select models in practice?

## 2.2 Evaluation Methods

In general, we can evaluate the generalization error through testing experiments. To do so, we use a *testing set* to estimate the learner's ability to classify the new samples, and use the *testing error* as an approximation to the generalization error. Generally, we assume that the testing samples are independent and identically sampled from the ground-truth sample distribution. Note that the testing set and the training set should be mutually exclusive as much as possible, that is, testing samples should avoid appearing in the training set or be used anyhow in the training process.

Here, we only consider the generalization error, but in real-world applications, we often consider more factors such as computational cost, memory cost, and interpretability.

Why should testing samples avoid appearing in the training set? To understand this, let us consider the following scenario. Suppose we use the same set of ten questions for both the exercise and exam, then does the exam reflect students' learning outcomes? The answer is "no" because some students can get good grades even if they only know how to solve those ten questions. Analogously, the generalization ability we wish the model to have is the same as we want students to study and master the knowledge. Accordingly, the training examples correspond to the exercises, and the testing samples correspond to the exam. Hence, the estimation could be too optimistic if the testing samples are already seen in the training process.

However, given the only data set of $m$ samples $D = \{(\boldsymbol{x}_1, y_1), (\boldsymbol{x}_2, y_2), \ldots, (\boldsymbol{x}_m, y_m)\}$, how can we do both training and testing? The answer is to produce both a training set $S$ and a testing set $T$ from the data set $D$. We discuss a few commonly used methods as follows.

### 2.2.1 Hold-Out

The *hold-out* method splits the data set $D$ into two disjoint subsets: one as the training set $S$ and the other as the testing set $T$, where $D = S \cup T$ and $S \cap T = \varnothing$. We train a model on the training set $S$ and then calculate the testing error on the testing set $T$ as an estimation of the generalization error.

Taking binary classification problems as an example, let $D$ be a data set with 1000 samples, and we split it into a training set $S$ with 700 samples and a testing set $T$ with 300 samples. After being trained on $S$, suppose the model misclassified 90 samples on $T$, then we have the error rate $(90/300) \times 100\% = 30\%$, and accordingly, the accuracy $1 - 30\% = 70\%$.

It is worth mentioning that the splitting should maintain the original data distribution to avoid introducing additional bias. Taking classification problems as an example, we should try to preserve the class ratio in different subsets, and the sampling methods that maintain the class ratio are called *stratified sampling*. For example, suppose we have a data set $D$ containing 500 positive examples and 500 negative examples, and we wish to split it into a training set $S$ with 70% of the examples and a testing set $T$ with 30% of the examples. Then, a stratified sampling method will ensure that $S$ contains 350 positive examples and 350 negative examples, and $T$ contains 150 positive examples and 150 negative examples. Without stratified sampling, the different class ratios in $S$ and $T$ can lead to biased error estimation since the data distributions are changed.

See Exercise 2.1.

However, even if the class ratios match, there still exist different ways of splitting the original data set $D$. For example, we can sort the samples in $D$ and then use the first 350 samples for training with the rest for testing. Different ways of splitting will result in different training and testing sets, and accordingly, different model evaluation results. Therefore, a single trial of hold-out testing usually leads to unreliable error estimation. In practice, we often perform the hold-out testing multiple times, where each trial splits the data randomly, and we use the average error as the final estimation. For example, we can randomly split the data set 100 times to produce 100 evaluation results and then take the average as the hold-out error estimation.

We can also check other statistical quantities such as standard deviation.

The hold-out method splits $D$ into a training set and a testing set, but the model we wish to evaluate is the one trained on $D$. Hence, we have a dilemma. If we place most samples in the training set $S$, then the trained model is an excellent approximation to the model trained on $D$. However, the evaluation is less reliable due to the small size of $T$. On the other hand, if we place more samples in the testing set $T$, then the difference between the model trained on $S$ and the model trained on $D$ becomes substantial, that is, the fidelity of evaluation becomes lower. There is no perfect solution to this dilemma, and we must make a trade-off. One routine is to use around $2/3$ to $4/5$ of the examples for training and the rest for testing.

The dilemma can be explained with *bias-variance* which will be discussed in Sect. 2.5. The variance of the evaluation result is large when the testing set is small, and the bias of the evaluation result is large when the training set is small.

Generally speaking, a testing set should contain at least 30 samples (Mitchell 1997).

### 2.2.2 Cross-Validation

*Cross-validation* splits data set $D$ into $k$ disjoint subsets with similar sizes, that is, $D = D_1 \cup D_2 \cup \cdots \cup D_k$, $D_i \cap D_j = \varnothing (i \neq j)$. Typically, each subset $D_i$ tries to maintain the original data distribution via stratified sampling. In each trial of cross-validation, we use the union of $k - 1$ subsets as the training set to train a model and then use the remaining subset as the testing set to evaluate the model. We repeat this process $k$ times and use each subset as the testing set precisely once. Finally, we average over $k$ trials to obtain the evaluation result. Since the stability and fidelity of cross-validation largely depend on the value of $k$, it is also known as $k$-fold cross-validation. The most commonly used value of $k$ is 10, and the corresponding method is called 10-fold cross-validation. Other common values of $k$ include 5 and 20. ◘ Figure 2.2 illustrates the idea of 10-fold cross-validation.

There are special cases, such as Leave-One-Out, which will be discussed shortly.

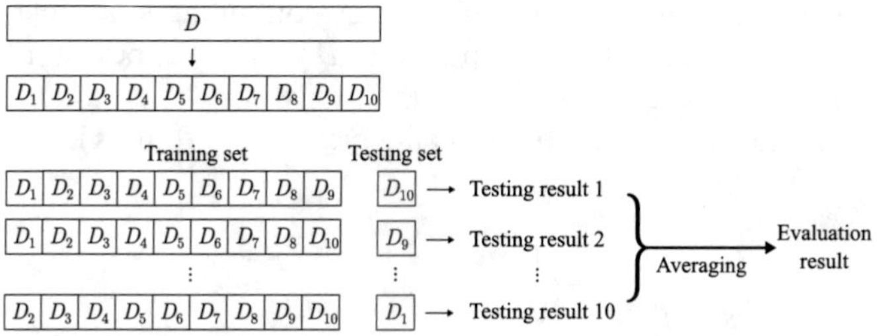


**Fig. 2.2** 10-fold cross-validation

Like hold-out, there are different ways of splitting the data set $D$ into $k$ subsets. To decrease the error introduced by splitting, we often repeat the random splitting $p$ times and average the evaluation results of $p$ times of $k$-fold cross-validation. For example, a common case is 10-time 10-fold cross-validation.

Both "10-time 10-fold cross-validation" and "100-time hold-out" run 100 evaluation experiments.

For a data set $D$ with $m$ samples, a special case of cross-validation is Leave-One-Out (LOO), which lets $k = m$. In such a case, the random splitting does not matter since there is only one way of splitting the $m$ samples into $m$ subsets. In LOO, each subset contains a single sample, and the training set is only one sample less than the original data set $D$; thus in most cases, the evaluation from LOO is very close to the ideal evaluation of training the model on $D$. Therefore, the results of LOO evaluations are often considered accurate. However, LOO has a flaw that the computational cost of training $m$ models could be prohibitive for large data sets (e.g., 1 million samples imply 1 million models), and it can be even worse if we take parameter tuning into consideration. Besides, LOO is not necessarily more accurate than other evaluation methods since the NFL theorem also applies to evaluation methods.

See Exercise 2.2.

See Sect. 1.4 for the NFL theorem.

### 2.2.3 Bootstrapping

What we want to evaluate is the model trained with $D$. However, no matter we use hold-out or cross-validation, the training set is always smaller than $D$. Hence, the estimation bias is unavoidable due to the size difference between the training set and $D$. We can reduce the bias by using LOO, but its computational complexity is often prohibitive. However, is it possible to reduce the impact of the small training set while still be computational efficient?

See Chap. 12 for more information about the relationship between the complexity of samples and the generalization ability.

One solution is *bootstrapping*, which employs the bootstrap sampling technique (Efron and Tibshirani 1993). Given a data set $D$ containing $m$ samples, bootstrapping samples a data set $D'$ by randomly picking one sample from $D$, copying it to $D'$, and then placing it back to $D$ so that it still has a chance to be picked next time. Repeating this process $m$ times results in the bootstrap sampling data set $D'$ containing $m$ samples. Due to replacement, some samples in $D$ may not appear in $D'$, while others may appear more than once. Let us do a quick estimation: the chance of not being picked in $m$ rounds is $(1 - \frac{1}{m})^m$, and hence taking the limit gives

The original meaning of bootstrap is to remove the strap of boots. The term comes from a story in the eighteenth century book *Baron Munchausen's Narrative of his Marvellous Travels and Campaigns in Russia*, in which Baron Munchausen pulls himself out of a swamp with his straps. Bootstrapping is also called *repeatable sampling* or *sampling with replacement*.

$e$ is Euler's number.

$$\lim_{m \to \infty} \left(1 - \frac{1}{m}\right)^m = \frac{1}{e} \approx 0.368, \tag{2.1}$$

which means that roughly 36.8% of the original samples do not appear in the data set $D'$. Then, we can use $D'$ as the training set and $D \backslash D'$ as the testing set such that both the evaluated model and the actual model that we wish to evaluate on $D$ are using $m$ training examples. Besides, we still have a separate testing set containing about 1/3 of the original examples that are not used for training. The evaluation result obtained via this approach is called *out-of-bag estimate*.

"\" is the subtraction of sets.

Bootstrapping is particularly useful when the data set is small, or when there is no effective way of splitting training and testing sets. Besides, bootstrapping can create multiple data sets, which can be useful for methods such as ensemble learning. Nevertheless, since the original data distribution has changed by bootstrapping, the estimation is also biased. Therefore, when we have abundant data, hold-out and cross-validation are often used instead.

See Chap. 8 for ensemble learning.

### 2.2.4 Parameter Tuning and Final Model

Most learning algorithms have parameters to set, and different parameter settings often lead to models with significantly different performance. Hence, the model evaluation and selection is not just about selecting the learning algorithms but also about the configuration of parameters. The process of finding the right parameters is called *parameter tuning*.

Readers may think there is no essential difference between parameter tuning and algorithm selection: each parameter setting leads to one model, and we select the one that produces the best results as the final model. This idea is basically sound; however, there is one issue: since parameters are often real-valued, it is impossible to try all parameter settings. Therefore, in practice, we usually set a range and a step size for each parameter, e.g., a range of [0, 0.2] and a step size of 0.05, which lead to only five candidate parameter settings. Such a trade-off between computational cost and quality of estimation makes the learning feasible, though the selected parameter setting is usually not optimal. In reality, even after making such a trade-off, parameter tuning can still be quite challenging. We can make a simple estimation. Suppose that the algorithm has three parameters and each considers only five candidate values, then we need to assess $5^3 = 125$ models for each pair of training and testing sets. Powerful learning algorithms often have quite many parameters to be configured, resulting in a heavy workload of parameter tuning. The quality of parameter tuning is often vital in real-world applications.

Machine learning typically involves two types of parameters. The first one is the algorithm parameters, also known as *hyper-parameters*, which are usually less than 10. The other one is the model parameters, which can be many, e.g., large-scale deep learning models can have more than 10 billion parameters. Both types of parameters are tuned similarly, that is, one generates candidate models and then selects via an evaluation method. The difference is that hyper-parameters are usually configured manually, whereas candidate models are generated by learning, e.g., parameters of neural networks that stop training at different iterations.

We should note that the training process does not use all data since part of the data is hold-out for model evaluation and selection. Therefore, after we have determined the algorithm and parameters via model selection, the entire data set should be used to re-train a model as the final delivery.

Last but not least, we should distinguish the data used for model selection from the testing data encountered after model selection. We often call the data set used in the model selection a *validation set*. For example, we may split data into a training set for training models, a validation set for model selection and parameter tuning, and a testing set for estimating the generalization ability of models.

## 2.3 Performance Measure

In order to evaluate the generalization ability of models, we need not only practical and effective estimation methods but also some performance measures that can quantify the generalization ability. Different performance measures reflect the varied demands of tasks and produce different evaluation results. In other words, the quality of a model is a relative concept that depends on the algorithm and data as well as the task requirement.

See Chap. 9 for the performance measures of clustering.

In prediction problems, we are given a data set $D = \{(\boldsymbol{x}_1, y_1), (\boldsymbol{x}_2, y_2), \ldots, (\boldsymbol{x}_m, y_m)\}$, where $y_i$ is the ground-truth label of the sample $\boldsymbol{x}_i$. To evaluate the performance of a learner $f$, we compare its prediction $f(\boldsymbol{x})$ to the ground-truth label $y$.

For regression problems, the most commonly used performance measure is the Mean Squared Error (MSE):

$$E(f; D) = \frac{1}{m}\sum_{i=1}^{m}(f(\boldsymbol{x}_i) - y_i)^2. \tag{2.2}$$

More generally, for a data distribution $\mathcal{D}$ and a probability density function $p(\cdot)$, the MSE is written as

$$E(f; \mathcal{D}) = \int_{\boldsymbol{x}\sim\mathcal{D}} (f(\boldsymbol{x}) - y)^2 p(\boldsymbol{x}) d\boldsymbol{x}. \tag{2.3}$$

The rest of this section will introduce some common performance measures for classification problems.

### 2.3.1 Error Rate and Accuracy

At the beginning of this chapter, we discussed error rate and accuracy, which are the most commonly used performance measures in classification problems, including both binary classification and multiclass classification. Error rate is the proportion of misclassified samples to all samples, whereas accuracy is the proportion of correctly classified samples instead. Given a data set $D$, we define error rate as

$$E(f; D) = \frac{1}{m} \sum_{i=1}^{m} \mathbb{I}(f(\boldsymbol{x}_i) \neq y_i), \tag{2.4}$$

and accuracy as

$$\begin{aligned} \operatorname{acc}(f; D) &= \frac{1}{m} \sum_{i=1}^{m} \mathbb{I}(f(\boldsymbol{x}_i) = y_i) \qquad (2.5) \\ &= 1 - E(f; D). \end{aligned}$$

More generally, for a data distribution $\mathcal{D}$ and a probability density function $p(\cdot)$, error rate and accuracy can be, respectively, written as

$$E(f; \mathcal{D}) = \int_{\boldsymbol{x} \sim \mathcal{D}} \mathbb{I}(f(\boldsymbol{x}) \neq y) p(\boldsymbol{x}) d\boldsymbol{x}, \tag{2.6}$$

$$\begin{aligned} \operatorname{acc}(f; \mathcal{D}) &= \int_{\boldsymbol{x} \sim \mathcal{D}} \mathbb{I}(f(\boldsymbol{x}) = y) p(\boldsymbol{x}) d\boldsymbol{x} \qquad (2.7) \\ &= 1 - E(f; \mathcal{D}). \end{aligned}$$

### 2.3.2 Precision, Recall, and *F*1

Error rate and accuracy are frequently used, but they are not suitable for all tasks. Taking our watermelon problem as an example, suppose we use a learned model to classify a new batch of watermelons. The error rate tells us the proportion of misclassified watermelons to all watermelons in this batch. However, we may want to know "What percentage of the picked watermelons are ripe?" or "What percentage of all ripe watermelons were picked out?" Unfortunately, the error rate is unable to answer such questions, and hence we need other performance measures.

Such questions often arise in applications like information retrieval and web search. For example, in information retrieval, we often want to know "What percentage of the retrieved information is of interest to users?" and "How much of the information the user is interested in is retrieved?" For such questions, *precision* and *recall* are better choices.

In binary classification problems, there are four combinations of the ground-truth class and the predicted class, namely true positive, false positive, true negative, and false negative, and we denote the number of samples in each case as $TP$, $FP$, $TN$, and $FN$, respectively. Then, $TP + FP + TN + FN =$ total number of samples. The four combinations can be displayed in a *confusion matrix*, as shown in ◘ Table 2.1. Then, the precision $P$ and the recall $R$ are, respectively, defined as

$$P = \frac{TP}{TP + FP}, \tag{2.8}$$

$$R = \frac{TP}{TP + FN}. \tag{2.9}$$

◘ **Tab. 2.1** The confusion matrix of binary classification

| Ground-truth class | Predicted class | |
|---|---|---|
| | Positive | Negative |
| Positive | $TP$ | $FN$ |
| Negative | $FP$ | $TN$ |

Precision and recall are contradictory. Generally speaking, the recall is often low when the precision is high, and the precision is often low when the recall is high. For example, to pick more ripe watermelons, we can increase the number of picked watermelons because, in an extreme case, if we pick all watermelons, then all ripe watermelons are picked as well. However, by doing so, the precision would be very low. On the other hand, if we wish the proportion of ripe watermelons to be high, then we should only pick watermelons that we are sure of. However, doing so could miss many ripe watermelons, and hence the recall becomes low. Typically, we can achieve high precision and high recall at the same time only in simple problems.

Quite often, we can use the learner's predictions to sort the samples by how likely they are positive. That is, the samples that are most likely to be positive are at the top of the ranking list, and the samples that are least likely to be positive are at the bottom. Starting from the top of the ranking list, we can incrementally label the samples as positive to calculate the precision and recall at each increment. Then, plotting the precisions as y-axis and the recalls as x-axis gives the Precision-Recall Curve (P-R curve). The plots of P-R curves are called P-R plots. ◘ Figure 2.3 gives an example of P-R curve.

Taking information retrieval as an example, the precision and recall can be calculated by sequentially returning each piece of information that the user might be interested in.

Also called *PR curve* or *PR plot*.

P-R plots intuitively show the overall precision and recall of learners. When comparing two learners, if the P-R curve of one learner entirely encloses the curve of another learner, then the performance of the first learner is superior. For example, in ◘ Figure 2.3, learner A is better than learner C. However, when the P-R curves intersect, such as curve A and curve B, we cannot

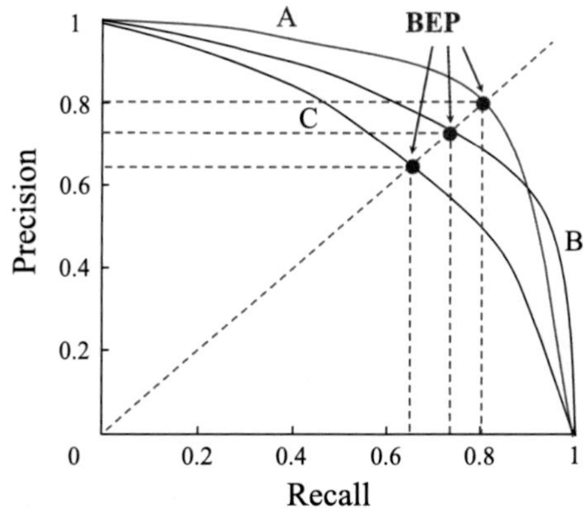


**Fig. 2.3** P-R curve and break-even points

say which learner is generally better and can only compare them at a specific precision or recall. Nevertheless, people often insist on finding out the best learner even if there exist intersections. A reasonable solution is to compare the areas under the P-R curves, which, to some extent, represent the proportion of cases when both precision and recall are relatively high. However, the areas are not easy to compute, and hence we often seek alternative performance measures that consider precision and recall simultaneously.

One alternative is Break-Even Point (BEP), which is the value when precision and recall are equal. For example, in ▪ Figure 2.3, the BEP of learner C is 0.64, and learner A is better than learner B, according to the BEP.

However, BEP could be oversimplified, and a more commonly used alternative is $F1$-measure:

$$F1 = \frac{2 \times P \times R}{P + R} = \frac{2 \times TP}{\text{total number of samples} + TP - TN}. \tag{2.10}$$

$F1$ is the harmonic mean of precision and recall: $\frac{1}{F1} = \frac{1}{2} \cdot \left(\frac{1}{P} + \frac{1}{R}\right)$.

In some applications, the importance of precision and recall are different. For example, precision is more critical in recommender systems since it is more desirable that the recommended content is of interest to the user and disturbs the user as little as possible. On the other hand, recall is more critical in criminal information retrieval systems since we wish to miss as few criminals as possible. The general form of $F1$-measure is $F_\beta$, which allows us to specify our preference over precision and recall, and is defined as

$F_\beta$ is the weighted harmonic mean: $\frac{1}{F_\beta} = \frac{1}{1+\beta^2} \cdot \left(\frac{1}{P} + \frac{\beta^2}{R}\right)$.

The Harmonic mean emphasizes more on smaller values compared to the arithmetic mean ($\frac{P+R}{2}$) and the geometric mean ($\sqrt{P \times R}$).

$$F_\beta = \frac{(1+\beta^2) \times P \times R}{(\beta^2 \times P) + R}, \tag{2.11}$$

where $\beta > 0$ gives the relative importance of recall to precision (Van Rijsbergen 1979). When $\beta = 1$, it reduces to the standard $F1$; when $\beta > 1$, recall is more important; when $\beta < 1$, precision is more important.

Sometimes we may have multiple confusion matrices in binary classification problems. For example, there is one confusion matrix for each round of training and testing. Also, there are multiple confusion matrices when we do training and testing on multiple data sets to estimate the overall performance. Besides, there is one confusion matrix for every class in multi-class classification problems. In all of these cases, we need to investigate the overall precision and recall on $n$ binary confusion matrices.

A straightforward approach is to calculate the precision and the recall for each confusion matrix, denoted by $(P_1, R_1)$, $(P_2, R_2), \ldots, (P_n, R_n)$. By taking the averages, we have the macro-$P$, the macro-$R$, and the macro-$F1$:

$$\text{macro-}P = \frac{1}{n}\sum_{i=1}^{n} P_i, \tag{2.12}$$

$$\text{macro-}R = \frac{1}{n}\sum_{i=1}^{n} R_i, \tag{2.13}$$

$$\text{macro-}F1 = \frac{2 \times \text{macro-}P \times \text{macro-}R}{\text{macro-}P + \text{macro-}R}. \tag{2.14}$$

We can also calculate element-wise averages across the confusion matrices to get $\overline{TP}, \overline{FP}, \overline{TN}, \overline{FN}$, and then take the averages to obtain the micro-$P$, the micro-$R$, and the micro-$F1$:

$$\text{micro-}P = \frac{\overline{TP}}{\overline{TP} + \overline{FP}}, \tag{2.15}$$

$$\text{micro-}R = \frac{\overline{TP}}{\overline{TP} + \overline{FN}}, \tag{2.16}$$

$$\text{micro-}F1 = \frac{2 \times \text{micro-}P \times \text{micro-}R}{\text{micro-}P + \text{micro-}R}. \tag{2.17}$$

### 2.3.3 ROC and AUC

Since the predictions from learners are often in the form of real values or probabilities, we can compare the predicted values against a classification threshold, that is, classify a sample as positive if the prediction value is greater than the threshold and classify it as negative otherwise. For example, typical neural networks predict real values in the interval [0.0, 1.0] for testing samples. We can compare the predicted values with 0.5, and classify a sample as positive if its predicted value is greater than 0.5, and negative otherwise. Hence, the predicted real values or probabilities directly determine the generalization ability. In practice, we sort the testing samples by the predicted real values or probabilities in descending order such that potential positive samples are at the top of the list. After that, we put a *cut point* in the sorted list and classify the samples above it as positive and the rest as negative.

See Chap. 5 for neural networks.

The position of the cut point depends on the specific application. For example, we move the cut point toward the top of the list if precision is more critical than recall, and move it toward the bottom otherwise. Consequently, the ranking quality reflects the learner's "expected generalization ability" for different tasks or the generalization ability for "typical cases". The Receiver Operating Characteristics (ROC) curve follows the above idea to measure the generalization ability of learners.

The ROC curve was initially developed for radar detection of enemy aircraft in World War II and then introduced to psychology and medical applications in the 1960s–1970s. Later on, it was introduced to machine learning (Spackman 1989). Similar to the P-R curve discussed in Sect. 2.3.2, we sort the samples by the predictions and then obtain two measures by gradually moving the cut point from the top toward the bottom of the ranked list. Using those two measures as x-axis and y-axis gives the ROC curve. Unlike precision and recall in P-R curves, the y-axis in ROC curves is True Positive Rate (TPR), and the x-axis is False Positive Rate (FPR). Reusing the notations in ◘ Table 2.1, these two measures are, respectively, defined as

$$\mathrm{TPR} = \frac{TP}{TP + FN}, \tag{2.18}$$

$$\mathrm{FPR} = \frac{FP}{TN + FP}. \tag{2.19}$$

The plot showing ROC curves is called an ROC plot. ◘ Figure 2.4a gives an example of an ROC plot in which the diagonal corresponds to the "random guessing" model, and the point (0, 1) corresponds to the "ideal model" that places all positive samples before negative samples.

In practice, we only have finite pairs of (FPR, TPR) coordinates for drawing the ROC plot since the testing samples are finite. Hence, the ROC curve may not look smooth like the one in ◘ Figure 2.4a but is only an approximation, like the one shown in ◘ Figure 2.4b. The plotting process is as follows: given $m^+$ positive samples and $m^-$ negative samples, we first

The same problem occurs when drawing P-R plots, but we deferred the discussion until now to facilitate the introduction of calculating AUC.

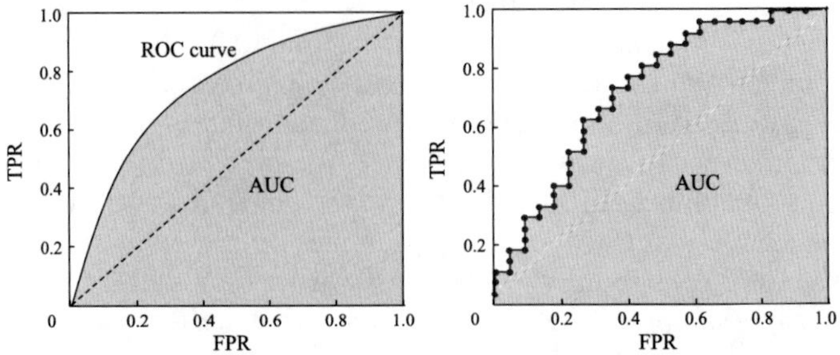


(a) ROC curve and AUC. (b) ROC curve and AUC with finite samples.

**Fig. 2.4** An illustration of ROC curve and AUC

sort all samples by the learner's predictions, and then set the threshold to maximum, that is, predicting all samples as negative. At this moment, both TPR and FPR are 0, so we mark at coordinate (0, 0). Then, we gradually decrease the threshold to the predicted value of each sample along the sorted list, that is, the samples are classified as positive successively. Let $(x, y)$ denote the previous coordinate, we put a mark at $(x, y + \frac{1}{m^+})$ if the current samples are true positive, and we put a mark at $(x + \frac{1}{m^-}, y)$ if the current samples are false positive. By connecting all adjacent marked points, we have the ROC curve.

Like P-R plots, we say learner A is better than learner B if A's ROC curve entirely encloses B's ROC curve. However, when there exist intersections, no learner is generally better than the other. One way of comparing intersected ROC curves is to calculate the areas under the ROC curves, that is, Area Under ROC Curve (AUC), as shown in ◘ Figure 2.4.

By its definition, AUC can be calculated by integrating the areas under the steps of ROC curve. Suppose that the ROC curve is obtained by sequentially connecting the points $\{(x_1, y_1), (x_2, y_2), \ldots, (x_m, y_m)\}$, where $x_1 = 0$ and $x_m = 1$. Then, as illustrated in ◘ Figure 2.4b, the AUC is estimated as

$$\mathrm{AUC} = \frac{1}{2}\sum_{i=1}^{m-1}(x_{i+1} - x_i) \cdot (y_i + y_{i+1}). \tag{2.20}$$

AUC is closely related to ranking errors since it considers the ranking quality of predictions. Let $m^+$ denote the number of positive samples, $m^-$ denote the number of negative samples, $D^+$ denote the set of positive samples, and $D^-$ denote the set of negative samples. Then, the ranking *loss* is defined as

$$\ell_{\mathrm{rank}} = \frac{1}{m^+m^-}\sum_{\boldsymbol{x}^+\in D^+}\sum_{\boldsymbol{x}^-\in D^-}\left(\mathbb{I}(f(\boldsymbol{x}^+) < f(\boldsymbol{x}^-)) + \frac{1}{2}\mathbb{I}(f(\boldsymbol{x}^+) = f(\boldsymbol{x}^-))\right). \tag{2.21}$$

For each pair of positive sample $\boldsymbol{x}^+$ and negative sample $\boldsymbol{x}^-$, the ranking loss applies a penalty of 1 if the predicted value of the positive sample is lower than that of the negative sample, and a penalty of 0.5 applies when the predicted values are equal. Suppose $(x, y)$ is the coordinate of a positive sample on the ROC curve, then $x$ is the proportion of negative samples ranked above this positive sample (i.e., FPR). Hence, the ranking loss $\ell_{\text{rank}}$ corresponds to the area above the ROC curve, that is,

$$\text{AUC} = 1 - \ell_{\text{rank}}. \tag{2.22}$$

### 2.3.4 Cost-Sensitive Error Rate and Cost Curve

In some problems, the consequences of making different errors are not the same. Taking medical diagnosis as an example, according to our earlier discussions, we receive the same amount of penalty for misclassifying someone as healthy or unhealthy. However, it turns out that misclassifying a sick patient as healthy is more serious since it risks the life of the patient. Another example is the access control system in which denying the access of normal users leads to unpleasant user experience while allowing intruders to enter causes security breach. In such cases, we need to assign *unequal costs* to different errors.

For binary classification problems, we can leverage domain knowledge to design a *cost matrix*, as shown in ◘ Table 2.2, where $\text{cost}_{ij}$ represents the cost of misclassifying a sample of class $i$ as class $j$. In general, $\text{cost}_{ii} = 0$, and $\text{cost}_{01} > \text{cost}_{10}$ if misclassifying class 0 as class 1 costs more than the other way around. The larger the difference between the costs is, the larger the difference between $\text{cost}_{01}$ and $\text{cost}_{10}$ will be.

Normally, we care more about the cost ratios rather than the absolute values, e.g., $\text{cost}_{01} : \text{cost}_{10} = 5 : 1$ is equivalent to $\text{cost}_{01} : \text{cost}_{10} = 50 : 10$.

◘ **Tab. 2.2** Cost matrix of binary classification

| Ground-truth class | Predicted class | |
|---|---|---|
| | Class 0 | Class 1 |
| Class 0 | 0 | $\text{cost}_{01}$ |
| Class 1 | $\text{cost}_{10}$ | 0 |

Almost all performance measures we discussed so far implicitly assumed equal-cost. For example, error rate (2.4) counts the number of errors without considering the different consequences. With unequal costs, however, we no longer minimize the counts but the *total cost*. For binary classification problems, we can call class 0 as the positive class and class 1 as the negative class. Let $D^+$ and $D^-$ denote, respectively, the set of positive samples and the set of negative samples. Then, based on Table 2.2, the *cost-sensitive* error rate is defined as

$$E(f; D; \text{cost}) = \frac{1}{m}\left(\sum_{\boldsymbol{x}_i \in D^+} \mathbb{I}(f(\boldsymbol{x}_i) \neq y_i) \times \text{cost}_{01} + \sum_{\boldsymbol{x}_i \in D^-} \mathbb{I}(f(\boldsymbol{x}_i) \neq y_i) \times \text{cost}_{10}\right). \tag{2.23}$$

Similarly, we can also define the distribution-based cost-sensitive error rate and the cost-sensitive version of accuracy. It is also possible to define cost-sensitive performance measures for multiclass cases by allowing $i$ and $j$ of $\text{cost}_{ij}$ to take values other than 0 and 1.

With unequal costs, we find the expected total costs of learners from *cost curves* rather than ROC curves. The x-axis of cost curves is the probability cost of positive class:

See Exercise 2.7.

$$P(+)\text{cost} = \frac{p \times \text{cost}_{01}}{p \times \text{cost}_{01} + (1-p) \times \text{cost}_{10}}, \tag{2.24}$$

where $p \in [0, 1]$ is the probability of a sample being positive. The y-axis is the normalized cost which takes values from [0, 1]:

*Normalization* is the process of mapping values from different ranges to a fixed range, e.g., [0, 1]. See Exercise 2.8.

$$\text{cost}_{\text{norm}} = \frac{\text{FNR} \times p \times \text{cost}_{01} + \text{FPR} \times (1-p) \times \text{cost}_{10}}{p \times \text{cost}_{01} + (1-p) \times \text{cost}_{10}}, \tag{2.25}$$

where FPR is the false positive rate defined in (2.19) and FNR = 1 − TPR is the false negative rate. We can draw a cost curve as follows: since every point (FPR, TPR) on the ROC curve corresponds to a line segment on the cost plane, we

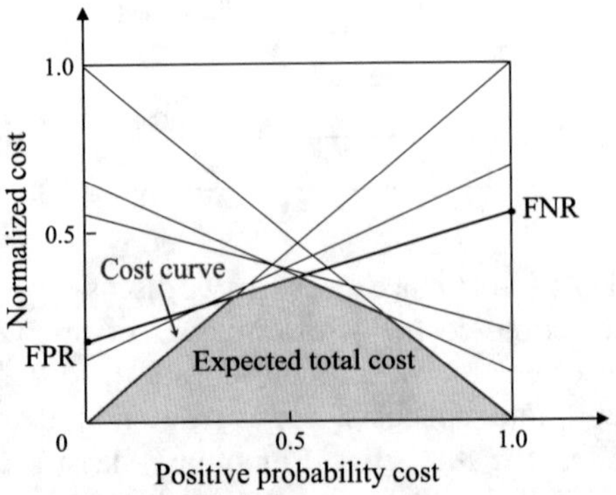


**Fig. 2.5** The cost curve and expected total cost

can calculate the FNR and draw a line segment from (0, FPR) to (1, FNR). Then, the area under the line segment represents the expected total cost for the given $p$, FPR, and TPR. By converting all points on the ROC curve to line segments on the cost plane, the expected total cost is given by the area under the lower bound of all line segments, as shown in ◘ Figure 2.5.

## 2.4 Comparison Test

It seems straightforward to compare learners using evaluation methods and performance measures. For example, we use an evaluation method to measure the performance of learners and then compare them. However, how should we make the "comparison"? Should we check which of the measured values is better? Performance comparisons are indeed far more complicated than we thought due to the following reasons. Firstly, we wish to compare the generalization performance of learners, but evaluation methods only measure performance on testing sets, that is, the comparisons may not reflect the actual generalization performance. Secondly, testing performance depends on the choice of the testing set, e.g., the results on two different-sized testing sets, or two equal-sized sets but with different samples, could be different. Finally, many machine learning algorithms have some build-in random behavior, which means that we may obtain different results even for the same parameter settings and testing set. Then, what is the appropriate way of comparing the performance of learners?

*Hypothesis testing* is one of the techniques to compare the performance of learners. Suppose that we observe learner A outperforms learner B on a testing set. Then, hypothesis testing can help us check whether the generalization performance of learner A is better than that of learner B in the statistical sense and how significant it is. In the following discussions, we introduce two basic hypothesis tests and several methods to compare learners' performance. For ease of discussion, the rest of this section assumes error rate, denoted by $\epsilon$, to be the default performance measure.

See Wellek (2010) for more information about hypothesis testing.

### 2.4.1 Hypothesis Testing

In hypothesis testing, a hypothesis is a statement or assumption about the learner's generalization error rate distribution, e.g., "$\epsilon = \epsilon_0$". In practice, however, we only have the testing error rate $\hat{\epsilon}$ but not the generalization error rate $\epsilon$. Though $\hat{\epsilon}$ and $\epsilon$ may not be identical, they are, intuitively, likely to be close.

Hence, we can use the testing error rate distribution to infer the generalization error rate distribution.

A generalization error rate of $\epsilon$ means that the learner has a probability of $\epsilon$ to make an incorrect prediction. A testing error rate of $\hat{\epsilon}$ means that the learner misclassified $\hat{\epsilon} \times m$ samples in a testing set of $m$ samples. Suppose the testing samples are drawn *i.i.d.* from the population distribution. Then, the probability that a learner with a generalization error rate of $\epsilon$ misclassifies $m'$ samples and correctly classifies the rest is $\binom{m}{m'}\epsilon^{m'}(1-\epsilon)^{m-m'}$. Consequently, for a learner with a generalization error rate of $\epsilon$, the probability of misclassifying $\hat{\epsilon} \times m$ samples, which is also the probability that the testing error rate being $\hat{\epsilon}$ on a testing set of $m$ samples, is

$$P(\hat{\epsilon}; \epsilon) = \binom{m}{\hat{\epsilon} \times m}\epsilon^{\hat{\epsilon}\times m}(1-\epsilon)^{m-\hat{\epsilon}\times m}. \tag{2.26}$$

By solving $\partial P(\hat{\epsilon}; \epsilon)/\partial\epsilon = 0$ with the testing error rate, we observe that $P(\hat{\epsilon}; \epsilon)$ is maximized when $\epsilon = \hat{\epsilon}$, and $P(\hat{\epsilon}; \epsilon)$ decreases as $|\epsilon - \hat{\epsilon}|$ increases. The observation follows the binomial distribution, and, as shown in ◼ Figure 2.6, the learner is most likely to misclassify 3 samples out of 10 samples when $\epsilon = 0.3$.

We can use *binomial test* to verify hypotheses such as "$\epsilon \leqslant 0.3$", that is, the generalization error rate is not greater than 0.3. More generally, for the hypothesis "$\epsilon \leqslant \epsilon_0$", (2.27) gives the maximum observable error rate within a probability of $1-\alpha$. The probability is also known as *confidence*, corresponding to the non-shaded part of ◼ Figure 2.6.

Common values of $\alpha$ include 0.05 and 0.1. We use a large $\alpha$ in ◼ Figure 2.6 for illustration purposes.

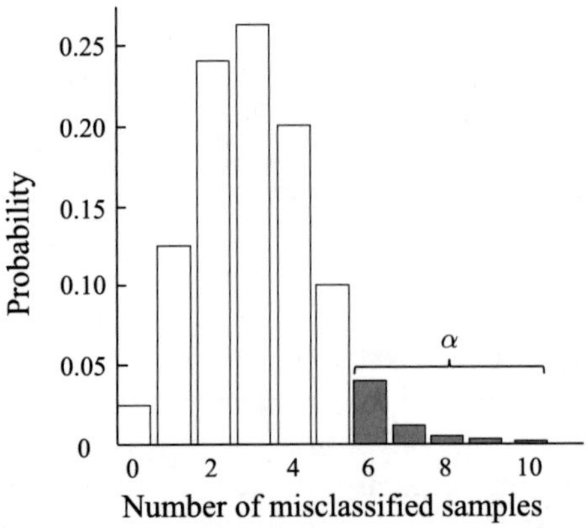


**Fig. 2.6** Binomial distribution ($m = 10$, $\epsilon = 0.3$)

$$\bar{\epsilon} = \min \epsilon \quad \text{s.t.} \quad \sum_{i=\epsilon \times m+1}^{m} \binom{m}{i} \epsilon_0^i (1-\epsilon_0)^{m-i} < \alpha. \tag{2.27}$$

"s.t." stands for "subject to", indicating that the expression on the right-hand side must be met while solving the expression on the left-hand side.

If the testing error rate $\hat{\epsilon}$ is greater than the critical value $\bar{\epsilon}$, then, according to the binomial test, the hypothesis "$\epsilon \leqslant \epsilon_0$" cannot be rejected at the significance level of $\alpha$, that is, the learner's generalization error rate is not greater than $\epsilon_0$ at the confidence level of $1-\alpha$; otherwise, we reject the hypothesis, that is, the learner's generalization error rate is greater than $\epsilon_0$ at the significance level of $\alpha$.

We can compute the critical value with the assistance of qbinom$(1-\alpha, m, \epsilon_0)$ in R or icdf('Binomial',$1-\alpha, m, \epsilon_0$) in MATLAB.

We often obtain multiple testing error rates from cross-validation or by doing multiple hold-out evaluations. In such cases, we can use $t$-test. Let $\hat{\epsilon}_1, \hat{\epsilon}_2, \ldots, \hat{\epsilon}_k$ denote the $k$ testing error rates, then the average testing error rate $\mu$ and variance $\sigma^2$ are, respectively,

R is an open-source scripting language for statistical computing. See ► http://www.r-project.org.

$$\mu = \frac{1}{k} \sum_{i=1}^{k} \hat{\epsilon}_i, \tag{2.28}$$

$$\sigma^2 = \frac{1}{k-1} \sum_{i=1}^{k} (\hat{\epsilon}_i - \mu)^2. \tag{2.29}$$

We can regard these $k$ testing error rates as *i.i.d.* samples of the generalization error rate $\epsilon_0$, and hence the variable

$$\tau_t = \frac{\sqrt{k}(\mu - \epsilon_0)}{\sigma} \tag{2.30}$$

follows a $t$-distribution with $k-1$ degrees of freedom, as shown in ■ Figure 2.7.

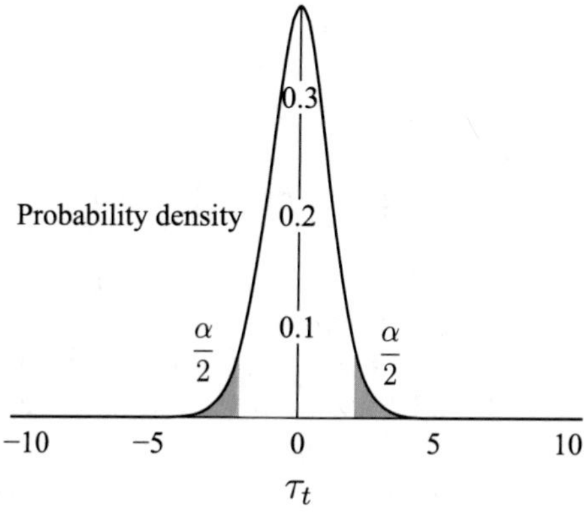


**Fig. 2.7** $t$-distribution ($k = 10$)

For the hypothesis "$\mu = \epsilon_0$" and significance level $\alpha$, we can calculate the maximum observable error rate (i.e., the critical value) within a probability of $1 - \alpha$, where $\epsilon_0$ is the average testing error rate. Here, we employ a two-tailed hypothesis, and there are $\alpha/2$ shaded areas at both tails of the distribution, as shown in ▪ Figure 2.7. Let $(-\infty, t_{-\alpha/2}]$ and $[t_{\alpha/2}, \infty)$ denote the ranges of the two shaded areas, respectively. If $\tau_t$ is within the critical value range $[t_{-\alpha/2}, t_{\alpha/2}]$, then the hypothesis "$\mu = \epsilon_0$" cannot be rejected, that is, the generalization error rate is $\epsilon_0$ at the confidence level of $1 - \alpha$; otherwise, we reject the hypothesis, that is, the generalization error rate is significantly different from $\epsilon_0$ at this confidence level. 0.05 and 0.1 are commonly used significance levels, and ▪ Table 2.3 shows some commonly used critical values for $t$-test.

The critical values $t_{\alpha/2}$ can be computed by qt$(1 - \alpha/2, k - 1)$ in R or icdf$('T', 1 - \alpha/2, k - 1)$ in MATLAB.

▪ **Tab. 2.3** Commonly used critical values for two-tailed $t$-test

| $\alpha$ | $k$ | | | | |
|---|---|---|---|---|---|
| | 2 | 5 | 10 | 20 | 30 |
| 0.05 | 12.706 | 2.776 | 2.262 | 2.093 | 2.045 |
| 0.10 | 6.314 | 2.132 | 1.833 | 1.729 | 1.699 |

Both methods introduced above compare the generalization performance of a single learner. In the following section, we discuss several hypothesis testing methods for comparing the generalization performance of multiple learners.

### 2.4.2 Cross-Validated *t*-Test

For two learners A and B, let $\epsilon_1^A, \epsilon_2^A, \ldots, \epsilon_k^A$ and $\epsilon_1^B, \epsilon_2^B, \ldots,$ $\epsilon_k^B$ denote their testing error rates obtained from $k$-fold cross-validation, where $i$ indicates the $i$th fold. Then, we can use $k$-fold cross-validated paired $t$-tests to compare the two learners. The basic idea is that if the performance of the two learners is the same, then the testing error rates should be the same on the same training and testing sets, that is, $\epsilon_i^A = \epsilon_i^B$.

To be specific, for the $k$ pairs of testing error rates obtained from the $k$-fold cross-validation, we calculate the difference of each pair of results as $\Delta_i = \epsilon_i^A - \epsilon_i^B$. Then, the mean of the differences should be zero if the two learners have the same performance. Consequently, based on the differences $\Delta_1, \Delta_2, \ldots, \Delta_k$, we perform a $t$-test on the hypothesis "learner A and learner B have the same performance". We calculate the mean $\mu$ and variance $\sigma^2$ of the differences, and if

$$\tau_t = \left| \frac{\sqrt{k}\mu}{\sigma} \right| \tag{2.31}$$

is less than the critical value $t_{\alpha/2,k-1}$ at the significance level of $\alpha$, then the hypothesis cannot be rejected, that is, there is no significant difference in the learners' performance; otherwise, these two learners have significantly different performance, and the one with the lower mean error rate is superior. Here, $t_{\alpha/2,k-1}$ is the critical value of a $t$-distribution with $k - 1$ degrees of freedom and a tail of $\alpha/2$.

The above hypothesis test assumes the testing error rates are *i.i.d.* samples of the generalization error rate. However, due to the finite training data, the training sets of different rounds are often overlapped in evaluation methods such as cross-validation. Therefore, the testing error rates are indeed not independent, resulting in an overestimated probability for the hypothesis to be true. To alleviate the problem, we can use "5 × 2 cross-validation" (Dietterich 1998).

As the name suggests, 5 × 2 cross-validation repeats two-fold cross-validation five times, where the data is randomly shuffled before each two-fold cross-validation such that the data splitting is different in the five rounds of cross-validations. For example, for two learners A and B, we obtain their testing error rates of the $i$th two-fold cross-validation. Then, we calculate the difference between their error rates of the first fold, denoted by $\Delta_i^1$, and the difference between their error rates of the second fold, denoted by $\Delta_i^2$. To alleviate the dependency of testing error rates, we calculate the variance of each two-fold cross-validation as $\sigma_i^2 = \left(\Delta_i^1 - \frac{\Delta_i^1+\Delta_i^2}{2}\right)^2 + \left(\Delta_i^2 - \frac{\Delta_i^1+\Delta_i^2}{2}\right)^2$; however, only the mean of the first two-fold cross-validation is calculated as $\mu = 0.5(\Delta_1^1 + \Delta_1^2)$. The variable

$$\tau_t = \frac{\mu}{\sqrt{0.2\sum_{i=1}^{5}\sigma_i^2}} \tag{2.32}$$

follows a $t$-distribution with five degrees of freedom, where its two-tailed critical value $t_{\alpha/2,5}$ is 2.5706 when $\alpha = 0.05$, and 2.0150 when $\alpha = 0.1$.

### 2.4.3 McNemar's Test

For binary classification problems, the hold-out method estimates not only the testing error rates of both learner A and learner B, but also the classification difference of the two learners, that is, the numbers of both correct, both incorrect, and one correct while the other incorrect. These numbers form a *contingency table*, as shown in Table 2.4.



**Tab. 2.4** The contingency table of two learners

| Algorithm B | Algorithm A | |
|---|---|---|
| | Correct | Incorrect |
| Correct | $e_{00}$ | $e_{01}$ |
| Incorrect | $e_{10}$ | $e_{11}$ |

If the performance of the two learners are the same, then we should have $e_{01} = e_{10}$. The variable $|e_{01} - e_{10}|$ follows a Gaussian distribution. McNemar's test considers the variable

Since $e_{01} + e_{10}$ is often small, we need the continuity correction, that is, $-1$ in the numerator.

$$\tau_{\chi^2} = \frac{(|e_{01} - e_{10}| - 1)^2}{e_{01} + e_{10}}, \tag{2.33}$$

which follows a chi-square distribution with one degree of freedom, that is, the distribution of the sum of squared standard normal random variables. At the significance level of $\alpha$, the hypothesis cannot be rejected if the variable is less than the critical value $\chi^2_\alpha$, that is, there is no significant difference between the performance of those two learners; otherwise, the hypothesis is rejected, that is, the performance of those two learners is significantly different, and the learner with smaller average error rate is superior. The critical value of $\chi^2$ test with one degree of freedom is 3.8415 when $\alpha = 0.05$ and 2.7055 when $\alpha = 0.1$.

The critical values $\chi^2_\alpha$ can be computed by $\texttt{qchisq}(1-\alpha, k-1)$ in R or $\texttt{icdf}('\texttt{Chisquare}', 1-\alpha, k-1)$ in MATLAB, where $k = 2$ is the number of algorithms being compared.

## 2.4.4 Friedman Test and Nemenyi Post-hoc Test

Both the cross-validated $t$-test and McNemar's test compare two algorithms on a single data set. However, in some cases, comparisons are made for multiple algorithms on multiple data sets. In such cases, we can compare each pair of algorithms on each data set using a cross-validated $t$-test or a McNemar's test. Alternatively, we can use the following ranking-based Friedman test to compare all algorithms on all data sets at once.

Suppose that we are comparing algorithms A, B, and C on four data sets $D_1$, $D_2$, $D_3$, and $D_4$. We first use either hold-out or cross-validation to obtain each algorithm's testing result on each data set. Then, we sort the algorithms on each data set by their testing performance and assign the ranks 1, 2, . . ., accordingly, where the algorithms with the same testing performance share the averaged rank. For example, as shown in Table 2.5, on data sets $D_1$ and $D_3$, A is the best, B is the second, and

C is the last; on data set $D_2$, A is the best, and B and C have the same performance. After collecting all the ranks, we calculate the average rank of each algorithm as the last row of ◘ Table 2.5.

**◘ Tab. 2.5** The ranking table of algorithms

| Data set | Algorithm A | Algorithm B | Algorithm C |
|---|---|---|---|
| $D_1$ | 1 | 2 | 3 |
| $D_2$ | 1 | 2.5 | 2.5 |
| $D_3$ | 1 | 2 | 3 |
| $D_4$ | 1 | 2 | 3 |
| Average rank | 1 | 2.125 | 2.875 |

According to the Friedman test, the algorithms with the same performance should have the same average rank. Let $k$ denote the number of algorithms, $N$ denote the number of data sets, and $r_i$ denote the average rank of the $i$th algorithm. Here, we ignore the ties to simplify our discussion. Then, the mean and the variance of $r_i$ are $(k+1)/2$ and $(k^2-1)/12N$, respectively. The variable

$$\begin{aligned}\tau_{\chi^2} &= \frac{k-1}{k} \cdot \frac{12N}{k^2-1} \sum_{i=1}^{k} \left(r_i - \frac{k+1}{2}\right)^2 \\ &= \frac{12N}{k(k+1)} \left(\sum_{i=1}^{k} r_i^2 - \frac{k(k+1)^2}{4}\right)\end{aligned} \tag{2.34}$$

follows a $\chi^2$ distribution with $k-1$ degrees of freedom when $k$ and $N$ are large.

The "original Friedman test" described above is too conservative, and hence the following variable is often used instead:

The "original Friedman test" requires a large $k$ (e.g., > 30), and tends to return no significant difference when $k$ is small.

$$\tau_F = \frac{(N-1)\tau_{\chi}^2}{N(k-1)-\tau_{\chi}^2}, \tag{2.35}$$

where $\tau_{\chi^2}$ is given by (2.34). $\tau_F$ follows a $F$-distribution with $k-1$ and $(k-1)(N-1)$ degrees of freedom. ◘ Table 2.6 shows some commonly used critical values for $F$-test.

2

The critical values for $F$-test can be computed by qf($1-\alpha, k-1, (k-1)(N-1)$) in R or icdf($'\mathrm{F}'$, $1-\alpha, k-1, (k-1)*(N-1)$) in MATLAB.

**Tab. 2.6** Commonly used critical values for $F$-test

$\alpha = 0.05$

| $N$ | $k$ = 2 | 3 | 4 | 5 | 6 | 7 | 8 | 9 | 10 |
|---|---|---|---|---|---|---|---|---|---|
| 4 | 10.128 | 5.143 | 3.863 | 3.259 | 2.901 | 2.661 | 2.488 | 2.355 | 2.250 |
| 5 | 7.709 | 4.459 | 3.490 | 3.007 | 2.711 | 2.508 | 2.359 | 2.244 | 2.153 |
| 8 | 5.591 | 3.739 | 3.072 | 2.714 | 2.485 | 2.324 | 2.203 | 2.109 | 2.032 |
| 10 | 5.117 | 3.555 | 2.960 | 2.634 | 2.422 | 2.272 | 2.159 | 2.070 | 1.998 |
| 15 | 4.600 | 3.340 | 2.827 | 2.537 | 2.346 | 2.209 | 2.104 | 2.022 | 1.955 |
| 20 | 4.381 | 3.245 | 2.766 | 2.492 | 2.310 | 2.179 | 2.079 | 2.000 | 1.935 |

$\alpha = 0.1$

| $N$ | $k$ = 2 | 3 | 4 | 5 | 6 | 7 | 8 | 9 | 10 |
|---|---|---|---|---|---|---|---|---|---|
| 4 | 5.538 | 3.463 | 2.813 | 2.480 | 2.273 | 2.130 | 2.023 | 1.940 | 1.874 |
| 5 | 4.545 | 3.113 | 2.606 | 2.333 | 2.158 | 2.035 | 1.943 | 1.870 | 1.811 |
| 8 | 3.589 | 2.726 | 2.365 | 2.157 | 2.019 | 1.919 | 1.843 | 1.782 | 1.733 |
| 10 | 3.360 | 2.624 | 2.299 | 2.108 | 1.980 | 1.886 | 1.814 | 1.757 | 1.710 |
| 15 | 3.102 | 2.503 | 2.219 | 2.048 | 1.931 | 1.845 | 1.779 | 1.726 | 1.682 |
| 20 | 2.990 | 2.448 | 2.182 | 2.020 | 1.909 | 1.826 | 1.762 | 1.711 | 1.668 |

The performance of algorithms is significantly different if the hypothesis "algorithms' performance is the same" is rejected. Then, we use a *post-hoc test* to further distinguish the algorithms. A common choice is the Nemenyi post-hoc test, which calculates the critical difference *CD* of the average rank difference as

$$CD = q_\alpha\sqrt{\frac{k(k+1)}{6N}}. \quad (2.36)$$

$q_\alpha$ is the critical value of Tukey distribution, which can be computed by qtukey($1-\alpha, k$, inf) / sqrt(2) in R.

Table 2.7 shows some commonly used values of $q_\alpha$ for $\alpha = 0.05$ and $\alpha = 0.1$. If the average rank difference of two algorithms is greater than the critical difference, then the hypothesis "algorithms' performance is the same" is rejected at the corresponding confidence level.

**Tab. 2.7** Commonly used values of $q_\alpha$ for Nemenyi test

| $\alpha$ | $k$ = 2 | 3 | 4 | 5 | 6 | 7 | 8 | 9 | 10 |
|---|---|---|---|---|---|---|---|---|---|
| 0.05 | 1.960 | 2.344 | 2.569 | 2.728 | 2.850 | 2.949 | 3.031 | 3.102 | 3.164 |
| 0.10 | 1.645 | 2.052 | 2.291 | 2.459 | 2.589 | 2.693 | 2.780 | 2.855 | 2.920 |

Taking the data in Table 2.5 as an example, we first calculate $\tau_F = 24.429$ according to (2.34) and (2.35). Then, from Table 2.6, we realize $\tau_F$ is greater than the critical value 5.143 when $\alpha = 0.05$. Hence, the hypothesis "algorithms' perfor-

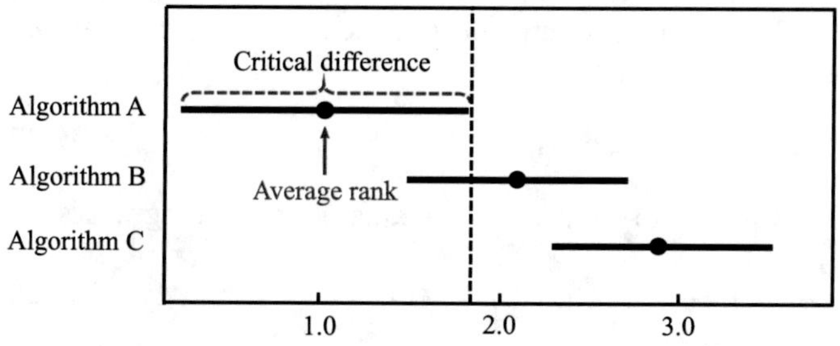


**Fig. 2.8** The plot of the Friedman test

mance is the same" is rejected. We proceed with the Nemenyi post-hoc test. From ◘ Table 2.7, we find $q_{0.05} = 2.344$ for $k = 3$, and hence the critical difference is $CD = 1.657$ according to (2.36). Based on the average ranks in ◘ Table 2.5, neither the difference between algorithms A and B nor the difference between algorithms B and C is greater than the critical difference, that is, there is no significant difference between their performance. However, the test confirms that the performance of algorithms A and C are significantly different since their difference is greater than the critical difference.

We can use a plot to illustrate the Friedman test, e.g., ◘ Figure 2.8 illustrates the Friedman test for ◘ Table 2.5, where the y-axis shows the algorithms, and the x-axis shows the average ranks. The dots mark the average ranks of algorithms, and the line segments centered at the dots are the corresponding critical difference. The performance of the two algorithms is not significantly different if their line segments overlap; otherwise, their performance is significantly different. From ◘ Figure 2.8, we can easily observe that there is no significant difference between algorithms A and B since their line segments overlap. On the other hand, algorithm A is better than algorithm C since their line segments do not overlap while A has a higher rank.

## 2.5 Bias and Variance

In addition to estimating the generalization performance of learning algorithms, people often wish to understand "why" learning algorithms have such performance. An essential tool for understanding the generalization performance of algorithms is the *bias-variance decomposition*, which decomposes the expected generalization error of learning algorithms.

For different training sets, the learning outcomes are often different, although the training samples are drawn from the same distribution. Let $\boldsymbol{x}$ be a testing sample, $y_D$ be the label of $\boldsymbol{x}$ in the data set $D$, $y$ be the ground-truth label of $\boldsymbol{x}$, and

Potential noise may lead to $y_D \neq y$.

$f(\boldsymbol{x}; D)$ be the output of $\boldsymbol{x}$ predicted by the model $f$ trained on $D$. Then, in regression problems, the expected prediction of a learning algorithm is

$$\bar{f}(\boldsymbol{x}) = \mathbb{E}_D[f(\boldsymbol{x}; D)]. \tag{2.37}$$

The variance of using different equal-sized training sets is

$$var(\boldsymbol{x}) = \mathbb{E}_D\left[(f(\boldsymbol{x}; D) - \bar{f}(\boldsymbol{x}))^2\right]. \tag{2.38}$$

The noise is

$$\varepsilon^2 = \mathbb{E}_D\left[(y_D - y)^2\right]. \tag{2.39}$$

The difference between the expected output and the ground-truth label is called bias, that is,

$$bias^2(\boldsymbol{x}) = \left(\bar{f}(\boldsymbol{x}) - y\right)^2. \tag{2.40}$$

For ease of discussion, we assume the expectation of noise is zero, i.e., $\mathbb{E}_D[y_D - y] = 0$. By expanding and combining the polynomial, we can decompose the expected generalization error as follows:

Since the noise does not rely on $f$, the last term equals to 0 according to (2.37).

The last term equals to 0 since the expectation of noise is 0.

$$\begin{aligned}
E(f; D) &= \mathbb{E}_D\left[(f(\boldsymbol{x}; D) - y_D)^2\right] \\
&= \mathbb{E}_D\left[\left(f(\boldsymbol{x}; D) - \bar{f}(\boldsymbol{x}) + \bar{f}(\boldsymbol{x}) - y_D\right)^2\right] \\
&= \mathbb{E}_D\left[\left(f(\boldsymbol{x}; D) - \bar{f}(\boldsymbol{x})\right)^2\right] + \mathbb{E}_D\left[\left(\bar{f}(\boldsymbol{x}) - y_D\right)^2\right] \\
&\quad + \mathbb{E}_D\left[2\left(f(\boldsymbol{x}; D) - \bar{f}(\boldsymbol{x})\right)\left(\bar{f}(\boldsymbol{x}) - y_D\right)\right] \\
&= \mathbb{E}_D\left[\left(f(\boldsymbol{x}; D) - \bar{f}(\boldsymbol{x})\right)^2\right] + \mathbb{E}_D\left[\left(\bar{f}(\boldsymbol{x}) - y_D\right)^2\right] \\
&= \mathbb{E}_D\left[\left(f(\boldsymbol{x}; D) - \bar{f}(\boldsymbol{x})\right)^2\right] + \mathbb{E}_D\left[\left(\bar{f}(\boldsymbol{x}) - y + y - y_D\right)^2\right] \\
&= \mathbb{E}_D\left[\left(f(\boldsymbol{x}; D) - \bar{f}(\boldsymbol{x})\right)^2\right] + \mathbb{E}_D\left[\left(\bar{f}(\boldsymbol{x}) - y\right)^2\right] \\
&\quad + \mathbb{E}_D\left[(y - y_D)^2\right] + 2\mathbb{E}_D\left[\left(\bar{f}(\boldsymbol{x}) - y\right)(y - y_D)\right] \\
&= \mathbb{E}_D\left[\left(f(\boldsymbol{x}; D) - \bar{f}(\boldsymbol{x})\right)^2\right] + \left(\bar{f}(\boldsymbol{x}) - y\right)^2 + \mathbb{E}_D\left[(y_D - y)^2\right].
\end{aligned} \tag{2.41}$$

That is,

$$E(f; D) = bias^2(\boldsymbol{x}) + var(\boldsymbol{x}) + \varepsilon^2, \tag{2.42}$$

which means the generalization error can be decomposed into the sum of bias, variance, and noise.

Bias (2.40) measures the difference between the learning algorithm's expected prediction and the ground-truth label, that is, expressing the fitting ability of the learning algorithm. Variance (2.38) measures the change of learning performance caused by changes to the equal-sized training set, that is, expressing the impact of data disturbance on the learning outcome. Noise (2.39) represents the lower bound of the expected generalization error that can be achieved by any learning algorithms for the given task, that is, the inherent difficulty of the learning problem. The bias-variance decomposition tells us that the generalization performance is jointly determined by the learning algorithm's ability, data sufficiency, and the inherent difficulty of the learning problem. In order to achieve excellent generalization performance, a small bias is needed by adequately fitting the data, and the variance should also be kept small by minimizing the impact of data disturbance.

Generally speaking, bias and variance are conflicted with each other, and this is known as the bias-variance dilemma. ◘ Figure 2.9 gives an illustrating example. Given a learning problem and a learner, suppose we can control the degree of training. If we limit the degree of training such that the learner is undertrained, its fitting ability is limited, and hence the data disturbances have a limited impact on the learner, that is, bias dominates the generalization error. As the training proceeds, the learner's fitting ability improves, and hence the learner starts to learn the data disturbances, that is, variance starts to dominate the generalization error. After a large amount of training, the fitting ability of the learner becomes very strong, and hence slight disturbances in the training data will cause significant changes to the learner. At this point, the learner may start to learn the peculiarities of the training data, and hence overfitting occurs.

Many learning algorithms allow users to control the degree of training, such as the number of levels in decision trees, the number of training epochs in neural networks, and the number of base learners in ensemble learning methods.

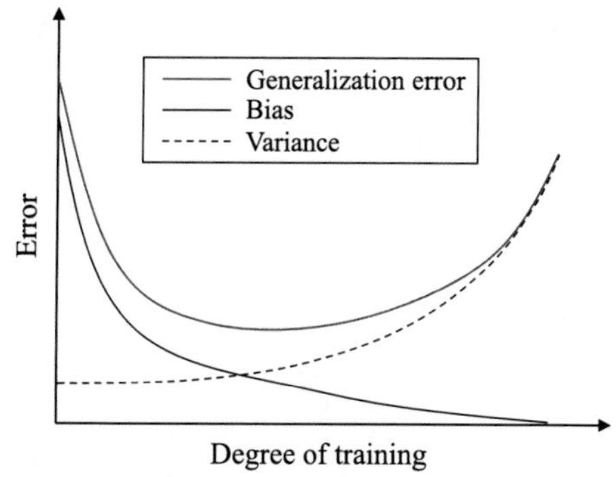


**Fig. 2.9** Relationships between generalization error, bias, and variance

2

## 2.6 Further Reading

Bootstrap sampling has crucial applications in machine learning, and a detailed discussion can be found in Efron and Tibshirani (1993).

ROC curve was introduced to machine learning in the late 1980s (Spackman 1989), and AUC started to be widely used in the field of machine learning since the middle 1990s (Bradley 1997). However, using the area under the ROC curve to evaluate the expected performance of models has already been done much earlier in medical diagnosis (Hanley and McNeil 1983). Hand and Till (2001) extended the ROC curve from binary classification problems to multiclass classification problems. Fawcett (2006) surveyed the use of the ROC curve.

Drummond and Holte (2006) invented the cost curve. Other than the misclassification cost, there are many costs involved in the machine learning process, such as the testing cost, labeling cost, and feature cost. The misclassification cost can also be further divided into the class-based misclassification cost and sample-based misclassification cost. Cost-sensitive learning (Elkan 2001; Zhou and Liu 2006) is a research topic for learning under unequal cost settings.

Section 2.3.4 only discussed the class-based misclassification cost.

Dietterich (1998) pointed out the risk of using the regular $k$-fold cross-validation method, and proposed the $5 \times 2$ cross-validation method. Demsar (2006) discussed the hypothesis testing methods for comparing multiple algorithms.

Geman et al. (1992) proposed the bias-variance-covariance decomposition for regression problems, which was later shortened as bias-variance decomposition. Though bias and variance reveal the internal factors of errors, we can only derive the elegant form of (2.42) for regression problems based on MSE. For classification problems, however, deriving the bias-variance decomposition is difficult since the 0/1 loss function is discontinuous. There exist many empirical methods for estimating bias and variance (Kong and Dietterich 1995; Kohavi and Wolpert 1996; Breiman 1996; Friedman 1997; Domingos 2000).

## Exercises

**2.1** Given a data set of 1000 samples, where 500 samples are positive and 500 samples are negative. To perform a hold-out evaluation, we split the data set into a training set with 70% of the samples and a testing set with 30% of the samples. Estimate the total number of possible splittings.

**2.2** Given a data set of 100 samples, where the positive and negative samples are half-half. Suppose that the model produced by a learning algorithm predicts every new sample as the majority class in the training set (random guessing if different classes have the same number of samples). Calculate the error rates of this model evaluated by 10-fold cross-validation and hold-out, respectively.

**2.3** Given that the $F1$ value of learner A is greater than that of learner B, find out whether the BEP value of A is also greater than that of B.

**2.4** Describe the relationships among TPR, FPR, Precision, and Recall.

**2.5** Prove (2.22).

**2.6** Describe the relationship between error rate and ROC curve.

**2.7** Prove that every ROC curve has a corresponding cost curve, and vice versa.

**2.8** The min-max normalization and $z$-score normalization are two commonly used normalization methods. Let $x$ and $x'$ denote the variable value before and after normalization, $x_{\min}$ and $x_{\max}$ denote the minimum and maximum value before normalization, $x'_{\min}$ and $x'_{\max}$ denote the minimum and maximum value after normalization, $\overline{x}$ denote the unnormalized mean, and $\sigma_x$ denote the standard deviation. Then, the min-max normalization and $z$-score normalization are, respectively, defined in (2.43) and (2.44). Discuss the pros and cons of each method.

$$x' = x'_{\min} + \frac{x - x_{\min}}{x_{\max} - x_{\min}} \times (x'_{\max} - x'_{\min}), \tag{2.43}$$

$$x' = \frac{x - \bar{x}}{\sigma_x}. \tag{2.44}$$

**2.9** Describe the process of $\chi^2$ test.

**2.10** * Describe the difference between using (2.34) and using (2.35) in the Friedman test.

2

## Break Time

### Short Story: *t*-Test, Beer, "Student", and William Gosset

In 1899, William Gosset (1876–1937), who majored in chemistry at the University of Oxford, joined Guinness Brewery in Dublin, Ireland after graduation and wished to apply his biology and chemistry knowledge to the brewing process. Gosset proposed *t*-test to reduce the cost of quality control of brewing, and published this work in *Biometrika* in 1908. In order to prevent the leak of the trade secret, the paper was published under the pseudonym of "Student", and this leads to the method's name "Student's *t*-test".

In 1954, the Guinness corporation started to publish *Guinness World Records*.

As a visionary corporation, Guinness Brewery grants its technical staff "sabbatical leave" just like in universities such that its staff can maintain a high level of technical skills. For this reason, Gosset had a chance to visit the lab led by Professor Karl Pearson (1857–1936) at University College London (UCL) in 1906–1907. Since *t*-test was published shortly after the visit, it is hard to tell whether it was developed at Guinness Brewery or during the visit at UCL. Nevertheless, the connection between "Student" and Gosset was found by statisticians from UCL, and this is not a surprise since Professor Pearson happened to be the editor-in-chief of *Biometrika*.

## References

Bradley AP (1997) The use of the area under the roc curve in the evaluation of machine learning algorithms. Pattern Recognit 30(7):1145–1159

Breiman L (1996) Bias, variance, and arcing classifiers. Technical Report 460, Statistics Department, University of California, CA

Demsar J (2006) Statistical comparison of classifiers over multiple data sets. J Mach Learn Res 7:1–30

Dietterich TG (1998) Approximate statistical tests for comparing supervised classification learning algorithms. Neural Comput 10(7):1895–1923

Domingos P (2000) A unified bias-variance decomposition. In: Proceedings of the 17th international conference on machine learning (ICML), pp 231–238. Stanford, CA

Drummond C, Holte RC (2006) Cost curves: an improved method for visualizing classifier performance. Mach Learn 65(1):95–130

Efron B, Tibshirani R (1993) An introduction to the bootstrap. Chapman & Hall, New York

Elkan C (2001) The foundations of cost-sensitive learning. In: Proceedings of the 17th international joint conference on artificial intelligence (IJCAI), pp 973–978. Seattle, WA

Fawcett T (2006) An introduction to roc analysis. Pattern Recognit Lett 27(8):861–874

Friedman JH (1997) On bias, variance, 0/1-loss, and the curse-of-dimensionality. Data Mining Knowl Disc 1(1):55–77

Geman S, Bienenstock E, Doursat R (1992) Neural networks and the bias/variance dilemma. Neural Comput 4(1):1–58

Hand DJ, Till RJ (2001) A simple generalisation of the area under the ROC curve for multiple class classification problems. Mach Learn 45(2):171–186

Hanley JA, McNeil BJ (1983) A method of comparing the areas under receiver operating characteristic curves derived from the same cases. Radiology 148(3):839–843

Kohavi R, Wolpert DH (1996) Bias plus variance decomposition for zero-one loss functions. In: Proceedings of the 13th international conference on machine learning (ICML), pp 275–283. Bari, Italy

Kong EB, Dietterich TG (1995) Error-correcting output coding corrects bias and variance. In: Proceedings of the 12th international conference on machine learning (ICML), pp 313–321. Tahoe City, CA

Mitchell T (1997) Machine learning. McGraw Hill, New York

Spackman KA (1989) Signal detection theory: valuable tools for evaluating inductive learning. In: Proceedings of the 6th international workshop on machine learning (IWML), pp 160–163. Ithaca, NY

Van Rijsbergen CJ (1979) Information retrieval, 2nd edn. Butterworths, London

Wellek S (2010) Testing statistical hypotheses of equivalence and noninferiority, 2nd edn. Chapman & Hall/CRC, Boca Raton

Zhou Z-H, Liu X-Y (2006) On multi-class cost-sensitive learning. In: Proceedings of the 21st national conference on artificial intelligence (AAAI), pp 567–572. Boston, WA

# Linear Models

## Table of Contents




Z.-H. Zhou, *Machine Learning*,
https://doi.org/10.1007/978-981-15-1967-3_3

## 3.1 Basic Form

Let $\boldsymbol{x} = (x_1; x_2; \ldots; x_d)$ be a sample described by $d$ variables, where $\boldsymbol{x}$ takes the value $x_i$ on the $i$th variable. A linear model aims to learn a function that makes predictions by a linear combination of the input variables, that is,



$$f(\boldsymbol{x}) = w_1 x_1 + w_2 x_2 + \cdots + w_d x_d + b, \tag{3.1}$$

or commonly written in the vector form

$$f(\boldsymbol{x}) = \boldsymbol{w}^\top \boldsymbol{x} + b, \tag{3.2}$$

where $\boldsymbol{w} = (w_1; w_2; \ldots; w_d)$. The model is determined once $\boldsymbol{w}$ and $b$ are learned.

Despite its simple form and ease of modeling, the basic linear model covers some important and fundamental ideas of machine learning. In fact, many powerful nonlinear models can be derived from linear models by introducing multi-layer structures or high-dimensional mapping. Besides, the learned weights $\boldsymbol{w}$ transparently indicate the importance of each input variable, and provide the linear model with excellent comprehensibility. For example, suppose the linear model learned in our watermelon problem is $f_{\mathtt{ripe}}(\boldsymbol{x}) = 0.2 \cdot x_{\mathtt{color}} + 0.5 \cdot x_{\mathtt{root}} + 0.3 \cdot x_{\mathtt{sound}} + 1$, which indicates that the ripeness of a watermelon can be determined by considering its color, root, and sound information. From the coefficients, we know that root is the most important variable, and sound is more important than color.

The rest of this chapter introduces some classic linear models, starting with the regression problems followed by binary classification and multiclass classification problems.

## 3.2 Linear Regression

Given a data set $D = \{(\boldsymbol{x}_1, y_1), (\boldsymbol{x}_2, y_2), \ldots, (\boldsymbol{x}_m, y_m)\}$, where $\boldsymbol{x}_i = (x_{i1}; x_{i2}; \ldots; x_{id})$ and $y_i \in \mathbb{R}$. Linear regression aims to learn a linear model that can accurately predict the real-valued output labels.

We start our discussion with the simplest case of a single input variable. To simplify the notation, we omit the subscript of variables, that is, $D = \{(x_i, y_i)\}_{i=1}^m$, where $x_i \in \mathbb{R}$. For discrete variables, they can be converted into real-valued variables when an ordinal relationship naturally exists between values. For example, the values tall and short of height can be converted into $\{1.0, 0.0\}$; the values high, medium and low of altitude can be converted into $\{1.0, 0.5, 0.0\}$. When no ordinal relationship exists, we often convert the discrete variable with $k$ possible values into a $k$-dimensional vector, e.g., for the variable `cucurbits`, its values `watermelon`, `pumpkin`, and `cucumber` can be converted into $(0, 0, 1)$, $(0, 1, 0)$, and $(1, 0, 0)$, respectively.

We should avoid converting categorical variables into continuous variables; otherwise, the incorrectly introduced ordinal relationships can mislead subsequent calculations such as distance calculations. See Sect. 9.3.

Linear regression aims to learn the function

$$f(x) = wx + b, \text{ such that } f(x_i) \simeq y_i, i = 1, \ldots, m. \tag{3.3}$$

To determine $w$ and $b$, the key is to measure the difference between $f(x)$ and $y$. For this purpose, the MSE (2.2) introduced in Sect. 2.3 is one of the most commonly used metrics. We can minimize MSE, that is,

Also known as *square loss*.

$$\begin{aligned}(w^*, b^*) &= \underset{(w,b)}{\arg\min} \sum_{i=1}^{m} (f(x_i) - y_i)^2 \\ &= \underset{(w,b)}{\arg\min} \sum_{i=1}^{m} (y_i - wx_i - b)^2 .\end{aligned} \tag{3.4}$$

$w^*$ and $b^*$ denote the solutions to $w$ and $b$, respectively.

MSE corresponds to the Euclidean distance and has an intuitive geometrical interpretation. A general method to minimize MSE is the *least squares method*. For linear regression, the least squares method attempts to find a straight line such that the total Euclidean distance from all samples to the line is minimized.

The least squares method has a wide range of applications other than linear regression.

The process of searching for $w$ and $b$ that minimize $E_{(w,b)} = \sum_{i=1}^{m}(y_i - wx_i - b)^2$ is called *least squares parameter estimation* of linear regression. To be specific, we can calculate the derivatives of $E_{(w,b)}$ with respect to $w$ and $b$, respectively:

$$\frac{\partial E_{(w,b)}}{\partial w} = 2\left(w\sum_{i=1}^{m} x_i^2 - \sum_{i=1}^{m}(y_i - b)x_i\right), \tag{3.5}$$

$$\frac{\partial E_{(w,b)}}{\partial b} = 2\left(mb - \sum_{i=1}^{m}(y_i - wx_i)\right). \tag{3.6}$$

Here, $E_{(w,b)}$ is a convex function of $w$ and $b$. The optimal solutions of $w$ and $b$ are obtained when the derivatives of $E_{(w,b)}$ with respect to both $w$ and $b$ are zero.
A function $f$ is said to be a convex function on the interval $[a, b]$ if there is $f(\frac{x_1+x_2}{2}) \leqslant \frac{f(x_1)+f(x_2)}{2}$ for any $x_1$ and $x_2$ on the interval. Functions with a U-shaped curve are usually convex functions, e.g., $f(x) = x^2$.
We can determine the convexity of a function defined over the real numbers by its second derivative: the function is convex on an interval if its second derivative is non-negative on the interval; the function is strictly convex on an interval if its second derivative is greater than 0 at all points on the interval.

By setting (3.5) and (3.6) equal to 0, we have the closed-form solutions of $w$ and $b$:

$$w = \frac{\sum_{i=1}^{m} y_i(x_i - \bar{x})}{\sum_{i=1}^{m} x_i^2 - \frac{1}{m}\left(\sum_{i=1}^{m} x_i\right)^2}, \tag{3.7}$$

$$b = \frac{1}{m}\sum_{i=1}^{m}(y_i - wx_i), \tag{3.8}$$

where $\bar{x} = \frac{1}{m}\sum_{i=1}^{m} x_i$ is the arithmetic mean of $x$.

3

More generally, samples are described by $d$ attributes, like the data set $D$ at the beginning of this section. In such cases, the model becomes

$$f(\boldsymbol{x}) = \boldsymbol{w}^\top \boldsymbol{x} + b, \text{ such that } f(\boldsymbol{x}_i) \simeq y_i, i = 1, \ldots, m,$$

which is known as *multivariate linear regression.*

Similarly, the parameters $\boldsymbol{w}$ and $b$ can be estimated using the least squares method. For ease of discussion, we rewrite $\boldsymbol{w}$ and $b$ as $\hat{\boldsymbol{w}} = (\boldsymbol{w}; b)$. Accordingly, the data set $D$ is represented as an $m$ by $(d+1)$ matrix $\mathbf{X}$, where each row corresponds to one sample with the first $d$ elements to be the values of the $d$ variables and the last element always to be 1, that is,

$$\mathbf{X} = \begin{pmatrix} x_{11} & x_{12} & \ldots & x_{1d} & 1 \\ x_{21} & x_{22} & \ldots & x_{2d} & 1 \\ \vdots & \vdots & \ddots & \vdots & \vdots \\ x_{m1} & x_{m2} & \ldots & x_{md} & 1 \end{pmatrix} = \begin{pmatrix} \boldsymbol{x}_1^\top & 1 \\ \boldsymbol{x}_2^\top & 1 \\ \vdots & \vdots \\ \boldsymbol{x}_m^\top & 1 \end{pmatrix}.$$

Vectorizing the labels to $\boldsymbol{y} = (y_1; y_2; \ldots; y_m)$, then, similar to (3.4), we have

$$\hat{\boldsymbol{w}}^* = \underset{\hat{\boldsymbol{w}}}{\arg\min} (\boldsymbol{y} - \mathbf{X}\hat{\boldsymbol{w}})^\top (\boldsymbol{y} - \mathbf{X}\hat{\boldsymbol{w}}). \tag{3.9}$$

Letting $E_{\hat{\boldsymbol{w}}} = (\boldsymbol{y} - \mathbf{X}\hat{\boldsymbol{w}})^\top (\boldsymbol{y} - \mathbf{X}\hat{\boldsymbol{w}})$ and finding the derivative with respect to $\hat{\boldsymbol{w}}$, we have

$$\frac{\partial E_{\hat{\boldsymbol{w}}}}{\partial \hat{\boldsymbol{w}}} = 2\mathbf{X}^\top (\mathbf{X}\hat{\boldsymbol{w}} - \boldsymbol{y}). \tag{3.10}$$

The closed-form solution of $\hat{\boldsymbol{w}}$ can be obtained by making (3.10) equal to 0. However, due to the matrix inverse operation, the calculation is more complicated than that of the single variable case. We provide a brief discussion as follows.

When $\mathbf{X}^\top\mathbf{X}$ is a full-rank matrix or a positive definite matrix, letting (3.10) equal to 0 gives

$$\hat{\boldsymbol{w}}^* = (\mathbf{X}^\top\mathbf{X})^{-1}\mathbf{X}^\top \boldsymbol{y}, \tag{3.11}$$

where $(\mathbf{X}^\top\mathbf{X})^{-1}$ is the inverse of $\mathbf{X}^\top\mathbf{X}$. Letting $\hat{\boldsymbol{x}}_i = (\boldsymbol{x}_i; 1)$, the learned multivariate linear regression model is

$$f(\hat{\boldsymbol{x}}_i) = \hat{\boldsymbol{x}}_i^\top (\mathbf{X}^\top\mathbf{X})^{-1}\mathbf{X}^\top \boldsymbol{y}. \tag{3.12}$$

Nevertheless, $\mathbf{X}^\top\mathbf{X}$ is often not full-rank in real-world applications. In practice, the number of variables can be large or even larger than the number of samples, that is, more columns than rows in $\mathbf{X}$. In such cases, $\mathbf{X}^\top\mathbf{X}$ is not full-rank, which means that there is more than one $\hat{\boldsymbol{w}}$ that can minimize the MSE. Then,

For example, genetic circuit data in bioinformatics are often with thousands or even more attributes but only hundreds of samples.

the choice of $\hat{w}$ is a matter of the inductive bias of the learning algorithm, e.g., some algorithms introduce a regularization term.

See Sect. 1.4 for inductive bias. See Sects. 6.4 and 11.4 for regularization.

Linear models are simple but diverse. Taking $(\boldsymbol{x}, y)$, $y \in \mathbb{R}$ as an example, by approximating the ground-truth label $y$ with a linear model, we obtain the linear regression model, which can be compactly written as

$$y = \boldsymbol{w}^{\top}\boldsymbol{x} + b. \tag{3.13}$$

Is it possible to let the predicted value approximate a variable derived from $y$? For instance, suppose the output label changes on the exponential scale, then the logarithm of the output label can be used for approximation, that is,

$$\ln y = \boldsymbol{w}^{\top}\boldsymbol{x} + b. \tag{3.14}$$

This is called *log-linear regression* which approximates $y$ with $e^{\boldsymbol{w}^{\top}\boldsymbol{x}+b}$. Though (3.14) is still in the form of linear regression, it is indeed searching for a nonlinear mapping from the input space to the output space, as shown in ◘ Figure 3.1. The logarithm function links the predictions of linear regression to the ground-truth labels.

More generally, for a monotonic differentiable function $g(\cdot)$,

$g(\cdot)$ is continuous and smooth.

$$y = g^{-1}(\boldsymbol{w}^{\top}\boldsymbol{x} + b) \tag{3.15}$$

is called *generalized linear model*, where the function $g(\cdot)$ is the link function. We see that log-linear regression is a special case of generalized linear models when $g(\cdot) = \ln(\cdot)$.

Parameter estimation of generalized linear models is usually performed through weighted least squares or maximum likelihood methods.

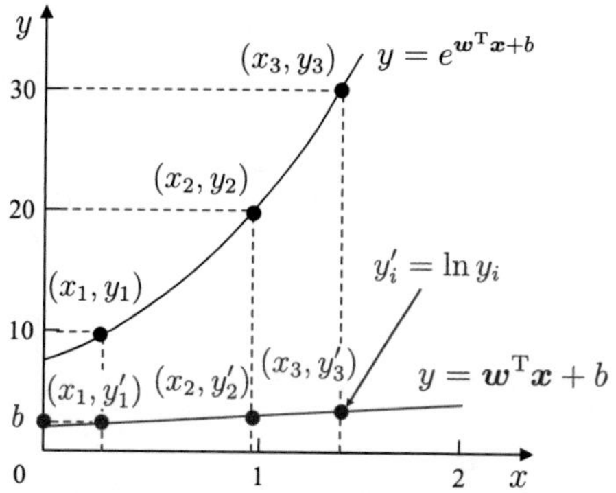


**Fig. 3.1** Log-linear regression

3

## 3.3 Logistic Regression

The previous section discussed the use of linear models in regression problems, but how can we solve classification problems? The answer lies in the generalized linear model (3.15): we just need to find a monotonic differentiable function $g(\cdot)$ that links the predictions of linear regression to the ground-truth labels of the classification problem.

For binary classification with output label $y \in \{0, 1\}$, the real-valued predictions of the linear regression model $z = \boldsymbol{w}^\top \boldsymbol{x} + b$ need to be converted into 0/1. Ideally, the unit-step function is desired:

Also called Heaviside function.

$$y = \begin{cases} 0, & z < 0; \\ 0.5, & z = 0; \\ 1, & z > 0, \end{cases} \tag{3.16}$$

which predicts positive for $z$ greater than 0, negative for $z$ smaller than 0, and an arbitrary output when $z$ equals to 0. The unit-step function is plotted in ◘ Figure 3.2.

Nevertheless, ◘ Figure 3.2 shows that the unit-step function is not continuous, and hence it cannot be used as $g^{-1}(\cdot)$ in (3.15). Therefore, we need to find a monotonic differentiable surrogate function to approximate the unit-step function, and a common choice is the logistic function:

$$y = \frac{1}{1 + e^{-z}}. \tag{3.17}$$

From ◘ Figure 3.2, we can see that the logistic function is a type of sigmoid function which converts $z$ to $y$ that is either close to 0 or 1, and the output value has a steep change near $z = 0$. Substituting $g^{-1}(\cdot)$ into (3.15), we have

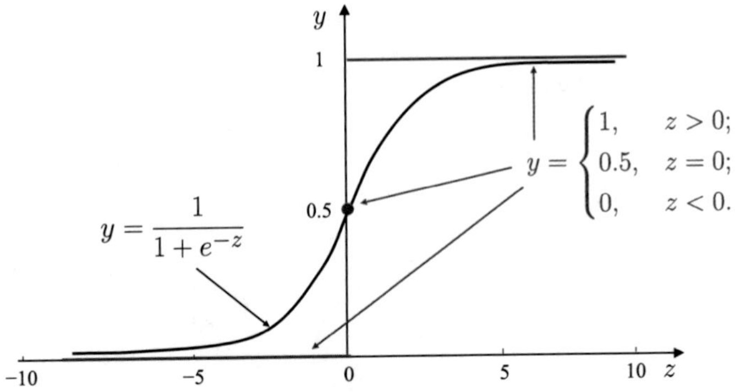


**Fig. 3.2** Unit-step function and logistic function

$$y = \frac{1}{1 + e^{-(\boldsymbol{w}^\top \boldsymbol{x}+b)}}. \tag{3.18}$$

Similar to (3.14), (3.18) can be transformed into

$$\ln \frac{y}{1-y} = \boldsymbol{w}^\top \boldsymbol{x} + b. \tag{3.19}$$

Let $y$ be the likelihood of $x$ being a positive sample and $1 - y$ be the likelihood of being a negative sample, then the ratio

$$\frac{y}{1-y} \tag{3.20}$$

is called the *odds*, indicating the relative likelihood of $\boldsymbol{x}$ being a positive sample. Taking the logarithm of odds gives the *log odds* (i.e., *logit*)

$$\ln \frac{y}{1-y}. \tag{3.21}$$

It turns out that (3.18) is using linear regression predictions to approximate the log odds of true labels. As such, the corresponding model is called *logistic regression*, also known as *logit regression*. It should be noted that logistic regression is indeed a classification model despite the term "regression" in its name. Logistic regression has several nice properties. For example, it directly models the label probability without requiring any prior assumptions on the data distribution and hence avoids issues such as inappropriate hypothetical data distributions. Also, it predicts labels together with associated probabilities, which is essential for tasks that use probability to aid decision-making. Finally, the objective function of logistic regression, as we will see later on, is a convex function having derivatives of all orders with many useful mathematical properties, and convexity makes it solvable with numerical optimization methods.

Now we turn our attention to the estimation of $\boldsymbol{w}$ and $b$ in (3.18). If we consider $y$ in (3.18) as the posterior probability $p(y = 1 \mid \boldsymbol{x})$, then (3.19) can be rewritten as

$$\ln \frac{p(y=1 \mid \boldsymbol{x})}{p(y=0 \mid \boldsymbol{x})} = \boldsymbol{w}^\top \boldsymbol{x} + b, \tag{3.22}$$

and consequently,

$$p(y=1 \mid \boldsymbol{x}) = \frac{e^{\boldsymbol{w}^\top \boldsymbol{x}+b}}{1 + e^{\boldsymbol{w}^\top \boldsymbol{x}+b}}, \tag{3.23}$$

$$p(y=0 \mid \boldsymbol{x}) = \frac{1}{1 + e^{\boldsymbol{w}^\top \boldsymbol{x}+b}}. \tag{3.24}$$

To maximize the posterior probability, we can apply the maximum likelihood method to estimate $\boldsymbol{w}$ and $b$. Given a data set $\{(\boldsymbol{x}_i, y_i)\}_{i=1}^m$, the log-likelihood to be maximized is

See Sect. 7.2 for the maximum likelihood method.

$$\ell(\boldsymbol{w}, b) = \sum_{i=1}^{m} \ln p(y_i \mid \boldsymbol{x}_i; \boldsymbol{w}, b), \tag{3.25}$$

3

i.e., maximizing the probability of each sample being predicted as the ground-truth label. For ease of discussion, we rewrite $\boldsymbol{w}^\top \boldsymbol{x} + b$ as $\boldsymbol{\beta}^\top \hat{\boldsymbol{x}}$, where $\boldsymbol{\beta} = (\boldsymbol{w}; b)$ and $\hat{\boldsymbol{x}} = (\boldsymbol{x}; 1)$. Also, letting $p_1(\hat{\boldsymbol{x}}; \boldsymbol{\beta}) = p(y = 1 \mid \hat{\boldsymbol{x}}; \boldsymbol{\beta})$ and $p_0(\hat{\boldsymbol{x}}; \boldsymbol{\beta}) = p(y = 0 \mid \hat{\boldsymbol{x}}; \boldsymbol{\beta}) = 1 - p_1(\hat{\boldsymbol{x}}; \boldsymbol{\beta})$, then the likelihood term in (3.25) can be rewritten as

$$p(y_i \mid \boldsymbol{x}_i; \boldsymbol{w}, b) = y_i p_1(\hat{\boldsymbol{x}}_i; \boldsymbol{\beta}) + (1 - y_i) p_0(\hat{\boldsymbol{x}}_i; \boldsymbol{\beta}). \tag{3.26}$$

Substituting (3.26) into (3.25), then, from (3.23) and (3.24), we know that maximizing (3.25) is equivalent to minimizing

Considering $y_i \in \{0, 1\}$.

$$\ell(\boldsymbol{\beta}) = \sum_{i=1}^{m} \left( -y_i \boldsymbol{\beta}^\top \hat{\boldsymbol{x}}_i + \ln \left( 1 + e^{\boldsymbol{\beta}^\top \hat{\boldsymbol{x}}_i} \right) \right). \tag{3.27}$$

Because (3.27) is a higher order differentiable convex function with respect to $\boldsymbol{\beta}$, the solutions, according to the convex optimization theory (Boyd and Vandenberghe 2004), can be found via classic numerical optimization methods such as the gradient descent method or even Newton's method. Hence, we have

See Appendix B.4 for the gradient descent method.

$$\boldsymbol{\beta}^* = \underset{\boldsymbol{\beta}}{\arg\min}\, \ell(\boldsymbol{\beta}). \tag{3.28}$$

Taking Newton's method as an example, the update rule at the $(t + 1)$th iteration is

$$\boldsymbol{\beta}^{t+1} = \boldsymbol{\beta}^t - \left( \frac{\partial^2 \ell(\boldsymbol{\beta})}{\partial \boldsymbol{\beta} \partial \boldsymbol{\beta}^\top} \right)^{-1} \frac{\partial \ell(\boldsymbol{\beta})}{\partial \boldsymbol{\beta}}, \tag{3.29}$$

where the first- and second-order derivatives with respect to $\boldsymbol{\beta}$ are, respectively,

$$\frac{\partial \ell(\boldsymbol{\beta})}{\partial \boldsymbol{\beta}} = - \sum_{i=1}^{m} \hat{\boldsymbol{x}}_i \left( y_i - p_1(\hat{\boldsymbol{x}}_i; \boldsymbol{\beta}) \right), \tag{3.30}$$

$$\frac{\partial^2 \ell(\boldsymbol{\beta})}{\partial \boldsymbol{\beta} \partial \boldsymbol{\beta}^\top} = \sum_{i=1}^{m} \hat{\boldsymbol{x}}_i \hat{\boldsymbol{x}}_i^\top p_1(\hat{\boldsymbol{x}}_i; \boldsymbol{\beta}) \left( 1 - p_1(\hat{\boldsymbol{x}}_i; \boldsymbol{\beta}) \right). \tag{3.31}$$

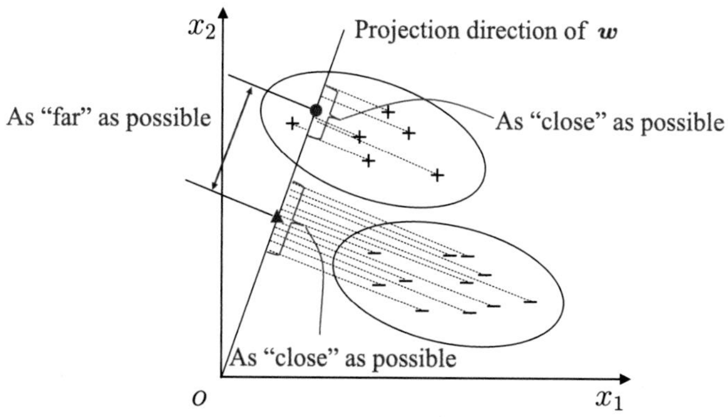


**Fig. 3.3** A two-dimensional illustration of LDA. "+" and "−" denote positive samples and negative samples, respectively. The ellipses are the boundaries of clusters; the dashed lines represent projections; the solid red dot and triangle are the centers of the projections

## 3.4 Linear Discriminant Analysis

Linear Discriminant Analysis (LDA) is a classic linear method, also known as Fisher's Linear Discriminant (FLD) since it was initially proposed by Fisher (1936) for binary classification problems.

Strictly speaking, LDA and FLD are slightly different, where LDA assumes equal and full-rank class covariances.

The idea of LDA is straightforward: projecting the training samples onto a line such that samples of the same class are close to each other, while samples of different classes are far away from each other. When classifying new samples, they are projected onto the same line and their classes are determined by their projected locations. ◘ Figure 3.3 gives a two-dimensional illustration.

Given a data set $D = \{(\boldsymbol{x}_i, y_i)\}_{i=1}^m, y_i \in \{0, 1\}$, let $X_i$, $\boldsymbol{\mu}_i$, and $\boldsymbol{\Sigma}_i$ denote, respectively, the sample set, mean vector, and covariance matrix of the $i$th class ($i \in \{0, 1\}$). After projecting data onto the line $\boldsymbol{w}$, the centers of those two classes samples are $\boldsymbol{w}^\top\boldsymbol{\mu}_0$ and $\boldsymbol{w}^\top\boldsymbol{\mu}_1$, respectively. The covariances of the two classes samples are $\boldsymbol{w}^\top\boldsymbol{\Sigma}_0\boldsymbol{w}$ and $\boldsymbol{w}^\top\boldsymbol{\Sigma}_1\boldsymbol{w}$, respectively. Since the line is a one-dimensional space, $\boldsymbol{w}^\top\boldsymbol{\mu}_0$, $\boldsymbol{w}^\top\boldsymbol{\mu}_1$, $\boldsymbol{w}^\top\boldsymbol{\Sigma}_0\boldsymbol{w}$, and $\boldsymbol{w}^\top\boldsymbol{\Sigma}_1\boldsymbol{w}$ are all real numbers.

To make the projection points of similar samples as close as possible, we can make the covariance of the projection points of similar samples as small as possible, that is, minimizing $\boldsymbol{w}^\top\boldsymbol{\Sigma}_0\boldsymbol{w} + \boldsymbol{w}^\top\boldsymbol{\Sigma}_1\boldsymbol{w}$. To make the projection points of examples from different classes as far away as possible, we can make the distance between the class centers as large as possible, that is, maximizing $\|\boldsymbol{w}^\top\boldsymbol{\mu}_0 - \boldsymbol{w}^\top\boldsymbol{\mu}_1\|$. Putting them together, we have the objective to be maximized:

$$J = \frac{\left\| \boldsymbol{w}^\top \boldsymbol{\mu}_0 - \boldsymbol{w}^\top \boldsymbol{\mu}_1 \right\|_2^2}{\boldsymbol{w}^\top \boldsymbol{\Sigma}_0 \boldsymbol{w} + \boldsymbol{w}^\top \boldsymbol{\Sigma}_1 \boldsymbol{w}} = \frac{\boldsymbol{w}^\top (\boldsymbol{\mu}_0 - \boldsymbol{\mu}_1)(\boldsymbol{\mu}_0 - \boldsymbol{\mu}_1)^\top \boldsymbol{w}}{\boldsymbol{w}^\top (\boldsymbol{\Sigma}_0 + \boldsymbol{\Sigma}_1) \boldsymbol{w}}. \tag{3.32}$$

By defining the *within-class scatter matrix*

$$\mathbf{S}_w = \boldsymbol{\Sigma}_0 + \boldsymbol{\Sigma}_1 = \sum_{\boldsymbol{x} \in X_0} (\boldsymbol{x} - \boldsymbol{\mu}_0)(\boldsymbol{x} - \boldsymbol{\mu}_0)^\top + \sum_{\boldsymbol{x} \in X_1} (\boldsymbol{x} - \boldsymbol{\mu}_1)(\boldsymbol{x} - \boldsymbol{\mu}_1)^\top \tag{3.33}$$

and the *between-class scatter matrix*

$$\mathbf{S}_b = (\boldsymbol{\mu}_0 - \boldsymbol{\mu}_1)(\boldsymbol{\mu}_0 - \boldsymbol{\mu}_1)^\top, \tag{3.34}$$

we can rewrite (3.32) as

$$J = \frac{\boldsymbol{w}^\top \mathbf{S}_b \boldsymbol{w}}{\boldsymbol{w}^\top \mathbf{S}_w \boldsymbol{w}}, \tag{3.35}$$

which is the objective of LDA to be maximized, that is, the *generalized Rayleigh quotient* of $\mathbf{S}_b$ and $\mathbf{S}_w$.

Then, how can we determine $\boldsymbol{w}$? Since the numerator and denominator in (3.35) are both quadratic terms of $\boldsymbol{w}$, the solution of (3.35) is independent of the magnitude of $\boldsymbol{w}$ but only about its direction. Without loss of generality, letting $\boldsymbol{w}^\top \mathbf{S}_w \boldsymbol{w} = 1$, then maximizing (3.35) is equivalent to

If $\boldsymbol{w}$ is a solution, then $\alpha\boldsymbol{w}$ is also a solution of (3.35) for any constant $\alpha$.

$$\begin{aligned} \min_{\boldsymbol{w}} \quad & -\boldsymbol{w}^\top \mathbf{S}_b \boldsymbol{w} \\ \text{s.t.} \quad & \boldsymbol{w}^\top \mathbf{S}_w \boldsymbol{w} = 1. \end{aligned} \tag{3.36}$$

Using the method of the Lagrange multipliers, which helps find the extremum of a function subject to equality constraints, the above equation is equivalent to

See Appendix B.1 for the method of the Lagrange multipliers.

$$\mathbf{S}_b \boldsymbol{w} = \lambda \mathbf{S}_w \boldsymbol{w}, \tag{3.37}$$

where $\lambda$ is the Lagrange multiplier. Since the direction of $\mathbf{S}_b \boldsymbol{w}$ is always $\boldsymbol{\mu}_0 - \boldsymbol{\mu}_1$, we can let

$(\boldsymbol{\mu}_0 - \boldsymbol{\mu}_1)^\top \boldsymbol{w}$ is a scalar.

$$\mathbf{S}_b \boldsymbol{w} = \lambda (\boldsymbol{\mu}_0 - \boldsymbol{\mu}_1), \tag{3.38}$$

and substitute it into (3.37), which gives

$$\boldsymbol{w} = \mathbf{S}_w^{-1} (\boldsymbol{\mu}_0 - \boldsymbol{\mu}_1). \tag{3.39}$$

In order to achieve the stability of numerical solutions, singular value decomposition is often applied to $\mathbf{S}_w$ in practice, that is, $\mathbf{S}_w = \mathbf{U}\boldsymbol{\Sigma}\mathbf{V}^\top$, where $\boldsymbol{\Sigma}$ is a real diagonal matrix and its diagonal elements are the singular values of $\mathbf{S}_w$. Then, we calculate $\mathbf{S}_w^{-1}$ as $\mathbf{S}_w^{-1} = \mathbf{V}\boldsymbol{\Sigma}^{-1}\mathbf{U}^\top$.

See Appendix A.3 for singular value decomposition.

It is worth mentioning that LDA can also be explained from the aspect of Bayesian decision theory, and it can be proved that we have the optimal solution of LDA when both classes follow Gaussian distribution with the same prior and covariance.

We can extend LDA to multiclass classification problems. Suppose that there are $N$ classes and the $i$th class has $m_i$ samples. First, we define the *global scatter matrix*

See Exercise 7.5.

$$\begin{aligned}\mathbf{S}_t &= \mathbf{S}_b + \mathbf{S}_w \\ &= \sum_{i=1}^{m} (\boldsymbol{x}_i - \boldsymbol{\mu})(\boldsymbol{x}_i - \boldsymbol{\mu})^\top,\end{aligned} \tag{3.40}$$

where $\boldsymbol{\mu}$ is the mean vector of all samples. We redefine the within-class scatter matrix $\mathbf{S}_w$ as the sum of scatter matrices of each class, that is,

$$\mathbf{S}_w = \sum_{i=1}^{N} \mathbf{S}_{w_i}, \tag{3.41}$$

where

$$\mathbf{S}_{w_i} = \sum_{\boldsymbol{x} \in X_i} (\boldsymbol{x} - \boldsymbol{\mu}_i)(\boldsymbol{x} - \boldsymbol{\mu}_i)^\top. \tag{3.42}$$

From (3.40) to (3.42), we have

$$\begin{aligned}\mathbf{S}_b &= \mathbf{S}_t - \mathbf{S}_w \\ &= \sum_{i=1}^{N} m_i (\boldsymbol{\mu}_i - \boldsymbol{\mu})(\boldsymbol{\mu}_i - \boldsymbol{\mu})^\top.\end{aligned} \tag{3.43}$$

Multiclass LDA can be implemented in different ways by choosing any two from $\mathbf{S}_b$, $\mathbf{S}_w$, and $\mathbf{S}_t$. A common implementation is to optimize the objective

$$\max_{\mathbf{W}} \frac{\operatorname{tr}(\mathbf{W}^\top \mathbf{S}_b \mathbf{W})}{\operatorname{tr}(\mathbf{W}^\top \mathbf{S}_w \mathbf{W})}, \tag{3.44}$$

where $\mathbf{W} \in \mathbf{R}^{d \times (N-1)}$, and $\operatorname{tr}(\cdot)$ is the trace of matrix. Equation (3.44) can be solved as a generalized eigenvalue problem:

$$\mathbf{S}_b \mathbf{W} = \lambda \mathbf{S}_w \mathbf{W}. \tag{3.45}$$

Concatenating the eigenvectors corresponding to the $d'$ largest non-zero eigenvalues of $\mathbf{S}_w^{-1}\mathbf{S}_b$ leads to the closed-form solution of $\mathbf{W}$, where $d' \leqslant N - 1$.

There are at most $N - 1$ non-zero eigenvalues.

If we consider $\mathbf{W}$ as a projection matrix, then multiclass LDA projects samples onto an $d'$-dimensional space, where $d'$ is often much smaller than the number of original features $d$. Since the projection reduces the data dimension while considering the class information, LDA is also considered as a classic supervised dimensionality reduction technique.



See Chap. 10 for dimensionality reduction.

## 3.5 Multiclass Classification

In practice, we often encounter multiclass classification problems. Some binary classification methods can be directly extended to accommodate multiclass cases. However, a more general approach is to apply some strategies to solve multiclass classification problems with any existing binary classification methods.

For example, the extension of LDA discussed in the previous section.

Without loss of generality, given $N$ classes $C_1, C_2, \ldots, C_N$, the basic idea of multiclass learning is decomposition, that is, dividing the multiclass classification problem into several binary classification problems. We begin by decomposing the problem, and then train a binary classifier for each divided binary classification problem. In the testing phase, we ensemble the outputs collected from all binary classifiers into the final multiclass predictions. In this process, the key questions are how to divide multiclass classification problems and how to ensemble multiple classifiers. The rest of this section focuses on introducing three classic dividing strategies, namely One versus One (OvO), One versus Rest (OvR), and Many versus Many (MvM).

Classification learners are often called *classifiers*.

See Chap. 8 for the ensemble of multiple classifiers.

Given a data set $D = \{(\boldsymbol{x}_1, y_1), (\boldsymbol{x}_2, y_2), \ldots, (\boldsymbol{x}_m, y_m)\}$, where $y_i \in \{C_1, C_2, \ldots, C_N\}$. OvO puts the $N$ classes into pairs, resulting in $N(N-1)/2$ binary classification problems. For example, OvO trains a classifier to distinguish class $C_i$ and $C_j$, where it regards $C_i$ as positive and $C_j$ as negative. During testing, a new sample is classified by all classifiers, resulting in $N(N-1)/2$ classification outputs. The final prediction can be made via voting, that is, the predicted class is the one that received the most votes. ■ Figure 3.4 gives an illustration of OvO.

OvR is also known as One versus All (OvA), but calling it OvA is not very accurate since we should not consider "all classes" as the negative class.

OvR trains $N$ classifiers by considering each class as positive in turn, and the rest classes are considered as negative. During testing, if there is only one classifier that predicts the new sample as positive, then it is the final classification result, as shown in ■ Figure 3.4. However, if multiple classifiers predict the new sample as positive, then the prediction confidences

We can also ensemble the classifiers based on information such as their confidence of predictions. See Sect. 8.4.

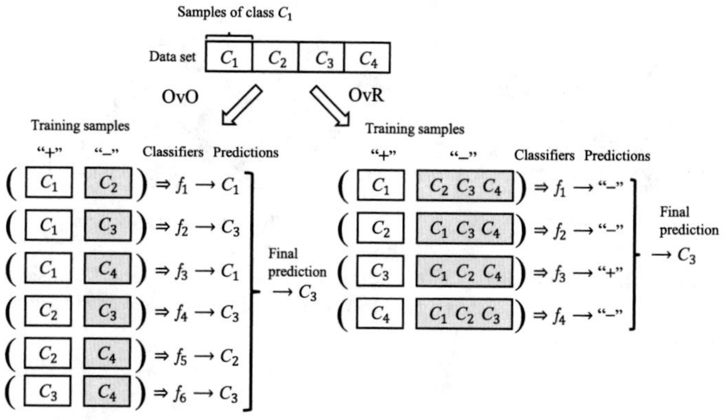


**Fig. 3.4** Illustration of OvO and OvR

are usually assessed, and the class with the highest confidence is used as the classification result.

Since OvR needs $N$ classifiers while OvO needs $N(N-1)/2$ classifiers, the memory and testing time costs of OvO are often higher compared to that of OvR. However, each OvR classifier uses all training samples, whereas each OvO classifier uses only samples of two classes. Hence, the computational cost of training OvO is lower compared to that of OvR, especially when there are many classes. As for the prediction performance, it depends on the specific data distribution, and in most cases, the two methods have similar performance.

MvM conducts multiple trials, and each trial puts several classes as positive and several classes as negative. Note that both OvO and OvR are special cases of MvM. The construction of positive and negative classes in MvM should be carefully designed. Here we introduce one of the most commonly used MvM techniques: Error Correcting Output Codes (ECOC).

ECOC (Dietterich and Bakiri 1995) introduces the idea of encoding into the dividing of classes and maintains error tolerance in the decoding step. ECOC has two main steps:

- Encoding: split the $N$ classes $M$ times, where each time splits some classes as positive and some classes as negative. In this way, a total of $M$ training sets are generated, and $M$ classifiers can be trained.
- Decoding: use the $M$ classifiers to predict a testing sample and combine the predicted labels into a codeword. Then, the distances between the codeword and the base codeword of each class are calculated. The class with the shortest distance is returned as the final prediction.

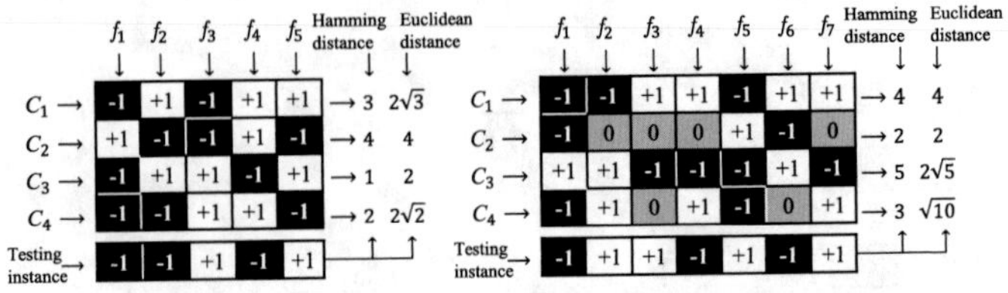


(a) Binary ECOC coding. (b) Ternary ECOC coding.

3

**Fig. 3.5** Illustration of ECOC encoding. "+" and "−" represent the positive and negative classes predicted by the learner $f_i$. "0" in ternary coding indicates that the class is not used by $f_i$

The *coding matrix* determines how classes are partitioned. There are different designs of the coding matrix, and commonly used designs are binary coding (Dietterich and Bakiri 1995) and ternary coding (Allwein et al. 2000). Binary coding puts each class as either positive or negative, while ternary coding adds an additional "excluded class". ◘ Figure 3.5 gives an example. In ◘ Figure 3.5a, classifier $f_2$ considers $C_1$ and $C_3$ as positive and considers $C_2$ and $C_4$ as negative. On the other hand, classifier $f_4$ in ◘ Figure 3.5b considers $C_1$ and $C_4$ as positive and $C_3$ as negative. In the decoding step, the predictions from all classifiers jointly generate the codeword for the testing sample. Then, the distances between the codeword and the base codeword of each class are calculated. The class with the shortest distance is returned as the final prediction. For example, in ◘ Figure 3.5a, the prediction is $C_3$ when using the Euclidean distance.

Why is it called "Error Correcting Output Codes"? Because the ECOC codeword has the error tolerance and correction ability in the testing phase. For example, in ◘ Figure 3.5a, the correct codeword for the testing sample is $(-1, +1, +1, -1, +1)$. Suppose that classifier $f_2$ has made a mistake, and the codeword becomes $(-1, -1, +1, -1, +1)$, but this codeword can still make the correct prediction $C_3$. In general, for the same problem, a longer ECOC codeword produces better correction ability. Nevertheless, a longer codeword implies more classifiers to be trained and consequently increased computation and memory costs. Besides, since class combinations are finite for finite classes, the extra length of codeword becomes meaningless when it reaches the limit.

In theory, the correction ability of a fixed length codeword increases as the distances between classes increase. Following this principle, the theoretically optimal codeword can be calculated when the codeword length is short. However, it becomes an NP-hard problem to find the optimal for long codewords. Fortunately, non-optimal codewords are often sufficient in practice, and the optimal codeword is rarely necessary. Besides, better theoretical properties are not necessarily associated with better classification performance since machine learning involves many factors. For example, when multiple classes are divided into two "class subsets", the different dividing methods lead to different class subsets with different classification difficulties. Therefore, we have two codewords: one has a nice theoretical property of error correction but leads to difficult binary classification problems, and the other one has the weaker error correction ability but leads to easier binary classification problems. It is hard to tell which codeword is better.

## 3.6 Class Imbalance Problem

The classification methods introduced so far have made a common assumption: there is no significant difference in the number of samples in each class. Generally, the impact of a small difference is limited, but a large difference becomes trouble in the learning process. For example, suppose there are 998 negative samples but only two positive samples, then a learner can easily achieve 99.8% accuracy by predicting every new sample as negative. Apparently, such a learner is useless since it cannot identify any positive samples.

The scenario described above is called class imbalance, which refers to the classification problems with a significantly different number of samples for each class. Without loss of generality, this section assumes that the positive class is the minority, and the negative class is the majority. Class imbalance is common in practice. For example, even if the original data set is class-balanced, it is still possible that the binary classification problems made by OvR or MvM are class-imbalanced. Therefore, it is necessary to understand the basic approach to class imbalance issues.

It is easy to understand from the perspective of linear classifiers. When we use $y = \boldsymbol{w}^{\top}\boldsymbol{x} + b$ to classify a new sample $\boldsymbol{x}$, we are actually comparing the predicted value with a threshold value, e.g., positive if $y > 0.5$ and negative otherwise. The value $y$ represents the likelihood of being positive, and the odds $\frac{y}{1-y}$ represent the ratio of likelihoods for being positive over negative. Setting the threshold to 0.5 implies that the classifier assumes the probabilities of a sample being positive or negative are equal. The decision rule of the classifier is

Since OvR and MvM do the same process for each class, the effects of class imbalance in binary classifications will cancel each other, and no special treatment is needed.

$$\text{If } \frac{y}{1-y} > 1 \text{ then predict as a positive.} \tag{3.46}$$

When the classes are imbalanced, let $m^+$ and $m^-$ denote the number of positive and negative samples, respectively. Then, the observed class ratio $\frac{m^+}{m^-}$ represents the ground-truth class ratio since the training set is assumed to be an unbiased sampling. Therefore, a new sample is classified as positive if the predicted odds are higher than the observed odds, i.e,

3

Unbiased sampling means the ground-truth class ratio is maintained in the training set.

$$\text{If } \frac{y}{1-y} > \frac{m^+}{m^-} \text{ then predict as a positive.} \tag{3.47}$$

However, since our classifier makes decisions via (3.46), it is necessary to adjust its prediction value so that when making a decision based on (3.46), it is actually executing (3.47). To do this, we can use

$$\frac{y'}{1-y'} = \frac{y}{1-y} \times \frac{m^-}{m^+}. \tag{3.48}$$

This gives a basic strategy for handling class imbalance learning—*rescaling*.

Also known as *rebalance*.

Though the idea of rescaling is simple, its implementation is non-trivial since the assumption "the training set is an unbiased sampling" often does not hold in practice. In other words, the ratio inferred from the training set may not be accurate. Overall, there are three major rescaling approaches. The first approach is to perform *undersampling* on the negative class, that is, some negative samples are selectively dropped so that the classes are balanced. The second approach is to perform *oversampling* on the positive class, that is, increase the number of positive samples. The third approach is *threshold-moving*, which uses the original training set for learning but uses (3.48) in the decision process.

Undersampling is also known as *downsampling* and oversampling is also known as *upsampling*.

Since undersampling discards negative samples, its computational cost is much lower compared to oversampling, which increases the number of positive samples. It is worth mentioning that oversampling is not simply duplicating existing samples; otherwise, serious overfitting will happen. A representative oversampling method is SMOTE (Chawla et al. 2002), which generates synthetic samples by interpolating neighborhood samples of the positive class. For undersampling, we may lose valuable information if the negative samples are discarded randomly. EasyEnsemble (Liu and Zhou 2009) is a representative undersampling algorithm, which utilizes the ensemble learning mechanism. EasyEnsemble divides negative samples into several smaller subsets for different learners such that undersampling is performed for each learner, but overall there is little loss of information.

Rescaling is also the basis for cost-sensitive learning, in which $m^-/m^+$ in (3.48) is replaced with $\mathrm{cost}^+/\mathrm{cost}^-$, where $\mathrm{cost}^+$ is the cost of misclassifying positive as negative and $\mathrm{cost}^-$ is the cost of misclassifying negative as positive.

## 3.7 Further Reading

*Sparse representation* has gained attention in recent years. However, finding the solution with optimal *sparsity* is not easy, even for simple models like multivariate linear regressions. Essentially, the sparsity problem corresponds to the $\mathrm{L}_0$ norm optimization, which is typically NP-hard. LASSO (Tibshirani 1996) approximates the $\mathrm{L}_0$ norm with the $\mathrm{L}_1$ norm, and is an important technique for finding the sparse solution. See Chap. 11.

Allwein et al. (2000) showed that OvO and OvR are special cases of ECOC. It was hoped that there exists a general codeword for all problems, but Crammer and Singer (2002) argued that codeword design should be problem-dependent and proved that the searching for the optimal discrete coding matrix is an NP-complete problem. Since then, many problem-dependent ECOC coding methods were proposed, generally by identifying representative binary classification problems to encode (Pujol et al. 2006, 2008). An open-source ECOC library was developed by Escalera et al. (2010).

ECOC is not the only implementation of MvM. For example, (Platt et al. 2000) used Directed Acyclic Graph (DAG) to divide classes into a tree structure, where each node corresponds to a binary classifier. Some efforts have been made on solving the multiclass problem directly without converting it into binary classifications, e.g., multiclass support vector machines (Crammer and Singer 2001; Lee et al. 2004).

The class-based *misclassification cost* (e.g., the cost matrix in ◘ Table 2.2) is the most widely studied topic in cost-sensitive learning. In this book, cost-sensitive learning is used on default to refer to the studies on misclassification cost, except as otherwise stated. Elkan (2001) proved that the optimal solution of binary classification problems could be obtained via rescaling. However, (Zhou and Liu 2006a) showed that closed-form solutions only exist under certain conditions for multiclass classification problems. Though both cost-sensitive learning and class imbalance learning can leverage the rescaling technique, they are essentially different (Zhou and Liu 2006b). Note that the cost of the minority class is often higher; otherwise, no special treatment is needed.

Though there are multiple classes in multiclass classification, each sample belongs to a single class. If more than one label is to be assigned, then it turns into *multi-label learning*. For example, a picture can be labeled as `blue sky`, `cloud`, `sheep`, and `natural scene` at the same time. Multi-label learning is a vigorous research area in recent years. Readers interested in this topic can find more information in Zhang and Zhou (2014).

## Exercises

**3.1** Analyze in what situations the bias term $b$ is not needed in (3.2).

**3.2** Prove that with respect to the parameter $\boldsymbol{w}$, the objective function (3.18) of logistic regression is non-convex, but its log-likelihood function (3.27) is convex.

**3.3** Implement and run logistic regression on the watermelon data set 3.0$\alpha$.

The watermelon data set 3.0$\alpha$ is in ◘ Table 4.5.

**3.4** Choose any two data sets from UCI, and compare the error of logistic regression obtained from 10-fold cross-validation and hold-out.

UCI data sets can be found at ► http://archive.ics.uci.edu/ml/.

**3.5** Implement and run linear discriminant analysis on the watermelon data set 3.0$\alpha$.

**3.6** Linear discriminant analysis only works well for linearly separable data. Design an improved version that can perform reasonably well on nonlinearly separable data.

Linearly separable means there exists at least one linear hyperplane that can separate samples of different classes. See Sect. 6.3.

**3.7** Let the length of codeword be nine and the number of classes be four. Find the theoretically optimal ECOC binary coding under the Hamming distance and prove its optimality.

**3.8** * The correction function of ECOC makes the important assumption that errors are independently incurring at each codeword position with similar probabilities. For the binary classifier obtained from ECOC coding, analyze the possibility that it satisfies the above assumption and the impact of such a possibility.

**3.9** OvR and MvM decompose multiclass problems into binary problems. Analyze why no treatment is needed for them even when the binary problems are class imbalance.

**3.10** * Derive the conditions for obtaining the theoretically optimal solution via rescaling in multiclass cost-sensitive learning (consider only class-based cost).

3

## Break Time

### Short Story: About the Least Squares Method

(Gauss on a 1993 Deutsche Mark banknote)

In 1801, Ceres, the first discovered asteroid, was observed by the Italian astronomer Piazzi. However, after being tracked for 40 days, Ceres disappeared behind the sun. Many astronomers have tried to recover its position but all failed. This drew the attention of the German mathematician Gauss (1777–1855), who developed a method and managed to calculate the position of Ceres using the observation data from Piazzi. With the predicted position and time, Gauss and the German astronomer Olbers recovered Ceres. In 1809, Gauss published his method, which is the method of least squares, in his book *Theory of the Motion of the Heavenly Bodies Moving about the Sun in Conic Sections.*

In 1805, *New Methods for the Determination of Comet Orbits* was published by the French mathematician Legendre (1752–1833), who have made numerous contributions to elliptic integral, number theory, and geometry. In this book, the least squares method was found in the appendix. In the eighteenth and nineteenth centuries, Legendre was one of the three pioneers in the French mathematics community, and was a fellow of French Academy of Sciences. However, Legendre's book did not discuss the error analysis of the least squares method, which was covered in Gauss's book in 1809. The error analysis is of great importance for statistics and even machine learning. In addition to this contribution, Gauss claimed that he started using least squares in 1799, and therefore the invention of the least squares method is often attributed to Gauss. Those two mathematicians had some debates at that time, and there is yet no conclusion after the efforts made by historians of mathematics.

The other two are Lagrange and Laplace. Since the surnames of them all start with the letter "L", they are called "3L" at that time.

## References

Allwein EL, Schapire RE, Singer Y (2000) Reducing multiclass to binary: a unifying approach for margin classifiers. J Mach Learn Res 1:113–141

Boyd S, Vandenberghe L (2004) Convex optimization. Cambridge University Press, Cambridge

Chawla NV, Bowyer KW, Hall LO, Kegelmeyer WP (2002) SMOTE: synthetic minority over-sampling technique. J Mach Learn Res 16:321–357

Crammer K, Singer Y (2001) On the algorithmic implementation of multiclass kernel-based vector machines. J Mach Learn Res 2:265–292

Crammer K, Singer Y (2002) On the learnability and design of output codes for multiclass problems. Mach Learn 47(2–3):201–233

Dietterich TG, Bakiri G (1995) Solving multiclass learning problems via error-correcting output codes. J Artif Intell Res 2:263–286

Elkan C (2001) The foundations of cost-sensitive learning. In: Proceedings of the 17th International Joint Conference on Artificial Intelligence (IJCAI), pp 973–978. Seattle, WA

Escalera S, Pujol O, Radeva P (2010) Error-correcting output codes library. J Artif Intell Res 11:661–664

Fisher RA (1936) The use of multiple measurements in taxonomic problems. Ann Eugen 7(2):179–188

Lee Y, Lin Y, Wahba G (2004) Multicategory support vector machines, theory, and application to the classification of microarray data and satellite radiance data. J Amer Stat Ass 99(465):67–81

Liu X-Y, Zhou Z-H (2009) Exploratory undersampling for class-imbalance learning. IEEE Trans Syst Man Cybern - Part B: Cybern 39(2):539–550

Platt JC, Cristianini N, Shawe-Taylor J (2000) Large margin DAGs for multiclass classification. In: Solla SA, Leen TK, Müller K (eds) Advances in neural information processing systems 12. MIT Press, Cambridge, pp 547–553

Pujol O, Escalera S, Radeva P (2008) An incremental node embedding technique for error correcting output codes. Pattern Recognit 41(2):713–725

Pujol O, Radeva P, Vitrià J (2006) Discriminant ECOC: a heuristic method for application dependent design of error correcting output codes. IEEE Trans Pattern Anal Mach Intell 28(6):1007–1012

Tibshirani R (1996) Regression shrinkage and selection via the LASSO. J R Stat Soc: Ser B 58(1):267–288

Zhang M-L, Zhou Z-H (2014) A review on multi-label learning algorithms. IEEE Trans Know Data Eng 26(8):1819–1837

Zhou Z-H, Liu X-Y (2006a) On multi-class cost-sensitive learning. In: Proceedings of the 21st national conference on artificial intelligence (AAAI), pp 567–572. Boston, WA

Zhou Z-H, Liu X-Y (2006b) Training cost-sensitive neural networks with methods addressing the class imbalance problem. IEEE Trans Know Data Eng 18(1):63–77

# Decision Trees

## Table of Contents




Z.-H. Zhou, *Machine Learning*,
https://doi.org/10.1007/978-981-15-1967-3_4

4

## 4.1 Basic Process

Decision trees are a popular class of machine learning methods. Taking binary classification as an example, we can regard the task as deciding the answer to the question "Is this instance positive?" As the name suggests, a decision tree makes decisions based on tree structures, which is also a common decision-making mechanism used by humans. For example, in order to answer the question "Is this watermelon ripe?" we usually go through a series of judgments or sub-decisions: we first consider "What is the color?" If it is green then "What is the shape of root?" If it is curly then "What is the knocking sound?" Finally, based on the observations, we decide whether the watermelon is ripe or not. Such a decision process is illustrated in ◘ Figure 4.1.

The conclusions at the end of the decision process correspond to the possible classifications, e.g., ripe or unripe. Every question asked in the decision process is a test on one feature, e.g.,color =? or root =?. Every test leads to either the conclusion or an additional test conditioned on the current answer. For example, if the current decision is color = green, the next test root =? considers only green watermelons.

Typically, a decision tree consists of one root node, multiple internal nodes, and multiple leaf nodes. The leaf nodes correspond to the decision outcomes, and every other node corresponds to a feature test. The samples in each node are divided into child nodes according to the splitting results of features. Each path from the root node to the leaf node is a decision sequence. The goal is to produce a tree that can generalize to predict unseen samples. The construction of decision trees follows the *divide-and-conquer* strategy, as shown in ◘ Algorithm 4.1.

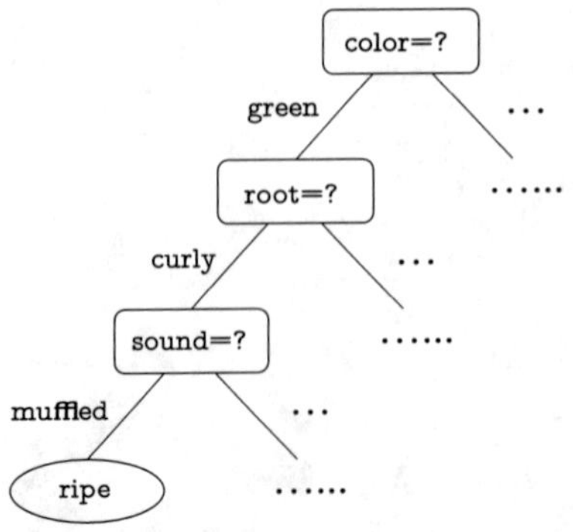


**Fig. 4.1** A decision tree of the watermelon problem

**Algorithm 4.1** Decision Tree Learning

**Input:** Training set $D = \{(\boldsymbol{x}_1, y_1), (\boldsymbol{x}_2, y_2), \ldots, (\boldsymbol{x}_m, y_m)\}$;
Feature set $A = \{a_1, a_2, \ldots, a_d\}$.
**Process:** Function TreeGenerate($D$, $A$)
1: Generate node $i$;
2: **if** All samples in $D$ belong to the same class $C$ **then**
3: Mark node $i$ as a class $C$ leaf node; **return**

Recursive return, case (1).

4: **end if**
5: **if** $A = \varnothing$ **OR** all samples in $D$ take the same value on $A$ **then**
6: Mark node $i$ as a leaf node, and its class label is the majority class in $D$; **return**

Recursive return, case (2).

7: **end if**
8: Select the optimal splitting feature $a_*$ from $A$;

We will discuss the optimal split selection in the next section.

9: **for** each value $a_*^v$ in $a_*$ **do**
10: Generate a branch for node $i$; Let $D_v$ be the subset of samples taking value $a_*^v$ on $a_*$;
11: **if** $D_v$ is empty **then**
12: Mark this child node as a leaf node, and label it with the majority class in $D$; **return**

Recursive return, case (3).

13: **else**
14: Use TreeGenerate($D_v$, $A \backslash \{a_*\}$) as the child node.

Exclude $a_*$ from $A$.

15: **end if**
16: **end for**
**Output:** A decision tree with root node $i$.

As shown in ◘ Algorithm 4.1, the tree is generated recursively, and the recursion stops in any of the following three cases: (1) all samples in the current node belong to the same class, that is, no further splitting is needed; (2) the current feature set is empty, or all samples have the same feature values, that is, not splittable; (3) there is no sample in the current node, that is, not splittable.

In case (2), we mark the current node as a leaf node and set its label to the majority class of its samples. In case (3), we mark the current node as a leaf node but set its label to the majority class of the samples in its parent node. Note that the two cases are different: case (2) uses the posterior probability of the current node, whereas case (3) uses the class probability of the parent node as the prior probability of the current node.

## 4.2 Split Selection

The core of the decision tree learning algorithm is the line 8 of ◘ Algorithm 4.1, that is, selecting the optimal splitting feature. Generally speaking, as the splitting process proceeds, we wish more samples within each node to belong to a single class, that is, increasing the *purity* of each node.

### 4.2.1 Information Gain

One of the most commonly used measures for purity is *information entropy*, or simply *entropy*. Let $p_k$ denotes the proportion of the $k$th class in the current data set $D$, where $k = 1, 2, \ldots, |\mathcal{Y}|$. Then, the entropy is defined as

In the calculation of entropy, $p \log_2 p = 0$ when $p = 0$.



$$\operatorname{Ent}(D) = -\sum_{k=1}^{|\mathcal{Y}|} p_k \log_2 p_k. \tag{4.1}$$

The minimum of Ent($D$) is 0 and the maximum is $\log_2 |\mathcal{Y}|$.

The lower the Ent($D$), the higher the purity of $D$.

Suppose that the discrete feature $a$ has $V$ possible values $\{a^1, a^2, \ldots, a^V\}$. Then, splitting the data set $D$ by feature $a$ produces $V$ child nodes, where the $v$th child node $D^v$ includes all samples in $D$ taking the value $a^v$ for feature $a$. Then, the entropy of $D^v$ can be calculated using (4.1). Since there are different numbers of samples in the child nodes, a weight $|D^v| / |D|$ is assigned to reflect the importance of each node, that is, the greater the number of samples, the greater the impact of the branch node. Then, the *information gain* of splitting the data set $D$ with feature $a$ is calculated as

$$\operatorname{Gain}(D, a) = \operatorname{Ent}(D) - \sum_{v=1}^{V} \frac{|D^v|}{|D|} \operatorname{Ent}(D^v). \tag{4.2}$$

In general, the higher the information gain, the more purity improvement we can expect by splitting $D$ with feature $a$. Therefore, information gain can be used for split selection, that is, using $a_* = \underset{a \in A \operatorname{Gain}(D,a)}{\arg\max}$ as the splitting feature on the line 8 of ◘ Algorithm 4.1. The well-known decision tree algorithm ID3 Quinlan (1986) takes information gain as the guideline for selecting the splitting features.

The term ID in ID3 stands for Iterative Dichotomiser.

Let us see a more concrete example with the watermelon data set 2.0 in ◘ Table 4.1. This data set includes 17 training samples, which are used to train a decision tree classifier for predicting the ripeness of uncut watermelons, where $|\mathcal{Y}| = 2$. In the beginning, the root node includes all samples in $D$, where $p_1 = \frac{8}{17}$ of them are positive and $p_2 = \frac{9}{17}$ of them are negative. According to (4.1), the entropy of the root node is

$$\operatorname{Ent}(D) = -\sum_{k=1}^{2} p_k \log_2 p_k = -\left(\frac{8}{17}\log_2\frac{8}{17} + \frac{9}{17}\log_2\frac{9}{17}\right) = 0.998.$$

Then, we need to calculate the information gain of each feature in the current feature set {`color`, `root`, `sound`, `texture`, `umbilicus`, `surface`}. Suppose that we have selected `color`,

**Tab. 4.1** The watermelon data set 2.0

| ID | color | root | sound | texture | umbilicus | surface | ripe |
|---|---|---|---|---|---|---|---|
| 1 | green | curly | muffled | clear | hollow | hard | true |
| 2 | dark | curly | dull | clear | hollow | hard | true |
| 3 | dark | curly | muffled | clear | hollow | hard | true |
| 4 | green | curly | dull | clear | hollow | hard | true |
| 5 | light | curly | muffled | clear | hollow | hard | true |
| 6 | green | slightly curly | muffled | clear | slightly hollow | soft | true |
| 7 | dark | slightly curly | muffled | slightly blurry | slightly hollow | soft | true |
| 8 | dark | slightly curly | muffled | clear | slightly hollow | hard | true |
| 9 | dark | slightly curly | dull | slightly blurry | slightly hollow | hard | false |
| 10 | green | straight | crisp | clear | flat | soft | false |
| 11 | light | straight | crisp | blurry | flat | hard | false |
| 12 | light | curly | muffled | blurry | flat | soft | false |
| 13 | green | slightly curly | muffled | slightly blurry | hollow | hard | false |
| 14 | light | slightly curly | dull | slightly blurry | hollow | hard | false |
| 15 | dark | slightly curly | muffled | clear | slightly hollow | soft | false |
| 16 | light | curly | muffled | blurry | flat | hard | false |
| 17 | green | curly | dull | slightly blurry | slightly hollow | hard | false |

which has three possible values {green, dark, light}. If $D$ is split by color, then there are three subsets: $D^1$ (color = green), $D^2$ (color = dark), and $D^3$ (color = light).

Subset $D^1$ includes six samples {1, 4, 6, 10, 13, 17}, in which $p_1 = \frac{3}{6}$ of them are positive and $p_2 = \frac{3}{6}$ of them are negative. Subset $D^2$ includes six samples {2, 3, 7, 8, 9, 15}, in which $p_1 = \frac{4}{6}$ of them are positive and $p_2 = \frac{2}{6}$ of them are negative. Subset $D^3$ includes five samples {5, 11, 12, 14, 16}, in which $p_1 = \frac{1}{5}$ of them are positive and $p_2 = \frac{4}{5}$ of them are negative. According to (4.1), the entropy of the three child nodes are

$$\begin{aligned}
\mathrm{Ent}(D^1) &= -\left(\frac{3}{6}\log_2\frac{3}{6} + \frac{3}{6}\log_2\frac{3}{6}\right) = 1.000,\\
\mathrm{Ent}(D^2) &= -\left(\frac{4}{6}\log_2\frac{4}{6} + \frac{2}{6}\log_2\frac{2}{6}\right) = 0.918,\\
\mathrm{Ent}(D^3) &= -\left(\frac{1}{5}\log_2\frac{1}{5} + \frac{4}{5}\log_2\frac{4}{5}\right) = 0.722.
\end{aligned}$$

Then, we use (4.2) to calculate the information gain of splitting by color as

$$\begin{aligned}
\mathrm{Gain}(D, \text{color}) &= \mathrm{Ent}(D) - \sum_{v=1}^{3}\frac{|D^v|}{|D|}\mathrm{Ent}(D^v)\\
&= 0.998 - \left(\frac{6}{17}\times 1.000 + \frac{6}{17}\times 0.918 + \frac{5}{17}\times 0.722\right)\\
&= 0.109.
\end{aligned}$$

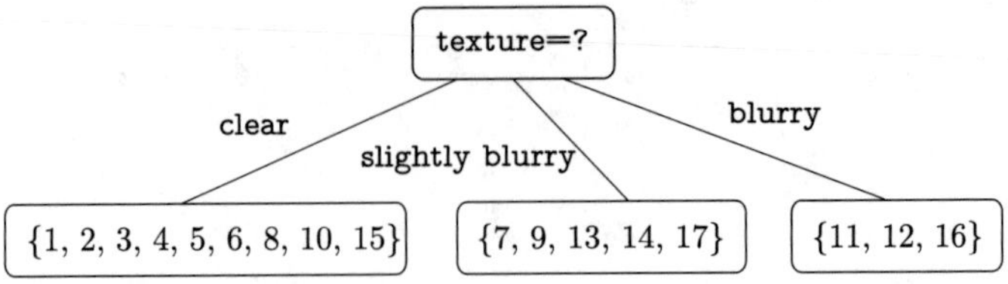


**Fig. 4.2** Splitting the root node by texture



Similarly, we calculate the information gain of other features:

$$\text{Gain}(D, \texttt{root}) = 0.143; \quad \text{Gain}(D, \texttt{sound}) = 0.141;$$
$$\text{Gain}(D, \texttt{texture}) = 0.381; \quad \text{Gain}(D, \texttt{umbilicus}) = 0.289;$$
$$\text{Gain}(D, \texttt{surface}) = 0.006.$$

Since splitting by texture gives the highest information gain, it is chosen as the splitting feature. ◘ Figure 4.2 shows the result of splitting the root node by texture.

Then, each child node is further split by the decision tree algorithm. For example, the first child node (i.e., texture = clear) includes nine samples: $D^1 = \{1, 2, 3, 4, 5, 6, 8, 10, 15\}$, and the available feature set is {color, root, sound, umbilicus, surface}. We calculate the information gains of these candidate features on $D^1$:

texture is no longer a candidate splitting feature.

$$\text{Gain}(D^1, \texttt{color}) = 0.043; \quad \text{Gain}(D^1, \texttt{root}) = 0.458;$$
$$\text{Gain}(D^1, \texttt{sound}) = 0.331; \quad \text{Gain}(D^1, \texttt{umbilicus}) = 0.458;$$
$$\text{Gain}(D^1, \texttt{surface}) = 0.458.$$

Since root, umbilicus, and surface lead to the highest information gains, any of them can be chosen as the splitting feature. Repeating this process for every node, we can obtain the final decision tree, as shown in ◘ Figure 4.3.

### 4.2.2 Gain Ratio

The process described above intentionally ignored the column ID. If we consider ID as a candidate splitting feature, then, from (4.2), we know its information gain is 0.998, which is much higher than that of any other features. This is reasonable since ID produces 17 child nodes, and each node has only a single sample with maximum purity. However, such a decision

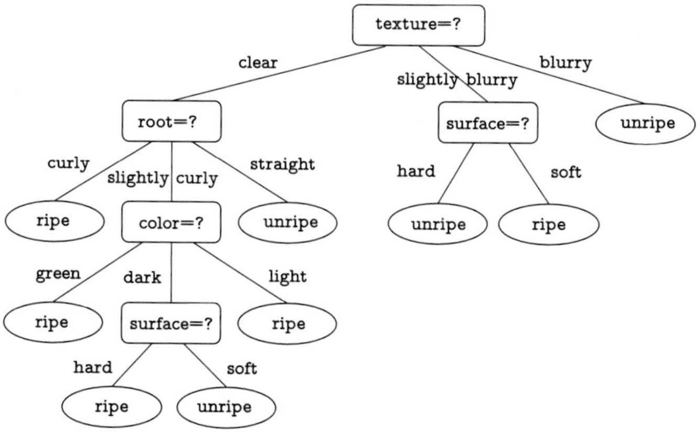


**Fig. 4.3** The information gain-based decision tree generated from ◘ Table 4.1

tree does not have generalization ability and cannot effectively predict new samples.

It turns out that the information gain criterion is biased toward features with more possible values. To reduce this bias, the renowned decision tree algorithm C4.5 (Quinlan 1993) employs *gain ratio* to select features instead of employing information gain. Using a notation similar to (4.2), the gain ratio of feature $a$ is defined as

$$\text{Gain\_ratio}(D, a) = \frac{\text{Gain}(D, a)}{\text{IV}(a)}, \tag{4.3}$$

where

$$\text{IV}(a) = -\sum_{v=1}^{V} \frac{|D^v|}{|D|} \log_2 \frac{|D^v|}{|D|} \tag{4.4}$$

is called the *intrinsic value* of feature $a$ (Quinlan 1993). $\text{IV}(a)$ is large when feature $a$ has many possible values (i.e., large $V$). Taking the watermelon data set 2.0 as an example, we have: $\text{IV}(\texttt{surface}) = 0.874$ ($V = 2$), $\text{IV}(\texttt{color}) = 1.580$ ($V = 3$), and $\text{IV}(\text{ID}) = 4.088$ ($V = 17$).

It should be noted that, in contrast to information gain, the gain ratio is biased toward features with fewer possible values. For this reason, the C4.5 algorithm does not use gain ratio directly for selecting the splitting feature, but uses a heuristic method (Quinlan 1993): selecting the feature with the highest gain ratio from the set of candidate features with an information gain above the average.

### 4.2.3 Gini Index

CART stands for Classification and Regression Tree, which is a well-known decision tree algorithm applicable to both classification and regression.

CART Breiman et al. (1984) employs the *Gini index* for selecting the splitting feature. Using a notation similar to (4.1), the Gini value of data set $D$ is defined as

$$\begin{aligned}\mathrm{Gini}(D) &= \sum_{k=1}^{|\mathcal{Y}|}\sum_{k'\neq k} p_k p_{k'}\\ &= 1-\sum_{k=1}^{|\mathcal{Y}|} p_k^2. \end{aligned} \tag{4.5}$$

Intuitively, Gini($D$) represents the likelihood of two samples we randomly selected from data set $D$ belonging to different classes. The lower the Gini($D$), the higher the purity of data set $D$.

Using a notation similar to (4.2), the Gini index of feature $a$ is defined as

$$\mathrm{Gini\_index}(D,a) = \sum_{v=1}^{V}\frac{|D^v|}{|D|}\mathrm{Gini}(D^v). \tag{4.6}$$

Given a candidate feature set $A$, we select the feature with the lowest Gini index as the splitting feature, that is, $a_* = \arg\min_{a\in A}\mathrm{Gini\_index}(D,a)$.

## 4.3 Pruning

See Sect. 2.1 for overfitting.

*Pruning* is the primary strategy of decision tree learning algorithms to deal with overfitting. To correctly classify the training samples, the learner repeats the split procedure. However, if there are too many branches, then the learner may be misled by the peculiarities of the training samples and incorrectly consider them as the underlying truth. Hence, we can prune some of the branches to reduce the risk of overfitting.

The general pruning strategies include *pre-pruning* and *post-pruning* (Quinlan 1993). Pre-pruning evaluates the improvement of the generalization ability of each split and cancels a split if the improvement is small, that is, the node is marked as a leaf node. In contrast, post-pruning re-examines the non-leaf nodes of a fully grown decision tree, and a node is replaced with a leaf node if the replacement leads to improved generalization ability.

How do we know if the generalization ability has been improved? We can use the performance evaluation methods introduced in Sect. 2.2. For example, we can use the hold-out

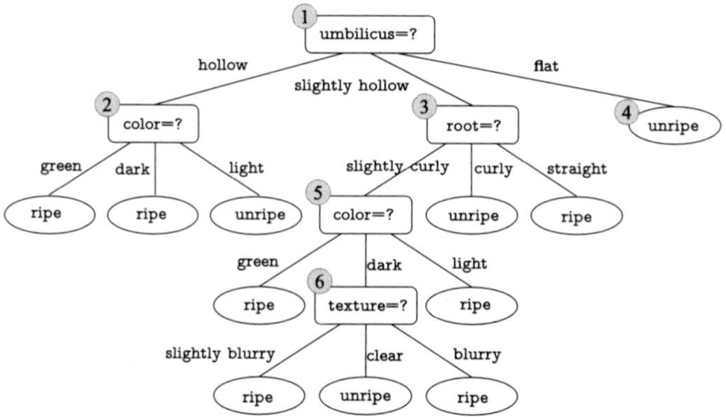


**Fig. 4.4** The unpruned decision tree generated from ◘ Table 4.2

method to reserve part of the data as a validation set for performance evaluation. Given the watermelon data set 2.0 in ◘ Table 4.1, suppose the samples are randomly partitioned into a training set {1, 2, 3, 6, 7, 10, 14, 15, 16, 17} and a validation set {4, 5, 8, 9, 11, 12, 13}, as shown in ◘ Table 4.2.

◘ **Tab. 4.2** Splitting the watermelon data set 2.0 into a training set (above the double dividing line) and a validation set (below the double dividing line)

| ID | color | root | sound | texture | umbilicus | surface | ripe |
|---|---|---|---|---|---|---|---|
| 1 | green | curly | muffled | clear | hollow | hard | true |
| 2 | dark | curly | dull | clear | hollow | hard | true |
| 3 | dark | curly | muffled | clear | hollow | hard | true |
| 6 | green | slightly curly | muffled | clear | slightly hollow | soft | true |
| 7 | dark | slightly curly | muffled | slightly blurry | slightly hollow | soft | true |
| 10 | green | straight | crisp | clear | flat | soft | false |
| 14 | light | slightly curly | dull | slightly blurry | hollow | hard | false |
| 15 | dark | slightly curly | muffled | clear | slightly hollow | soft | false |
| 16 | light | curly | muffled | blurry | flat | hard | false |
| 17 | green | curly | dull | slightly blurry | slightly hollow | hard | false |

| ID | color | root | sound | texture | umbilicus | surface | ripe |
|---|---|---|---|---|---|---|---|
| 4 | green | curly | dull | clear | hollow | hard | true |
| 5 | light | curly | muffled | clear | hollow | hard | true |
| 8 | dark | slightly curly | muffled | clear | slightly hollow | hard | true |
| 9 | dark | slightly curly | dull | slightly blurry | slightly hollow | hard | false |
| 11 | light | straight | crisp | blurry | flat | hard | false |
| 12 | light | curly | muffled | blurry | flat | soft | false |
| 13 | green | slightly curly | muffled | slightly blurry | hollow | hard | false |

Suppose we use the information gain criterion described in Sect. 4.2.1 for deciding the splitting features, then ◘ Figure 4.4 shows the decision tree trained on the data set in ◘ Table 4.2. For ease of discussion, we numbered some nodes in the figures.

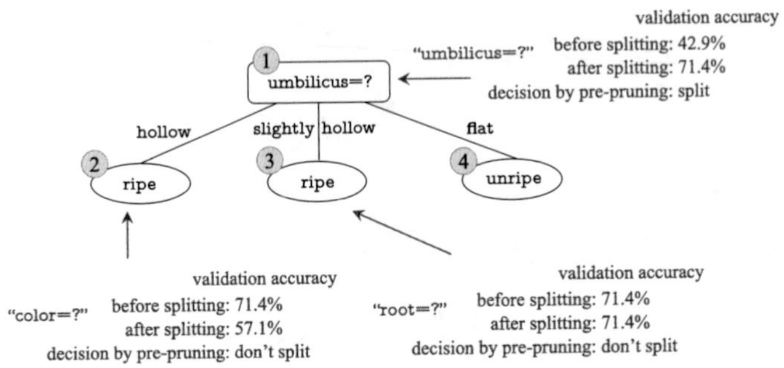


**Fig. 4.5** The pre-pruned decision tree generated from ◘ Table 4.2

### 4.3.1 Pre-pruning

Let us take a look at pre-pruning first. According to the information gain criterion, umbilicus should be chosen to split the training set into three branches, as shown in ◘ Figure 4.5. However, shall we proceed with this split? Pre-pruning decides by comparing the generalization abilities before and after splitting.

Prior to splitting, all samples are in the root node. When no splitting is performed, this node is marked as a leaf node according to line 6 of ◘ Algorithm 4.1, and its label is set to the majority class (i.e., ripe). By evaluating this single-node decision tree using the validation set in ◘ Table 4.2, we have the samples {4, 5, 8} correctly classified and the other four samples misclassified. Then, the validation accuracy is $\frac{3}{7} \times 100\% = 42.9\%$.

When there is more than one class with the largest number of samples, we randomly select one of the classes.

After splitting the root node by umbilicus, the samples are placed into three child nodes, as shown in ◘ Figure 4.5: node ② with the samples {1, 2, 3, 14}, node ③ with the samples {6, 7, 15, 17}, and node ④ with the samples {10, 16}. We mark these 3 nodes as leaf nodes and set the labels to the majority classes, that is, ② is ripe, ③ is ripe, and ④ is unripe. Then, the validation accuracy is $\frac{5}{7} \times 100\% = 71.4\% > 42.9\%$. Since the validation accuracy is improved, the splitting using umbilicus is adopted.

After that, the decision tree algorithm moves on to split node ②, and color is chosen based on the information gain criterion. However, since the sample {5} in the validation set is misclassified, the validation accuracy drops to 57.1%. Hence, the pre-pruning strategy stops splitting node ②. For node ③, the best feature to split on is root. However, since the validation accuracy after splitting remains the same as 71.4%, pre-pruning strategy stops splitting node ③. For node ④, no splitting is needed since all samples belong to the same class.

Finally, the pre-pruning decision tree constructed based on the data in ◘ Table 4.2 is given in ◘ Figure 4.5, and its validation accuracy is 71.4%. Because there is only one splitting, such a decision tree is also called a *decision stump*.

By comparing ◘ Figures 4.5 and ◘ 4.4, we can see that applying pre-pruning reduces the branches of the decision tree, which reduces not only the risk of overfitting but also the computational cost of training and testing. On the other hand, although some branches are prevented by pre-pruning due to little or even negative improvement on generalization ability, it is still possible that their subsequent splits can lead to significant improvement. These branches are pruned due to the greedy nature of pre-pruning, and it may introduce the risk of underfitting.

### 4.3.2 Post-pruning

Post-pruning allows a decision tree to grow into a complete tree, e.g., ◘ Figure 4.4 shows a fully grown decision tree based on data in ◘ Table 4.2. The validation accuracy of this decision tree is 42.9%.

In ◘ Figure 4.4, node ⑥ is the first one examined by post-pruning. If the subtree led by node ⑥ is pruned and replaced with a leaf node, then it includes the samples {7, 15} and its label is set to the majority class ripe. Since the validation accuracy increases to 57.1%, the pruning is performed, resulting in the decision tree, and the result is shown in ◘ Figure 4.6.

Next, post-pruning examines node ⑤. If the subtree led by node ⑤ is replaced by a leaf node, then it includes the samples {6, 7, 15} and its label is set to the majority class ripe. Since the validation accuracy remains at 57.1%, no pruning is performed.

If the subtree led by node ② is replaced by a leaf node, then it includes the samples {1, 2, 3, 14} and its label is set to the majority class ripe. Since the validation accuracy increases to 71.4%, the pruning is performed.

For nodes ③ and ①, replacing them as leaf nodes gives the validation accuracies 71.4% and 42.9%, respectively. Since there is no improvement in both cases, the nodes remain unchanged.

Although the accuracy of the validation set is not improved in this case, according to Occam's razor principle, the model would be better after pruning. In fact, the actual decision tree algorithm usually needs pruning in this case. For the convenience of drawing, this book adopts a conservative strategy of not pruning.

Finally, the post-pruning decision tree constructe using data in ◘ Table 4.2 is given in ◘ Figure 4.6, and its validation accuracy is 71.4%.

By comparing ◘ Figs. 4.6 and ◘ 4.5, we can see that post-pruning keeps more branches than pre-pruning. In general, post-pruning is less prone to underfitting and leads to better generalization ability compared to pre-pruning. However, the training time of post-pruning is much longer since it takes a bottom-up strategy to examine every non-leaf node in a completely grown decision tree.

## 4.4 Continuous and Missing Values

### 4.4.1 Handling Continuous Values

Our discussions so far are limited to discrete features. However, since continuous features are also common in practice, it is necessary to know how to incorporate continuous features into decision trees.

We cannot directly split nodes with continuous features since their values are infinite. The discretization techniques come in handy in such cases. The most straightforward discretization strategy is bi-partition, which is used by C4.5 decision tree (Quinlan 1993).

Given a data set $D$ and a continuous feature $a$, suppose $n$ values of $a$ are observed in $D$, and we sort these values in ascending order, denoted by $\{a^1, a^2, \ldots, a^n\}$. With a split point $t$, $D$ is partitioned into the subsets $D_t^-$ and $D_t^+$, where $D_t^-$ includes the samples with the value of $a$ not greater than $t$, and $D_t^+$ includes the samples with the value of $a$ greater than $t$. For adjacent feature values $a^i$ and $a^{i+1}$, the partitions are identical for choosing any $t$ in the interval $[a^i, a^{i+1})$. As a result, for continuous feature $a$, there are $n-1$ elements in the following set of candidate split points:

$$T_a = \left\{ \frac{a^i + a^{i+1}}{2} \mid 1 \leqslant i \leqslant n-1 \right\}, \tag{4.7}$$

where the midpoint $\frac{a^i+a^{i+1}}{2}$ is used as the candidate split point for the interval $[a^i, a^{i+1})$. Then, the split points are examined in the same way as discrete features, and the optimal split points

The split point can be set to the maximum observed value of this feature in $\left[a^i, \frac{a^i+a^{i+1}}{2}\right]$. Doing so ensures that all split points are the values appeared in the training set (Quinlan 1993).

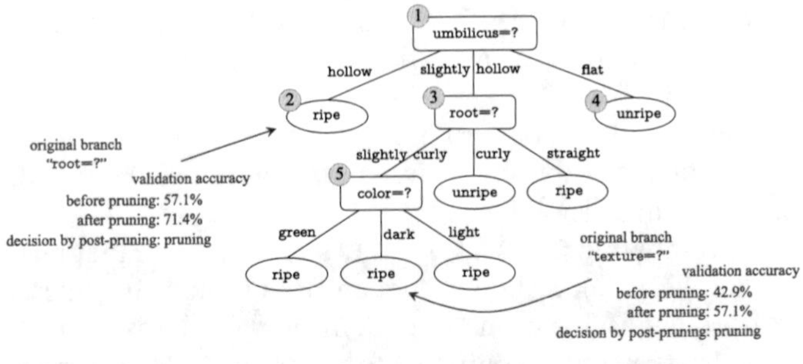


**Fig. 4.6** The post-pruned decision tree generated from ◘ Table 4.2

are selected for splitting nodes. For example, we can modify (4.2) as

$$\begin{aligned}\mathrm{Gain}(D, a) &= \max_{t\in T_a} \mathrm{Gain}(D, a, t)\\ &= \max_{t\in T_a} \mathrm{Ent}(D) - \sum_{\lambda\in\{-,+\}} \frac{|D_t^\lambda|}{|D|}\mathrm{Ent}(D_t^\lambda),\end{aligned} \quad (4.8)$$

where $\mathrm{Gain}(D, a, t)$ is the information gain of bi-partitioning $D$ by $t$, and the split point with the largest $\mathrm{Gain}(D, a, t)$ is selected.

For illustration, we create the watermelon data set 3.0 in ◘ Table 4.3 by adding two continuous features density and sugar to the watermelon data set 2.0. Now, we build a decision tree using this new data set.

◘ **Tab. 4.3** The watermelon data set 3.0

| ID | color | root | sound | texture | umbilicus | surface | density | sugar | ripe |
|---|---|---|---|---|---|---|---|---|---|
| 1 | green | curly | muffled | clear | hollow | hard | 0.697 | 0.460 | true |
| 2 | dark | curly | dull | clear | hollow | hard | 0.774 | 0.376 | true |
| 3 | dark | curly | muffled | clear | hollow | hard | 0.634 | 0.264 | true |
| 4 | green | curly | dull | clear | hollow | hard | 0.608 | 0.318 | true |
| 5 | light | curly | muffled | clear | hollow | hard | 0.556 | 0.215 | true |
| 6 | green | slightly curly | muffled | clear | slightly hollow | soft | 0.403 | 0.237 | true |
| 7 | dark | slightly curly | muffled | slightly blurry | slightly hollow | soft | 0.481 | 0.149 | true |
| 8 | dark | slightly curly | muffled | clear | slightly hollow | hard | 0.437 | 0.211 | true |
| 9 | dark | slightly curly | dull | slightly blurry | slightly hollow | hard | 0.666 | 0.091 | false |
| 10 | green | straight | crisp | clear | flat | soft | 0.243 | 0.267 | false |
| 11 | light | straight | crisp | blurry | flat | hard | 0.245 | 0.057 | false |
| 12 | light | curly | muffled | blurry | flat | soft | 0.343 | 0.099 | false |
| 13 | green | slightly curly | muffled | slightly blurry | hollow | hard | 0.639 | 0.161 | false |
| 14 | light | slightly curly | dull | slightly blurry | hollow | hard | 0.657 | 0.198 | false |
| 15 | dark | slightly curly | muffled | clear | slightly hollow | soft | 0.360 | 0.370 | false |
| 16 | light | curly | muffled | blurry | flat | hard | 0.593 | 0.042 | false |
| 17 | green | curly | dull | slightly blurry | slightly hollow | hard | 0.719 | 0.103 | false |

At the beginning, all 17 training samples have different density values. According to (4.7), the candidate split point set includes 16 values: $T_{\text{density}} = \{0.244, 0.294, 0.351, 0.381, 0.420, 0.459, 0.518, 0.574, 0.600, 0.621, 0.636, 0.648, 0.661, 0.681, 0.708, 0.746\}$. According to (4.8), the information gain of density is 0.262, and the corresponding split point is 0.381.

For the feature sugar, its candidate split point set includes 16 values: $T_{\text{sugar}} = \{0.049, 0.074, 0.095, 0.101, 0.126, 0.155, 0.179, 0.204, 0.213, 0.226, 0.250, 0.265, 0.292, 0.344, 0.373, 0.418\}$. Similarly, the information gain of sugar is 0.349 according to (4.8), and the corresponding split point is 0.126.

Combining the results from Sect. 4.2.1, the information gains of features in ◘ Table 4.3 are

$$\begin{aligned}
&\text{Gain}(D, \texttt{color}) = 0.109; && \text{Gain}(D, \texttt{root}) = 0.143;\\
&\text{Gain}(D, \texttt{sound}) = 0.141; && \text{Gain}(D, \texttt{texture}) = 0.381;\\
&\text{Gain}(D, \texttt{umbilicus}) = 0.289; && \text{Gain}(D, \texttt{surface}) = 0.006;\\
&\text{Gain}(D, \texttt{density}) = 0.262; && \text{Gain}(D, \texttt{sugar}) = 0.349.
\end{aligned}$$

Since splitting by texture has the largest information gain, it is selected as the splitting feature for the root node. The splitting process proceeds recursively, and the final decision tree is shown in ◘ Figure 4.7.

Unlike discrete features, a continuous feature can be used as a splitting feature more than once in a decision sequence.

For example, using density $\leqslant 0.381$ in a parent node does not forbid the use of density $\leqslant 0.294$ in a child node.

### 4.4.2 Handling Missing Values

In practice, data is often incomplete, that is, some feature values are missing in some samples. Taking medical diagnosis data as an example, feature values such as HIV test results could be unavailable due to privacy concerns. Sometimes we may have a large number of incomplete samples, especially when there are many features. Though we can simply discard the incomplete samples, it is a huge waste of data. For example, ◘ Table 4.4 shows a watermelon data set with missing values. If we discard the incomplete samples, then we will have only four samples {4, 7, 14, 16} left for training. Apparently, we need a method to utilize incomplete samples.

Learning from incomplete samples raises two problems: (1) how to choose the splitting features when there are missing values? (2) how to split a sample with the splitting feature value missing?

Given a training set $D$ and a feature $a$, let $\tilde{D}$ be the subset of samples in $D$ that has values of $a$. For problem (1), we can simply use $\tilde{D}$ to evaluate $a$. Let $\{a^1, a^2, \ldots, a^V\}$ denote the $V$ possible values of $a$, $\tilde{D}^v$ denote the subset of samples in $\tilde{D}$

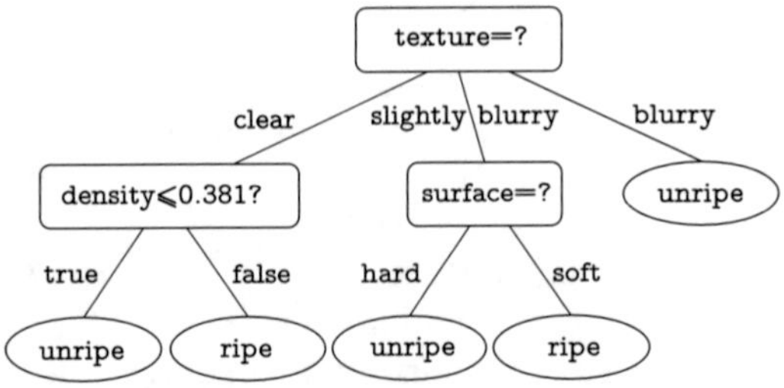


**Fig. 4.7** The information gain-based decision tree generated from ◘ Table 4.3

**Tab. 4.4** The watermelon data set 2.0$\alpha$

| ID | color | root | sound | texture | umbilicus | surface | ripe |
|---|---|---|---|---|---|---|---|
| 1 | - | curly | muffled | clear | hollow | hard | true |
| 2 | dark | curly | dull | clear | hollow | - | true |
| 3 | dark | curly | - | clear | hollow | hard | true |
| 4 | green | curly | dull | clear | hollow | hard | true |
| 5 | - | curly | muffled | clear | hollow | hard | true |
| 6 | green | slightly curly | muffled | clear | - | soft | true |
| 7 | dark | slightly curly | muffled | slightly blurry | slightly hollow | soft | true |
| 8 | dark | slightly curly | muffled | - | slightly hollow | hard | true |
| 9 | dark | - | dull | slightly blurry | slightly hollow | hard | false |
| 10 | green | straight | crisp | - | flat | soft | false |
| 11 | light | straight | crisp | blurry | flat | - | false |
| 12 | light | curly | - | blurry | flat | soft | false |
| 13 | - | slightly curly | muffled | slightly blurry | hollow | hard | false |
| 14 | light | slightly curly | dull | slightly blurry | hollow | hard | false |
| 15 | dark | slightly curly | muffled | clear | - | soft | false |
| 16 | light | curly | muffled | blurry | flat | hard | false |
| 17 | green | - | dull | slightly blurry | slightly hollow | hard | false |

taking the value $a^v$, and $\tilde{D}_k$ denote the subset of samples in $\tilde{D}$ belonging to the $k$th class, where $k = 1, 2, \ldots, |\mathcal{Y}|$. Then, we have $\tilde{D} = \bigcup_{k=1}^{|\mathcal{Y}|} \tilde{D}_k$ and $\tilde{D} = \bigcup_{v=1}^{V} \tilde{D}^v$. We assign a weight $w_{\boldsymbol{x}}$ to each sample $\boldsymbol{x}$, and define

At the beginning of decision tree learning, the weights are initialized to 1 for all samples in the root node.

$$\rho = \frac{\sum_{\boldsymbol{x}\in\tilde{D}} w_{\boldsymbol{x}}}{\sum_{\boldsymbol{x}\in D} w_{\boldsymbol{x}}}, \tag{4.9}$$

$$\tilde{p}_k = \frac{\sum_{\boldsymbol{x}\in\tilde{D}_k} w_{\boldsymbol{x}}}{\sum_{\boldsymbol{x}\in\tilde{D}} w_{\boldsymbol{x}}} \quad (1 \leqslant k \leqslant |\mathcal{Y}|), \tag{4.10}$$

$$\tilde{r}_v = \frac{\sum_{\boldsymbol{x}\in\tilde{D}^v} w_{\boldsymbol{x}}}{\sum_{\boldsymbol{x}\in\tilde{D}} w_{\boldsymbol{x}}} \quad (1 \leqslant v \leqslant V). \tag{4.11}$$

Intuitively, for the feature $a$, $\rho$ represents the proportion of samples without missing values, $\tilde{p}_k$ represents the proportion of the $k$th class in all samples without missing values, and $\tilde{r}_v$ represents the proportion of samples taking the feature value $a^v$ in all samples without missing values. Then, we have $\sum_{k=1}^{|\mathcal{Y}|} \tilde{p}_k = 1$ and $\sum_{v=1}^{V} \tilde{r}_v = 1$.

With the above definitions, we extend the information gain (4.2) to

$$\begin{aligned}\mathrm{Gain}(D, a) &= \rho \times \mathrm{Gain}(\tilde{D}, a)\\ &= \rho \times \left(\mathrm{Ent}\left(\tilde{D}\right) - \sum_{v=1}^{V} \tilde{r}_v \mathrm{Ent}\left(\tilde{D}^v\right)\right),\end{aligned} \tag{4.12}$$

where, according to (4.1),

$$\text{Ent}(\tilde{D}) = -\sum_{k=1}^{|\mathcal{Y}|} \tilde{p}_k \log_2 \tilde{p}_k.$$

For problem (2), when the value of $a$ is known, we place the sample $\boldsymbol{x}$ into the corresponding child node without changing its weight $w_{\boldsymbol{x}}$. When the value of $a$ is unknown, we place the sample $\boldsymbol{x}$ into all child nodes, and set its weight in the child node of value $a^v$ to $\tilde{r}_v \cdot w_{\boldsymbol{x}}$. In other words, we place the same sample into different child nodes with different probabilities.

4

The above solution is employed by the C4.5 algorithm (Quinlan 1993), and we will use it to construct a decision tree for ◘ Table 4.4.

In the beginning, the root node includes all of the 17 samples in $D$, and all samples have the weight of 1. Taking color as an example, the set of samples without missing values of this feature includes 14 samples {2, 3, 4, 6, 7, 8, 9, 10, 11, 12, 14, 15, 16, 17}, denoted by $\tilde{D}$. The entropy of $\tilde{D}$ is calculated as

$$\begin{aligned}\text{Ent}(\tilde{D}) &= -\sum_{k=1}^{2} \tilde{p}_k \log_2 \tilde{p}_k \\ &= -\left(\frac{6}{14}\log_2\frac{6}{14} + \frac{8}{14}\log_2\frac{8}{14}\right) = 0.985.\end{aligned}$$

Let $\tilde{D}^1$, $\tilde{D}^2$, and $\tilde{D}^3$ be the subsets of samples with color = green, color = dark, and color = light, respectively. Then, we have

$$\begin{aligned}\text{Ent}(\tilde{D}^1) &= -\left(\frac{2}{4}\log_2\frac{2}{4} + \frac{2}{4}\log_2\frac{2}{4}\right) = 1.000, \\ \text{Ent}(\tilde{D}^2) &= -\left(\frac{4}{6}\log_2\frac{4}{6} + \frac{2}{6}\log_2\frac{2}{6}\right) = 0.918, \\ \text{Ent}(\tilde{D}^3) &= -\left(\frac{0}{4}\log_2\frac{0}{4} + \frac{4}{4}\log_2\frac{4}{4}\right) = 0.000.\end{aligned}$$

The information gain of color for subset $\tilde{D}$ is

$$\begin{aligned}\text{Gain}(\tilde{D}, \text{color}) &= \text{Ent}(\tilde{D}) - \sum_{v=1}^{3} \tilde{r}_v \text{Ent}(\tilde{D}^v) \\ &= 0.985 - \left(\frac{4}{14} \times 1.000 + \frac{6}{14} \times 0.918 + \frac{4}{14} \times 0.000\right) \\ &= 0.306.\end{aligned}$$

The information gain of color for data set $D$ is

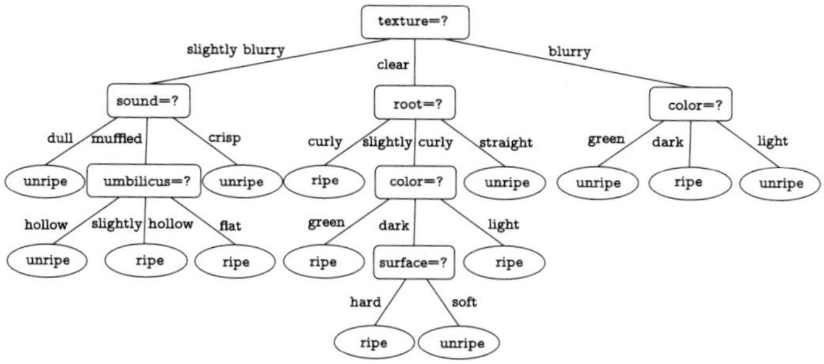


**Fig. 4.8** The information gain-based decision tree generated from ◘ Table 4.4

$$\text{Gain}(D, \texttt{color}) = \rho \times \text{Gain}(\tilde{D}, \texttt{color}) = \frac{14}{17} \times 0.306 = 0.252.$$

Similarly, we can calculate the information gain of all features for $D$:

$$\text{Gain}(D, \texttt{color}) = 0.252; \qquad \text{Gain}(D, \texttt{root}) = 0.171;$$
$$\text{Gain}(D, \texttt{sound}) = 0.252; \qquad \text{Gain}(D, \texttt{texture}) = 0.424;$$
$$\text{Gain}(D, \texttt{umbilicus}) = 0.289; \quad \text{Gain}(D, \texttt{texture}) = 0.006.$$

Since splitting by texture has the largest information gain, it is selected for splitting the root node. Specifically, the samples $\{1, 2, 3, 4, 5, 6, 15\}$ are placed into the child node of $\texttt{texture} = \texttt{clear}$, the samples $\{7, 9, 13, 14, 17\}$ are placed into the child node $\texttt{texture} = \texttt{slightly blurry}$, and the samples $\{11, 12, 16\}$ are placed into the child node of $\texttt{texture} = \texttt{blurry}$. The weights of these samples (i.e., 1) remain unchanged in the child nodes. However, since the value of texture is missing for the sample $\{8\}$, the sample is placed into all of the three child nodes with different weights: $\frac{7}{15}$, $\frac{5}{15}$, and $\frac{3}{15}$. The sample $\{10\}$ is processed similarly. The splitting process proceeds recursively, and the final constructed decision tree is shown in ◘ Figure 4.8.

## 4.5 Multivariate Decision Trees

If we regard each feature as a coordinate axis in the coordinate space, a sample with $d$ features corresponds to a point in the $d$-dimensional space. Classifying samples is then about finding the decision boundaries in this space to separate the samples of different classes. For decision trees, the decision boundaries have a distinct characteristic: axis-parallel, that is, the decision boundaries are multiple segments parallel to the axes.

The watermelon data set 3.0$\alpha$ is a copy of the watermelon data set 3.0 excluding discrete features.

**Tab. 4.5** The watermelon data set 3.0$\alpha$

| ID | density | sugar | ripe |
|---|---|---|---|
| 1 | 0.697 | 0.460 | true |
| 2 | 0.774 | 0.376 | true |
| 3 | 0.634 | 0.264 | true |
| 4 | 0.608 | 0.318 | true |
| 5 | 0.556 | 0.215 | true |
| 6 | 0.403 | 0.237 | true |
| 7 | 0.481 | 0.149 | true |
| 8 | 0.437 | 0.211 | true |
| 9 | 0.666 | 0.091 | false |
| 10 | 0.243 | 0.267 | false |
| 11 | 0.245 | 0.057 | false |
| 12 | 0.343 | 0.099 | false |
| 13 | 0.639 | 0.161 | false |
| 14 | 0.657 | 0.198 | false |
| 15 | 0.360 | 0.370 | false |
| 16 | 0.593 | 0.042 | false |
| 17 | 0.719 | 0.103 | false |

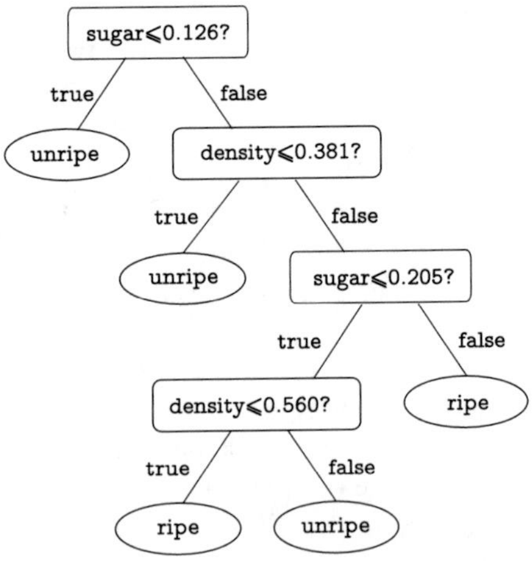


**Fig. 4.9** The decision tree generated from Table 4.5

For example, Figure 4.9 shows the decision tree trained on the watermelon data set 3.0$\alpha$ in Table 4.5, and the corresponding decision boundaries are shown in Figure 4.10.

From Figure 4.10, we can observe that every segment is parallel to the axis. Since every segment corresponds to a specific value of a feature, such decision boundaries make the learning outcome easy to interpret. In practice, the decision boundaries often need many segments for good approximations, e.g., Figure 4.11. However, such complex decision trees are often slow to make predictions since they contain many feature tests.

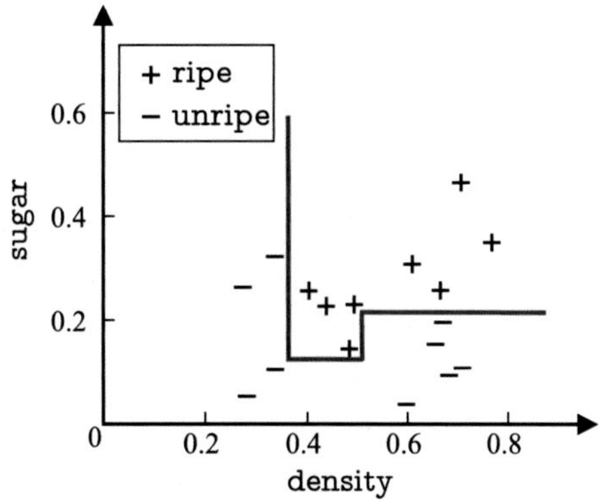


**Fig. 4.10** The decision boundaries of the decision tree in Figure 4.9

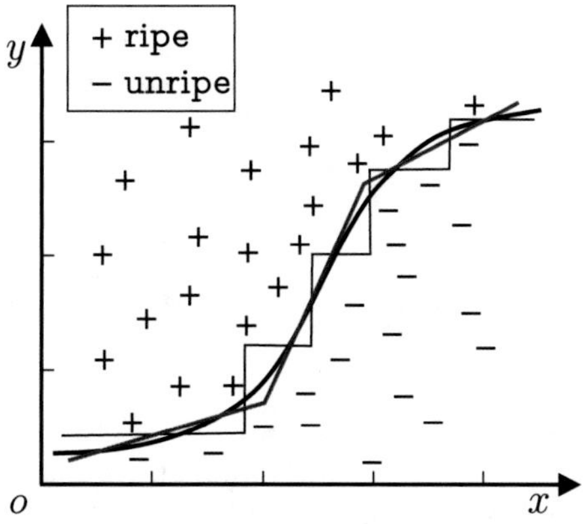


**Fig. 4.11** The piecewise approximation of complex decision boundaries

If we can make the decision boundaries oblique, as shown by the red line in Figure 4.11, then the decision tree model can be significantly simplified. *Multivariate decision tree* enables oblique partitions or even more complicated decision boundaries. With oblique boundaries, each non-leaf node is no longer a test for a particular feature but a linear combination of features. In other words, each non-leaf node is a linear classifier in the form of $\sum_{i=1}^{d} w_i a_a = t$, where $w_i$ is the weight of

Also known as *oblique decision tree*.

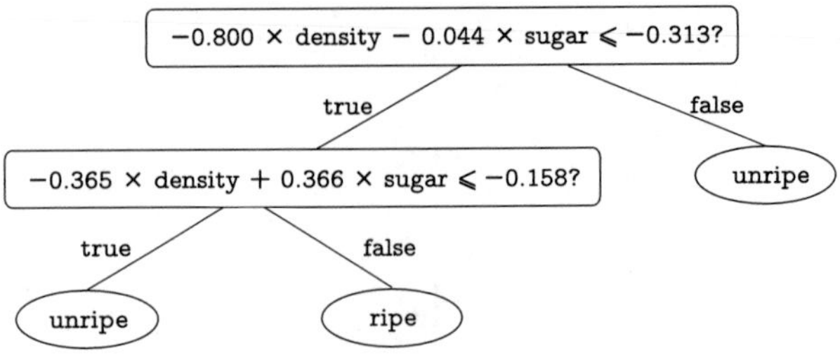


**Fig. 4.12** The multivariate decision tree generated from ◘ Table 4.5

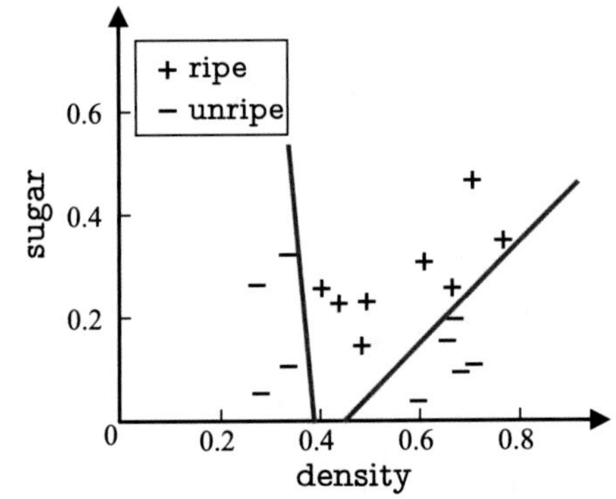


**Fig. 4.13** The decision boundaries of the decision tree in ◘ Figure 4.12

feature $a_i$, and $w_i$ and $t$ are learned from the data set and feature set of the node. Unlike the traditional univariate decision tree, the learning process of multivariate decision tree does not look for an optimal splitting feature but tries to establish a suitable linear classifier. ◘ Figure 4.12 shows the multivariate decision tree learned from the watermelon data set 3.0$\alpha$, and the corresponding decision boundaries are shown in ◘ Figure 4.13.

See Chap. 3 for linear classifier.

## 4.6 Further Reading

Representative decision tree learning algorithms include ID3 (Quinlan 1979, 1986), C4.5 (Quinlan 1993), and CART (Breiman et al. 1984). Murthy (1989) surveyed decision tree techniques. C4.5Rule (Quinlan 1993) is an algorithm that converts C4.5 decision trees into symbolic rules by rewriting each split as a rule. The converted rule set may possess even better generalization ability compared to the original decision tree due to the merge, addition, and subtraction operations on the rules during the conversion process.

Apart from the information gain, gain ratio, and Gini index, many criteria have been developed for split selection. Empirical studies (Mingers 1989b) showed that different criteria have a limited impact on the generalization ability, though they often lead to decision trees with different sizes. A theoretical study (Raileanu and Stoffel 2004) also showed that the information gain and Gini index are only different in 2% of cases. The general strategies for pruning decision trees were discussed in Sect. 4.3. Empirical studies (Mingers 1989a) showed that pruning could lead to even a 25% improvement of the generalization ability for noisy data.

Any feature selection methods can be used for selecting the splitting features. See Chap. 11 for feature selection.

Representative multivariate decision tree learning algorithms include OC1 (Murthy et al. 1994) and a series of algorithms proposed in Brodley and Utgoff (1995). The OC1 algorithm starts with a greedy search of the optimal weight for each feature, and then, on top of local optimization, performs random manipulation on the decision boundaries to find potentially better decision boundaries; by contrast, Brodley and Utgoff (1995) directly introduced the least squares method of linear classifiers. There are also algorithms trying to embed neural networks into leaf nodes to take advantage of both learning mechanisms. For example, *Perceptron tree* (Utgoff 1989b) trained a perceptron at each leaf node, and (Guo and Gelfand 1992) embedded multi-layer neural networks into leaf nodes.

See Chap. 5 for Perceptron and Neural Networks.

Some decision tree learning algorithms support *incremental learning*, that is, adjusting the learned model using newly received samples without re-training with the whole data set. The main idea is to partially restructure the decision tree by reordering the features along paths, and representative algorithms include ID4 (Schlimmer and Fisher 1986), ID5R (Utgoff 1989a), and ITI (Utgoff et al. 1997). Incremental learning can effectively reduce the computational cost of training upon receiving new samples, but the model after multiple steps of incremental learning could be considerably different from a model re-trained with the whole data set.

## Exercises

**4.1** Prove that if the training set contains no conflicting data (i.e., same feature vector but different labels), then there must exist a decision tree that is consistent with the training set (i.e., training error is 0).

**4.2** Analyze the disadvantages of using minimizing training error as the splitting criterion for decision tree learning.



**4.3** Implement a decision tree algorithm with entropy as its splitting criterion, and generate a decision tree using the data set in ◘ Table 4.3.

**4.4** Implement a decision tree algorithm with the Gini index as its splitting criterion, and generate a decision tree using the data set in ◘ Table 4.2.

**4.5** Implement a decision tree algorithm with logistic regression as its splitting criterion, and generate a decision tree using the data set in ◘ Table 4.3.

UCI data sets are available at ► http://archive.ics.uci.edu/ml/.

**4.6** Choose four data sets from UCI, and conduct empirical comparisons on the unpruned, pre-pruned, and post-pruned decision trees generated by the decision tree algorithms in the above three exercises. Apply an appropriate statistical hypothesis test.

See Sect. 2.4 for statistical hypothesis test.

**4.7** Since ◘ Algorithm 4.1 is a recursive algorithm, the depth of decision trees learned from massive data can easily cause stack overflow. Try to use the queue data structure to implement the decision tree algorithm, and add a parameter *MaxDepth* to control the depth of decision trees.

**4.8** * Rewrite the decision tree algorithm in Exercise 4.7 such that breadth-first search is used instead of depth-first search, and add a parameter *MaxNode* to control the number of nodes. Analyze which of the two decision tree algorithms is easier to control the memory cost.

**4.9** Extend the method of handling missing values in Sect. 4.4.2 to the calculation of the Gini index.

The watermelon data set 3.0 is in ◘ Table 4.3.

**4.10** Download or implement a multivariate decision tree learning algorithm, and investigate its results on the watermelon data set 3.0.

## Break Time

### Short Story: Decision Tree and John Ross Quinlan

Speaking of decision trees, we must talk about the Australian computer scientist John Ross Quinlan (1943–).

The initial decision tree algorithm originated from the Concept Learning System (CLS) proposed by the American psychologist and computer scientist E. B. Hunt in 1962. The CLS algorithm, which was developed for studying the concept learning process of humans, established the divide-and-conquer strategy in decision tree learning. Under the supervision of Hunt, Quinlan obtained his doctoral degree from the University of Washington in 1968 and started a career at the University of Sydney. In 1978, he visited Stanford University during his sabbatical leave and enrolled in an engrossing graduate course taught by D. Michie, who was the assistant of A. Turing. In this course, there was an assignment asking students to implement a program to determine whether a given chess endgame will finish in two moves. Quinlan developed a program that is similar to the CLS algorithm but introduced the information gain criterion. In 1979, he published this work, which is the ID3 algorithm.

In 1986, Quinlan was invited to republish the ID3 algorithm in the first issue of *Machine Learning* journal, and this started a research trend of decision tree learning. In a few years, numerous decision tree algorithms were proposed, and names like ID4 and ID5 have soon been taken. Quinlan had to name his own successor of ID3 as C4.0, and later on the well-known C4.5. Quinlan modestly claimed that C4.5 is only a slightly improved version of C4.0, and hence is called the 4.5th generation classifier. The commercialized successor is called C5.0.

C4.0 stands for Classifier 4.0.

The implementation of C4.5 in WEKA is called J4.8.

4

## References

Breiman L, Friedman J, Stone CJ, Olshen RA (1984) Classification and regression trees. Chapman & Hall/CRC, Boca Raton, FL

Brodley CE, Utgoff PE (1995) Multivariate decision trees. Mach Learn 19(1):45–77

Guo H, Gelfand SB (1992) Classification trees with neural network feature extraction. IEEE Trans Neural Netw 3(6):923–933

Mingers J (1989a) An empirical comparison of pruning methods for decision tree induction. Mach Learn 4(2):227–243

Mingers J (1989b) An empirical comparison of selection measures for decision-tree induction. Mach Learn 3(4):319–342

Murthy SK (1989) Automatic construction of decision trees from data: a multi-disciplinary survey. Data Min Knowl Discov 2(4):345–389

Murthy SK, Kasif S, Salzberg S (1994) A system for induction of oblique decision trees. J Artif Intell Res 2:1–32

Quinlan JR (1979) Discovering rules by induction from large collections of examples. In: Michie D (ed) Expert systems in the micro-electronic age. Edinburgh University Press, Edinburgh, pp 168–201

Quinlan JR (1986) Induction of decision trees. Mach Learn 1(1):81–106

Quinlan JR.(1993) C4.5: program for machine learning. Morgan Kaufmann, San Mateo, CA

Raileanu LE, Stoffel K (2004) Theoretical comparison between the Gini index and information gain criteria. Ann Math Artif Intell 41(1):77–93

Schlimmer JC, Fisher D (1986) A case study of incremental concept induction. In: Proceedings of the 5th national conference on artificial intelligence (AAAI), Philadelphia, PA, pp 495–501

Utgoff PE (1989a) Incremental induction of decision trees. Mach Learn 4(2):161–186

Utgoff PE (1989b) Perceptron trees: a case study in hybrid concept representations. Connect Sci 1(4):377–391

Utgoff PE, Berkman NC, Clouse JA (1997) Decision tree induction based on efficient tree restructuring. Mach Learn 29(1):5–44

# Neural Networks

Table of Contents




Z.-H. Zhou, *Machine Learning*,
https://doi.org/10.1007/978-981-15-1967-3_5

## 5.1 Neuron Model

In this book, neural networks refer to artificial neural networks rather than biological neural networks.

Research on neural networks started quite a long time ago, and it has become a broad and interdisciplinary research field today. Though neural networks have various definitions across disciplines, this book uses a widely adopted one: "Artificial neural networks are massively parallel interconnected networks of simple (usually adaptive) elements and their hierarchical organizations which are intended to interact with the objects of the real world in the same way as biological nervous systems do" (Kohonen 1988). In the context of machine learning, neural networks refer to "neural networks learning", or in other words, the intersection of machine learning research and neural networks research.

This is the definition given by T. Kohonen in the first issue of *Neural Networks* journal in 1988.

5

Neuron is also known as unit.

The basic element of neural networks is neuron, which is the "simple element" in the above definition. In biological neural networks, the neurons, when "excited", send neurotransmitters to interconnected neurons to change their electric potentials. When the electric potential exceeds a *threshold*, the neuron is activated (i.e., "excited"), and it will send neurotransmitters to other neurons.

Threshold is also known as bias.

In 1943, (McCulloch and Pitts 1943) abstracted the above process into a simple model called the McCulloch–Pitts model (M-P neuron model), which is still in use today. As illustrated in ◘ Figure 5.1, each neuron in the M-P neuron model receives input signals from $n$ neurons via weighted connections. The weighted sum of received signals is compared against the threshold, and the output signal is produced by the *activation function*.

Also known as the transfer function.

Step function is a variant of the unit-step function. The logistic function is a representative sigmoid function. See Sect. 3.3.

The ideal activation function is the step function illustrated in ◘ Figure 5.2a, which maps the input value to the output value "0" (non-excited) or "1" (excited). Since the step function has some undesired properties such as being discontinuous and non-smooth, we often use the sigmoid function instead. ◘ Figure 5.2b illustrates a typical sigmoid function that squashes the input values from a large interval into the open unit interval (0, 1), and hence also is known as the *squashing function*.

For example, for 10 pairwise linked neurons, there are 100 parameters, including 90 connection weights and 10 thresholds.

"Simulation of biological neural networks" is an analogous interpretation of neural networks made by cognitive scientists.

A neural network is derived by connecting the neurons into a layered structure. From the perspective of computer science, we can simply regard a neural network as a mathematical model with many parameters , and put aside whether it simulates the biological neural networks or not. The model consists of multiple functions, e.g., nesting $y_j = f(\sum_i w_i x_i - \theta_j)$ multiple times. Effective neural network learning algorithms are often supported by mathematical proofs.

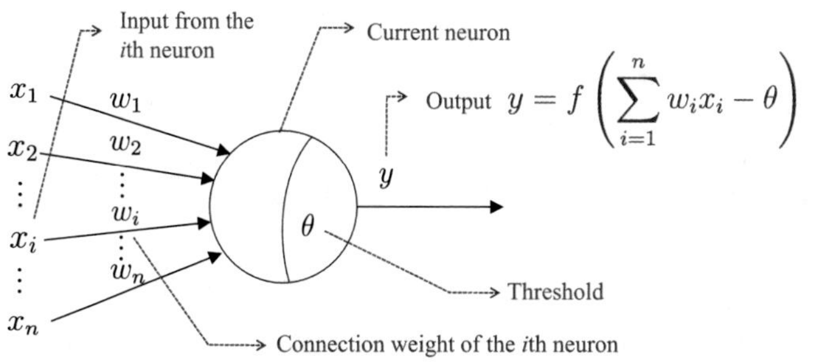


**Fig. 5.1** The M-P neuron model

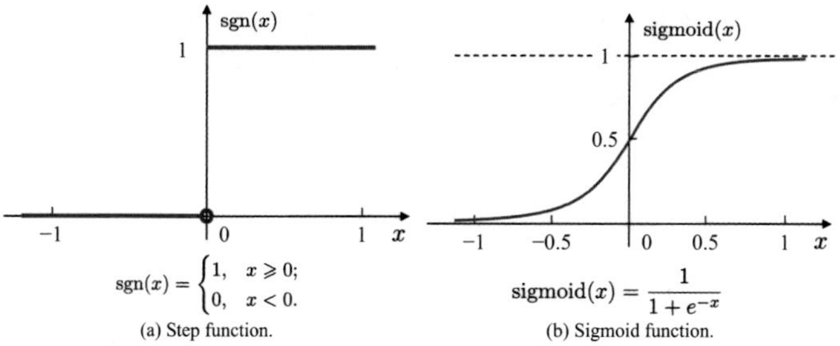


**Fig. 5.2** Typical neuron activation functions

## 5.2 Perceptron and Multi-layer Network

*Perceptron* is a binary classifier consisting of two layers of neurons, as illustrated in ◘ Figure 5.3. The input layer receives external signals and transmits them to the output layer, which is an M-P neuron, also known as *threshold logic unit*.

Perceptron can easily implement the logic operations "AND", "OR", and "NOT". Suppose the function $f$ in $y = f(\sum_i w_i x_i - \theta)$ is the step function shown in ◘ Figure 5.2, the logic operations can be implemented as follows:

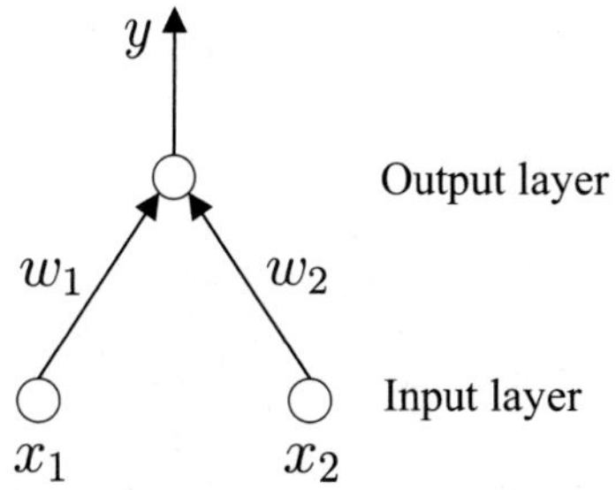


**Fig. 5.3** A perceptron with two input neurons

- "AND" ($x_1 \wedge x_2$): letting $w_1 = w_2 = 1, \theta = 2$, then $y = f(1 \cdot x_1 + 1 \cdot x_2 - 2)$, and $y = 1$ if and only if $x_1 = x_2 = 1$;
- "OR" ($x_1 \vee x_2$): letting $w_1 = w_2 = 1, \theta = 0.5$, then $y = f(1 \cdot x_1 + 1 \cdot x_2 - 0.5)$, and $y = 1$ when $x_1 = 1$ or $x_2 = 1$;
- "NOT" ($\neg x_1$): letting $w_1 = -0.6, w_2 = 0, \theta = -0.5$, then $y = f(-0.6 \cdot x_1 + 0 \cdot x_2 + 0.5)$, and $y = 0$ when $x_1 = 1$ and $y = 1$ when $x_1 = 0$.



More generally, the weight $w_i$ ($i = 1, 2, \ldots, n$) and threshold $\theta$ can be learned from training data. If we consider the threshold $\theta$ as a *dummy node* with the connection weight $w_{n+1}$ and fixed input $-1.0$, then the weight and threshold are unified as weight learning. The learning of perceptron is simple: for training sample $(\boldsymbol{x}, y)$, if the perceptron outputs $\hat{y}$, then the weight is updated by

$x_i$ corresponds to the value of the $i$th input neuron.

$$w_i \leftarrow w_i + \Delta w_i, \tag{5.1}$$

$$\Delta w_i = \eta(y - \hat{y})x_i, \tag{5.2}$$

$\eta$ is typically set to a small positive number, e.g., 0.1.

where $\eta \in (0, 1)$ is known as the *learning rate*. From (5.1) we can see that the perceptron remains unchanged if it correctly predicts the sample $(\boldsymbol{x}, y)$ (i.e., $\hat{y} = y$). Otherwise, the weight is updated based on the degree of error.

The learning ability of perceptrons is rather weak since only the output layer has activation functions, that is, only one layer of functional neurons. In fact, the "AND", "OR", and "NOT" problems are all linearly separable problems. Minsky and Papert (1969) showed that there must exist a linear hyperplane that can separate two classes if they are linearly separable. This means that the perceptron learning process is guaranteed to *converge* to an appropriate weight vector $\mathbf{w} = (w_1; w_2; \ldots; w_{n+1})$, as illustrated in ◘ Figure 5.4a–c. Otherwise, *fluctuation* will happen in the learning process, and no appropriate solution can be found since $\mathbf{w}$ cannot be stabilized. For example, perceptron cannot even solve simple nonlinearly separable problems like "XOR", as shown in ◘ Figure 5.4d.

"Nonlinearly separable" means that no linear hyperplane can separate data from different classes.

In order to solve nonlinearly separable problems, we can use multi-layer functional neurons. For example, the simple two-layer perceptron illustrated in ◘ Figure 5.5 can solve the "XOR" problem. In ◘ Figure 5.5a, the neuron layer between the input layer and the output layer is known as the hidden layer, which has activation functions like the output layer does.

◘ Figure 5.6 illustrates two typical multi-layer neural network structures, in which the neurons in each layer are fully connected with the neurons in the next layer. However, neurons within the same layer or from non-adjacent layers are

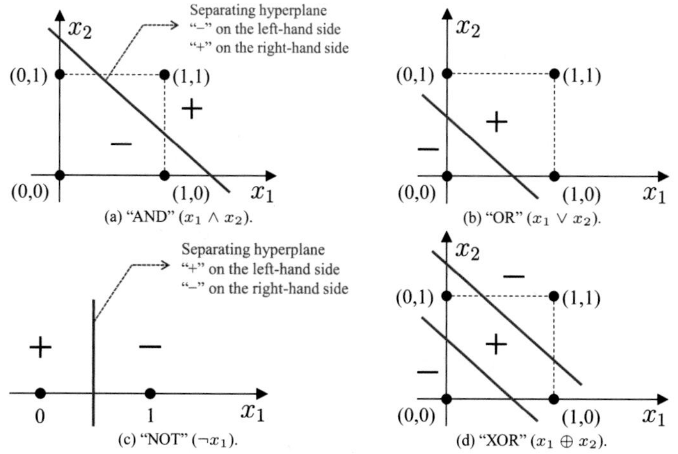


**Fig. 5.4** "AND", "OR", and "NOT" are linearly separable problems. "XOR" is a nonlinearly separable problem

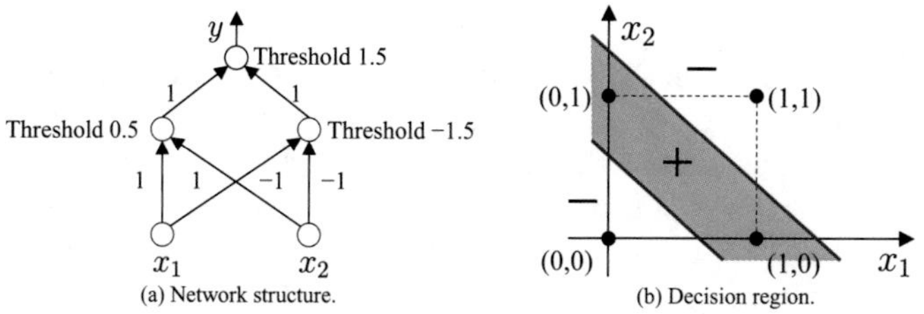


**Fig. 5.5** A two-layer perceptron that solves the "XOR" problem

not connected. Neural networks following such a structure are known as *multi-layer feedforward neural networks*, in which the input layer receives external signals, the hidden and output layers process the signals, and the output layer outputs the processed signals. In other words, the input layer has no processing function but only receives the input, whereas the hidden and output layers have functional neurons. Since only two layers are functional, the neural network in ◼ Figure 5.6a is often called "two-layer neural network". To avoid ambiguity, we call it "single hidden layer neural network" in this book. For neural networks with at least one hidden layer, we call them multi-layer neural networks. The learning process of neural networks is about learning from the training data to adjust the *connection weights* among neurons and the thresholds of functional neurons. In other words, the "knowledge" learned by neural networks is in the connection weights and thresholds.

"Feedforward" does not mean signals cannot be transmitted backward but refers to no recurrent or circular connections. See Sect. 5.5.5.

That is, the weights of the connections between neurons.

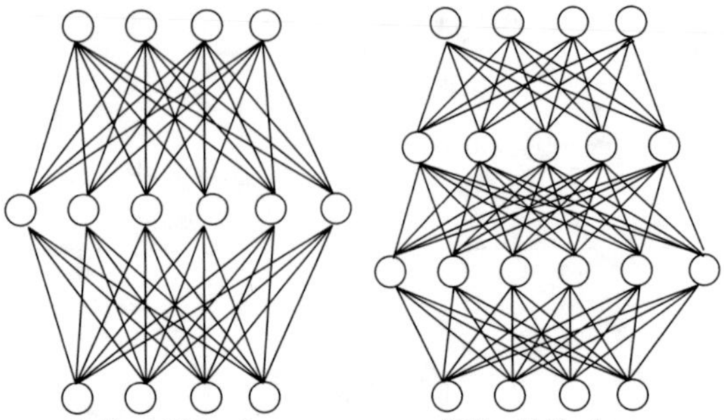


**Fig. 5.6** Muti-layer feedforward neural networks

5

## 5.3 Error Backpropagation Algorithm

The learning ability of multi-layer neural networks is much stronger than single-layer perceptrons. Nevertheless, strong learning ability requires more powerful learning algorithms rather than the simple method of (5.1). Among them, the *error Backpropagation* (BP) algorithm is a representative and by far the most successful neural network learning algorithm, which trained most neural networks in real-world applications. The BP algorithm can train not only feedforward neural networks but also other types of neural networks, such as recurrent neural networks (Pineda 1987). However, "BP neural networks" usually refer to feedforward neural networks trained with the BP algorithm.

Next, let us take a closer look at the BP algorithm. Given a training set $D = \{(\boldsymbol{x}_1, \boldsymbol{y}_1), (\boldsymbol{x}_2, \boldsymbol{y}_2), \ldots, (\boldsymbol{x}_m, \boldsymbol{y}_m)\}$, where $\boldsymbol{x}_i \in \mathbb{R}^d$, $\boldsymbol{y}_i \in \mathbb{R}^l$, that is, the input sample is described by $d$ attributes and the output is an $l$-dimensional real-valued vector. For ease of discussion, ◘ Figure 5.7 shows a multi-layer feedforward neural network with $d$ input neurons, $l$ output neurons, and $q$ hidden neurons. Let $\theta_j$ denote the threshold of the $j$th neuron in the output layer, $\gamma_h$ denote the threshold of the $h$th neuron in the hidden layer, $v_{ih}$ denote the connection weight between the $i$th neuron of the input layer and the $h$th neuron of the hidden layer, $w_{hj}$ denote the connection weight between the $h$th neuron of the hidden layer and the $j$th neuron of the output layer, $\alpha_h = \sum_{i=1}^{d} v_{ih}x_i$ denote the input received by the $h$th neuron in the hidden layer, and $\beta_j = \sum_{h=1}^{q} w_{hj}b_h$ denote the input received by the $j$th neuron in the output layer, where $b_h$ is the output of the $h$ neuron in the hidden layer. Suppose the neurons in both the hidden layer and output layer employ the sigmoid function, which was given in ◘ Figure 5.2b.

Discrete attributes require pre-processing. If there is an ordinal relationship between attribute values, the attribute can be easily converted to continuous values. Otherwise it is usually converted into a $k$-dimensional vector, where $k$ is the number of attribute values. See Sect. 3.2.

It is a logistic function. See Sect. 3.3.

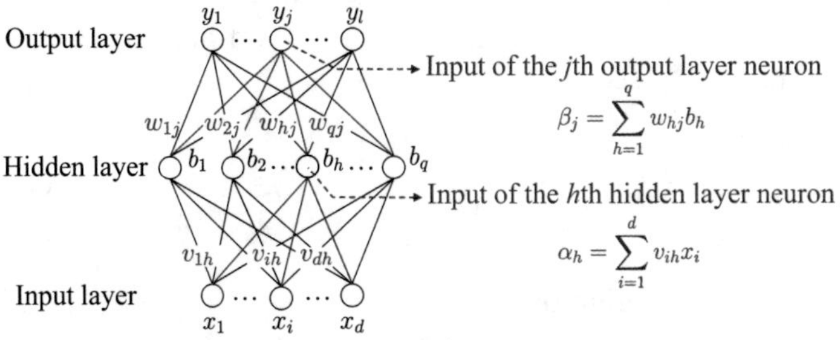


**Fig. 5.7** Notations of BP neural networks

For training sample $(\boldsymbol{x}_k, \boldsymbol{y}_k)$, suppose the neural network outputs $\hat{\boldsymbol{y}}_k = (\hat{y}_1^k, \hat{y}_2^k, \ldots, \hat{y}_l^k)$, that is,

$$\hat{y}_j^k = f(\beta_j - \theta_j), \tag{5.3}$$

then the MSE of the neural network on sample $(\boldsymbol{x}_k, \boldsymbol{y}_k)$ is

$$E_k = \frac{1}{2}\sum_{j=1}^{l}(\hat{y}_j^k - y_j^k)^2. \tag{5.4}$$

Adding 1/2 is for the convenience of subsequent calculations.

The neural network in ■ Figure 5.7 has $(d + l + 1)q + l$ parameters to be determined, including $d \times q$ connection weights from the input layer to the hidden layer, $q \times l$ connection weights from the hidden layer to the output layer, $q$ thresholds of the hidden layer neurons, and $l$ thresholds of the output layer neurons. BP is an iterative learning algorithm, and each iteration employs the general form of perceptron learning rule to estimate and update the parameter, that is, similar to (5.1), the update rule of any parameter $v$ is

$$v \leftarrow v + \Delta v. \tag{5.5}$$

Next, we take the connection weight $w_{hj}$ as an example to demonstrate the derivation.

The BP algorithm employs the *gradient descent* method and tunes the parameters toward the direction of the negative gradient of the objective. For the error $E_k$ in (5.4) and learning rate $\eta$, we have

See Appendix B.4 for the gradient descent method.

$$\Delta w_{hj} = -\eta\frac{\partial E_k}{\partial w_{hj}}. \tag{5.6}$$

Note that $w_{hj}$ first influences the input value $\beta_j$ of the $j$th output layer neuron, then the output value $\hat{y}_j^k$, and finally, the error $E_k$. Hence, we have

The "Chain rule".

$$\frac{\partial E_k}{\partial w_{hj}} = \frac{\partial E_k}{\partial \hat{y}_j^k} \cdot \frac{\partial \hat{y}_j^k}{\partial \beta_j} \cdot \frac{\partial \beta_j}{\partial w_{hj}}. \tag{5.7}$$

From the definition of $\beta_j$, we have

$$\frac{\partial \beta_j}{\partial w_{hj}} = b_h. \tag{5.8}$$

5

The sigmoid function in ◘ Figure 5.2 has the following nice property:

$$f'(x) = f(x)(1 - f(x)). \tag{5.9}$$

Hence, from (5.4) and (5.3), we have

$$\begin{aligned} g_j &= -\frac{\partial E_k}{\partial \hat{y}_j^k} \cdot \frac{\partial \hat{y}_j^k}{\partial \beta_j} \\ &= -(\hat{y}_j^k - y_j^k) f'(\beta_j - \theta_j) \\ &= \hat{y}_j^k (1 - \hat{y}_j^k)(y_j^k - \hat{y}_j^k). \end{aligned} \tag{5.10}$$

By substituting (5.10) and (5.8) into (5.7), and then into (5.6), we have the update rule of $w_{hj}$ as

$$\Delta w_{hj} = \eta g_j b_h. \tag{5.11}$$

Similarly we can derive

$$\Delta \theta_j = -\eta g_j, \tag{5.12}$$

$$\Delta v_{ih} = \eta e_h x_i, \tag{5.13}$$

$$\Delta \gamma_h = -\eta e_h, \tag{5.14}$$

where

$$\begin{aligned} e_h &= -\frac{\partial E_k}{\partial b_h} \cdot \frac{\partial b_h}{\partial \alpha_h} \\ &= -\sum_{j=1}^{l} \frac{\partial E_k}{\partial \beta_j} \cdot \frac{\partial \beta_j}{\partial b_h} f'(\alpha_h - \gamma_h) \end{aligned}$$

$$= \sum_{j=1}^{l} w_{hj} g_j f'(\alpha_h - \gamma_h)$$

$$= b_h(1 - b_h) \sum_{j=1}^{l} w_{hj} g_j . \tag{5.15}$$

The learning rate $\eta \in (0, 1)$ controls the step size of the update in each round. An overly large learning rate may cause fluctuations, whereas a too small value leads to slow convergence. For fine-tuning purposes, we can use $\eta_1$ for (5.11) and (5.12) and $\eta_2$ for (5.13) and (5.14), where $\eta_1$ and $\eta_2$ could take different values.

$\eta = 0.1$ is a typical choice.

The workflow of the BP algorithm is illustrated in ◘ Algorithm 5.1. For each training sample, the BP algorithm executes the following operations: feeding each input sample to the input layer neurons, and then forwarding the signals layer by layer until the output layer produces results; then, the error of the output layer is calculated (lines 4–5) and propagated back to the hidden layer neurons (line 6) for adjusting the connection weights and thresholds (line 7). The BP algorithm repeats the above operations until the termination condition is met, e.g., a small training error. ◘ Figure 5.8 shows the BP algorithm running on a watermelon data set with five samples and two features. It shows how the parameters and decision boundaries change in different rounds.

The termination condition relates to the strategy of dealing with overfitting.

**Algorithm 5.1** Error Backpropagation

**Input:** Training set $D = \{(\boldsymbol{x}_k, \boldsymbol{y}_k)\}_{k=1}^{m}$;
Learning rate $\eta$.

**Process:**

1: Randomly initialize connection weights and thresholds from $(0, 1)$
2: **repeat**
3:     **for all** $(\boldsymbol{x}_k, \boldsymbol{y}_k) \in D$ **do**
4:         Compute the output $\hat{\boldsymbol{y}}_k$ for the current sample according to the current parameters and (5.3);
5:         Compute the gradient term $g_j$ of the output layer neuron according to (5.10);
6:         Compute the gradient term $e_h$ of the hidden layer neurons according to (5.15);
7:         Update connection weights $w_{hj}$, $v_{ih}$ and thresholds $\theta_j$, $\gamma_h$ according to (5.11)–(5.14).
8:     **end for**
9: **until** The termination condition is met

**Output:** A feedforward neural network with determined connection weights and thresholds.

The BP algorithm aims to minimize the accumulated error on the training set $D$

$$E = \frac{1}{m} \sum_{k=1}^{m} E_k . \tag{5.16}$$

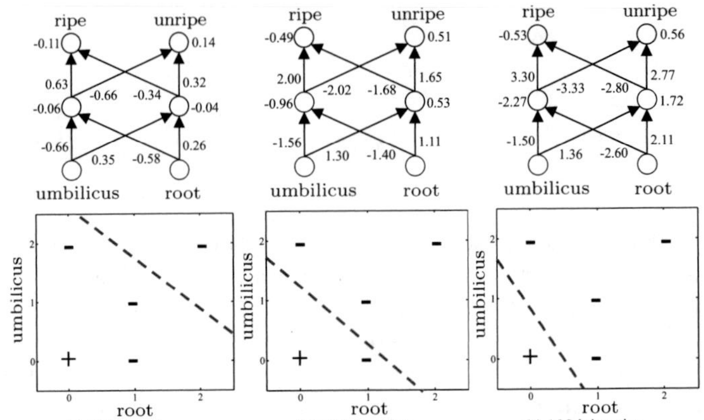


(a) 25th iteration. (b) 50th iteration. (c) 100th iteration.

**Fig. 5.8** The changes of parameters and decision boundaries of a BP neural network running on a watermelon data set with five samples and two features



However, the "standard BP algorithm" introduced above uses only one training sample at a time to update the connection weights and thresholds. In other words, the update rules in ◘ Algorithm 5.1 are derived from the error $E_k$ of individual samples. If we use a similar method to derive the update rules of minimizing the accumulated error, then we have the *accumulated error backpropagation* algorithm. Both the standard and accumulated BP algorithms are commonly used in practice. Generally speaking, the parameters in the standard BP algorithm are updated frequently since each update uses one sample, and hence the updates of different samples may "offset" each other. As a result, the standard BP algorithm often needs more iterations to achieve the same minimum error as the accumulated BP algorithm does. By contrast, the accumulated BP algorithm minimizes the accumulated error directly, and it tunes parameters less frequently since it tunes once after a full scan of the training set $D$. However, in some tasks, especially when the training set $D$ is large, the accumulated BP algorithm can become slow after the accumulated error decreases to a certain level. In contrast, the standard BP algorithm can achieve a reasonably good solution quicker.

A full scan of the training set is called *one round* of learning, also known as *one epoch*.

The difference between the standard BP algorithm and the accumulated BP algorithm is similar to the difference between the stochastic gradient descent (SGD) algorithm and the standard gradient descent algorithm.

Hornik et al. (1989) proved that a feedforward neural network consisting of a single hidden layer with sufficient neurons could approximate continuous functions of any complexity up to arbitrary accuracy. However, there is yet no principled method for setting the number of hidden layer neurons, and *trial-by-error* is usually used in practice.

Along with the strong expressive power, BP neural networks suffer from overfitting, that is, the training error decreases while the testing error increases. There are two general strategies to alleviate the overfitting problem of BP neural networks. The first strategy is *early stopping*: dividing the data into a training set and a validation set, where the training set is for calculating the gradient to update the connection weights and thresholds, and the validation set is for estimating the error. Once the training error decreases while the validation error increases, the training process stops and returns the connection weights and thresholds corresponding to the minimum validation error. The other strategy is *regularization* (Barron 1991, Girosi et al. 1995): the main idea is to add a regularization term to the objective function, describing the complexity of neural network (e.g., the sum of squared connection weights and thresholds). Let $E_k$ denote the error on the $k$th training sample and $w_i$ denote the connection weights and thresholds, then the error objective function (5.16) becomes

Neural networks with regularization are very similar to SVM, which will be introduced in Chap. 6.

The regularization term makes the training process biased toward smaller connection weights and thresholds so that the output becomes "smoother" and the model is less prone to overfitting.

$$E = \lambda \frac{1}{m} \sum_{k=1}^{m} E_k + (1 - \lambda) \sum_i w_i^2, \tag{5.17}$$

where $\lambda \in (0, 1)$ is a trade-off between the empirical error and the complexity of neural network. The value of $\lambda$ is usually estimated by cross-validation.

## 5.4 Global Minimum and Local Minimum

Since $E$ represents the training error of the neural network, it is a function of the connection weights $\mathbf{w}$ and thresholds $\boldsymbol{\theta}$. From this perspective, the training process of neural networks is a parameter optimization process, that is, searching for the set of parameters in the parameter space that minimizes $E$.

Our discussions here also apply to other machine learning models.

We often talk about two types of optimality: the *local minimum* and the *global minimum*. We say $(\mathbf{w}^*; \boldsymbol{\theta}^*)$ is a local minimum solution if there exists $\epsilon > 0$ such that

$$E(\mathbf{w}; \boldsymbol{\theta}) \geqslant E(\mathbf{w}^*; \boldsymbol{\theta}^*), \ \forall \ (\mathbf{w}; \boldsymbol{\theta}) \in \{(\mathbf{w}; \boldsymbol{\theta}) \mid \left\|(\mathbf{w}; \boldsymbol{\theta}) - (\mathbf{w}^*; \boldsymbol{\theta}^*)\right\| \leqslant \epsilon\}.$$

On the other hand, $(\mathbf{w}^*; \boldsymbol{\theta}^*)$ is the global minimum solution if $E(\mathbf{w}; \boldsymbol{\theta}) \geqslant E(\mathbf{w}^*; \boldsymbol{\theta}^*)$ holds for any $(\mathbf{w}; \boldsymbol{\theta})$ in the parameter space. Intuitively, a local minimum solution refers to a point in the parameter space that has an error smaller than the errors of the points in its neighborhood. By contrast, the global min-

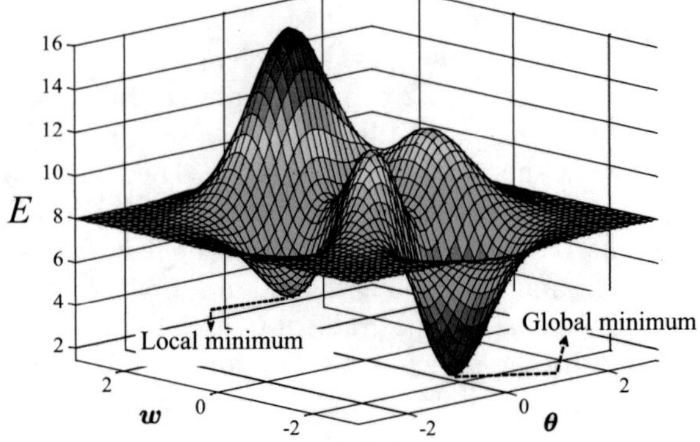


**Fig. 5.9** The global minimum and local minimum

5

imum solution refers to the point with an error smaller than that of any other points in the parameter space. The errors $E(\mathbf{w}^*; \boldsymbol{\theta}^*)$ of these two minimums are called the local minimum value and global minimum value of the error function.

In the parameter space, any point with a zero gradient is a local minimum if its error is smaller than the error of any point in its neighborhood. Though there may exist multiple local minimum values, the global minimum value is always unique. In other words, the global minimum must be a local minimum, but not vice versa. For example, ◘ Figure 5.9 shows two local minimums, but only one of them is the global minimum. The objective of parameter optimization is to find the global minimum.

The most widely used parameter optimization methods are gradient-based search methods. These methods start from an initial solution and search for the optimal parameters iteratively. In each round, the search direction at the current point is determined by the gradient of the error function. For example, the gradient descent method takes steps in the direction of the negative gradient since this is the direction that the function value decreases the most. When the gradient turns zero, the search reaches a local minimum and stops since the update term becomes zero. This local minimum is the global minimum if the error function has only one local minimum. On the other hand, if there is more than one local minimum, the solution we found might not be the global minimum, and the parameter optimization is stuck at the local minimum, which is undesirable.

The update rules of perceptron (5.1) and BP (5.11)–(5.14) are based on the gradient descent method.

In real-world applications, we often use the following strategies to "jump out" from the local minimum to get closer to the global minimum:

- We can use different sets of parameters to initialize multiple neural networks and take the one with the smallest error. Since the search starts from different initial points, we obtain multiple local minimums, and the smallest one among them is a closer estimation of the global minimum.
- We can use the *simulated annealing* technique (Aarts and Korst 1989), which accepts a worse solution at a certain probability, and hence it can jump out from the local minimum. To maintain the algorithm's stability, we progressively decrease the probability of accepting suboptimal solutions as the search proceeds.
- We can use the stochastic gradient descent method, which introduces random factors to the gradient calculations rather than the exact calculations used in the standard gradient descent method. With random factors, the gradient may not be zero even if it is a local minimum, that is, there is a chance to jump out from the local minimum.

However, it may also cause a jump out from the global minimum.

Besides, the *genetic algorithm* (Goldberg 1989) is also frequently used to train neural networks to better approximate the global minimum. Note that the above techniques for jumping out from local minimums are mostly heuristic methods without theoretical guarantees.

## 5.5 Other Common Neural Networks

Due to space limitations, we are unable to cover the numerous neural network models and algorithms, so only several commonly used neural networks are introduced in the rest of this section.

### 5.5.1 RBF Network

Radial Basis Function (RBF) networks (Broomhead and Lowe 1988) are feedforward neural networks with single hidden layer. It employs the radial basis function as the activation function for hidden layer neurons, and the output layer computes a linear combination of the outputs from the hidden layer neurons. Suppose that the input is a $d$-dimensional vector $\boldsymbol{x}$ and the output is a real value, then the RBF network can be expressed as

Using multiple hidden layers is also theoretically feasible, but typical RBF networks use single hidden layer.

$$\varphi(\boldsymbol{x}) = \sum_{i=1}^{q} w_i \rho(\boldsymbol{x}, \boldsymbol{c}_i), \tag{5.18}$$

where $q$ is the number of hidden layer neurons, and $\boldsymbol{c}_i$ and $w_i$ are, respectively, the center and the weight of the $i$th hidden layer neuron. $\rho(\boldsymbol{x}, \boldsymbol{c}_i)$ is the radial basis function, which is a radially symmetric scalar function, generally defined as a monotonic function based on the Euclidean distance between the sample $\boldsymbol{x}$ and the data centroid $\boldsymbol{c}_i$. The commonly used Gaussian radial basis function is in the form of

$$\rho(\boldsymbol{x}, \boldsymbol{c}_i) = e^{-\beta_i \|\boldsymbol{x}-\boldsymbol{c}_i\|^2}. \tag{5.19}$$

Park and Sandberg (1991) proved that an RBF network with sufficient hidden layer neurons can approximate continuous functions of any complexity up to arbitrary accuracy.

RBF networks are typically trained with two steps. The first step is to identify the neuron center $\boldsymbol{c}_i$ using methods such as random sampling and clustering, and the second step is to determine the parameters $w_i$ and $\beta_i$ using BP algorithms.



### 5.5.2 ART Network

*Competitive learning* is a commonly used unsupervised learning strategy in neural networks. In this strategy, the neurons compete with each other, and only one of them will be the winner who can be activated while others become inactivated. Such a mechanism is also known as the "winner-take-all" principle.

Adaptive Resonance Theory (ART) networks (Carpenter and Grossberg 1987) are an important representative of competitive learning. An ART network consists of four major components, namely the comparison layer, the recognition layer, the reset module, and the recognition threshold (a.k.a. vigilance parameter). Among them, the comparison layer accepts input samples and passes information to the recognition layer neurons. Each neuron in the recognition layer corresponds to a pattern class, and the number of neurons can be dynamically increased in the training process to incorporate new pattern classes.

Pattern class can be regarded as the "sub-class" of a class.

After receiving a signal, the recognition layer neurons compete with each other to become the winner neuron. The simplest way of competition is to calculate the distance between the input vector and the representation vector of the pattern class in each recognition layer neuron. The winner neuron is the one with the shortest distance, and it suppresses the activation of other neurons in the recognition layer by sending them signals. When the similarity between the input sample and the representation vector of the winner neuron exceeds the recognition threshold, the input sample is classified as the class of the representation vector. In the meantime, the connection weights are updated such that this winner neuron is more likely to win future input samples that are similar to the current sam-

The "winner-take-all" principle.

ple. However, if the similarity does not exceed the recognition threshold, then the reset module will create a new neuron in the recognition layer and initialize the representation vector of it to the input vector.

The recognition threshold has a significant impact on the performance of ART networks. When the recognition threshold is large, the input samples are divided into more fine-grained pattern classes. On the other hand, a small recognition threshold leads to fewer and coarse-grained pattern classes.

ART networks offer a solution to the *stability-plasticity dilemma* in competitive learning, where plasticity refers to the ability to learn new knowledge, and stability refers to the ability to maintain existing knowledge. This property gives ART networks an important advantage to be able to perform *incremental learning* and *online learning*.

Incremental learning refers to the learning of new samples after a model has already been trained. In other words, the model can update incrementally with the new samples without re-training the whole model, and the useful information that has been previously learned will not be overridden. Online learning refers to the learning of new samples one by one. Online learning is a special case of incremental learning, and incremental learning can be regarded as batch-mode online learning.

ART networks in the early days can only handle input data of Boolean type. Later on, ART networks evolved into a family of algorithms, such as ART2 networks that can handle real-valued inputs, FuzzyART networks that incorporate fuzzy processing, and ARTMAP networks that can perform supervised learning.

### 5.5.3 SOM Networks

Self-Organizing Map (SOM) networks (Kohonen 1982) are unsupervised neural networks based on competitive learning. An SOM network can map high-dimensional input data to low-dimensional space (typically two-dimensional) while preserving the topological structure of the data in high-dimensional space, that is, mapping samples that are similar in the high-dimensional space to neighborhood neurons in the output layer.

Also known as Self-Organizing Feature Map and Kohonen Network.

As illustrated in ◘ Figure 5.10, neurons in the output layer are placed like a matrix in the two-dimensional space. Each neuron has a weight vector that is used to determine the winner neuron based on the input vector, that is, the neuron with the

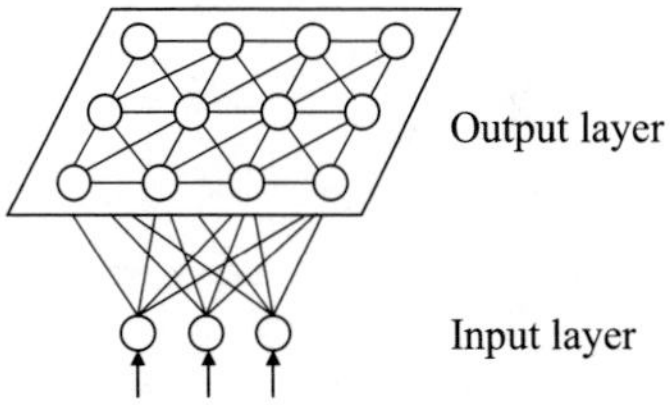


**Fig. 5.10** SOM network

highest similarity wins. The objective of SOM training is to find the appropriate weight vector for each neuron in the output layer such that the topological structure is preserved.

The training process of SOM is simple: upon the arrival of a new sample, each neuron in the output layer computes the distance from its weight vector to the sample vector, and the neuron with the shortest distance (i.e., the *best matching unit*) wins. Then, the weight vectors of the best matching unit and its surrounding neurons will be updated toward the current input sample. This process repeats until converge.



### 5.5.4 Cascade-Correlation Network

Typical neural networks have fixed network structures, and the purpose of training is to determine the appropriate parameters such as connection weights and thresholds. Constructive networks, however, consider the construction of network structure as one of the learning objectives, that is, identifying the best network structure that fits the training data. A representative constructive network is the Cascade-Correlation Network (Fahlman and Lebiere 1990).

The ART network introduced in Sect. 5.5.2 is also a constructive network since its recognition neurons are dynamically added.

Cascade-Correlation networks have two major components, namely "Cascading" and "Correlation". Cascading refers to the establishment of a hierarchy with hierarchical connections. At the beginning of training, the network has the minimum topology, that is, only an input layer and an output layer. As the training progresses, as shown in ◘ Figure 5.11, new hidden neurons are gradually added to create a hierarchy. When a new hidden layer neuron is added, its input connection weight will be fixed. Correlation refers to training-related parameters by maximizing the correlation between the output of new neurons and network errors.

Since there is no configuration of the number of layers and the number of hidden neurons, cascade-correction networks

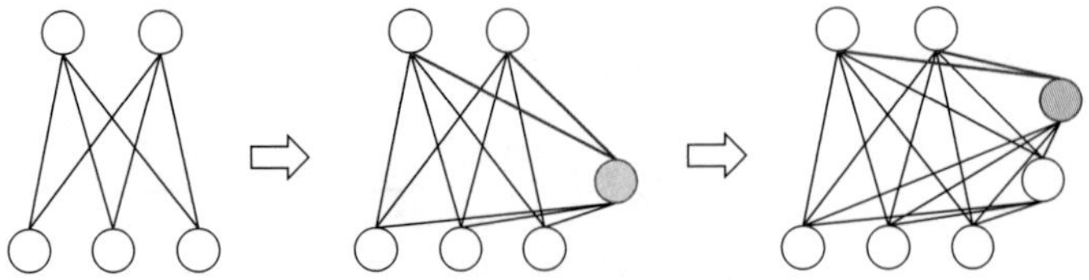

**Fig. 5.11** The training process of a Cascade-Correlation network starts with the initial state and gradually adds the first and the second hidden neurons. When adding a new hidden neuron, the weights shown in red are updated by maximizing the correlation between the new hidden neuron's output and the network error

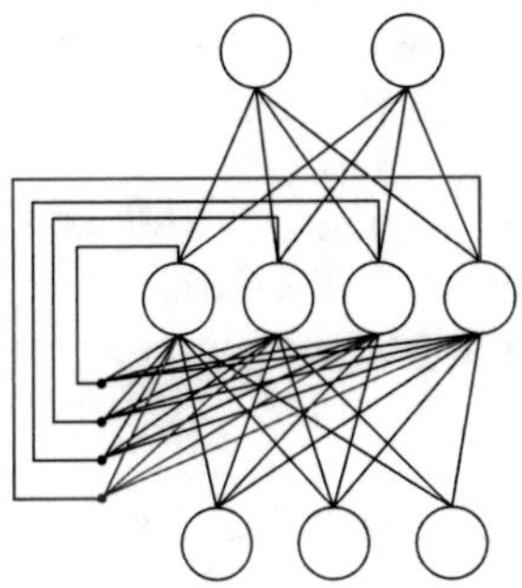

**Fig. 5.12** Elman network

are faster to train than typical feedforward neural networks. However, cascade-correction networks are prone to overfitting on small data sets.

### 5.5.5 Elman Network

Unlike feedforward neural networks, Recurrent Neural Networks (RNN) permit circular structures, in which the neurons can take output feedback as input signals. Hence, the network's output at time $t$ not only depends on the input at time $t$ but also the network status at time $t-1$. Due to this behavior, recurrent neural networks are capable of modeling temporal dynamics.

Also known as Recursive Neural Networks.

One of the most widely used recurrent neural networks is the Elman network (Elman 1990), which, as illustrated in ■ Figure 5.12, has a similar structure to multi-layer feedforward networks. However, the difference is that the outputs from hidden neurons are reused as the input to hidden neurons at the next iteration, together with signals from input neurons. Typically, hidden neurons in the Elman networks employ sigmoid activation functions, and the networks are trained with generalized BP algorithms (Pineda 1987).

### 5.5.6 Boltzmann Machine

One type of neural network model defines an *energy* for the state of the network. The network reaches the ideal state when the energy is minimized, and training such a network is to minimize the energy function. One representative of energy-based models is the Boltzmann Machine (Ackley et al. 1985). As illustrated in ■ Figure 5.13a, a typical structure of the Boltzmann machines has two layers: the visible layer and the hidden layer. The visible layer is for representing data input and output,

■ Figure 5.13a shows that the Boltzmann machines are recurrent neural networks.

while the hidden layer is understood as the intrinsic representation of the data. Neurons in the Boltzmann machines are Boolean-typed, that is, 0 for inactivated and 1 for activated. Let $\boldsymbol{s} \in \{0, 1\}^n$ denote the states of $n$ neurons, $w_{ij}$ denote the connection weight between neuron $i$ and neuron $j$, and $\theta_i$ denote the threshold of neuron $i$. Then, the energy of the Boltzmann machine corresponding to the state vector $\boldsymbol{s}$ is defined as

$$E(\boldsymbol{s}) = -\sum_{i=1}^{n-1} \sum_{j=i+1}^{n} w_{ij} s_i s_j - \sum_{i=1}^{n} \theta_i s_i. \tag{5.20}$$



The Boltzmann distribution is also known as equilibrium or stationary distribution.

If the neurons are updated in arbitrary order independent of the inputs, then the network will eventually reach a Boltzmann distribution, in which the probability of having the state vector $\boldsymbol{s}$ is solely determined by its energy and the energies of all possible state vectors:

$$P(\boldsymbol{s}) = \frac{e^{-E(\boldsymbol{s})}}{\sum_{\boldsymbol{t}} e^{-E(\boldsymbol{t})}}. \tag{5.21}$$

Then, the training of a Boltzmann machine is to regard each training sample as a state vector and maximize its probability. Since the standard Boltzmann machines are fully connected graphs with high complexity, they are impractical to use in real-world applications. In practice, we often use the restricted Boltzmann machines instead. As illustrated in ◘ Figure 5.13b, a restricted Boltzmann machine keeps only the connections between the visible layer and the hidden layer and simplifies the network structure to a bipartite graph.

The restricted Boltzmann machines are often trained using the Contrastive Divergence (CD) algorithm (Hinton 2010). Let $d$ denote the number of visible neurons, $q$ denote the number of hidden neurons, and $\mathbf{v}$ and $\boldsymbol{h}$ denote the state vectors of the visible layer and the hidden layer, respectively. Since there is

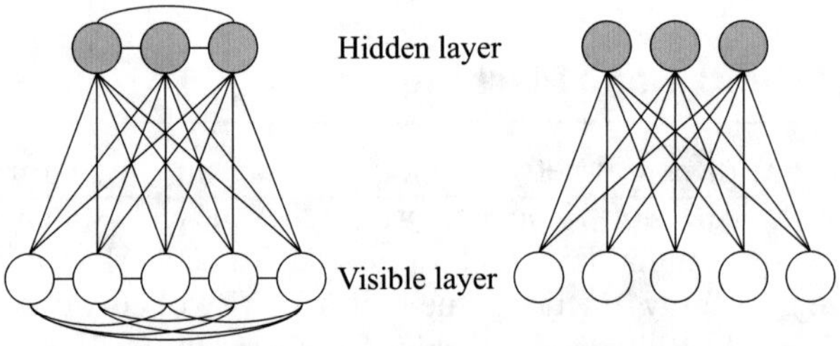


(a) Boltzmann machine. (b) Restricted Boltzmann machine.

**Fig. 5.13** Boltzmann machine and restricted Boltzmann machine no connection within the same layer, we have

$$P(\mathbf{v} \mid \boldsymbol{h}) = \prod_{i=1}^{d} P(v_i \mid \boldsymbol{h}), \tag{5.22}$$

$$P(\boldsymbol{h} \mid \mathbf{v}) = \prod_{j=1}^{q} P(h_j \mid \mathbf{v}). \tag{5.23}$$

For each training sample $\mathbf{v}$, the CD algorithm first computes the probability distribution of hidden neurons using (5.23) and then obtains $\boldsymbol{h}$ by sampling from this distribution. Thereafter, similarly, $\mathbf{v}'$ is generated from $\boldsymbol{h}$ according to (5.22), and $\boldsymbol{h}'$ is further generated from $\mathbf{v}'$. Then, the connection weight can be updated by

The threshold can be updated similarly.

$$\Delta w = \eta \left( \mathbf{v}\boldsymbol{h}^{\top} - \mathbf{v}'\boldsymbol{h}'^{\top} \right). \tag{5.24}$$

## 5.6 Deep Learning

In theory, a model with more parameters has a larger *capacity* and, consequently, is more capable of handling complex learning tasks. In practice, however, complex models are not favored since they are often slow to train and are prone to overfitting. Fortunately, since we have entered the age of cloud computing and big data, the significant improvement of computing power has enabled efficient training, while the big data reduces the risk of overfitting. Since the major barriers of complex models are alleviated, complex models, represented by *deep learning*, have begun to attract attention.

See Chap. 12 for the capacity of learners.

Large-scale deep learning model can contain tens of billions of parameters.

Typically, a deep learning model is a neural network with many layers. For neural networks, a simple method of increasing the model capacity is to add more hidden layers, which corresponds to more parameters such as connection weights and thresholds. The model complexity can also be increased by simply adding more hidden neurons since we have seen previously that a multi-layer feedforward network with a single hidden layer already has very strong learning ability. However, to increase model complexity, adding more hidden layers is more effective than adding hidden neurons since more layers not only imply more hidden neurons but also more nesting of activation functions. Multi-hidden layers neural networks are difficult to train directly with classic algorithms (such as the standard BP algorithm), because errors tend to diverge and fail to converge to a stable state when backpropagating within the multi-hidden layers.

Here, "multi-hidden layers" refer to three or more hidden layers. Deep learning models typically have eight or nine, or even more hidden layers.

5

One effective method of training neural networks with multi-hidden layers is *unsupervised layer-wise training*, in which the hidden layers are trained one by one, that is, the outputs of the previous hidden layer are used as the inputs to train the current hidden layer, and this process is known as *pre-training*. Following pre-training, *fine-tuning* is performed on the whole network. For example, in a Deep Belief Network (DBN) (Hinton et al. 2006), each layer is a restricted Boltzmann machine, that is, the network is a stack of multiple RBM models. In unsupervised layer-wise training, the first hidden layer is trained in the same way as in the standard RBM training. Then, the pre-trained neurons of the first hidden layer are used as the input neurons for the second hidden layer, and so on. Once pre-training is completed for each layer, the entire network is trained with BP algorithms.

We can view "pre-training + fine-tuning" as splitting the parameters into groups and then optimizing the parameters in each group locally. The locally optimized parameters are then joint together to search for the global optimum. This approach reduces the training cost without sacrificing the freedom offered by a large number of parameters.

Another strategy for saving the training cost is *weight sharing*, in which a set of neurons share the same connection weight. This strategy plays an important role in Convolutional Neural Networks (CNN) (LeCun and Bengio 1995, LeCun et al. 1998). To demonstrate the idea, let us take a look at how CNN is applied to the task of handwritten character recognition (LeCun et al. 1998). As illustrated in ◘ Figure 5.14, the input is a $32 \times 32$ image of a handwritten digit, and the output is the recognition result. A CNN combines multiple *convolutional layers* and *pooling layers* to process the input signals and ends the mapping to the output result with the fully connected layers. Each convolutional layer contains multiple *feature maps*, where each feature map is a "plane" consisting of multiple neurons. We can regard a feature map as a convolutional filter that performs convolution operations on the input matrix to extract convolved features. For example, the first convolutional layer in ◘ Figure 5.14 consists of six feature maps, where each feature map is a $28 \times 28$ matrix of neurons, and each neuron is responsible for extracting the local feature of a $5 \times 5$ region using the convolutional filter. The function of pooling layers is to use local correlations to perform sampling such that the data volume is reduced while useful information is preserved. For example, the first pooling layer in ◘ Figure 5.14 contains six feature maps with the size of $14 \times 14$, where each neuron on the feature map computes its output using a $2 \times 2$ region on the feature map of the previously convolutional layer. The original image is mapped into a 120-dimensional feature vector through

Recently, the sigmoid activation function in CNN is often replaced with the linear rectifier function

$$f(x) = \begin{cases} 0, & \text{if } x < 0; \\ x, & \text{otherwise,} \end{cases}$$

and such neurons are called Rectified Linear Unit (ReLU). Besides, the typical operations in the pooling layer are "max" and "mean", which are similar to the operations in ensemble learning. See Sect. 8.4.

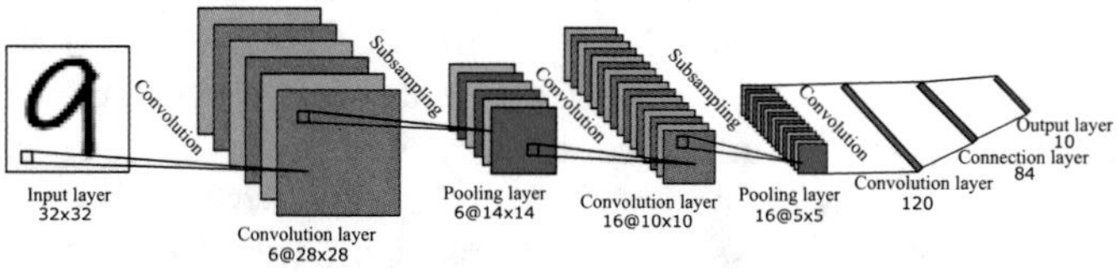


**Fig. 5.14** Handwritten character recognition using CNN (LeCun et al. 1998)

combinations of convolutional layers and pooling layers, and finally, a connection layer composed of 84 neurons and an output layer are connected to complete the recognition task. CNN can be trained using BP algorithms. However, in both convolutional layers and pooling layers, each set of neurons (i.e., each "plane" in ◘ Figure 5.14) shares the same connection weights so that the number of parameters is significantly reduced.

There is another perspective to understand deep learning. No matter it is a DBN or a CNN, the mechanisms are the same, that is, using multiple hidden layers to process information layer by layer. This mechanism can be viewed as progressively converting the original input representation, which is not closely related to the targeted output, into a representation that is closely related to the targeted output. By doing so, the mapping task of the final output layer, which used to be difficult, becomes feasible. In other words, through multi-layer processing, the low-level feature representations are converted into the high-level feature representations, which can be used by simple models for complex classifications. Therefore, deep learning can also be viewed as *feature learning* or *representation learning*.

If we consider the layers at the front are working on the feature representations, and only the last layer is working on the "classification", then the classification is using a simple model.

In the past, real-world applications often require human experts to design features, known as *feature engineering*. Though the features, good or bad, are vital for the generalization ability, it is often not easy to design good features even for human experts. As feature learning can produce good features via machine learning, we are one more step toward "fully automated data analytics".

## 5.7 Further Reading

Haykin (1998) is a good textbook for studying neural networks, and Bishop (1995) is an alternative which puts more emphasis on machine learning and pattern recognition. Notable journals in the field of neural networks include *Neural Computation*, *Neural Networks*, and *IEEE Transactions on Neural Networks and Learning Systems*. Notable international con-

Was named *IEEE Transactions on Neural Networks* prior to 2012.

ferences include International Conference on Neural Information Processing Systems (NeurIPS, or NIPS prior to 2018) and International Joint Conference on Neural Networks (IJCNN). Notable regional conferences include the Europe-based International Conference on Artificial Neural Networks (ICANN) and Asia Pacific-based International Conference on Neural Information Processing (ICONIP).

NeurIPS has put more emphasis on machine learning in recent years.

Though M-P neuron is the most widely used neuron model, there are other neuron models, such as the *spiking neuron* model (Gerstner and Kistler 2002) which considers the temporal information of spike potentials rather than only the accumulated electric potentials.



The BP algorithm was initially proposed by Werbos (1974) and then reinvented by Rumelhart et al. (1986a, b). Essentially, the BP algorithm is a generalization of the Least Mean Square (LMS) algorithm. LMS aims at minimizing the MSE of the network output and can be used in Perceptron learning when the neuron activation function is differentiable. Generalizing LMS to differentiable nonlinear activation function leads to the BP algorithm. For this reason, the BP algorithm is also called generalized $\delta$ rule (Chauvin and Rumelhart 1995).

LMS is also known as the Widrow–Hoff rule or $\delta$ rule.

MacKay (1992) proposed a method that can automatically determine the regularization parameter under the Bayesian framework. Gori and Tesi (1992) provided a detailed discussion on the local minimum problem of BP networks. Yao (1999) surveyed the use of evolutionary computation techniques, such as the genetic algorithm, to generate neural networks. There are many studies on the improvement of BP algorithms. For example, in order to speed up, the learning rate can be adaptively reduced during training, that is, the larger learning rate is used first and then gradually reduced. More "tricks" can be found in Reed and Marks (1998), Orr and Müller (1998).

Schwenker et al. (2001) provided detailed information about the training of RBF networks. Carpenter and Grossberg (1991) provided an introduction to the family of ART algorithms. SOM networks (Kohonen 2001) have been widely used in applications like clustering, high-dimensional data visualization, and image segmentation. Goodfellow et al. (2016) discussed recent progress on deep learning.

Neural networks are *black boxes* which are difficult to interpret. Some attempts have been made to improve the interpretability of neural networks by extracting easy-to-understand symbolic rules (Tickle et al. 1998, Zhou 2004).

## Exercises

**5.1** Describe the drawbacks of using the linear function $f(\boldsymbol{x}) = \boldsymbol{w}^\top \boldsymbol{x}$ as the activation function for neurons.

**5.2** Describe the relationship between the logistic regression and the neuron using the activation function as shown in ◘ Figure 5.2b.

**5.3** For $v_{ih}$ in ◘ Figure 5.7, derive its update rule (5.13) in the BP algorithm.

**5.4** Describe the impact of the learning rate in (5.6) on the training process of neural networks.

**5.5** Implement the standard BP algorithm and the accumulated BP algorithm. Use the algorithms to train single hidden layer networks on the watermelon data set 3.0 and make a comparison.

The watermelon data set 3.0 is in ◘ Table 4.3.

**5.6** Design and implement an improved BP algorithm that can dynamically adjust the learning rate to speed up the convergence. Compare the improved BP algorithm against the standard BP algorithm on two UCI data sets.

UCI data sets can be found at ► http://archive.ics.uci.edu/ml/.

**5.7** Based on (5.18) and (5.19), construct a single-layer RBF neural network that can solve the XOR problem.

**5.8** Download or implement an SOM network, and then investigate its results on the watermelon data set 3.0$\alpha$.

The watermelon data set 3.0$\alpha$ is in ◘ Table 4.5.

**5.9** * Derive the BP algorithm for the Elman network.

**5.10** Download or implement a convolutional neural network, and then test it on the MNIST handwritten digit data set.

The MNIST data set can be found at ► http://yann.lecun.com/exdb/mnist/.

## Break Time

### Short Story: The Rises and Falls of Neural Networks

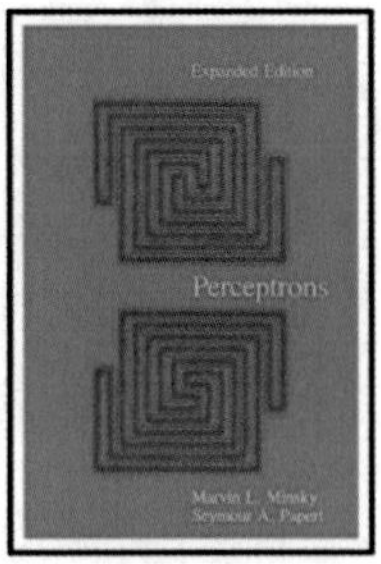


Following the development of the M-P neuron model and the Hebb learning rule in the 1940s, a series of researches, represented by Perceptron and Adaline, have emerged in the 1950s, and this is the first golden age of neural networks. Unfortunately, in the 1969 book *Perceptrons*, the MIT computer science pioneers Marvin Minsky (1927–2016) and Seymour Papert pointed out that single-layer neural networks cannot solve nonlinear problems, and it is unknown whether it is possible to have effective algorithms to train multi-layer networks. This conclusion directly pushed neural network research into an "ice age", and both the United States and the former Soviet Union stopped funding neural network research. Neural network researchers around the world had to change their research topics, and only a few research groups insisted. In 1974, when Paul Werbos from Harvard University presented the BP algorithm, it did not receive the deserved attention since neural network research was still in the ice age.

The book provides many insights about neural networks, but its important conclusion hindered the subsequent development of neural networks and even artificial intelligence. For this reason, the book has been criticized after the come back of neural networks. When the book was reprinted in 1988, Minsky added some clarifications as an additional chapter.

Minsky received the Turing Award in 1969.

In 1983, John Hopfield, a physicist from Caltech, has made a sensation by tackling the NP-hard "traveling salesman problem" using neural networks. Later on, the PDP group, led by David Rumelhart and James McClelland, published the book *Parallel Distributed Processing: Explorations in the Microstructure of Cognition*, in which the BP algorithm was reinvented. Following the excitement of Hopfield's work, the BP algorithm has soon become popular, and it marks the second golden age of neural networks. In the middle 1990s, statistical learning theory and Support Vector Machine have emerged, while neural networks were suffering from the lacking of theories, heavily relying on trial-and-error, and full of "tricks". As a result, researchers turned their attention to statistical learning, and neural network research entered another "winter". Ironically, the NeurIPS conference, which was named with neural networks, accepted few neural network research papers for several years in that period.

Around 2010, neural networks took advantage of the significant advancements of computing power and big data and came back with a new name of "deep learning". Neural networks regained its reputation by winning competitions (e.g., ImageNet) and attracted massive research funding from industry, entering its third golden age.

## References

Aarts E, Korst J (1989) Simulated annealing and Boltzmann machines: a stochastic approach to combinatorial optimization and neural computing. Wiley, New York

Ackley DH, Hinton GE, Sejnowski TJ (1985) A learning algorithm for Boltzmann machines. Cogn Sci 9(1):147–169

Barron AR (1991) Complexity regularization with application to artificial neural networks. In: Roussas G (ed) Nonparametric functional estimation and related topics; NATO ASI series, vol 335. Kluwer, Amsterdam, pp 561–576

Bishop CM (1995) Neural networks for pattern recognition. Oxford University Press, New York

Broomhead DS, Lowe D (1988) Multivariate functional interpolation and adaptive networks. Complex Syst 2(3):321–355

Carpenter GA, Grossberg S (1987) A massively parallel architecture for a self-organizing neural pattern recognition machine. Comput Vis 37(1):54–115

Carpenter GA, Grossberg S (eds) (1991) Pattern recognition by self-organizing neural networks. MIT Press, Cambridge

Chauvin Y, Rumelhart DE (eds) (1995) Backpropagation: theory, architecture, and applications. Lawrence Erlbaum Associates, Hillsdale

Elman JL (1990) Finding structure in time. Cogn Sci 14(2):179–211

Fahlman SE, Lebiere C (1990) The cascade-correlation learning architecture. Technical report CMU-CS-90-100, School of Computer Sciences, Carnegie Mellon University, Pittsburgh, PA

Gerstner W, Kistler W (2002) Spiking neuron models: single neurons, populations, plasticity. Cambridge University Press, Cambridge

Girosi F, Jones M, Poggio T (1995) Regularization theory and neural networks architectures. Neural Comput 7(2):219–269

Goldberg DE (1989) Genetic algorithms in search, optimization and machine learning. Addison-Wesley, Boston

Goodfellow I, Bengio Y, Courville A (2016) Deep learning. MIT Press, Cambridge

Gori M, Tesi A (1992) On the problem of local minima in backpropagation. IEEE Trans Pattern Anal Mach Intell 14(1):76–86

Haykin S (1998) Neural networks: a comprehensive foundation, 2nd edn. Prentice-Hall, Upper Saddle River

Hinton G (2010) A practical guide to training restricted Boltzmann machines. Technical report UTML TR 2010-003, Department of Computer Science, University of Toronto

Hinton G, Osindero S, Teh Y-W (2006) A fast learning algorithm for deep belief nets. Neural Comput 18(7):1527–1554

Hornik K, Stinchcombe M, White H (1989) Multilayer feedforward networks are universal approximators. Neural Netw 2(5):359–366

Kohonen T (1982) Self-organized formation of topologically correct feature maps. Biol Cybern 43(1):59–69

Kohonen T (1988) An introduction to neural computing. Neural Netw 1(1):3–16

Kohonen T (2001) Self-organizing maps, 3rd edn. Springer, Berlin

LeCun Y, Bengio Y (1995) Convolutional networks for images, speech, and time-series. In: Arbib MA (ed) The handbook of brain theory and neural networks. MIT Press, Cambridge

LeCun Y, Bottou L, Bengio Y, Haffner P (1998) Gradient-based learning applied to document recognition. Proc IEEE 86(11):2278–2324

MacKay DJC (1992) A practical Bayesian framework for backpropagation networks. Neural Comput 4(3):448–472

McCulloch WS, Pitts W (1943) A logical calculus of the ideas immanent in nervous activity. Bull Math Biophys 5(4):115–133

Minsky M, Papert S (1969) Perceptrons. MIT Press, Cambridge

Orr GB, Müller K-R (eds) (1998) Neural networks: tricks of the trade. Springer, London

Park J, Sandberg IW (1991) Universal approximation using radial-basis-function networks. Neural Comput 3(2):246–257

Pineda FJ (1987) Generalization of back-propagation to recurrent neural networks. Phys Rev Lett 59(19):2229–2232

Reed RD, Marks RJ (1998) Neural smithing: supervised learning in feed-forward artificial neural networks. MIT Press, Cambridge

Rumelhart DE, Hinton GE, Williams RJ (1986a) Learning internal representations by error propagation. In: Rumelhart DE, McClelland JL (eds) Parallel distributed processing: explorations in the microstructure of cognition, vol 1. MIT Press, Cambridge, pp 318–362

Rumelhart DE, Hinton GE, Williams RJ (1986b) Learning representations by backpropagating errors. Nature 323(9):533–536

Schwenker F, Kestler HA, Palm G (2001) Three learning phases for radial-basis-function networks. Neural Netw 14(4–5):439–458

Tickle AB, Andrews R, Golea M, Diederich J (1998) The truth will come to light: directions and challenges in extracting the knowledge embedded within trained artificial neural networks. IEEE Trans Neural Netw 9(6):1057–1067

Werbos P (1974) Beyond regression: new tools for prediction and analysis in the behavior science. PhD thesis, Harvard University, Cambridge, MA

Yao X (1999) Evolving artificial neural networks. Proc IEEE 87(9):1423–1447

Zhou Z-H (2004) Rule extraction: using neural networks or for neural networks? J Comput Sci Technol 19(2):249–253

# Support Vector Machine

## Table of Contents




Z.-H. Zhou, *Machine Learning*,
https://doi.org/10.1007/978-981-15-1967-3_6

6

## 6.1 Margin and Support Vector

Given a training set $D = \{(\boldsymbol{x}_1, y_1), (\boldsymbol{x}_2, y_2), \ldots, (\boldsymbol{x}_m, y_m)\}$, where $y_i \in \{-1, +1\}$. The basic idea of classification is to utilize the training set $D$ to find a separating hyperplane in the sample space that can separate samples of different classes. However, there could be multiple qualified separating hyperplanes, as shown in ◘ Figure 6.1, which one should be chosen?

Intuitively, we should choose the one right in the middle of two classes, that is, the red one in ◘ Figure 6.1, since this separating hyperplane has the best "tolerance" to local data perturbation. For example, the samples not in the training set could be closer to the decision boundary due to the noises or limitations of the training set. As a result, many separating hyperplanes that perform well on the training set will make mistakes, whereas the red hyperplane is less likely to be affected. In other words, this separating hyperplane has the strongest generalization ability and the most robust classification results.

A separating hyperplane in the sample space can be expressed as the following linear function:

$$\boldsymbol{w}^\top \boldsymbol{x} + b = 0, \tag{6.1}$$

where $\boldsymbol{w} = (w_1; w_2; \ldots; w_d)$ is the normal vector which controls the direction of the hyperplane, and $b$ is the bias which controls the distance between the hyperplane and the origin. The normal vector $\boldsymbol{w}$ and the bias $b$ determine the separating hyperplane, denoted by $(\boldsymbol{w}, b)$. The distance from any point $\boldsymbol{x}$ in the sample space to the hyperplane $(\boldsymbol{w}, b)$ can be written as

See Exercise 6.1.

$$r = \frac{|\boldsymbol{w}^\top \boldsymbol{x} + b|}{\|\boldsymbol{w}\|}. \tag{6.2}$$

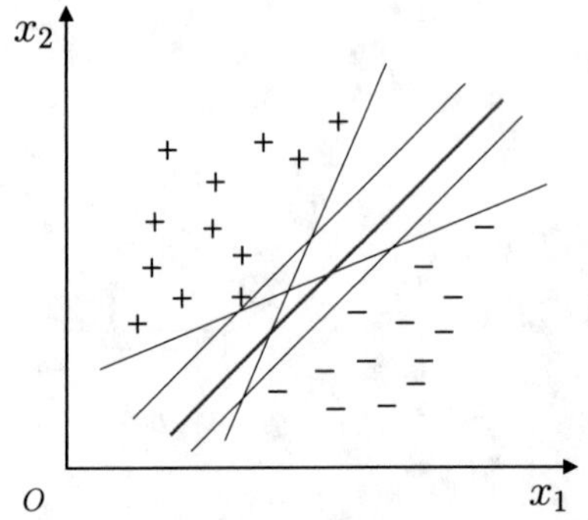


**Fig. 6.1** More than one hyperplanes can separate the training samples

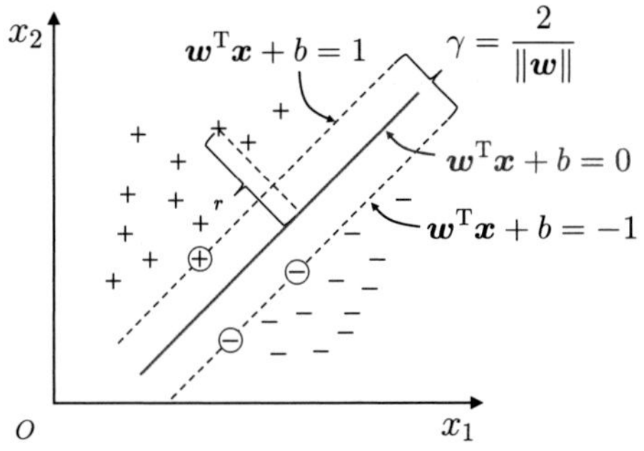


**Fig. 6.2** Support vectors and margin

Suppose the hyperplane $(\boldsymbol{w}, b)$ can correctly classify the training samples, that is, for $(\boldsymbol{x}_i, y_i) \in D$, there is $\boldsymbol{w}^\top \boldsymbol{x}_i + b > 0$ when $y_i = +1$, and $\boldsymbol{w}^\top \boldsymbol{x}_i + b < 0$ when $y_i = -1$. Let

$$\begin{cases} \boldsymbol{w}^\top \boldsymbol{x}_i + b \geqslant +1, & y_i = +1, \\ \boldsymbol{w}^\top \boldsymbol{x}_i + b \leqslant -1, & y_i = -1. \end{cases} \tag{6.3}$$

If the hyperplane $(\boldsymbol{w}', b')$ can correctly classify the training samples, then there always exists a scaling transformations $\varsigma\boldsymbol{w} \mapsto \boldsymbol{w}'$ and $\varsigma b \mapsto b'$ such that (6.3) holds.

As illustrated in ◘ Figure 6.2, the equality in (6.3) holds for the sample points closest to the hyperplane, and these sample points are called *support vectors*. The total distance from two support vectors of different classes to the hyperplane is

Each sample point corresponds to a feature vector.

$$\gamma = \frac{2}{\|\boldsymbol{w}\|}, \tag{6.4}$$

which is called *margin*.

Finding the separating hyperplane with the *maximum margin* is equivalent to finding the parameters $\boldsymbol{w}$ and $b$ that maximize $\gamma$ subject to the constraints in (6.3), that is

$$\begin{aligned} &\max_{\boldsymbol{w},b} \ \frac{2}{\|\boldsymbol{w}\|} \\ &\text{s.t. } y_i(\boldsymbol{w}^\top \boldsymbol{x}_i + b) \geqslant 1, \quad i = 1, 2, \ldots, m. \end{aligned} \tag{6.5}$$

The margin can be optimized by maximizing $\|\boldsymbol{w}\|^{-1}$, which is equivalent to minimizing $\|\boldsymbol{w}\|^2$. Hence, we can rewrite (6.5) as

It appears that the margin only depends on $\boldsymbol{w}$, but $b$ also implicitly changes the margin by influencing $\boldsymbol{w}$ through the constraints.

$$\begin{aligned} &\min_{\boldsymbol{w},b} \ \frac{1}{2}\|\boldsymbol{w}\|^2 \\ &\text{s.t. } y_i(\boldsymbol{w}^\top \boldsymbol{x}_i + b) \geqslant 1, \quad i = 1, 2, \ldots, m. \end{aligned} \tag{6.6}$$

This is the primal form of Support Vector Machine (SVM).

## 6.2 Dual Problem

We wish to solve (6.6) to obtain the maximum margin separating hyperplane model

$$f(\boldsymbol{x}) = \boldsymbol{w}^\top \boldsymbol{x} + b, \tag{6.7}$$

where $\boldsymbol{w}$ and $b$ are the model parameters. (6.6) is a convex quadratic programming problem, which can be solved with existing optimization packages. However, there are more efficient methods.

See Appendix B.1 for the Lagrange multiplier method.

Applying Lagrange multipliers to (6.6) leads to its *dual problem*. To be specific, introducing a Lagrange multiplier $\alpha_i \geqslant 0$ to each constraint in (6.6) gives the Lagrange function



$$L(\boldsymbol{w}, b, \boldsymbol{\alpha}) = \frac{1}{2}\|\boldsymbol{w}\|^2 + \sum_{i=1}^{m} \alpha_i (1 - y_i(\boldsymbol{w}^\top \boldsymbol{x}_i + b)), \tag{6.8}$$

where $\boldsymbol{\alpha} = (\alpha_1; \alpha_2; \ldots; \alpha_m)$. Setting the partial derivatives of $L(\boldsymbol{w}, b, \boldsymbol{\alpha})$ with respect to $\boldsymbol{w}$ and $b$ to 0 gives

$$\boldsymbol{w} = \sum_{i=1}^{m} \alpha_i y_i \boldsymbol{x}_i, \tag{6.9}$$

$$0 = \sum_{i=1}^{m} \alpha_i y_i, \tag{6.10}$$

Substituting (6.9) into (6.8) eliminates $\boldsymbol{w}$ from $L(\boldsymbol{w}, b, \boldsymbol{\alpha})$. Then, with the constraint (6.10), we have the dual problem of (6.6) as

$$\begin{aligned} \max_{\boldsymbol{\alpha}} \ & \sum_{i=1}^{m} \alpha_i - \frac{1}{2} \sum_{i=1}^{m} \sum_{j=1}^{m} \alpha_i \alpha_j y_i y_j \boldsymbol{x}_i^\top \boldsymbol{x}_j \\ \text{s.t.} \ & \sum_{i=1}^{m} \alpha_i y_i = 0, \\ & \alpha_i \geqslant 0, \quad i = 1, 2, \ldots, m. \end{aligned} \tag{6.11}$$

By solving this optimization problem, we obtain $\boldsymbol{\alpha}$, and subsequently $\boldsymbol{w}$ and $b$. Then, we have the desired model

$$\begin{aligned} f(\boldsymbol{x}) &= \boldsymbol{w}^\top \boldsymbol{x} + b \\ &= \sum_{i=1}^{m} \alpha_i y_i \boldsymbol{x}_i^\top \boldsymbol{x} + b. \end{aligned} \tag{6.12}$$

The variable $\alpha_i$ solved from the dual problem (6.11) is the Lagrange multiplier in (6.8), corresponding to the training sample $(\boldsymbol{x}_i, y_i)$. Since (6.6) is an optimization problem with inequality constraints, it must satisfy the Karush–Kuhn–Tucker (KKT) conditions

See Appendix B.1.

$$\begin{cases} \alpha_i \geqslant 0; \\ y_i f(\boldsymbol{x}_i) - 1 \geqslant 0; \\ \alpha_i (y_i f(\boldsymbol{x}_i) - 1) = 0. \end{cases} \tag{6.13}$$

Hence, for any training sample $(\boldsymbol{x}_i, y_i)$, we either have $\alpha_i = 0$ or $y_i f(\boldsymbol{x}_i) = 1$. When $\alpha_i = 0$, the sample is not included in the summation in (6.12) and consequently has no impact on $f(\boldsymbol{x})$. On the other hand, when $\alpha_i > 0$, we have $y_i f(\boldsymbol{x}_i) = 1$, and the sample point lies on the maximum-margin hyperplanes, that is, it is a support vector . This observation reveals an important property of support vector machines: once the training completed, most training samples are no longer needed since the final model only depends on the support vectors.

As pointed out in (Vapnik 1999), the naming of support vector reflects the fact that finding the solution using support vectors is the key to such learners. This also implies that the model complexity mainly depends on the number of support vectors.

How can we solve (6.11)? This quadratic programming problem can be solved with quadratic programming algorithms. However, the computational cost is often high since the complexity is seriously affected by the number of training samples. To overcome this limitation, researchers proposed many efficient algorithms by exploiting the structure of the optimization problem. Among them, Sequential Minimal Optimization (SMO) (Platt 1998) is a celebrated representative.

See Appendix B.2 for quadratic programming.

The basic idea of SMO is to iteratively find the local optimal solutions of $\alpha_i$ by fixing all the other parameters as constants. Due to the constraint $\sum_{i=1}^{m} \alpha_i y_i = 0$, $\alpha_i$ can be derived from other fixed variables. Hence, in each iteration, SMO selects two variables $\alpha_i$ and $\alpha_j$ and fixes other parameters. After initializing the parameters, SMO repeats the following two steps until convergence:

- Select two variables to be updated: $\alpha_i$ and $\alpha_j$;
- Fix all the parameters and solve (6.11) to update $\alpha_i$ and $\alpha_j$.

The objective value will increase iteratively once any of the selected $\alpha_i$ and $\alpha_j$ violates the KKT conditions (6.13) (Osuna et al. 1997). Intuitively, the larger the magnitude of the KKT conditions being violated, the larger the possible magnitude of the increase of the objective value after updating. Following this idea, SMO should select the most violated variable as the first variable and select the one leading to the fastest increase of the objective value as the second variable. However, since it is computationally expensive to compare the increases of the objective values of variables, SMO takes a heuristic approach: select the two variables whose corresponding samples have the largest distance. Intuitively, since these two variables are very different from each other, they are more likely to lead to a significant update to the objective value compared to those two variables that are similar to each other.

The efficiency of the SMO algorithm is a result of the efficient optimization of two variables by fixing the others. To be specific, if we only consider $\alpha_i$ and $\alpha_j$, then we can rewrite the constraints in (6.11) as

$$\alpha_i y_i + \alpha_j y_j = c, \quad \alpha_i \geqslant 0, \;\; \alpha_j \geqslant 0, \tag{6.14}$$

where

$$c = -\sum_{k \neq i,j} \alpha_k y_k \tag{6.15}$$

is a constant such that $\sum_{i=1}^{m} \alpha_i y_i = 0$ holds. Eliminating the variable $\alpha_j$ from (6.11) with

$$\alpha_i y_i + \alpha_j y_j = c \tag{6.16}$$

leads to a univariate quadratic programming problem, in which the only constraint is $\alpha_i \geqslant 0$. Since this kind of quadratic programming problems have closed-form solutions, no numerical optimization algorithm is needed to get the new $\alpha_i$ and $\alpha_j$.

How can we determine the bias term $b$? We notice that there is $y_s f(\boldsymbol{x}_s) = 1$ for every support vector $(\boldsymbol{x}_s, y_s)$, that is

$$y_s \left( \sum_{i \in S} \alpha_i y_i \boldsymbol{x}_i^\top \boldsymbol{x}_s + b \right) = 1, \tag{6.17}$$

where $S = \{i \mid \alpha_i > 0, i = 1, 2, \ldots, m\}$ is the index set of all support vectors. Theoretically, we can find $b$ by substituting any support vectors to (6.17). In practice, a more robust method is taking the average of $b$ obtained from all support vectors

$$b = \frac{1}{|S|} \sum_{s \in S} \left( \frac{1}{y_s} - \sum_{i \in S} \alpha_i y_i \boldsymbol{x}_i^\top \boldsymbol{x}_s \right). \tag{6.18}$$

## 6.3 Kernel Function

Previous discussions in this chapter assumed the training samples are linearly separable, that is, there exist hyperplanes that can classify all training samples correctly. However, this assumption often does not hold in practice. For example, the exclusive disjunction (i.e., XOR) problem, as shown in ▫ Figure 6.3, is not linearly separable.

In such cases, we can map the samples from the original feature space to a higher dimensional feature space. That

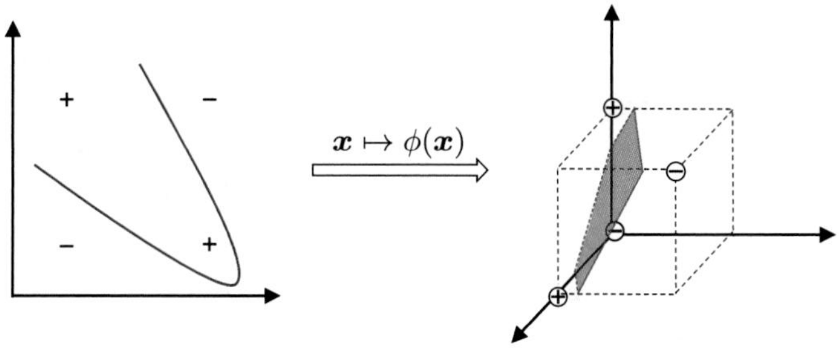


**Fig. 6.3** The XOR problem and non-linear mapping

way the samples become linearly separable. For example, in ◘ Figure 6.3, a qualified hyperplane can be found after mapping the 2-dimensional space to a 3-dimensional space. Fortunately, if the original feature space has a finite number of features, then there must exist a higher dimensional feature space in which the samples are linearly separable.

Let $\phi(\boldsymbol{x})$ denote the mapped feature vector of $\boldsymbol{x}$, then the separating hyperplane model in the feature space can be expressed as

See Chap. 12.

$$f(\boldsymbol{x}) = \boldsymbol{w}^\top \phi(\boldsymbol{x}) + b, \tag{6.19}$$

where $\boldsymbol{w}$ and $b$ are the model parameters. Similar to (6.6), we have

$$\begin{aligned} &\min_{\boldsymbol{w},b} \ \frac{1}{2}\|\boldsymbol{w}\|^2 \\ &\text{s.t. } y_i(\boldsymbol{w}^\top \phi(\boldsymbol{x}_i) + b) \geqslant 1, \quad i = 1, 2, \ldots, m. \end{aligned} \tag{6.20}$$

Its dual problem is

$$\begin{aligned} &\max_{\boldsymbol{\alpha}} \ \sum_{i=1}^m \alpha_i - \frac{1}{2}\sum_{i=1}^m \sum_{j=1}^m \alpha_i \alpha_j y_i y_j \phi(\boldsymbol{x}_i)^\top \phi(\boldsymbol{x}_j) \\ &\text{s.t. } \sum_{i=1}^m \alpha_i y_i = 0, \\ &\quad\ \ \alpha_i \geqslant 0, \quad i = 1, 2, \ldots, m. \end{aligned} \tag{6.21}$$

Solving (6.21) involves the calculation of $\phi(\boldsymbol{x}_i)^\top \phi(\boldsymbol{x}_j)$, which is the inner product of the mapped feature vectors of $\boldsymbol{x}_i$ and $\boldsymbol{x}_j$. Since the mapped feature space can have very high or even infinite dimensionality, it is often difficult to calculate $\phi(\boldsymbol{x}_i)^\top \phi(\boldsymbol{x}_j)$ directly. To avoid this difficulty, we suppose there exists a function in the following form:

$$\kappa(\boldsymbol{x}_i, \boldsymbol{x}_j) = \langle \phi(\boldsymbol{x}_i), \phi(\boldsymbol{x}_j) \rangle = \phi(\boldsymbol{x}_i)^\top \phi(\boldsymbol{x}_j), \tag{6.22}$$

This is called *kernel trick*.

which says the inner product of $\boldsymbol{x}_i$ and $\boldsymbol{x}_j$ in the feature space can be calculated in the sample space using the function $\kappa(\cdot, \cdot)$. With such a function, we no longer need to calculate the inner product in the feature space and can rewrite (6.21) as

$$\begin{aligned} \max_{\boldsymbol{\alpha}} \ & \sum_{i=1}^{m} \alpha_i - \frac{1}{2} \sum_{i=1}^{m} \sum_{j=1}^{m} \alpha_i \alpha_j y_i y_j \kappa(\boldsymbol{x}_i, \boldsymbol{x}_j) \\ \text{s.t.} \ & \sum_{i=1}^{m} \alpha_i y_i = 0, \\ & \alpha_i \geqslant 0, \quad i = 1, 2, \ldots, m. \end{aligned} \tag{6.23}$$

6

Solving it gives

$$\begin{aligned} f(\boldsymbol{x}) &= \boldsymbol{w}^\top \phi(\boldsymbol{x}) + b \\ &= \sum_{i=1}^{m} \alpha_i y_i \phi(\boldsymbol{x}_i)^\top \phi(\boldsymbol{x}) + b \\ &= \sum_{i=1}^{m} \alpha_i y_i \kappa(\boldsymbol{x}, \boldsymbol{x}_i) + b, \end{aligned} \tag{6.24}$$

where $\kappa(\cdot, \cdot)$ is the *kernel function*. From (6.24), we see that the optimal solution can be expanded by training samples with the kernel functions, and this is known as the *support vector expansion*.

We can derive the kernel function $\kappa(\cdot, \cdot)$ if we know the details of the mapping $\phi(\cdot)$. However, $\phi(\cdot)$ is often unknown in practice. Then, how do we know if there is a proper kernel function? What kind of functions are valid kernel functions? Let us see the following theorem:

See (Schölkopf and Smola 2002) for the proof.

**Theorem 6.1** (Kernel Function) *Let $\mathcal{X}$ denote the input space, and $\kappa(\cdot, \cdot)$ denote a symmetric function defined in $\mathcal{X} \times \mathcal{X}$. Then, $\kappa$ is a kernel function if and only if the kernel matrix* $\mathbf{K}$ *is positive semidefinite for any data set $D = \{\boldsymbol{x}_1, \boldsymbol{x}_2, \ldots, \boldsymbol{x}_m\}$*

$$\mathbf{K} = \begin{bmatrix} \kappa(\boldsymbol{x}_1, \boldsymbol{x}_1) & \ldots & \kappa(\boldsymbol{x}_1, \boldsymbol{x}_j) & \ldots & \kappa(\boldsymbol{x}_1, \boldsymbol{x}_m) \\ \vdots & \ddots & \vdots & \ddots & \vdots \\ \kappa(\boldsymbol{x}_i, \boldsymbol{x}_1) & \ldots & \kappa(\boldsymbol{x}_i, \boldsymbol{x}_j) & \ldots & \kappa(\boldsymbol{x}_i, \boldsymbol{x}_m) \\ \vdots & \ddots & \vdots & \ddots & \vdots \\ \kappa(\boldsymbol{x}_m, \boldsymbol{x}_1) & \ldots & \kappa(\boldsymbol{x}_m, \boldsymbol{x}_j) & \ldots & \kappa(\boldsymbol{x}_m, \boldsymbol{x}_m) \end{bmatrix}$$

Theorem 6.1 says that every symmetric function with a positive semidefinite kernel matrix is a valid kernel function. For every positive semidefinite kernel matrix, there is always a corresponding mapping $\phi$. In other words, every kernel function implicitly defines a feature space known as the Reproducing Kernel Hilbert Space (RKHS).

Since we wish the samples to be linearly separable in the feature space, the quality of the feature space is vital to the performance of support vector machines. However, we do not know which kernel functions are good because we do not know the feature mapping. Therefore, the "choice of kernel" is the biggest uncertainty of support vector machines. A poor kernel will map the samples to a poor feature space, resulting in poor performance. Some common kernel functions are listed in ◘ Table 6.1.

There are some rules of thumb, such as using linear kernels for textual data since the data is already high-dimensional, and using the Gaussian kernel when the situation is unclear.

**◘ Tab. 6.1** Some common kernel functions

| Name | Expression | Parameters |
|---|---|---|
| Linear kernel | $\kappa(\boldsymbol{x}_i, \boldsymbol{x}_j) = \boldsymbol{x}_i^\top \boldsymbol{x}_j$ | |
| Polynomial kernel | $\kappa(\boldsymbol{x}_i, \boldsymbol{x}_j) = (\boldsymbol{x}_i^\top \boldsymbol{x}_j)^d$ | $d \geqslant 1$ is the degree of the polynomial |
| Gaussian kernel | $\kappa(\boldsymbol{x}_i, \boldsymbol{x}_j) = \exp\left(-\frac{\lVert \boldsymbol{x}_i - \boldsymbol{x}_j \rVert^2}{2\sigma^2}\right)$ | $\sigma > 0$ is the width of the Gaussian kernel |
| Laplacian kernel | $\kappa(\boldsymbol{x}_i, \boldsymbol{x}_j) = \exp\left(-\frac{\lVert \boldsymbol{x}_i - \boldsymbol{x}_j \rVert}{\sigma}\right)$ | $\sigma > 0$ |
| Sigmoid kernel | $\kappa(\boldsymbol{x}_i, \boldsymbol{x}_j) = \tanh(\beta \boldsymbol{x}_i^\top \boldsymbol{x}_j + \theta)$ | tanh is the hyperbolic tangent function, $\beta > 0$, $\theta < 0$ |

It reduces to linear kernel when $d = 1$.

Gaussian kernel is also called RBF kernel.

Besides, we can construct kernel functions by function composition, e.g.

- For any positive numbers $\gamma_1$ and $\gamma_2$, if $\kappa_1$ and $\kappa_2$ are kernel functions, then their linear combination

$$\gamma_1 \kappa_1 + \gamma_2 \kappa_2 \tag{6.25}$$

  is also a kernel function;
- If $\kappa_1$ and $\kappa_2$ are kernel functions, then their direct product

$$\kappa_1 \otimes \kappa_2(\boldsymbol{x}, \boldsymbol{z}) = \kappa_1(\boldsymbol{x}, \boldsymbol{z}) \kappa_2(\boldsymbol{x}, \boldsymbol{z}) \tag{6.26}$$

  is also a kernel function;
- If $\kappa_1$ is a kernel function, then, for any function $g(\boldsymbol{x})$,

$$\kappa(\boldsymbol{x}, \boldsymbol{z}) = g(\boldsymbol{x}) \kappa_1(\boldsymbol{x}, \boldsymbol{z}) g(\boldsymbol{z}) \tag{6.27}$$

  is also a kernel function.

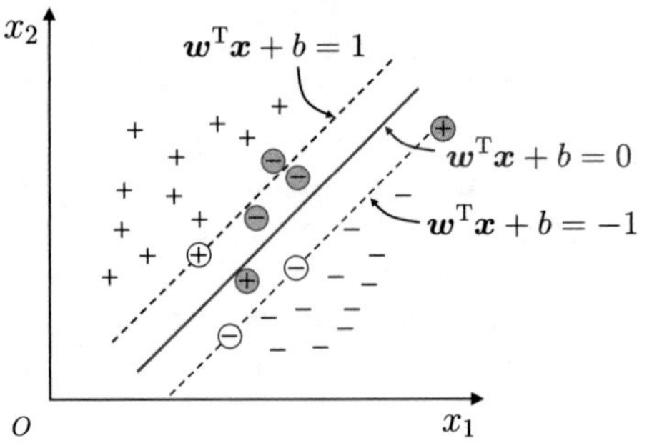


**Fig. 6.4** Soft margin. The samples in red violate the constraints



## 6.4 Soft Margin and Regularization

Our discussions so far assumed the samples to be linearly separable in either the sample space or the feature space. However, it is often difficult to find an appropriate kernel function to make the training samples linearly separable in the feature space. Even if we do find such a kernel function, it is hard to tell if it is a result of overfitting.

One way of alleviating this situation is to allow a support vector machine to make mistakes on a few samples. This idea is implemented by the concept of *soft margin*, as shown in ▪ Figure 6.4.

To be specific, the previously introduced support vector machines are subject to the constraints (6.3), that is, the *hard margin* requires all samples to be correctly classified. The soft margin, however, allows the violation of the constraint

$$y_i(\boldsymbol{w}^\top \boldsymbol{x}_i + b) \geqslant 1. \tag{6.28}$$

Of course, the number of samples violating the constraint should be minimized while maximizing the margin. Hence, the optimization objective can be written as

$$\min_{\boldsymbol{w},b} \ \frac{1}{2} \|\boldsymbol{w}\|^2 + C \sum_{i=1}^{m} \ell_{0/1}(y_i(\boldsymbol{w}^\top \boldsymbol{x}_i + b) - 1), \tag{6.29}$$

where $C > 0$ is a constant, and $\ell_{0/1}$ is the 0/1 loss function

$$\ell_{0/1}(z) = \begin{cases} 1, & \text{if } z < 0; \\ 0, & \text{otherwise.} \end{cases} \tag{6.30}$$

When $C$ is infinitely large, (6.29) forces all samples to obey the constraint (6.28), and (6.29) is equivalent to (6.6), that is, the

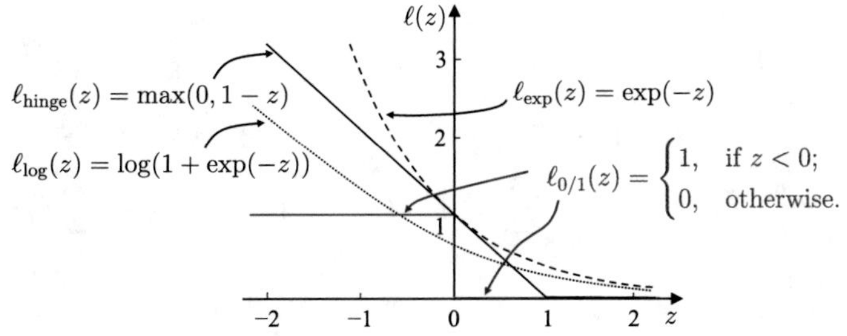


**Fig. 6.5** Three surrogate losses, namely, hinge loss, exponential loss, and logistic loss

hard margin. On the other hand, some samples may violate the constraint when $C$ takes a finite value.

Solving (6.29) directly is difficult since $\ell_{0/1}$ has poor mathematical properties, that is, non-convex and discontinuous. Therefore, we often replace $\ell_{0/1}$ with some other functions, known as *surrogate loss* functions, which have nice mathematical properties, e.g., convex, continuous, and are upper bound of $\ell_{0/1}$. ◘ Figure 6.5 illustrates three commonly used surrogate loss functions:

The logistic loss is a transformation of the logistic function. See Sect. 3.3.

$$\text{hinge loss: } \ell_{\text{hinge}}(z) = \max(0, 1 - z); \tag{6.31}$$
$$\text{exponential loss: } \ell_{\exp}(z) = \exp(-z); \tag{6.32}$$
$$\text{logistic loss: } \ell_{\log}(z) = \log(1 + \exp(-z)). \tag{6.33}$$

Since the logistic loss is often written as $\ell_{\log}(\cdot)$, (3.15) rewrites $\ln(\cdot)$ in (6.33) to $\log(\cdot)$.

When hinge loss is used, (6.29) becomes

$$\min_{\boldsymbol{w},b} \ \frac{1}{2}\|\boldsymbol{w}\|^2 + C\sum_{i=1}^{m}\max(0, 1 - y_i(\boldsymbol{w}^\top \boldsymbol{x}_i + b)). \tag{6.34}$$

By introducing *slack variables* $\xi_i \geqslant 0$, (6.34) can be rewritten as

$$\begin{aligned} \min_{\boldsymbol{w},b} \ & \frac{1}{2}\|\boldsymbol{w}\|^2 + C\sum_{i=1}^{m}\xi_i \\ \text{s.t. } & y_i(\boldsymbol{w}^\top \boldsymbol{x}_i + b) \geqslant 1 - \xi_i \\ & \xi_i \geqslant 0, \quad i = 1, 2, \ldots, m \end{aligned} \tag{6.35}$$

which is the commonly used Soft Margin Support Vector Machine.

In (6.35), each sample has a corresponding slack variable indicating the degree to which the constraint (6.28) is violated. Similar to (6.6), this is again a quadratic programming problem. Hence, we apply the Lagrange multipliers to (6.35) to obtain the Lagrange function

$$L(\boldsymbol{w}, b, \boldsymbol{\alpha}, \boldsymbol{\xi}, \boldsymbol{\mu}) = \frac{1}{2}\|\boldsymbol{w}\|^2 + C\sum_{i=1}^{m}\xi_i + \sum_{i=1}^{m}\alpha_i(1 - \xi_i - y_i(\boldsymbol{w}^\top \boldsymbol{x}_i + b)) - \sum_{i=1}^{m}\mu_i\xi_i, \tag{6.36}$$

where $\alpha_i \geqslant$ and $\mu_i \geqslant 0$ are the Lagrange multipliers.

Setting the partial derivatives of $L(\boldsymbol{w}, b, \boldsymbol{\alpha}, \boldsymbol{\xi}, \boldsymbol{\mu})$ with respect to $\boldsymbol{w}$, $b$, and $\xi_i$ to 0 gives

6

$$\boldsymbol{w} = \sum_{i=1}^{m}\alpha_i y_i \boldsymbol{x}_i, \tag{6.37}$$

$$0 = \sum_{i=1}^{m}\alpha_i y_i, \tag{6.38}$$

$$C = \alpha_i + \mu_i. \tag{6.39}$$

Substituting (6.37)–(6.39) into (6.36) gives the dual problem of (6.35)

$$\begin{aligned} \max_{\boldsymbol{\alpha}} \quad & \sum_{i=1}^{m}\alpha_i - \frac{1}{2}\sum_{i=1}^{m}\sum_{j=1}^{m}\alpha_i\alpha_j y_i y_j \boldsymbol{x}_i^\top \boldsymbol{x}_j \\ \text{s.t.} \quad & \sum_{i=1}^{m}\alpha_i y_i = 0, \\ & 0 \leqslant \alpha_i \leqslant C, \quad i = 1, 2, \ldots, m. \end{aligned} \tag{6.40}$$

Comparing (6.40) with the dual problem (6.11) of hard margin, we observe that the only difference is the constraint on dual variables: $0 \leqslant \alpha_i \leqslant C$ for soft margin and $\alpha_i \geqslant 0$ for hard margin. Hence, (6.40) can be solved in the same way as we did in Sect. 6.2. By introducing kernel function, we obtain the support vector expansion the same as we had in (6.24).

Similar to (6.13), the KKT conditions for soft margin support vector machines are

$$\begin{cases} \alpha_i \geqslant 0, \quad \mu_i \geqslant 0, \\ y_i f(\boldsymbol{x}_i) - 1 + \xi_i \geqslant 0, \\ \alpha_i(y_i f(\boldsymbol{x}_i) - 1 + \xi_i) = 0, \\ \xi_i \geqslant 0, \quad \mu_i\xi_i = 0. \end{cases} \tag{6.41}$$

Hence, for each training sample $(\boldsymbol{x}_i, y_i)$, we either have $\alpha_i = 0$ or $y_i f(\boldsymbol{x}_i) = 1 - \xi_i$. When $\alpha_i = 0$, this sample has no impact on $f(\boldsymbol{x})$. When $\alpha_i > 0$, $y_i f(\boldsymbol{x}_i) = 1 - \xi_i$, that is, this sample is a support vector. From (6.39), we know that if $\alpha_i < C$ then $\mu_i > 0$ and subsequently $\xi_i = 0$, that is, this sample point lies on the maximum-margin hyperplanes. When $\alpha_i = C$, we have $\mu_i = 0$, which means the sample falls inside the margin if $\xi_i \leqslant 1$ and it is incorrectly classified if $\xi_i > 1$. This shows that the final model of soft margin support vector machine only depends on the support vectors, that is, the sparseness is preserved after using the hinge loss.

Can we use other surrogate loss functions for (6.29)? In fact, if we replace the 0/1 loss function of (6.29) by the logistic loss function $\ell_{\log}$, then we end up with a model that is almost the same as the logistic regression (3.27). Since the optimization objectives of support vector machine and logistic regression are similar, their performance is also similar in many cases. The main advantage of logistic regression is its output naturally carries probability meanings, that is, each predicted label comes with a probability. By contrast, the predictions of support vector machines do not associate with probabilities, and probabilities can only be obtained with additional processing (Platt 2000). Besides, the logistic regression can be directly applied to multiclass classifications, whereas the support vector machine requires extensions (Hsu and Lin 2002). On the other hand, from ◘ Figure 6.5, we can see there is a "flat zero region" for hinge loss, which makes the solution of support vector machines sparse. The logistic loss, which is a smooth and monotonically decreasing function, cannot derive the concept like support vectors, and hence relies on more training samples, and its prediction cost is also higher.

We can obtain other learning models by replacing the 0/1 loss function of (6.29) with other surrogate functions. The obtained models depend on the choice of the surrogate functions, but they have one thing in common: the first term in the objective function represents the "margin" size of the separating hyperplane and the other term $\sum_{i=1}^{m} \ell(f(\boldsymbol{x}_i), y_i)$ represents the error on the training set. Hence, we can rewrite the loss in a more general form as

$$\min_{f} \ \Omega(f) + C \sum_{i=1}^{m} \ell(f(\boldsymbol{x}_i), y_i), \tag{6.42}$$

where $\Omega(f)$ is known as *structural risk*, representing some properties of the model $f$. The second term $\sum_{i=1}^{m} \ell(f(\boldsymbol{x}_i), y_i)$ is known as the *empirical risk*, which describes how well the model matches the training data. The constant $C$ makes a trade-off

Typically, *structural risk* refers to the whole risk after incorporating the model structure factor. In this book, however, *structural risk* refers to the part corresponding to model structure in the total risk so that its meaning is more intuitive, and its connection to other mechanisms of machine learning becomes more apparent. See Page 169.

between these two risks. From the perspective of minimizing the empirical risk, $\Omega(f)$ represents what kind of properties we would like the model to have (e.g., prefers a model with low complexity), which provides a way for incorporating domain knowledge and user's requirements. On the other hand, $\Omega(f)$ is also helpful for reducing the hypothesis space, which reduces the overfitting risk of minimizing the training error. From this point of view, (6.42) is a *regularization* problem, where $\Omega(f)$ is the regularization term, and $C$ is the regularization constant. A typical regularization term is the $L_p$ norm, where the $L_2$ norm $\|\boldsymbol{w}\|_2$ is biased towards balanced weights $\boldsymbol{w}$, i.e., a dense solution with many non-zero weights. On the other hand, the $L_0$ norm $\|\boldsymbol{w}\|_0$ and the $L_1$ norm $\|\boldsymbol{w}\|_1$ are biased towards making $\boldsymbol{w}$ have sparse elements, that is, with only a few non-zero elements.

We can regard regularization as a *penalty function method* which applies penalties to undesired outcomes such that the optimization is biased towards the desired outcome. From the Bayesian inference point of view, the regularization term can be seen as a *prior* to the model.

See Sect. 11.4 for $L_1$ and $L_2$ regularization.



## 6.5 Support Vector Regression

Now let us consider regression problems. Given a training set $D = \{(\boldsymbol{x}_1, y_1), (\boldsymbol{x}_2, y_2), \ldots, (\boldsymbol{x}_m, y_m)\}$, where $y_i \in \mathbb{R}$. Suppose we wish to learn a regression model in the form of (6.7) such that $f(\boldsymbol{x})$ and $y$ are as close as possible, where $\boldsymbol{w}$ and $b$ are the parameters to be learned from data.

The loss in traditional regression models is calculated from the difference between the model output $f(\boldsymbol{x})$ and the ground-truth output $y$, and the loss is 0 if and only if $f(\boldsymbol{x})$ equals to $y$. By contrast, Support Vector Regression (SVR) allows an error $\epsilon$ between $f(\boldsymbol{x})$ and $y$, that is, a loss incurs only if the difference between $f(\boldsymbol{x})$ and $y$ exceeds $\epsilon$. We can consider it as establishing a band buffer region surrounding $f(\boldsymbol{x})$ with a width of $2\epsilon$. Training samples falling within this buffer region are considered as correctly predicted, that is, no loss (◘ Figure 6.6).

Formally, the SVR problem can be written as

$$\min_{\boldsymbol{w},b} \frac{1}{2}\|\boldsymbol{w}\|^2 + C\sum_{i=1}^{m} \ell_\epsilon(f(\boldsymbol{x}_i) - y_i), \tag{6.43}$$

where $C$ is the regularization constant, and $\ell_\epsilon$, as illustrated in ◘ Figure 6.7, is the $\epsilon$-insensitive loss function

The slackness of two sides can be different.

$$\ell_\epsilon(z) = \begin{cases} 0, & \text{if } |z| \leqslant \epsilon; \\ |z| - \epsilon, & \text{otherwise.} \end{cases} \tag{6.44}$$

With slack variables $\xi_i$ and $\hat{\xi}_i$, (6.43) can be rewritten as

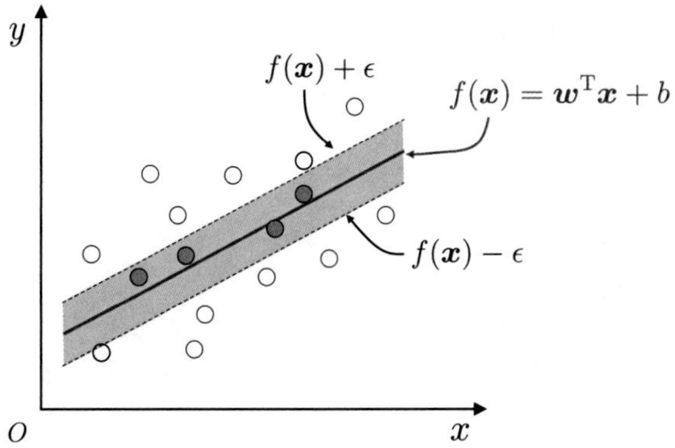


**Fig. 6.6** Support vector regression. The samples that fall into the red $\epsilon$-region do not incur any loss

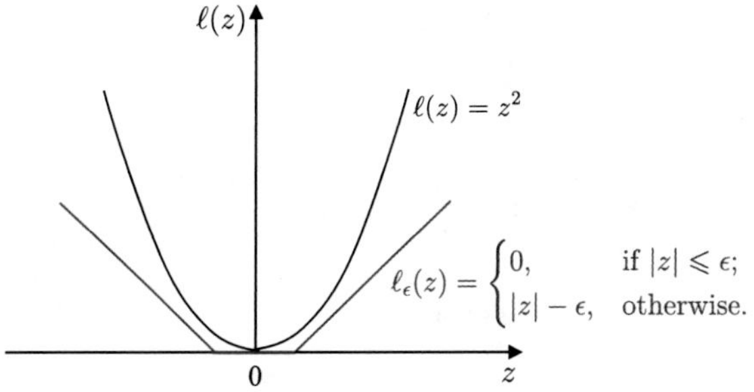


**Fig. 6.7** The $\epsilon$-insensitive loss function

$$\begin{aligned} \min_{\boldsymbol{w},b,\xi_i,\hat{\xi}_i} \quad & \frac{1}{2}\|\boldsymbol{w}\|^2 + C\sum_{i=1}^{m}(\xi_i + \hat{\xi}_i) \\ \text{s.t.} \quad & f(\boldsymbol{x}_i) - y_i \leqslant \epsilon + \xi_i, \\ & y_i - f(\boldsymbol{x}_i) \leqslant \epsilon + \hat{\xi}_i, \\ & \xi_i \geqslant 0, \hat{\xi}_i \geqslant 0, \quad i = 1, 2, \ldots, m. \end{aligned} \tag{6.45}$$

Similar to (6.36), applying the Lagrange multipliers $\mu_i \geqslant 0$, $\hat{\mu}_i \geqslant 0$, $\alpha_i \geqslant 0$, and $\hat{\alpha}_i \geqslant 0$ to (6.45) gives its Lagrange function

$$\begin{aligned} & L(\boldsymbol{w}, b, \boldsymbol{\alpha}, \hat{\boldsymbol{\alpha}}, \boldsymbol{\xi}, \hat{\boldsymbol{\xi}}, \boldsymbol{\mu}, \hat{\boldsymbol{\mu}}) \\ &= \frac{1}{2}\|\boldsymbol{w}\|^2 + C\sum_{i=1}^{m}(\xi_i + \hat{\xi}_i) - \sum_{i=1}^{m}\mu_i\xi_i - \sum_{i=1}^{m}\hat{\mu}_i\hat{\xi}_i \\ &+ \sum_{i=1}^{m}\alpha_i(f(\boldsymbol{x}_i) - y_i - \epsilon - \xi_i) + \sum_{i=1}^{m}\hat{\alpha}_i(y_i - f(\boldsymbol{x}_i) - \epsilon - \hat{\xi}_i). \end{aligned} \tag{6.46}$$

Substituting (6.7) and setting the partial derivatives of $L(\boldsymbol{w}, b, \boldsymbol{\alpha}, \hat{\boldsymbol{\alpha}}, \boldsymbol{\xi}, \hat{\boldsymbol{\xi}}, \boldsymbol{\mu}, \hat{\boldsymbol{\mu}})$ with respect to $\boldsymbol{w}$, $b$, $\xi_i$, and $\hat{\xi}_i$ to 0, gives

$$\boldsymbol{w} = \sum_{i=1}^{m} (\hat{\alpha}_i - \alpha_i)\boldsymbol{x}_i, \tag{6.47}$$

$$0 = \sum_{i=1}^{m} (\hat{\alpha}_i - \alpha_i), \tag{6.48}$$

$$C = \alpha_i + \mu_i, \tag{6.49}$$

$$C = \hat{\alpha}_i + \hat{\mu}_i. \tag{6.50}$$

6

Substituting (6.47)–(6.50) into (6.46) gives the dual problem of SVR as

$$\begin{aligned} \max_{\boldsymbol{\alpha}, \hat{\boldsymbol{\alpha}}} \quad & \sum_{i=1}^{m} y_i(\hat{\alpha}_i - \alpha_i) - \epsilon(\hat{\alpha}_i + \alpha_i) \\ & - \frac{1}{2} \sum_{i=1}^{m} \sum_{j=1}^{m} (\hat{\alpha}_i - \alpha_i)(\hat{\alpha}_j - \alpha_j)\boldsymbol{x}_i^{\top} \boldsymbol{x}_j \\ \text{s.t.} \quad & \sum_{i=1}^{m} (\hat{\alpha}_i - \alpha_i) = 0, \\ & 0 \leqslant \alpha_i, \hat{\alpha}_i \leqslant C. \end{aligned} \tag{6.51}$$

For SVR, the KKT conditions are

$$\begin{cases} \alpha_i(f(\boldsymbol{x}_i) - y_i - \epsilon - \xi_i) = 0, \\ \hat{\alpha}_i(y_i - f(\boldsymbol{x}_i) - \epsilon - \hat{\xi}_i) = 0, \\ \alpha_i \hat{\alpha}_i = 0, \quad \xi_i \hat{\xi}_i = 0, \\ (C - \alpha_i)\xi_i = 0, \quad (C - \hat{\alpha}_i)\hat{\xi}_i = 0. \end{cases} \tag{6.52}$$

From (6.52), we see that $\alpha_i$ is non-zero if and only if $f(\boldsymbol{x}_i) - y_i - \epsilon - \xi_i = 0$, and $\hat{\alpha}_i$ is non-zero if and only if $y_i - f(\boldsymbol{x}_i) - \epsilon - \hat{\xi}_i = 0$. In other words, $\alpha_i$ and $\hat{\alpha}_i$ take non-zero values if and only if the sample $(\boldsymbol{x}_i, y_i)$ falls outside of the $\epsilon$-insensitive region. Moreover, since the constraints $f(\boldsymbol{x}_i) - y_i - \epsilon - \xi_i = 0$ and $y_i - f(\boldsymbol{x}_i) - \epsilon - \hat{\xi}_i = 0$ cannot be true at the same time, $\alpha_i$ or $\hat{\alpha}_i$ must be zero.

Substituting (6.47) into (6.7), we have the solution of SVR as

$$f(\boldsymbol{x}) = \sum_{i=1}^{m} (\hat{\alpha}_i - \alpha_i)\boldsymbol{x}_i^{\top} \boldsymbol{x} + b. \tag{6.53}$$

In (6.53), the samples are the support vectors of SVR if $(\hat{\alpha}_i - \alpha_i) \neq 0$, and they fall outside the $\epsilon$-insensitive region. The solution of SVR is sparse since the support vectors are only a subset of the training samples.

All samples fall inside of the $\epsilon$-insensitive region satisfy $\alpha_i = 0$ and $\hat{\alpha}_i = 0$.

The KKT conditions in (6.52) show that each sample $(\boldsymbol{x}_i, y_i)$ satisfies $(C - \alpha_i)\xi_i = 0$ and $\alpha_i(f(\boldsymbol{x}_i) - y_i - \epsilon - \xi_i) = 0$. Hence, $\xi_i = 0$ if $0 < \alpha_i < C$, and consequently

$$b = y_i + \epsilon - \sum_{j=1}^{m}(\hat{\alpha}_j - \alpha_j)\boldsymbol{x}_j^\top \boldsymbol{x}_i. \tag{6.54}$$

Therefore, after obtaining $\alpha_i$ by solving (6.51), we can find $b$ by substituting any $\alpha_i$ satisfying $0 < \alpha_i < C$ into (6.54). In practice, a more robust method is taking the average of $b$ obtained from multiple or all $\alpha_i$ satisfying $0 < \alpha_i < C$.

Similar to (6.19), after mapping to the feature space, (6.47) becomes

$$\boldsymbol{w} = \sum_{i=1}^{m}(\hat{\alpha}_i - \alpha_i)\phi(\boldsymbol{x}_i). \tag{6.55}$$

By substituting (6.55) into (6.19), SVR can be expressed as

$$f(\boldsymbol{x}) = \sum_{i=1}^{m}(\hat{\alpha}_i - \alpha_i)\kappa(\boldsymbol{x}, \boldsymbol{x}_i) + b. \tag{6.56}$$

where $\kappa(\boldsymbol{x}_i, \boldsymbol{x}_j) = \phi(\boldsymbol{x}_i)^\top \phi(\boldsymbol{x}_j)$ is the kernel function.

## 6.6 Kernel Methods

By revisiting (6.24) and (6.56), we see that if we do not consider the bias term $b$, then both the SVM and the SVR models can be expressed as linear combinations of the kernel functions $\kappa(\boldsymbol{x}, \boldsymbol{x}_i)$. In fact, there is a more generalized conclusion known as the *representer theorem*.

**Theorem 6.2** (Representer Theorem) *Let $\mathbb{H}$ denote the reproducing kernel Hilbert space associated to the kernel function $\kappa$, and $\|h\|_{\mathbb{H}}$ denote the norm of $h$ in the space $\mathbb{H}$. Then, for any monotonically increasing function $\Omega : [0, \infty] \mapsto \mathbb{R}$ and any non-negative loss function $\ell : \mathbb{R}^m \mapsto [0, \infty]$, the solution to the optimization problem*

The proof can be found in (Schölkopf and Smola 2002), with the use of Mercer's theorem.

$$\min_{h \in \mathbb{H}} \; F(h) = \Omega(\|h\|_{\mathbb{H}}) + \ell\,(h(\boldsymbol{x}_1), h(\boldsymbol{x}_2), \ldots, h(\boldsymbol{x}_m)) \tag{6.57}$$

*can always be written in the form of*

$$h^*(\boldsymbol{x}) = \sum_{i=1}^{m} \alpha_i \kappa(\boldsymbol{x}, \boldsymbol{x}_i). \tag{6.58}$$

The representer theorem has few restrictions on the loss function, and the regularization term $\Omega$ is only required to be monotonically increasing, not even necessary to be convex. This implies that, for typical loss functions and regularization terms, the optimal solution $h^*(\boldsymbol{x})$ for the optimization problem (6.57) can always be expressed as a linear combination of the kernel functions $\kappa(\boldsymbol{x}, \boldsymbol{x}_i)$. This observation exhibits the great power of kernel functions.

By utilizing kernel functions, researchers developed a series of learning methods known as *kernel methods*. A typical approach is to extend linear learners to non-linear learners by *kernelization*, that is, introducing kernel functions. Next, we demonstrate how to apply kernelization to the linear discriminant analysis to obtain its non-linear extension, that is, Kernelized Linear Discriminant Analysis (KLDA).

See Sect. 3.4 for linear discriminant analysis.

Suppose there is a mapping $\phi : \mathcal{X} \mapsto \mathbb{F}$ that maps samples into a feature space $\mathbb{F}$, and we perform linear discriminant analysis in $\mathbb{F}$ to get

$$h(\boldsymbol{x}) = \boldsymbol{w}^\top \phi(\boldsymbol{x}). \tag{6.59}$$

Similar to (3.35), the learning objective of KLDA is

$$\max_{\boldsymbol{w}} \ J(\boldsymbol{w}) = \frac{\boldsymbol{w}^\top \mathbf{S}_b^\phi \boldsymbol{w}}{\boldsymbol{w}^\top \mathbf{S}_w^\phi \boldsymbol{w}}, \tag{6.60}$$

where $\mathbf{S}_b^\phi$ and $\mathbf{S}_w^\phi$ are, respectively, the between-class scatter matrix and the within-class scatter matrix of training samples in the feature space $\mathbb{F}$. Let $X_i$ denote the set of $m_i$ samples of the $i$th ($i \in \{0, 1\}$) class, and $m = m_0 + m_1$ denote the total number of samples. Then, the mean of the $i$th class samples in the feature space $\mathbb{F}$ is

$$\boldsymbol{\mu}_i^\phi = \frac{1}{m_i} \sum_{\boldsymbol{x} \in X_i} \phi(\boldsymbol{x}), \tag{6.61}$$

and the two scatter matrices are

$$\mathbf{S}_b^\phi = (\boldsymbol{\mu}_1^\phi - \boldsymbol{\mu}_0^\phi)(\boldsymbol{\mu}_1^\phi - \boldsymbol{\mu}_0^\phi)^\top, \tag{6.62}$$

$$\mathbf{S}_w^\phi = \sum_{i=0}^{1} \sum_{\boldsymbol{x} \in X_i} (\phi(\boldsymbol{x}) - \boldsymbol{\mu}_i^\phi)(\phi(\boldsymbol{x}) - \boldsymbol{\mu}_i^\phi)^\top. \tag{6.63}$$

Typically, the exact form of the mapping $\phi$ is unknown, and hence we use the kernel function $\kappa(\boldsymbol{x}, \boldsymbol{x}_i) = \phi(\boldsymbol{x}_i)^\top \phi(\boldsymbol{x})$ to implicitly express the mapping $\phi$ and the feature space $\mathbb{F}$. Using $J(\boldsymbol{w})$ as the loss function $\ell$ in (6.57) and letting $\Omega \equiv 0$, then, according to the representer theorem, the function $h(\boldsymbol{x})$ can be written as

$$h(\boldsymbol{x}) = \sum_{i=1}^{m} \alpha_i \kappa(\boldsymbol{x}, \boldsymbol{x}_i), \tag{6.64}$$

and, from (6.59), we have

$$\boldsymbol{w} = \sum_{i=1}^{m} \alpha_i \phi(\boldsymbol{x}_i). \tag{6.65}$$

Let $\mathbf{K} \in \mathbb{R}^{m \times m}$ denote the kernel matrix associated to the kernel function $\kappa$, where $(\mathbf{K})_{ij} = \kappa(\boldsymbol{x}_i, \boldsymbol{x}_j)$. Let $\mathbf{1}_i \in \{1, 0\}^{m \times 1}$ denote the indicator vector of the $i$th class samples, that is, the $j$th element of $\mathbf{1}_i$ is 1 if and only if $\boldsymbol{x}_j \in X_i$, and the $j$th element of $\mathbf{1}_i$ is 0 otherwise. Letting

$$\hat{\boldsymbol{\mu}}_0 = \frac{1}{m_0} \mathbf{K} \mathbf{1}_0, \tag{6.66}$$

$$\hat{\boldsymbol{\mu}}_1 = \frac{1}{m_1} \mathbf{K} \mathbf{1}_1, \tag{6.67}$$

$$\mathbf{M} = (\hat{\boldsymbol{\mu}}_0 - \hat{\boldsymbol{\mu}}_1)(\hat{\boldsymbol{\mu}}_0 - \hat{\boldsymbol{\mu}}_1)^\top, \tag{6.68}$$

$$\mathbf{N} = \mathbf{K}\mathbf{K}^\top - \sum_{i=0}^{1} m_i \hat{\boldsymbol{\mu}}_i \hat{\boldsymbol{\mu}}_i^\top. \tag{6.69}$$

Then, (6.60) is equivalent to

$$\max_{\boldsymbol{\alpha}} \; J(\boldsymbol{\alpha}) = \frac{\boldsymbol{\alpha}^\top \mathbf{M} \boldsymbol{\alpha}}{\boldsymbol{\alpha}^\top \mathbf{N} \boldsymbol{\alpha}}. \tag{6.70}$$

which can be solved in the same way as in the linear discriminant analysis. With the solution $\boldsymbol{\alpha}$, the mapping function $h(\boldsymbol{x})$ can be obtained using (6.64).

See Sect. 3.4 for linear discriminant analysis.

## 6.7 Further Reading

SVM was officially published in 1995 (Cortes and Vapnik 1995). Because of its outstanding performance in text classification tasks (Joachims 1998), SVM has soon become the mainstream of machine learning and has contributed to the boom of the *statistical learning* around 2000. However, the concept of support vectors already appeared in the 1960s, and the foundation of the statistical learning theory was set in the 1970s. Studies on kernel functions even go back a longer time. For example, Mercer's theorem (Cristianini and Shawe-Taylor 2000) can be traced back to 1909, and the RKHS has already been studied in the 1940s. However, the kernel trick did not become a general machine learning technique until the statistical learning gained its popularity. There are many books and introductory articles for support vector machines and kernel methods, including (Burges 1998, Cristianini and Shawe-Taylor 2000, Schölkopf et al. 1999, Schölkopf and Smola 2002). The statistical learning theory can refer to (Vapnik 1995, 1998, 1999).

Till now, SVM with a linear kernel is still the preferred method for text classifications. An important possible reason is that it is already a very high dimensionality feature space with high information redundancy if we consider each word as a feature. The representation power is strong enough to "shatter" different text documents. See Sect. 12.4 for shattering.



Solutions to SVM are often found via convex optimization techniques (Boyd and Vandenberghe 2004). A critical research topic on SVM is how to improve its efficiency for large-scale data. There are already considerable research outcomes for linear kernels. For example, the $\text{SVM}^{\text{perf}}$, which is based on the cutting plane algorithm, has achieved linear complexity (Joachims 2006); the Pegasos method (Shalev-Shwartz et al. 2011), which is based on stochastic gradient descent, is even faster, and the coordinate descent method is highly efficient on sparse data (Hsieh et al. 2008). Theoretically, the time complexity of SVM with non-linear kernels is no less than $O(m^2)$, and hence researchers mainly focus on designing efficient approximation methods, such as the CVM model (Tsang et al. 2006) based on sampling, the Nyström method (Williams and Seeger 2001) based on low-rank approximation, and the random Fourier features method (Rahimi and Recht 2007). Recent studies show that the Nyström method is superior to the random Fourier features method when a large difference between the eigenvalues of the kernel matrices exists (Yang et al. 2012).

$m$ is the number of samples.

Since support vector machines are originally designed for binary classifications, an extension is required for multiclass classifications (Hsu and Lin 2002) and structured output variables (Tsochantaridis et al. 2005). The study of support vector regressions started in (Drucker et al. 1997), and a comprehensive introduction is given in (Smola and Schölkopf 2004).

The choice of kernel functions directly decides the performance of support vector machines and kernel methods. Unfortunately, the kernel function selection is still an unsolved problem. Multiple kernel learning employs multiple kernel functions and aims to find the best convex combination of the kernels as the final kernel function (Bach et al. 2004, Lanckriet et al. 2004), and this is indeed taking advantages of ensemble learning.

See Chap. 8 for ensemble learning.

Surrogate loss functions are widely used in machine learning. However, how do we know if the solution via a surrogate loss function is still the solution to the original problem? This is known as the "consistency" problem in theoretical studies. (Vapnik and Chervonenkis 1991) gave the necessary and sufficient conditions of consistency when using the surrogate loss for empirical risk minimization. (Zhang 2004) proved the consistency of several common convex surrogate loss functions.

There are many software packages implementing SVM, such as LIBSVM (Chang and Lin 2011) and LIBLINEAR (Fan et al. 2008).

## Exercises

LIBSVM is available at
► http://www.csie.ntu.edu.tw/~cjlin/libsvm/.
The watermelon data set 3.0$\alpha$ is in ■ Table 4.5.

UCI data sets are available at
► http://archive.ics.uci.edu/ml/.

6

**6.1** Prove that (6.2) is the distance from every point $\boldsymbol{x}$ in the sample space to the hyperplane $(\boldsymbol{w}, b)$.

**6.2** Run LIBSVM on the watermelon data set 3.0$\alpha$ and build two SVM models using the linear kernel and the Gaussian kernel. Compare their support vectors.

**6.3** Choose two UCI data sets and build two SVM models using the linear kernel and the Gaussian kernel. Compare the performance against BP neural networks and C4.5 decision trees.

**6.4** Discuss under what conditions the linear discriminant analysis and the support vector machine with linear kernel are equivalent.

**6.5** Discuss the relationship between the SVM with the Gaussian kernel and the RBF neural network.

**6.6** Analyze the reason of why the SVM is sensitive to noises.

**6.7** Write down the full KKT conditions of (6.52).

**6.8** Use LIBSVM to train an SVR on the watermelon data set 3.0$\alpha$, where the input is `density` and the output is `sugar`.

**6.9** Use kernel trick to extend the logistic regression to the kernelized logistic regression.

**6.10** * Design a method to significantly reduce the number of support vectors without sacrificing much generalization ability.

## Break Time

### Short Story: Vladimir N. Vapnik—father of the statistical learning theory

Vladimir N. Vapnik (1936–) is an outstanding mathematician, statistician, and computer scientist. He was born in the former Soviet Union and obtained his master's degree in mathematics from the Uzbek State University in 1958. In 1964, he obtained his Ph.D. degree in statistics at the Institute of Control Sciences in Moscow, and then worked there and eventually became the head of the computer science research department. In 1990, one year before the dissolution of the former Soviet Union, he moved to the United States and joined the AT&T Bell Labs in New Jersey. In 1995, he and his colleagues published the seminal paper of SVM. Since neural networks were groundbreaking at that time, the paper was published under the name "Support Vector Networks" as requested by the *Machine Learning* journal.

In fact, Vapnik has already proposed the concept of support vectors in 1963. In 1968, Vapnik and another former Soviet Union mathematician A. Chervonenkis proposed the concept of *VC dimension* named after their surnames. Later on, in 1974, the principle of structural risk minimization was proposed. These works established the statistical learning theory in the 1970s. However, most of these works were published in Russian, and hence relevant research did not gain attention from the western academia until Vapnik moved to the United States. At the end of the twentieth century, the statistical learning theory, support vector machines, and kernel methods became the superstars in machine learning.

In 2002, Vapnik left the AT&T Bell Labs and joined NEC Laboratories in Princeton. In 2014, Vapnik joined Facebook AI Research. He has also held professorship at the University of London and Columbia University since 1995. There is a very well-known quote of Vapnik: "*Nothing more practical than a good theory.*"

Actually, the SVM does have close connections to neural networks: setting the number of hidden neurons to the number of training samples and letting each sample corresponds to a neuron center, then, an RBF network (see Sect. 5.5.1) using Gaussian radial basis function has the same form as the prediction function of SVM with the Gaussian kernel.

## References


Bach, R. R., Lanckriet, G. R. G., and Jordan, M. I. (2004). Multiple kernel learning, conic duality, and the smo algorithm. In *Proceedings of the 21st International Conference on Machine Learning (ICML)*, pages 6–13. Banff, Canada

Boyd S, Vandenberghe L (2004) Convex Optimization. Cambridge University Press, Cambridge, UK

Burges CJC (1998) A tutorial on support vector machines for pattern recognition. Data Mining and Knowledge Discovery 2(1):121–167

Chang C-C, Lin C-J (2011) LIBSVM: A library for support vector machines. ACM Transactions on Intelligent Systems and Technology 2(3):27

Cortes C, Vapnik VN (1995) Support vector networks. Machine Learning 20(3):273–297

Cristianini N, Shawe-Taylor J (2000) An Introduction to Support Vector Machines and Other Kernel-Based Learning Methods. Cambridge University Press, Cambridge, UK

Drucker H, Burges CJC, Kaufman L, Srriola AJ, Vapnik V (1997) Support vector regression machines. In: Mozer MC, Jordan MI, Petsche T (eds) Advances in Neural Information Processing Systems 9 (NIPS). MIT Press, Cambridge, MA, pp 155–161

Fan R-E, Chang K-W, Hsieh C-J, Wang X-R, Lin C-J (2008) LIBLINEAR: A library for large linear classification. Journal of Machine Learning Research 9:1871–1874

Hsieh, C.-J., Chang, K.-W., Lin, C.-J., Keerthi, S. S., and Sundararajan, S. (2008). A dual coordinate descent method for large-scale linear SVM. In *Proceedings of the 25th International Conference on Machine Learning (ICML)*, pages 408–415. Helsinki, Finland

Hsu C-W, Lin C-J (2002) A comparison of methods for multi-class support vector machines. IEEE Transactions on Neural Networks 13(2):415–425

Joachims, T. (1998). Text classification with support vector machines: Learning with many relevant features. In *Proceedings of the 10th European Conference on Machine Learning (ECML)*, pages 137–142. Chemnitz, Germany

Joachims, T. (2006). Training linear SVMs in linear time. In *Proceedings of the 12th ACM SIGKDD International Conference on Knowledge Discovery and Data Mining (KDD)*, pages 217–226. Philadelphia, PA

Lanckriet GRG, Cristianini N, Bartlett MIJP, Ghaoui LE (2004) Learning the kernel matrix with semidefinite programming. Journal of Machine Learning Research 5:27–72

Osuna, E., Freund, R., and Girosi, F. (1997). An improved training algorithm for support vector machines. In *Proceedings of the IEEE Workshop on Neural Networks for Signal Processing (NNSP)*, pages 276–285. Amelia Island, FL

Platt, J. (1998). Sequential minimal optimization: A fast algorithm for training support vector machines. Technical Report MSR-TR-98-14, Microsoft Research

Platt J (2000) Probabilities for SV machines. In: Smola A, Bartlett P, Schölkopf B, Schuurmans D (eds) Advances in Large Margin Classifiers. MIT Press, Cambridge, MA, pp 61–74

Rahimi A, Recht B (2007) Random features for large-scale kernel machines. In: Platt JC, Koller D, Singer Y, Roweis S (eds) Advances in Neural Information Processing Systems 20 (NIPS). MIT Press, Cambridge, MA, pp 1177–1184


6

Schölkopf B, Burges CJC, Smola AJ (eds) (1999) Advances in Kernel Methods: Support Vector Learning. MIT Press, Cambridge, MA

Schölkopf B, Smola AJ (2002) Learning with Kernels: Support Vector Machines, Regularization. MIT Press, Cambridge, MA, Optimization and Beyond

Shalev-Shwartz S, Y. S., Srebro, N., and Cotter, A. (2011) Pegasos: Primal estimated sub-gradient solver for SVM. Mathematical Programming 127(1):3–30

Smola AJ, Schölkopf B (2004) A tutorial on support vector regression. Statistics and Computing 14(3):199–222

Tsang IW, Kwok JT, Cheung P (2006) Core vector machines: Fast SVM training on very large data sets. Journal of Machine Learning Research 6:363–392

Tsochantaridis I, Joachims T, Hofmann T, Altun., Y. (2005) Large margin methods for structured and interdependent output variables. Journal of Machine Learning Research 6:1453–1484

Vapnik VN (1995) The Nature of Statistical Learning Theory. Springer, New York, NY

Vapnik VN (1998) Statistical Learning Theory. Wiley, New York, NY

Vapnik VN (1999) An overview of statistical learning theory. IEEE Transactions on Neural Networks 10(5):988–999

Vapnik VN, Chervonenkis AJ (1991) The necessary and sufficient conditions for consistency of the method of empirical risk. Pattern Recognition and Image Analysis 1(3):284–305

Williams CK, Seeger M (2001) Using the nyström method to speed up kernel machines. In: Leen TK, Dietterich TG, Tresp V (eds) Advances in Neural Information Processing Systems 13 (NIPS). MIT Press, Cambridge, MA, pp 682–688

Yang T-B, Li Y-F, Mahdavi M, Jin R, Zhou Z-H (2012) Nyström method vs random fourier features: A theoretical and empirical comparison. In: Bartlett P, Pereira FCN, Burges CJC, Bottou L, Weinberger KQ (eds) Advances in Neural Information Processing Systems 25 (NIPS). MIT Press, Cambridge, MA, pp 485–493

Zhang T (2004) Statistical behavior and consistency of classification methods based on convex risk minimization (with discussion). Annals of Statistics 32(5):56–85

# Bayes Classifiers

Table of Contents




Z.-H. Zhou, *Machine Learning*,
https://doi.org/10.1007/978-981-15-1967-3_7

## 7.1 Bayesian Decision Theory

Bayesian decision theory is a fundamental decision-making approach under the probability framework. In an ideal situation when all relevant probabilities were known, Bayesian decision theory makes optimal classification decisions based on the probabilities and costs of misclassifications. In the following, we demonstrate the basic idea of Bayesian decision theory with multiclass classification.

Let us assume that there are $N$ distinct class labels, that is, $\mathcal{Y} = \{c_1, c_2, \ldots, c_N\}$. Let $\lambda_{ij}$ denote the cost of misclassifying a sample of class $c_j$ as class $c_i$. Then, we can use the posterior probability $P(c_i \mid \boldsymbol{x})$ to calculate the *expected loss* of classifying a sample $\boldsymbol{x}$ as class $c_i$, that is, the *conditional risk* of the sample $\boldsymbol{x}$:

In decision theory, the *expected loss* is known as the *risk*.



$$R(c_i \mid \boldsymbol{x}) = \sum_{j=1}^{N} \lambda_{ij} P(c_j \mid \boldsymbol{x}). \tag{7.1}$$

Our task is to find a decision rule $h : \mathcal{X} \mapsto \mathcal{Y}$ that minimizes the overall risk

$$R(h) = \mathbb{E}_{\boldsymbol{x}} \left[ R(h(\boldsymbol{x}) \mid \boldsymbol{x}) \right]. \tag{7.2}$$

The overall risk $R(h)$ is minimized when the conditional risk $R(h(\boldsymbol{x}) \mid \boldsymbol{x})$ of each sample $\boldsymbol{x}$ is minimized. This leads to the Bayes decision rule: to minimize the overall risk, classify each sample as the class that minimizes the conditional risk $R(c \mid \boldsymbol{x})$:

$$h^*(\boldsymbol{x}) = \underset{c \in \mathcal{Y}}{\arg\min}\, R(c \mid \boldsymbol{x}), \tag{7.3}$$

where $h^*$ is called the *Bayes optimal classifier*, and its associated overall risk $R(h^*)$ is called the *Bayes risk*. $1 - R(h^*)$ is the best performance that can be achieved by any classifiers, that is, the theoretically achievable upper bound of accuracy for any machine learning models.

To be specific, if the objective is to minimize the misclassification rate, then the misclassification loss $\lambda_{ij}$ can be written as

The misclassification rate corresponds to the 0/1 loss function in Chap. 6.

$$\lambda_{ij} = \begin{cases} 0, & \text{if } i = j; \\ 1, & \text{otherwise}, \end{cases} \tag{7.4}$$

and the conditional risk is

$$R(c \mid \boldsymbol{x}) = 1 - P(c \mid \boldsymbol{x}), \tag{7.5}$$

Then, the Bayes optimal classifier that minimizes the misclassification rate is

$$h^*(\boldsymbol{x}) = \underset{c \in \mathcal{Y}}{\arg\max}\, P(c \mid \boldsymbol{x}), \tag{7.6}$$

which classifies each sample $\boldsymbol{x}$ as the class that maximizes its posterior probability $P(c \mid \boldsymbol{x})$.

From (7.6), we can see that the Bayes decision rule relies on the posterior probability $P(c \mid \boldsymbol{x})$, which is often difficult to obtain. From this perspective, the task of machine learning is then to accurately estimate the posterior probability $P(c \mid \boldsymbol{x})$ from the finite training samples. Generally speaking, there are two strategies. The first strategy is to predict $c$ by estimating $P(c \mid \boldsymbol{x})$ directly, and the corresponding models are known as *discriminative models*. The second strategy is to estimate the joint probability $P(\boldsymbol{x}, c)$ first and then estimate $P(c \mid \boldsymbol{x})$, and the corresponding models are called *generative models*. The models introduced in earlier Chapters are all discriminative models, including decision trees, BP neural networks, and support vector machines. For generative models, we must evaluate

Note that this is just for the understanding of machine learning under the probabilistic framework. In fact, many machine learning techniques can perform accurate classifications without estimating the posterior probabilities.

$$P(c \mid \boldsymbol{x}) = \frac{P(\boldsymbol{x}, c)}{P(\boldsymbol{x})}. \tag{7.7}$$

According to Bayes' theorem, $P(c \mid \boldsymbol{x})$ can be written as

$$P(c \mid \boldsymbol{x}) = \frac{P(c)P(\boldsymbol{x} \mid c)}{P(\boldsymbol{x})}, \tag{7.8}$$

where $P(c)$ is the *prior* probability of $c$, $P(\boldsymbol{x} \mid c)$ is the *class-conditional probability*, also known as the *likelihood*, of the sample $\boldsymbol{x}$ with respect to class $c$, and $P(\boldsymbol{x})$ is the *evidence* factor for normalization. Given $\boldsymbol{x}$, the evidence factor $P(\boldsymbol{x})$ is independent of the class, and thus the estimation of $P(c \mid \boldsymbol{x})$ is transformed to estimating the prior $P(c)$ and the likelihood $P(\boldsymbol{x} \mid c)$ from the training set $D$.

$P(\boldsymbol{x})$ is the same for all classes.

The prior probability $P(c)$ represents the proportion of each class in the sample space. Based on the law of large numbers, $P(c)$ can be estimated by the frequency of each class in the training set as long as there are sufficient *i.i.d.* samples.

For ease of discussion, we assume all features to be discrete. For continuous features, we can replace the probability mass function $P(\cdot)$ by the probability density function $p(\cdot)$.

It is difficult to calculate the class-conditional probability $P(\boldsymbol{x} \mid c)$ directly since the calculation involves the joint probability of all features of $\boldsymbol{x}$. For example, suppose that there are $d$ binary features in each sample, then there are $2^d$ possible combinations in the sample space. In practice, the number of training samples $m$ is usually much smaller than $2^d$, and therefore, many combinations have never appeared in the training set. Hence, estimating $P(\boldsymbol{x} \mid c)$ directly by the frequencies in the

See Sect. 7.3.

training set is infeasible since "unobserved" and "zero probability" are generally different.

## 7.2 Maximum Likelihood Estimation

A general strategy of estimating the class-conditional probability is to hypothesize a fixed form of probability distribution, and then estimate the distribution parameters using the training samples. To be specific, let $P(\boldsymbol{x} \mid c)$ denote class-conditional probability of class $c$, and suppose $P(\boldsymbol{x} \mid c)$ has a fixed form determined by a parameter vector $\boldsymbol{\theta}_c$. Then, the task is to estimate $\boldsymbol{\theta}_c$ from a training set $D$. To be precise, we write $P(\boldsymbol{x} \mid c)$ as $P(\boldsymbol{x} \mid \boldsymbol{\theta}_c)$.

$p(\boldsymbol{x} \mid c)$ for continuous distribution.

The training process of probabilistic models is the process of parameter estimation. There are two different ways of thinking about parameters. On the one hand, the Bayesian school thinks that the parameters are unobserved random variables following some distribution, and hence we can assume prior distributions for the parameters and estimate posterior distribution from observed data. On the other hand, the Frequentist school supports the view that parameters have fixed values though they are unknown, and hence they can be determined by some approaches such as optimizing the likelihood function. The remaining of this section discusses the Maximum Likelihood Estimation (MLE), which comes from the Frequentist school and is a classic method of estimating the probability distribution from samples.

The ongoing debate between the Frequentist school and the Bayesian school started around the 1920s and the 1930s. These two schools have different views on many important questions and even the interpretation of probability itself. Readers who are interested in this can find more information in Efron (2005), Samaniego (2010).

Also known as the Maximum Likelihood Method.

Let $D_c$ denote the set of class $c$ samples in the training set $D$, and further suppose the samples are *i.i.d.* samples. Then, the likelihood of $D_c$ for a given parameter $\boldsymbol{\theta}_c$ is

$$P(D_c \mid \boldsymbol{\theta}_c) = \prod_{\boldsymbol{x} \in D_c} P(\boldsymbol{x} \mid \boldsymbol{\theta}_c). \tag{7.9}$$

Applying the MLE to $\boldsymbol{\theta}_c$ is about finding a parameter value $\hat{\boldsymbol{\theta}}_c$ that maximizes the likelihood $P(D_c \mid \boldsymbol{\theta}_c)$. Intuitively, the MLE aims to find a value of $\boldsymbol{\theta}_c$ that maximizes the "likelihood" that the data will present.

Since the product of a sequence in (7.9) can easily lead to underflow, we often use the *log-likelihood* instead:

$$\begin{aligned} LL(\boldsymbol{\theta}_c) &= \log P(D_c \mid \boldsymbol{\theta}_c) \\ &= \sum_{\boldsymbol{x} \in D_c} \log P(\boldsymbol{x} \mid \boldsymbol{\theta}_c), \end{aligned} \tag{7.10}$$

and the MLE of $\boldsymbol{\theta}_c$ is

$$\hat{\boldsymbol{\theta}}_c = \arg\max_{\boldsymbol{\theta}_c} LL(\boldsymbol{\theta}_c). \tag{7.11}$$

For example, suppose the features are continuous and the probability density function follows the Gaussian distribution $p(\boldsymbol{x} \mid c) \sim \mathcal{N}(\boldsymbol{\mu}_c, \boldsymbol{\sigma}_c^2)$, then the MLE of the parameters $\boldsymbol{\mu}_c$ and $\boldsymbol{\sigma}_c^2$ are

$\mathcal{N}$ is the Gaussian distribution. See Appendix C.1.7.

$$\hat{\boldsymbol{\mu}}_c = \frac{1}{|D_c|} \sum_{\boldsymbol{x} \in D_c} \boldsymbol{x}, \tag{7.12}$$

$$\hat{\boldsymbol{\sigma}}_c^2 = \frac{1}{|D_c|} \sum_{\boldsymbol{x} \in D_c} (\boldsymbol{x} - \hat{\boldsymbol{\mu}}_c)(\boldsymbol{x} - \hat{\boldsymbol{\mu}}_c)^\top. \tag{7.13}$$

In other words, the estimated mean of Gaussian distribution obtained by the MLE is the sample mean, and the estimated variance is the mean of $(\boldsymbol{x} - \hat{\boldsymbol{\mu}}_c)(\boldsymbol{x} - \hat{\boldsymbol{\mu}}_c)^\top$; this conforms to our intuition. Conditional probabilities can also be estimated similarly for discrete features.

Such kind of parametric methods simplify the estimation of posterior probabilities, but the accuracy of estimation heavily relies on whether the hypothetical probability distribution matches the unknown ground-truth data distribution. In practice, a "guessed" probability distribution could lead to misleading results, and hence we often need domain knowledge to help hypothesize a good approximation to the ground-truth data distribution.

## 7.3 Naïve Bayes Classifier

When estimating the posterior probability $P(c \mid \boldsymbol{x})$ with the Bayes rule (7.8), there is a difficulty: it is not easy to calculate the class-conditional probability $P(\boldsymbol{x} \mid c)$ from the finite training samples since $P(\boldsymbol{x} \mid c)$ is the joint probability on all attributes. To avoid this, the naïve Bayes classifier makes the "attribute conditional independence assumption": given any known class, assume all attributes are independent of each other. In other words, we assume each attribute influences the prediction result independently.

Computing the joint probability with finite training samples suffers from the combinatorial explosion problem in computation and the data sparsity problem in data. The more attributes there are, the severer the problems are.

With the attribute conditional independence assumption, we rewrite (7.8) as

$$P(c \mid \boldsymbol{x}) = \frac{P(c)P(\boldsymbol{x} \mid c)}{P(\boldsymbol{x})} = \frac{P(c)}{P(\boldsymbol{x})} \prod_{i=1}^{d} P(x_i \mid c), \tag{7.14}$$

where $d$ is the number of attributes and $x_i$ is the value taken on the $i$th attribute of $\boldsymbol{x}$

Strictly speaking, $x_i$ is an "attribute-value" pair, e.g., color=green. For ease of discussion, when the context is clear, $x_i$ may either represent the $i$th attribute (e.g., color) or the value of $\boldsymbol{x}$ on the $i$th attribute (e.g., green).

Since $P(\boldsymbol{x})$ is the same for all classes, from the Bayes decision rule (7.6), we have

$$h_{\text{nb}}(\boldsymbol{x}) = \underset{c \in \mathcal{Y}}{\arg\max}\, P(c) \prod_{i=1}^{d} P(x_i \mid c), \tag{7.15}$$

which is the formulation of the naïve Bayes classifier.

To train a naïve Bayes classifier, we compute the prior probability $P(c)$ from the training set $D$ and then compute the conditional probability $P(x_i \mid c)$ for each attribute.

Let $D_c$ denote a subset of $D$ containing all samples of class $c$. Then, given sufficient *i.i.d.* samples, the prior probability can be easily estimated by



$$P(c) = \frac{|D_c|}{|D|}. \tag{7.16}$$

For discrete attributes, let $D_{c,x_i}$ denote a subset of $D_c$ containing all samples taking the value $x_i$ on the $i$th attribute. Then, the conditional probability $P(x_i \mid c)$ can be estimated by

$$P(x_i \mid c) = \frac{|D_{c,x_i}|}{|D_c|}. \tag{7.17}$$

For continuous features, suppose $p(x_i \mid c) \sim \mathcal{N}(\mu_{c,i}, \sigma_{c,i}^2)$, where $\mu_{c,i}$ and $\sigma_{c,i}^2$ are, respectively, the mean and variance of the $i$th feature of class $c$ samples. Then, we have

$$p(x_i \mid c) = \frac{1}{\sqrt{2\pi}\sigma_{c,i}} \exp\left(-\frac{(x_i - \mu_{c,i})^2}{2\sigma_{c,i}^2}\right). \tag{7.18}$$

The watermelon data set 3.0 is in ▪ Table 4.3.

Now, let us train a naïve Bayes classifier using the watermelon data set 3.0 and classify the following watermelon T1:

| ID | color | root | sound | texture | umbilicus | surface | density | sugar | ripe |
|---|---|---|---|---|---|---|---|---|---|
| T1 | green | curly | muffled | clear | hollow | hard | 0.697 | 0.460 | ? |

First, we estimate the prior probability $P(c)$:

$$P(\text{ripe} = \text{true}) = \frac{8}{17} \approx 0.471,$$

$$P(\text{ripe} = \text{false}) = \frac{9}{17} \approx 0.529.$$

Then, we estimate the conditional probability of each feature $P(x_i \mid c)$:

The toy watermelon data set 3.0 is for demonstration purposes. In practice, we need sufficient samples to make meaningful probability estimations.

$$P_{\text{green}|\text{true}} = P(\text{color} = \text{green} \mid \text{ripe} = \text{true}) = \frac{3}{8} = 0.375,$$

$$P_{\text{green}|\text{false}} = P(\text{color} = \text{green} \mid \text{ripe} = \text{false}) = \frac{3}{9} \approx 0.333,$$

$$P_{\text{curly}|\text{true}} = P(\text{root} = \text{curly} \mid \text{ripe} = \text{true}) = \frac{5}{8} = 0.625,$$

$$P_{\text{curly}|\text{false}} = P(\text{root} = \text{curly} \mid \text{ripe} = \text{false}) = \frac{3}{9} \approx 0.333,$$

$$P_{\text{muffled}|\text{true}} = P(\text{sound} = \text{muffled} \mid \text{ripe} = \text{true}) = \frac{5}{8} = 0.625,$$

$$P_{\text{muffled}|\text{false}} = P(\text{sound} = \text{muffled} \mid \text{ripe} = \text{false}) = \frac{4}{9} \approx 0.444,$$

$$P_{\text{clear}|\text{true}} = P(\text{texture} = \text{clear} \mid \text{ripe} = \text{true}) = \frac{7}{8} = 0.875,$$

$$P_{\text{clear}|\text{false}} = P(\text{texture} = \text{clear} \mid \text{ripe} = \text{false}) = \frac{2}{9} \approx 0.222,$$

$$P_{\text{hollow}|\text{true}} = P(\text{umbilicus} = \text{hollow} \mid \text{ripe} = \text{true}) = \frac{5}{8} = 0.625,$$

$$P_{\text{hollow}|\text{false}} = P(\text{umbilicus} = \text{hollow} \mid \text{ripe} = \text{false}) = \frac{2}{9} \approx 0.222,$$

$$P_{\text{hard}|\text{true}} = P(\text{surface} = \text{hard} \mid \text{ripe} = \text{true}) = \frac{6}{8} = 0.750,$$

$$P_{\text{hard}|\text{false}} = P(\text{surface} = \text{hard} \mid \text{ripe} = \text{false}) = \frac{6}{9} \approx 0.667,$$

$$\begin{aligned} p_{\text{density}:0.697|\text{true}} &= p(\text{density} = 0.697 \mid \text{ripe} = \text{true}) \\ &= \frac{1}{\sqrt{2\pi} \cdot 0.129} \exp\left(-\frac{(0.697 - 0.574)^2}{2 \cdot 0.129^2}\right) \approx 1.959, \end{aligned}$$

$$\begin{aligned} p_{\text{density}:0.697|\text{false}} &= p(\text{density} = 0.697 \mid \text{ripe} = \text{false}) \\ &= \frac{1}{\sqrt{2\pi} \cdot 0.195} \exp\left(-\frac{(0.697 - 0.496)^2}{2 \cdot 0.195^2}\right) \approx 1.203, \end{aligned}$$

$$\begin{aligned} p_{\text{sugar}:0.460|\text{true}} &= p(\text{density} = 0.460 \mid \text{ripe} = \text{true}) \\ &= \frac{1}{\sqrt{2\pi} \cdot 0.101} \exp\left(-\frac{(0.460 - 0.279)^2}{2 \cdot 0.101^2}\right) \approx 0.788, \end{aligned}$$

$$\begin{aligned} p_{\text{sugar}:0.460|\text{false}} &= p(\text{density} = 0.460 \mid \text{ripe} = \text{false}) \\ &= \frac{1}{\sqrt{2\pi} \cdot 0.108} \exp\left(-\frac{(0.460 - 0.154)^2}{2 \cdot 0.108^2}\right) \approx 0.066. \end{aligned}$$

Hence, we have

$$\begin{aligned} P(\text{ripe} = \text{true}) &\times P_{\text{green}|\text{true}} \times P_{\text{curly}|\text{true}} \times P_{\text{muffled}|\text{true}} \\ &\times P_{\text{clear}|\text{true}} \times P_{\text{hollow}|\text{true}} \times P_{\text{hard}|\text{true}} \\ &\times p_{\text{density}:0.697|\text{true}} \times p_{\text{sugar}:0.460|\text{true}} \approx 0.052, \\ P(\text{ripe} = \text{false}) &\times P_{\text{green}|\text{false}} \times P_{\text{curly}|\text{false}} \times P_{\text{muffled}|\text{false}} \\ &\times P_{\text{clear}|\text{false}} \times P_{\text{hollow}|\text{false}} \times P_{\text{hard}|\text{false}} \\ &\times p_{\text{density}:0.697|\text{false}} \times p_{\text{sugar}:0.460|\text{false}} \approx 6.80 \times 10^{-5}. \end{aligned}$$

In practice, we often use logarithms to convert the multiplications into additions to avoid numerical underflow.

Since $0.052 > 6.80 \times 10^{-5}$, the naïve Bayes classifier classifies the testing sample T1 as ripe.

What if a feature value has never appeared together with a particular class? In such cases, it becomes problematic to use the probability (7.17) for predicting the class with (7.15). For instance, given a testing sample with sound=crisp, the naïve Bayes classifier trained on the watermelon data set 3.0 will predict

$$P_{\text{crisp}|\text{true}} = P(\text{sound} = \text{crisp} \mid \text{ripe} = \text{true}) = \frac{0}{8} = 0.$$

Since the product of the sequence in (7.15) gives a probability of zero, the classification result will always be ripe=false regardless of the values of other features, even if it is obviously a ripe watermelon. Such a behavior is unreasonable.



To avoid "removing" the information carried by other features, the probability estimation requires "smoothing", and a common choice is the Laplacian correction. To be specific, let $N$ denote the number of distinct classes in the training set $D$, $N_i$ denote the number of distinct values the $i$th feature can take. Then, (7.16) and (7.17) can be, respectively, corrected as

$$\hat{P}(c) = \frac{|D_c| + 1}{|D| + N}, \tag{7.19}$$

$$\hat{P}(x_i \mid c) = \frac{|D_{c,x_i}| + 1}{|D_c| + N_i}. \tag{7.20}$$

Taking the watermelon data set 3.0 as an example, the prior probabilities can be estimated as

$$\hat{P}(\text{ripe} = \text{true}) = \frac{8+1}{17+2} \approx 0.474.$$

$$\hat{P}(\text{ripe} = \text{false}) = \frac{9+1}{17+2} \approx 0.526.$$

Similarly, $P_{\text{green}|\text{true}}$ and $P_{\text{green}|\text{false}}$ can be estimated as

$$\hat{P}_{\text{green}|\text{true}} = \hat{P}(\text{color} = \text{green} \mid \text{ripe} = \text{true}) = \frac{3+1}{8+3} \approx 0.364,$$

$$\hat{P}_{\text{green}|\text{false}} = \hat{P}(\text{color} = \text{green} \mid \text{ripe} = \text{false}) = \frac{3+1}{9+3} \approx 0.333.$$

Also, the probability $P_{\text{crisp}|\text{true}}$, which was zero, is now estimated as

$$\hat{P}_{\texttt{crisp}|\texttt{true}} = \hat{P}(\texttt{crisp} = \texttt{true} \mid \texttt{ripe} = \texttt{true}) = \frac{0+1}{8+3} \approx 0.091.$$

The Laplacian correction avoids the problem of zero probabilities caused by insufficient training samples. The prior introduced by the correction will become negligible as the size of training set increases.

The Laplacian correction essentially assumes that the feature values and classes are evenly distributed. This is an extra prior on the data introduced in the process of naïve Bayesian learning.

There are different ways of using a naïve Bayes classifier in practice. For example, if the speed of prediction is important, then a naïve Bayes classifier can pre-calculate all relevant probabilities and save for later use. After that, the prediction can be made by looking up the saved probability table. On the other hand, if the training data changes frequently, then we can take a *lazy learning* approach, in which the probabilities are estimated once a prediction request is received, that is, no training before prediction. If we keep receiving new training samples, then we can enable incremental learning by updating only the probabilities that are related to the new samples.

See Sect. 10.1 for lazy learning.

See Sect. 5.5.2 for incremental learning.

## 7.4 Semi-Naïve Bayes Classifier

To overcome the difficulty of computing $P(c \mid \boldsymbol{x})$ in Bayes' theorem, the naïve Bayes classifier makes the attribute conditional independence assumption, which often does not hold in practice. Hence, semi-naïve Bayes classifiers are developed to relax this assumption to some extent.

The basic idea of semi-naïve Bayes classifiers is to consider some strong dependencies among features without calculating the complete joint probabilities. A general strategy used by semi-naïve Bayes classifiers is One-Dependent Estimator (ODE). As the name suggests, "one-dependent" means each feature can depend on at most one feature other than the class information, that is,

$$P(c \mid \boldsymbol{x}) \propto P(c) \prod_{i=1}^{d} P(x_i \mid c, pa_i), \tag{7.21}$$

where $x_i$ depends on $pa_i$, and $pa_i$ is called the parent feature of $x_i$. For each feature $x_i$, if the parent feature $pa_i$ is known, then $P(x_i \mid c, pa_i)$ can be estimated in a similar manner as in (7.20). Hence, the key problem becomes determining the parent features, and different approaches will lead to different one-dependent classifiers.

The most straightforward approach is called Super-Parent ODE (SPODE), which assumes that all features depend on just one feature called the *super-parent*. SPODE selects the super-

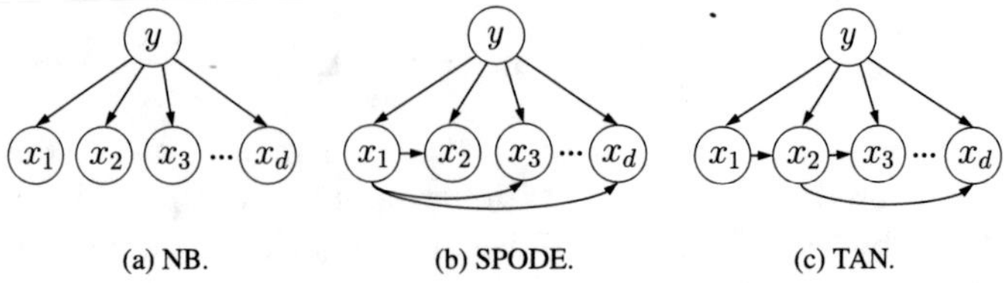

**Fig. 7.1** Feature dependencies of naïve Bayes and semi-naïve Bayes classifiers

parent feature using model selection methods such as cross-validation. For example, in Figure 7.1b, $x_1$ is a super-parent feature.

Tree Augmented naïve Bayes (TAN) (Friedman et al. 1997), which is based on maximum weighted spanning trees (Chow and Liu 1968), simplifies feature dependencies into a tree structure, as shown in Figure 7.1c, by the following steps:



(1) Compute the conditional mutual information for each pair of features

$$I(x_i, x_j \mid y) = \sum_{x_i, x_j; c \in \mathcal{Y}} P(x_i, x_j \mid c) \log \frac{P(x_i, x_j \mid c)}{P(x_i \mid c)P(x_j \mid c)}; \tag{7.22}$$

(2) Construct a complete undirected graph in which the nodes are features. Set $I(x_i, x_j \mid y)$ as the weight of the edge between $x_i$ and $x_j$;

(3) Construct a maximum weighted spanning tree and select a root feature. Set the direction of each edge outward from the root feature;

(4) Add a class node $y$ and add direct edges from $y$ to other feature nodes.

The mutual information $I(x_i, x_j \mid y)$ describes the correlation between $x_i$ and $x_j$ given the class information. Hence, via the maximum weighted spanning tree algorithm, TAN keeps only the dependencies among highly correlated features.

Averaged One-Dependent Estimator (AODE) (Webb et al. 2005) is a more powerful one-dependent classifier, which takes advantage of ensemble learning. Unlike SPODE, which selects a super-parent feature via model selection, AODE tries to use each feature as a super-parent to build multiple SPODE models and then integrates those supported by sufficient training data, that is,

See Chap. 8 for ensemble learning.

$$P(c \mid \boldsymbol{x}) \propto \sum_{\substack{i=1 \\ |D_{x_i}| \geqslant m'}}^{d} P(c, x_i) \prod_{j=1}^{d} P(x_j \mid c, x_i), \tag{7.23}$$

where $D_{x_i}$ is the subset of samples taking the value $x_i$ on the $i$th feature, and $m'$ is a threshold constant. AODE needs to estimate $P(c, x_i)$ and $P(x_j \mid c, x_i)$. Similar to (7.20), we have

By default, $m'$ is set to 30 (Webb et al. 2005).

$$\hat{P}(c, x_i) = \frac{|D_{c,x_i}| + 1}{|D| + N \times N_i}, \tag{7.24}$$

$$\hat{P}(x_j \mid c, x_i) = \frac{|D_{c,x_i,x_j}| + 1}{|D_{c,x_i}| + N_j}, \tag{7.25}$$

where $N$ is the number of distinct classes in $D$, $N_i$ is the number of distinct values the $i$th feature can take, $D_{c,x_i}$ is the subset of class $c$ samples taking the value $x_i$ on the $i$th feature, and $D_{c,x_i,x_j}$ is the subset of class $c$ samples taking the value $x_i$ on the $i$th feature while taking the value $x_j$ on the $j$th feature. For example, for the watermelon data set 3.0, we have

$$\begin{aligned}
\hat{P}_{\texttt{true,muffled}} &= \hat{P}(\texttt{ripe} = \texttt{true}, \texttt{sound} = \texttt{muffled}) \\
&= \frac{6+1}{17+3\times 2} = 0.304, \\
\hat{P}_{\texttt{hollow|true,muffled}} &= \hat{P}(\texttt{umbilicus} = \texttt{hollow} \mid \texttt{ripe} = \texttt{true}, \\
&\quad \texttt{sound} = \texttt{muffled}) \\
&= \frac{3+1}{6+3} = 0.444.
\end{aligned}$$

Similar to naïve Bayes classifiers, the training process of AODE is also about "counting" the number of training samples satisfying some conditions. AODE does not require model selection, and hence it supports quick predictions with pre-calculations, lazy learning, and incremental learning.

Since relaxing the attribute conditional independence assumption to one-dependent assumption leads to better generalization ability, is it possible to make further improvement by considering higher order dependencies? In other words, we extend ODE to the $k$DE by replacing the feature $pa_i$ in (7.21) with a feature set $\mathbf{pa}_i$ containing $k$ features. It is worth mentioning that the number of samples required for an accurate estimation of $P(x_i \mid y, \mathbf{pa}_i)$ increases exponentially as $k$ increases. Hence, improving the generalization performance requires abundant data; otherwise it would be difficult to calculate higher order joint probabilities.

"Higher order dependency" refers to dependency among multiple features.

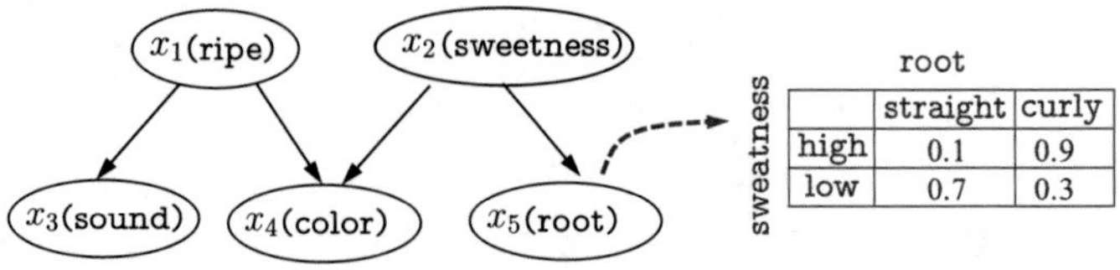


**Fig. 7.2** The Bayesian network structure of the watermelon problem together with the conditional probability table of root

## 7.5 Bayesian Network

*Bayesian networks*, also known as *belief networks*, utilize Directed Acyclic Graphs (DAG) to represent dependencies among features and build Conditional Probability Tables (CPT) to describe the joint probabilities of features.

The Bayesian network is a classic probabilistic graphical model. See Chap. 14 for probabilistic graphical models.

For ease of discussion, this section assumes discrete features. For continuous features, the CPT can be generalized to probability density functions.

To be specific, a Bayesian network $B$ consists of structure $G$ and parameter $\Theta$, denoted by $B = \langle G, \Theta \rangle$. The structure $G$ is a DAG in which each node corresponds to a feature. In the graph, two features with a direct dependency are connected by an edge, and, collectively, these dependencies are encoded by the parameter $\Theta$. Let $\pi_i$ denote the parent node set of the feature $x_i$ in $G$, then $\Theta$ includes the CPT $\theta_{x_i|\pi_i} = P_B(x_i \mid \pi_i)$ of each feature.

For example, ◘ Figure 7.2 shows a Bayesian network structure and the CPT of the feature root. From the network structure, we see that color directly depends on ripe and sweetness, while root directly depends on sweetness. From the CPT, we can further obtain a quantified dependency of root on sweetness, e.g., $P(\text{root} = \text{straight} \mid \text{sweetness} = \text{high}) = 0.1$.

The continuous feature sugar in the watermelon data set is converted to the discrete feature sweetness.

### 7.5.1 Network Structure

The Bayesian network structure effectively expresses the conditional independence among features. Given the sets of parent nodes, the Bayesian network assumes each feature is independent of its all non-descendant features. Hence, $B = \langle G, \Theta \rangle$ defines the joint probability of the features $x_1, x_2, \ldots, x_d$ as

$$P_B(x_1, x_2, \ldots, x_d) = \prod_{i=1}^{d} P_B(x_i \mid \pi_i) = \prod_{i=1}^{d} \theta_{x_i|\pi_i}. \tag{7.26}$$

Taking ◘ Figure 7.2 as an example, the joint probability is defined as

$$P(x_1, x_2, x_3, x_4, x_5) = P(x_1)P(x_2)P(x_3 \mid x_1)P(x_4 \mid x_1, x_2)P(x_5 \mid x_2).$$

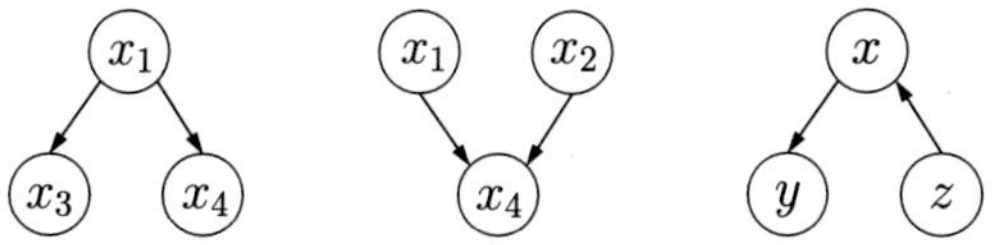

(a) Common parent. (b) V-structure. (c) Cascade.

**Fig. 7.3** Typical feature dependencies in Bayesian networks

We can see that $x_3$ and $x_4$ are independent given the value of $x_1$ (i.e., $x_3 \perp x_4 \mid x_1$). Also, $x_4$ and $x_5$ are independent given the value of $x_2$ (i.e., $x_4 \perp x_5 \mid x_2$).

▫ Figure 7.3 shows three typical cases of dependencies among three variables, and two of them have already appeared in (7.26).

Not all conditional independence relationships are enumerated here.

In the common parent structure, $x_3$ and $x_4$ are conditionally independent given the value of their parent node $x_1$. In the cascade structure, $y$ and $z$ are conditionally independent given the value of $x$. In the V-structure, $x_1$ and $x_2$ are dependent given the value of $x_4$. Interestingly, when the value of $x_4$ is unknown in the V-structure, $x_1$ and $x_2$ are independent of each other. We can make a simple validation as follows:

$$\begin{aligned} P(x_1, x_2) &= \sum_{x_4} P(x_1, x_2, x_4) \\ &= \sum_{x_4} P(x_4 \mid x_1, x_2) P(x_1) P(x_2) \\ &= P(x_1) P(x_2). \end{aligned} \tag{7.27}$$

Such independence is called *marginal independence*, denoted by $x_1 \perp\!\!\!\perp x_2$.

Calculating the integration or summation of a variable is called *marginalization*.

In the V-structure, the independence between two variables is subject to the presence of a third variable. However, this may also be the case in other structures. For example, the conditional independence $x_3 \perp x_4 \mid x_1$ exists in the common parent structure, but $x_3$ and $x_4$ are not independent when the value of $x_1$ is unknown, that is, $x_3 \perp\!\!\!\perp x_4$ does not hold. In the cascade structure, $y \perp z \mid x$, but $y \perp\!\!\!\perp z$ does not hold.

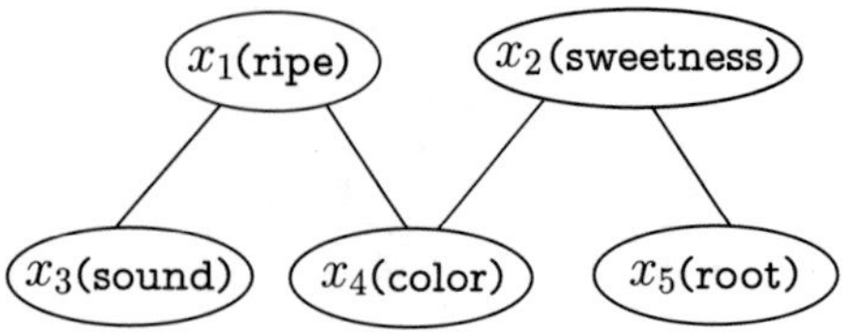


**Fig. 7.4** The moral graph of Fig. 7.2

"d" stands for "directed".

The discovery of common parent structure, cascade structure, and V-structure together with the proposal of d-separation have promoted the research on causal discovery. See Pearl (1988).

We usually prune the graph first and keep only $x$, $y$, $\mathbf{z}$, and all of their ancestor variables.

We can use *d-separation* to analyze the conditional independence among variables in a directed graph. First, we convert the directed graph into an undirected graph:

- Find all V-structures in the directed graph and then add an undirected edge between the common parents if they share the same child node;
- Change all directed edges to undirected edges.

The undirected graph produced by this process is called a *moral graph*, and the process of connecting common parents is called *moralization* (Cowell et al. 1999).

With a moral graph, the conditional independence between variables can be intuitively and efficiently located. Suppose we have a moral graph containing the variables $x$ and $y$ and a variable set $\mathbf{z} = \{z_i\}$. We say $x$ and $y$ are d-separated by $\mathbf{z}$ (i.e., $x \perp y \mid \mathbf{z}$) if $x$ and $y$ can be separated by $\mathbf{z}$, that is, $x$ and $y$ belong to separated subgraphs after dropping $\mathbf{z}$. For example, the moral graph corresponding to ◘ Figure 7.2 is given in ◘ Figure 7.4, from which we can easily find all conditional dependencies: $x_3 \perp x_4 \mid x_1$, $x_4 \perp x_5 \mid x_2$, $x_3 \perp x_2 \mid x_1$, $x_3 \perp x_5 \mid x_1$, $x_3 \perp x_5 \mid x_2$, and so on.

### 7.5.2 Learning

When the network structure is known (i.e., the dependencies between variables are given), learning a Bayesian network is relatively easy as we only need to count the training samples to estimate the CPT of each node. In practice, the network structure is rarely known in advance, and hence the primary task of Bayesian network learning is to find the most "appropriate" Bayesian network. A common method to solve this problem is *score-and-search*. To be specific, we first define a score function to evaluate the fitness between a Bayesian network and a training set, and then use this score function to find the most optimal Bayesian network. The score function, which can be designed in different ways, carries the inductive bias of a Bayesian network.

See Sect. 1.4 for inductive bias.

Typical score functions are based on the information theory principle, which considers the learning problem as a data compression task. Then, the task is to find a model that can describe the training data with the minimum coding length. The coding length includes the bit length required to describe the model itself and the bit length required by the model to describe the data. For Bayesian network learning, the model is a Bayesian network that describes a probability distribution of the training data. According to the Minimum Description Length (MDL) principle, we search for the Bayesian network with the shortest combined code length.

Given a training set $D = \{\boldsymbol{x}_1, \boldsymbol{x}_2, \ldots, \boldsymbol{x}_m\}$, the score function of a Bayesian network $B = \langle G, \Theta \rangle$ on $D$ can be written as

Here, the class is also considered as a feature, that is, $\boldsymbol{x}_i$ is a vector containing a sample and its class.

$$s(B \mid D) = f(\theta)\,|B| - LL(B \mid D), \tag{7.28}$$

where $|B|$ is the number of parameters of the Bayesian network, $f(\theta)$ is the coding length required to describe the parameter $\theta$, and the log-likelihood of the Bayesian network $B$ is

$$LL(B \mid D) = \sum_{i=1}^{m} \log P_B(\boldsymbol{x}_i). \tag{7.29}$$

In (7.28), the first term is about the coding length required for describing the Bayesian network, and the second term is about how well the probability distribution $P_B$ describes the data $D$. Hence, the learning task becomes an optimization task, that is, searching for a Bayesian network $B$ that minimizes the score function $s(B \mid D)$.

In the view of statistical learning, these two terms correspond to the structural risk and the empirical risk, respectively.

If $f(\theta) = 1$, that is, each parameter is described by 1 coding bit, then we have the Akaike Information Criterion (AIC) score function

$$\mathrm{AIC}(B \mid D) = |B| - LL(B \mid D). \tag{7.30}$$

If $f(\theta) = \frac{1}{2}\log m$, that is, each parameter is described by $\frac{1}{2}\log m$ coding bits, then we have the Bayesian Information Criterion (BIC) score function

$$\mathrm{BIC}(B \mid D) = \frac{\log m}{2}\,|B| - LL(B \mid D). \tag{7.31}$$

If $f(\theta) = 0$, that is, the number of bits for describing the Bayesian network is not considered in the calculation, then the score function is reduced to the negative log-likelihood, and consequently, the learning task reduces to the maximum likelihood estimation.

For a Bayesian network $B = \langle G, \Theta \rangle$, if the network structure $G$ is fixed, then the first term of the score function $s(B \mid D)$ is a constant. Hence, minimizing $s(B \mid D)$ is equivalent to the maximum likelihood estimation with respect to $\Theta$. From (7.29) and (7.26), we know that each parameter $\theta_{x_i|\pi_i}$ can be obtained by an empirical estimation based on the training set $D$, that is,

$$\theta_{x_i|\pi_i} = \hat{P}_D(x_i \mid \pi_i), \tag{7.32}$$

where $\hat{P}_D(\cdot)$ is the empirical distribution from $D$. Therefore, to minimize the score function $s(B \mid D)$, we only need to search in the candidate network structures instead of parameters, since the optimal parameters of a candidate can be directly calculated from the training set.

That is, the frequency of an event in a training set.

Unfortunately, searching for the optimal Bayesian network structure in all network structures is an NP-hard problem, which cannot be solved efficiently. There are two general strategies to find approximate solutions within finite time. The first strategy is to use greedy methods. For example, starting with a network structure and gradually updating each edge (add, delete, or change direction) until the score function does not decrease any more. The second strategy is adding constraints to reduce the size of the search space (e.g., limiting the network structure to a tree structure).

7

For example, TAN (Friedman et al. 1997) limits the network structure to be a tree structure (we can consider semi-naïve Bayes classifiers as a special case of Bayesian networks).

### 7.5.3 Inference

A trained Bayesian network can answer "queries", that is, inferring the values of other features from observed feature values. For example, we may want to know the ripeness and sweetness of a watermelon after observing color = green, sound = muffled, and root = curly. The process of inferring the value of a queried feature from the observed feature values is called *inference*, where the observed feature values are called *evidence*.

The class can also be regarded as a feature.

Ideally, we wish to use the joint probability distribution defined by the Bayesian network to compute the exact posterior probabilities. Unfortunately, such an *exact inference* is NP-hard (Cooper 1990). In other words, exact inferences are difficult when the network contains many nodes with dense connections. In such cases, we often leverage *approximate inference* methods to obtain approximate solutions within a finite time by sacrificing some accuracy. A typical approximate inference method for Bayesian networks is the *Gibbs sampling*, which is a type of random sampling method. Now, let us take a closer look at the Gibbs sampling.

Chapter 14 provides more discussions on inferences.

The variational inference is also frequently used. See Sect. 14.5.

Let $\mathbf{Q} = \{Q_1, Q_2, \ldots, Q_n\}$ denote the query variables, $\mathbf{E} = \{E_1, E_2, \ldots, E_k\}$ denote the evidence variables, whose values are $\mathbf{e} = \{e_1, e_2, \ldots, e_k\}$. The task is to compute the posterior probability $P(\mathbf{Q} = \mathbf{q} \mid \mathbf{E} = \mathbf{e})$, where $\mathbf{q} = \{q_1, q_2, \ldots, q_n\}$ is a set of values of the query variables. For instance, the query variables could be $\mathbf{Q} = \{\texttt{ripe}, \texttt{sweetness}\}$, and the evidence variables are $\mathbf{E} = \{\texttt{color}, \texttt{sound}, \texttt{root}\}$ with the values $\mathbf{e} = \{\texttt{green}, \texttt{muffled}, \texttt{curly}\}$. Then, the target values of the query are $\mathbf{q} = \{\texttt{true}, \texttt{high}\}$, that is, how likely this is a ripe watermelon with a high level of sweetness.

The Gibbs sampling algorithm, as shown in ◘ Algorithm 7.1, starts with a randomly generated sample $\mathbf{q}^0$ with random values assigned to non-evidence variables. Its evidence variables, however, are assigned the same values as the query sample (i.e., $\mathbf{E} = \mathbf{e}$). Then, at each step, it generates a new sample by modifying the non-evidence variables of the current sample, e.g., it generates $\mathbf{q}^1$ from $\mathbf{q}^0$. More generally, at the $t$th step, the algorithm first sets $\mathbf{q}^t = \mathbf{q}^{t-1}$ and then modifies the value of each non-evidence variable one by one via sampling. The sampling probability of each value is computed using the Bayesian network $B$ and the current values of other variables (i.e., $\mathbf{Z} = \mathbf{z}$). Suppose we sampled $T$ times and obtained $n_q$ samples that are consistent with the query $\mathbf{q}$, then the posterior probability can be approximated as

$$P(\mathbf{Q} = \mathbf{q} \mid \mathbf{E} = \mathbf{e}) \simeq \frac{n_q}{T}. \tag{7.33}$$

For the joint state space of all variables in the Bayesian network, the Gibbs sampling is actually a *random walk* in the subspace that is consistent with the evidence $\mathbf{E} = \mathbf{e}$. Since every step only depends on the state of the previous step, the sampling generates a *Markov chain*. Under certain conditions, the state distribution always converges to a stationary distribution as $t \to \infty$, regardless of the initial starting point. The converged distribution by Gibbs sampling happens to be $P(\mathbf{Q} \mid \mathbf{E} = \mathbf{e})$. Hence, the Gibbs sampling with a large $T$ approximates a sampling from $P(\mathbf{Q} \mid \mathbf{E} = \mathbf{e})$, and consequently ensures that (7.33) converges to $P(\mathbf{Q} = \mathbf{q} \mid \mathbf{E} = \mathbf{e})$.

See Sect. 14.5 for more information about Markov chain and Gibbs sampling.

Note that the convergence speed of the Gibbs sampling algorithm is usually slow since it often takes a long time for the Markov chain to converge to a stationary distribution. Besides, if the Bayesian network contains extreme probabilities like "0" or "1", then a stationary distribution is not guaranteed by the Markov chain, and consequently, Gibbs sampling will produce an incorrect estimation.

**Algorithm 7.1** Gibbs Sampling

**Input:** Bayesian network $B = \langle G, \Theta \rangle$;
Number of samplings $T$;
Evidence variables $\mathbf{E}$ and their values $\mathbf{e}$;
Query variables $\mathbf{Q}$ and their values $\mathbf{q}$.

**Process:**

$n_q = 0$;
$\mathbf{q}^0 =$ assign random initial values to $\mathbf{Q}$;
**for** $t = 1, 2, \ldots, T$ **do**
  **for** $Q_i \in \mathbf{Q}$ **do**
    $\mathbf{Z} = \mathbf{E} \cup \mathbf{Q} \backslash \{Q_i\}$;
    $\mathbf{z} = \mathbf{e} \cup \mathbf{q}^{t-1} \backslash \{q_i^{t-1}\}$;
    Compute distribution $P_B(Q_i \mid \mathbf{Z} = \mathbf{z})$ according to $B$;
    $q_i^t = Q_i$ sampled from $P_B(Q_i \mid \mathbf{Z} = \mathbf{z})$;
    $\mathbf{q}^t = \mathbf{q}^{t-1}$ with $q_i^{t-1}$ replaced by $q_i^t$;
  **end for**
  **if** $\mathbf{q}^t = \mathbf{q}$ **then**
    $n_q = n_q + 1$.
  **end if**
**end for**

**Output:** $P(\mathbf{Q} = \mathbf{q} \mid \mathbf{E} = \mathbf{e}) \simeq \frac{n_q}{T}$.

All variables excluding $Q_i$.

## 7.6 EM Algorithm

The previous discussions assumed the values of all features are observed for the training samples, that is, each training sample is *complete*. In practice, however, we often have *incomplete* training samples. For instance, the root of a watermelon may have been removed, so we do not know whether it is straight or curly, that is, the feature root is unknown. Then, in the presence of *unobserved* features, also known as *latent variables*, how can we estimate the model parameters?

Let $\mathbf{X}$ denote the set of observed variables, $\mathbf{Z}$ denote the set of latent variables, and $\Theta$ denote the model parameters. Then, the maximum likelihood estimation of $\Theta$ maximizes the log-likelihood

Since the "likelihood" is often based on exponential family functions, the natural logarithm $\ln(\cdot)$ is generally used for log-likelihood and also in the EM algorithm.

$$LL(\Theta \mid \mathbf{X}, \mathbf{Z}) = \ln P(\mathbf{X}, \mathbf{Z} \mid \Theta). \tag{7.34}$$

We cannot solve (7.34) directly because $\mathbf{Z}$ are latent variables. However, we can use the expectation of $\mathbf{Z}$ to maximize the log *marginal likelihood* of the observed data:

$$LL(\Theta \mid \mathbf{X}) = \ln P(\mathbf{X} \mid \Theta) = \ln \sum_{\mathbf{Z}} P(\mathbf{X}, \mathbf{Z} \mid \Theta). \tag{7.35}$$

The Expectation-Maximization (EM) algorithm (Dempster et al. 1977) is a powerful iterative method for estimating latent variables. Its basic idea is as follows: given the value of $\Theta$, from the training data, we can infer the expected value for each latent variable in $\mathbf{Z}$ (the E-step); given the values of the latent variables in $\mathbf{Z}$, we can estimate $\Theta$ with the maximum likelihood estimation (the M-step).

To implement the above idea, we initialize (7.35) with $\Theta^0$ and iteratively apply the following two steps until convergence:

This is only a brief description of the EM algorithm. See Sect. 9.4.3 for more concrete examples.

- Infer the expectation of $\mathbf{Z}$ with $\Theta^t$, denoted by $\mathbf{Z}^t$;
- Estimate $\Theta$ with the maximum likelihood estimation based on $\mathbf{Z}^t$ and the observed variables $\mathbf{X}$, denoted by $\Theta^{t+1}$.

The above iterative process is the prototype of the EM algorithm. Instead of using the expectation of $\mathbf{Z}$, we can compute the probability distribution $P(\mathbf{Z} \mid \mathbf{X}, \Theta^t)$ based on $\Theta^t$, and the above two steps become

- E-step (Expectation): infer the latent variable distribution $P(\mathbf{Z} \mid \mathbf{X}, \Theta^t)$ based on $\Theta^t$, and then compute the expectation of the log-likelihood $LL(\Theta \mid \mathbf{X}, \mathbf{Z})$ with respect to $\mathbf{Z}$ as

$$Q(\Theta \mid \Theta^t) = \mathbb{E}_{\mathbf{Z}|\mathbf{X},\Theta^t} LL(\Theta \mid \mathbf{X}, \mathbf{Z}). \tag{7.36}$$

- M-step (Maximization): find the parameters that maximize the expected log-likelihood, that is,

$$\Theta^{t+1} = \arg\max_{\Theta} Q(\Theta \mid \Theta^t). \tag{7.37}$$

In short, the EM algorithm alternatively computes the following two steps: the expectation (E) step, which uses the current parameters to compute the expected log-likelihood; the maximization (M) step, which finds the parameters that maximize the expected log-likelihood in the E-step. Iterating the above two steps until converging to a local optimum.

See Wu (1983) for the convergence analysis of the EM algorithm.

We can also use optimization algorithms to estimate the latent variables, such as gradient descent. However, computing the gradients could be difficult since the number of terms in the summation increases exponentially as the number of latent variables increases. In contrast, the EM algorithm can be seen as a non-gradient optimization method.

The EM algorithm can be seen as a coordinate descent method which maximizes the lower bound of the log-likelihood. See Appendix B.5 for the coordinate descent method.

## 7.7 Further Reading

Bayesian decision theory plays a crucial role in fields related to data analytics, such as machine learning and pattern recognition. Finding the approximate solution of Bayes' theorem is an effective approach for designing machine learning algorithms. Applying Bayes' theorem suffers from the problems of combinatorial explosion and data sparsity. To avoid these problems, the naïve Bayes classifier introduces the attribute conditional independence assumption. Though this assumption can rarely hold in practice, the performance of naïve Bayes classifiers is often surprisingly good (Domingos and Pazzani 1997, Ng and Jordan 2002). One possible explanation is that correct classification typically requires only the conditional probabilities to be ranked correctly rather than accurate probability values (Domingos and Pazzani 1997). Another possible explanation is that if the inter-attribute dependencies are the same for all classes, or if the effects of the dependency relationship can cancel each other out, then attribute conditional independence assumption reduces the calculation overhead without affecting the performance negatively (Zhang 2004). Naïve Bayes classifiers are especially popular in the field of information retrieval (Lewis 1991). (McCallum and Nigam 1998) compared two common ways of utilizing naïve Bayes classifiers in text classifications.



Bayes classifiers form a "spectrum" based on the extent to which inter-attribute dependencies are involved: naïve Bayes classifiers do not consider feature dependency at all, whereas Bayesian networks can represent any dependencies among features. These two are located at the two ends of the "spectrum", and the classifiers sit in the middle of them are a series of semi-naïve Bayes classifiers that consider partial feature dependencies with different assumptions and constraints. It is generally believed that (Kononenko 1991) initialized the research on semi-naïve classifiers. ODE considers dependency on a single parent feature, and representative one-dependent classifiers include TAN (Friedman et al. 1997), AODE (Webb et al. 2005), and LBR (Zheng and Webb 2000). On the other hand, $k$DE considers multiple dependencies on $k$ parent features, and representative $k$-dependent classifiers include KDB (Sahami 1996) and NBtree (Kohavi 1996).

Note that Bayes classifiers are different from the general *Bayesian learning*. Bayes classifiers make point estimates with maximum posterior probabilities, whereas Bayesian learning performs probability distribution estimations. More information about Bayesian learning can be found in Bishop (2006).

J. Pearl received the Turing Award in 2011 for his significant contributions to this field.

Bayesian networks provide an essential framework for learning and inference under uncertainty and have attracted wide attention because of its powerful representation ability and good interpretability (Pearl 1988). Bayesian network learning consists of structure learning and parameter learning. Parameter learning is relatively easy, while structure learning has been proven to be NP-hard (Cooper 1990; Chickering et al. 2004). Hence, many score-and-search methods were proposed (Friedman and Goldszmidt 1996). Though Bayesian networks are generally considered as generative models, there are also some recent studies on discriminative Bayesian network classifiers (Grossman and Domingos 2004). More information about Bayesian networks can be found in Jensen (1997), Heckerman (1998).

Bayesian networks are classic probabilistic graphical models. See Chap. 14.

The EM algorithm is the most commonly used method for estimating latent variables and has been extensively used in machine learning. For example, the parameter estimation of Gaussian Mixture Models (GMM) is based on the EM algorithm, and the $k$-means clustering algorithm, which will be introduced in Sect. 9.4, is also a typical EM algorithm. McLachlan and Krishnan (2008) provided detailed discussions on the EM algorithm, including its analysis, extensions, and applications.

The naïve Bayes algorithm and the EM algorithm are both in the "Top 10 algorithms in data mining" (Wu et al. 2007).

The "Top 10 algorithms in data mining" also includes the previously introduced C4.5, CART decision tree, and support vector machine algorithms. It also includes the AdaBoost, $k$-means, and $k$-nearest neighbors algorithms that will be introduced in later chapters.

## Exercises

The watermelon data set 3.0 is in Table 4.3.

**7.1** Use the maximum likelihood estimation to estimate the class-conditional probabilities of the first three features in the watermelon data set 3.0.

**7.2** * Prove that the naïve Bayes classifier may still produce a Bayes optimal classifier even if the attribute conditional independence assumption does not hold.

**7.3** Implement and train a Laplacian corrected naïve Bayes classifier on the watermelon data set 3.0 to classify the T1 sample in Sect. 7.3.

**7.4** In practice, when the data dimensionality is high, the product of the probabilities $\prod_{i=1}^{d} P(x_i \mid c)$ in (7.15) can be very close to 0 and underflow will happen. Discuss possible methods to prevent underflow.



Assuming their priors are the same. See Sect. 3.4.

**7.5** Prove that, for binary classification problems, the linear discriminant analysis produces a Bayes optimal classifier if the data of both classes follow the Gaussian distributions with the same variance.

**7.6** Implement and train an AODE classifier on the watermelon data set 3.0 to classify the T1 sample in Sect. 7.3.

**7.7** For a binary classification problem with $d$ binary features, suppose the empirical estimation of any prior probabilities requires at least 30 samples, then estimating the prior term $P(c)$ in the naïve Bayes classifier (7.15) requires $30 \times 2 = 60$ samples. How many samples are needed to estimate the prior term $P(c, x_i)$ in AODE (7.23)? (consider the best and the worst cases.)

**7.8** For Figure 7.3, prove for the common parent structure, $x_3 \perp\!\!\!\perp x_4$ does not hold if $x_1$ is unknown; for the cascade structure, $y \perp z \mid x$ but $y \perp\!\!\!\perp z$ does not hold.

The watermelon data set 2.0 is in Table 4.1.

**7.9** Use the watermelon data set 2.0 to construct a Bayesian network based on the BIC criterion.

**7.10** Use the watermelon data set 2.0 to construct a Bayesian network based on the EM algorithm by considering umbilicus as the latent variable.

## Break Time

### Short Story: The Mystery of Bayes

On February 23, 1763, the priest R. Price, who inherited the legacy of Thomas Bayes (1701?–1761), presented Bayes' unpublished work "An Essay toward solving a Problem in the Doctrine of Chances" to the Royal Society. This essay introduced the famous Bayes' theorem, and that day is considered as the birth date of Bayes' theorem. Though Bayes' theorem has become one of the most classic contents in probability and statistics today, Bayes' life is still a mystery.

It is known that Bayes was a clergy who has spent most of his adult life as a priest at a chapel in Tunbridge Wells, England. His devotion to mathematics research was motivated by the hope to prove the existence of god by mathematics. In 1742, he was elected as a Fellow of the Royal Society while he only published two works in his lifetime, one in theology and one in mathematics. A senior member signed his nomination, but it is still a mystery regarding why he was nominated and how could he be elected given his limited achievement at that time (Bayes' theorem was published posthumously). Bayes' life and work did not receive much attention while he was alive, and even Bayes' theorem was soon forgotten by the public. Bayes' theorem was later brought back by the great mathematician Laplace, but it had not become famous until its full applications in statistics in the twentieth century. The birth date of Bayes is still unclear. Moreover, whether the photo we see today is Bayes or not remains debatable.

## References

Bishop CM (2006) Pattern recognition and machine learning. Springer, New York, NY

Chickering DM, Heckerman D, Meek C (2004) Large-sample learning of Bayesian networks is NP-hard. J Mach Learn Res 5:1287–1330

Chow CK, Liu CN (1968) Approximating discrete probability distributions with dependence trees. IEEE Trans Inf Theory 14(3):462–467

Cooper GF (1990) The computational complexity of probabilistic inference using Bayesian belief networks. Artif Intell 42(2–3):393–405

Cowell RG, Dawid P, Lauritzen, SL, Spiegelhalter, DJ (1999) Probabilistic networks and expert systems. Springer, New York, NY

Dempster AP, Laird NM, Rubin DB (1977) Maximum likelihood from incomplete data via the EM algorithm. J R Stat Soc -Ser B 39(1):1–38

Domingos P, Pazzani M (1997) On the optimality of the simple Bayesian classifier under zero-one loss. Mach Learn 29(2–3):103–130

Efron B (2005) Bayesians, frequentists, and scientists. J Am Stat Assoc 100(469):1–5

Friedman N, Geiger D, Goldszmidt M (1997) Bayesian network classifiers. Mach Learn 29(2–3):131–163

Friedman N, Goldszmidt M (1996) Learning Bayesian networks with local structure. In: Proceedings of the 12th annual conference on uncertainty in artificial intelligence (UAI), Portland, OR, pp 252–262

Grossman D, Domingos P (2004) Learning Bayesian network classifiers by maximizing conditional likelihood. In: Proceedings of the 21st international conference on machine learning (ICML), Banff, Canada, pp 46–53

Heckerman D (1998) A tutorial on learning with Bayesian networks. In: Jordan MI (ed) Learning in graphical models. Kluwer, Dordrecht, Netherlands, pp 301–354

Jensen FV (1997) An introduction to Bayesian networks. Springer, New York, NY

Kohavi R (1996) Scaling up the accuracy of naive-Bayes classifiers: a decision-tree hybrid. In: Proceedings of the 2nd international conference on knowledge discovery and data mining (KDD), Portland, OR, pp 202–207

Kononenko I (1991) Semi-naive Bayesian classifier. In: Proceedings of the 6th European working session on learning (EWSL), Porto, Portugal, pp 206–219

Lewis DD (1991) Naive (Bayes) at forty: the independence assumption in information retrieval. In: Proceedings of the 10th European conference on machine learning (ECML), Chemnitz, Germany, pp 4–15

McCallum A, Nigam K (1998) A comparison of event models for naive Bayes text classification. In: Working notes of the AAAI'98 workshop on learning for text categorization, Madison, WI, pp 4–15

McLachlan G, Krishnan T (2008) The EM algorithm and extensions, 2nd edn. Wiley, Hoboken, NJ

Ng AY, Jordan MI (2002) On discriminative versus generative classifiers: a comparison of logistic regression and naive Bayes. In: Dietterich TG, Becker S, Ghahramani Z (eds), Advances in neural information processing systems 14 (NIPS), pp 841–848. MIT Press, Cambridge, MA

Pearl J (1988) Probabilistic reasoning in intelligent systems: networks of plausible inference. Morgan Kaufmanns, San Francisco, CA

Sahami M (1996) Learning limited dependence Bayesian classifiers. In: Proceedings of the 2nd international conference on knowledge discovery and data mining (KDD), Portland, OR, pp 335–338

Samaniego FJ (2010) A comparison of the Bayesian and frequentist approaches to estimation. Springer, New York, NY
Webb G, Boughton J, Wang Z (2005) Not so naive Bayes: aggregating one-dependence estimators. Mach Learn 58(1):5–24
Wu CFJ (1983) On the convergence properties of the EM algorithm. Ann Stat 11(1):95–103
Wu X, Kumar V, Quinlan JR, Ghosh J, Yang Q, Motoda H, McLachlan GJ, Ng A, Liu B, Yu PS, Zhou Z-H, Steinbach M, Hand DJ, Steinberg D (2007) Top 10 algorithms in data mining. Knowl Inf Syst 14(1):1–37
Zhang H (2004) The optimality of naive Bayes. In: Proceedings of the 17th international Florida artificial intelligence research society conference (FLAIRS), Miami, FL, pp 562–567
Zheng Z, Webb GI (2000) Lazy learning of Bayesian rules. Mach Learn 41(1):53–84

# Ensemble Learning

Table of Contents




Z.-H. Zhou, *Machine Learning*,
https://doi.org/10.1007/978-981-15-1967-3_8

## 8.1 Individual and Ensemble

Ensemble learning, also known as multiple classifier system and committee-based learning, trains and combines multiple learners to solve a learning problem.

As shown in ◘ Figure 8.1, the typical workflow of ensemble learning is training a set of individual learners first and then combining them via some strategies, where an individual learner is usually trained by an existing learning algorithm, such as the C4.5 algorithm and the BP neural network algorithm. An ensemble is said to be homogeneous if all individual learners are of the same type, e.g., a "decision tree ensemble" contains only decision trees, while a "neural network ensemble" contains only neural networks. For homogeneous ensembles, the individual learners are called *base learners*, and the corresponding learning algorithms are called *base learning algorithms*. In contrast, a heterogeneous ensemble contains different individual learners and learning algorithms, and there is no single base learner or base learning algorithm. For heterogeneous ensembles, the individual learners are unusually called *component learners* or simply individual learners.



By combining multiple learners, the generalization ability of an ensemble is often much stronger than that of an individual learner, and this is especially true for *weak learners*. Therefore, theoretical studies on ensemble learning often focus on weak learners, and hence base learners are sometimes called weak learners. In practice, however, despite that an ensemble of weak learners can theoretically obtain good performance, people still prefer strong learners for some reasons, such as reducing the number of individual learners and reusing existing knowledge about the strong learners.

Weak learners typically refer to learners with generalization ability just slightly better than random guessing, e.g., with an accuracy slightly above 50% in binary classification problems.

Intuitively, mixing things with different qualities will produce something better than the worst one but worse than the best one. Then, how can an ensemble produce better performance than the best individual learner?

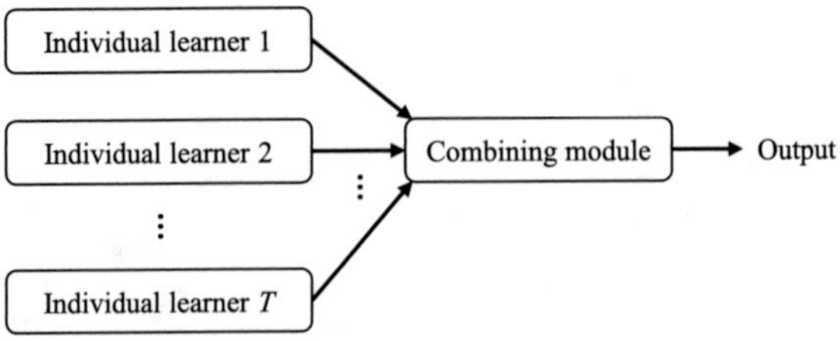


**Fig. 8.1** The workflow of ensemble learning

| | Testing sample 1 | Testing sample 2 | Testing sample 3 |
|---|---|---|---|
| $h_1$ | ✓ | ✓ | ✗ |
| $h_2$ | ✗ | ✓ | ✓ |
| $h_3$ | ✓ | ✗ | ✓ |
| Ensemble | ✓ | ✓ | ✓ |

(a) Ensemble helps.

| | Testing sample 1 | Testing sample 2 | Testing sample 3 |
|---|---|---|---|
| $h_1$ | ✓ | ✓ | ✗ |
| $h_2$ | ✓ | ✓ | ✗ |
| $h_3$ | ✓ | ✓ | ✗ |
| Ensemble | ✓ | ✓ | ✗ |

(b) Ensemble doesn't help.

| | Testing sample 1 | Testing sample 2 | Testing sample 3 |
|---|---|---|---|
| $h_1$ | ✓ | ✗ | ✗ |
| $h_2$ | ✗ | ✓ | ✗ |
| $h_3$ | ✗ | ✗ | ✓ |
| Ensemble | ✗ | ✗ | ✗ |

(c) Ensemble hurts.

**Fig. 8.2** Individual learners should be "accurate and diverse" ($h_i$ is the $i$th learner)

Taking binary classification as an example, suppose three classifiers are applied to three testing samples, as shown in ◘ Figure 8.2, where the ticks indicate the correct classifications, and the crosses indicate the incorrect classifications. The classification of ensemble learning is made by voting. In ◘ Figure 8.2a, each classifier achieves an accuracy of 66.6%, while the ensemble achieves an accuracy of 100%. In ◘ Figure 8.2b, the three classifiers have made identical decisions, and the ensemble does not improve the result. In ◘ Figure 8.2c, each classifier achieves an accuracy of 33.3%, while the ensemble gives an even worse result. From this example, we see that a good ensemble should contain individual learners that are "accurate and diverse". In other words, individual learners must be not bad and have *diversity* (i.e., the learners are different).

Let us do a simple analysis with binary classification, that is, $y \in \{-1, +1\}$. Suppose the ground-truth function is $f$, and the error rate of each base learner is $\epsilon$. Then, for each base learner $h_i$, we have

An individual learner should be no worse than a weak learner.

$$P(h_i(\boldsymbol{x}) \neq f(\boldsymbol{x})) = \epsilon. \tag{8.1}$$

Suppose ensemble learning combines the $T$ base learners by voting, then the ensemble will make an correct classification if more than half of the base learners are correct:

For ease of discussion, we assume $T$ is odd.

$$F(\boldsymbol{x}) = \operatorname{sign}\left(\sum_{i=1}^{T} h_i(\boldsymbol{x})\right). \tag{8.2}$$

Assuming the error rates of base learners are independent, then, from Hoeffding's inequality, the error rate of the ensemble is

See Exercise 8.1.

$$\begin{aligned} P(F(\boldsymbol{x}) \neq f(\boldsymbol{x})) &= \sum_{k=0}^{\lfloor T/2 \rfloor} \binom{T}{k} (1-\epsilon)^k \epsilon^{T-k} \\ &\leqslant \exp\left(-\frac{1}{2} T (1-2\epsilon)^2\right). \end{aligned} \tag{8.3}$$

The above equation shows that as the number of base learners $T$ in the ensemble increases, the error rate decreases exponentially and eventually approaches zero.

The above analysis makes a critical assumption that the error rates of base learners are independent. However, this assumption is invalid in practice since the learners are trained to solve the same problem and thus cannot be independent. In fact, accuracy and diversity are two conflicting properties of individual learners. Generally, when the accuracy is already high, we usually need to sacrifice some accuracy if we wish to increase diversity. It turns out that the generation and combination of "accurate and diverse" individual learners are the fundamental issues in ensemble learning.

Current ensemble learning methods can be roughly grouped into two categories, depending on how the individual learners are generated. The first category, represented by *Boosting*, creates individual learners with strong correlations and generates the learners sequentially. The second category, represented by *Bagging* and *Random Forest*, creates individual learners independently and can parallelize the generation process.



## 8.2 Boosting

Boosting is a family of algorithms that convert weak learners to strong learners. Boosting algorithms start with training a base learner and then adjust the distribution of the training samples according to the result of the base learner such that incorrectly classified samples will receive more attention by subsequent base learners. After training the first base learner, the second base learner is trained with the adjusted training samples, and the result is used to adjust the training sample distribution again. Such a process repeats until the number of base learners reaches a predefined value $T$, and finally, these base learners are weighted and combined.

The most well-known Boosting algorithm is AdaBoost (Freund and Schapire 1997), as shown in ◘ Algorithm 8.1, where $y_i \in \{-1, +1\}$ and $f$ is the ground-truth function.

There are multiple ways to derive the AdaBoost algorithm, but one that is easy to understand is based on the *additive model*, that is, using the linear combination of base learners

$$H(\boldsymbol{x}) = \sum_{t=1}^{T} \alpha_t h_t(\boldsymbol{x}) \tag{8.4}$$

to minimize the *exponential loss function* (Friedman et al. 2000)

**Algorithm 8.1** AdaBoost

**Input:** Training set $D = \{(\boldsymbol{x}_1, y_1), (\boldsymbol{x}_2, y_2), \ldots, (\boldsymbol{x}_m, y_m)\}$;
Base learner $\mathcal{L}$;
Number of training rounds $T$.

**Process:**

$\mathcal{D}_1(\boldsymbol{x}) = 1/m$;
**for** $t = 1, 2, \ldots, T$ **do**
$h_t = \mathcal{L}(D, \mathcal{D}_t)$;
$\epsilon_t = P_{\boldsymbol{x}\sim\mathcal{D}_t}(h_t(\boldsymbol{x}) \neq f(\boldsymbol{x}))$;
**if** $\epsilon_t > 0.5$ **then break**
$\alpha_t = \frac{1}{2}\ln\left(\frac{1-\epsilon_t}{\epsilon_t}\right)$;

$$\mathcal{D}_{t+1}(\boldsymbol{x}) = \frac{\mathcal{D}_t(\boldsymbol{x})}{Z_t} \times \begin{cases} \exp(-\alpha_t), & \text{if } h_t(\boldsymbol{x}) = f(\boldsymbol{x}); \\ \exp(\alpha_t), & \text{if } h_t(\boldsymbol{x}) \neq f(\boldsymbol{x}); \end{cases}$$
$$= \frac{\mathcal{D}_t(\boldsymbol{x})\exp(-\alpha_t f(\boldsymbol{x}) h_t(\boldsymbol{x}))}{Z_t};$$

**end for**

**Output:** $F(\boldsymbol{x}) = \text{sign}\left(\sum_{t=1}^{T} \alpha_t h_t(\boldsymbol{x})\right)$.

Initialize the sample weight distribution.

Train classifier $h_t$ using data set $D$ that follows distribution $\mathcal{D}_t$.

Estimate the error of $h_t$.

Determine the weight of classifier $h_t$.

Update the sample distribution, where $Z_t$ is the normalization factor that ensures $\mathcal{D}_{t+1}$ is a valid distribution.

$$\ell_{\exp}(H \mid \mathcal{D}) = \mathbb{E}_{\boldsymbol{x}\sim\mathcal{D}}\left[e^{-f(\boldsymbol{x})H(\boldsymbol{x})}\right]. \tag{8.5}$$

If $H(\boldsymbol{x})$ minimizes the exponential loss, then we consider the partial derivative of (8.5) with respect to $H(\boldsymbol{x})$

$$\frac{\partial \ell_{\exp}(H \mid \mathcal{D})}{\partial H(\boldsymbol{x})} = -e^{-H(\boldsymbol{x})}P(f(\boldsymbol{x}) = 1 \mid \boldsymbol{x}) + e^{H(\boldsymbol{x})}P(f(\boldsymbol{x}) = -1 \mid \boldsymbol{x}), \tag{8.6}$$

and setting it to zero gives

$$H(\boldsymbol{x}) = \frac{1}{2}\ln\frac{P(f(\boldsymbol{x}) = 1 \mid \boldsymbol{x})}{P(f(\boldsymbol{x}) = -1 \mid \boldsymbol{x})}. \tag{8.7}$$

Hence, we have

$$\begin{aligned} \text{sign}(H(\boldsymbol{x})) &= \text{sign}\left(\frac{1}{2}\ln\frac{P(f(\boldsymbol{x}) = 1 \mid \boldsymbol{x})}{P(f(\boldsymbol{x}) = -1 \mid \boldsymbol{x})}\right) \\ &= \begin{cases} 1, & P(f(\boldsymbol{x}) = 1 \mid \boldsymbol{x}) > P(f(\boldsymbol{x}) = -1 \mid \boldsymbol{x}) \\ -1, & P(f(\boldsymbol{x}) = 1 \mid \boldsymbol{x}) < P(f(\boldsymbol{x}) = -1 \mid \boldsymbol{x}) \end{cases} \\ &= \underset{y\in\{-1,1\}}{\arg\max}\, P(f(\boldsymbol{x}) = y \mid \boldsymbol{x}), \end{aligned} \tag{8.8}$$

Here, we ignore the case of $P(f(\boldsymbol{x}) = 1 \mid \boldsymbol{x}) = P(f(\boldsymbol{x}) = -1 \mid \boldsymbol{x})$.

See Sect. 6.7 for the "consistency" of surrogate functions.

which implies that $\text{sign}(H(\boldsymbol{x}))$ achieves the Bayes optimal error rate. In other words, the classification error rate is minimized when the exponential loss is minimized, and hence the exponential loss function is a consistent surrogate function of the original 0/1 loss function. Since this surrogate function has better mathematical properties, e.g., continuously differentiable, it is used as the optimization objective replacing the 0/1 loss function.

In the AdaBoost algorithm, the base learning algorithm generates the first base classifier $h_1$ from the original training data and then iteratively generates the subsequent base classifiers $h_t$ and associated weights $\alpha_t$. Once the classifier $h_t$ is generated from the distribution $\mathcal{D}_t$, its weight $\alpha_t$ is estimated by letting $\alpha_t h_t$ minimize the exponential loss function

$$\begin{aligned}\ell_{\exp}(\alpha_t h_t \mid \mathcal{D}_t) &= \mathbb{E}_{\boldsymbol{x}\sim\mathcal{D}_t}\left[e^{-f(\boldsymbol{x})\alpha_t h_t(\boldsymbol{x})}\right]\\ &= \mathbb{E}_{\boldsymbol{x}\sim\mathcal{D}_t}\left[e^{-\alpha_t}\mathbb{I}(f(\boldsymbol{x}) = h_t(\boldsymbol{x})) + e^{\alpha_t}\mathbb{I}(f(\boldsymbol{x}) \neq h_t(\boldsymbol{x}))\right]\\ &= e^{-\alpha_t}P_{\boldsymbol{x}\sim\mathcal{D}_t}(f(\boldsymbol{x}) = h_t(\boldsymbol{x})) + e^{\alpha_t}P_{\boldsymbol{x}\sim\mathcal{D}_t}(f(\boldsymbol{x}) \neq h_t(\boldsymbol{x}))\\ &= e^{-\alpha_t}(1-\epsilon_t) + e^{\alpha_t}\epsilon_t, \end{aligned} \tag{8.9}$$

where $\epsilon = P_{\boldsymbol{x}\sim\mathcal{D}_t}(h_t(\boldsymbol{x}) \neq f(\boldsymbol{x}))$. Setting the derivative of the exponential loss function

$$\frac{\partial \ell_{\exp}(\alpha_t h_t \mid \mathcal{D}_t)}{\partial \alpha_t} = -e^{-\alpha_t}(1-\epsilon_t) + e^{\alpha_t}\epsilon_t \tag{8.10}$$

to zero gives the optimal $\alpha_t$ as

$$\alpha_t = \frac{1}{2}\ln\left(\frac{1-\epsilon_t}{\epsilon_t}\right), \tag{8.11}$$

which is exactly the equation shown in line 6 of ◘ Algorithm 8.1.

The AdaBoost algorithm adjusts the sample distribution based on $H_{t-1}$ such that the base learner $h_t$ in the next round can correct some mistakes made by $H_{t-1}$. Ideally, we wish $h_t$ to correct all mistakes made by $H_{t-1}$. The minimization of $\ell_{\exp}(H_{t-1} + \alpha_t h_t \mid \mathcal{D})$ can be simplified to

$$\begin{aligned}\ell_{\exp}(H_{t-1} + h_t \mid \mathcal{D}) &= \mathbb{E}_{\boldsymbol{x}\sim\mathcal{D}}\left[e^{-f(\boldsymbol{x})(H_{t-1}(\boldsymbol{x})+h_t(\boldsymbol{x}))}\right]\\ &= \mathbb{E}_{\boldsymbol{x}\sim\mathcal{D}}\left[e^{-f(\boldsymbol{x})H_{t-1}(\boldsymbol{x})}e^{-f(\boldsymbol{x})h_t(\boldsymbol{x})}\right]. \end{aligned} \tag{8.12}$$

Since $f^2(\boldsymbol{x}) = h_t^2(\boldsymbol{x}) = 1$, (8.12) can be approximated using Taylor expansion of $e^{-f(\boldsymbol{x})h_t(\boldsymbol{x})}$ as

$$\begin{aligned}\ell_{\exp}(H_{t-1}+h_t \mid \mathcal{D}) &\simeq \mathbb{E}_{\boldsymbol{x}\sim\mathcal{D}}\left[e^{-f(\boldsymbol{x})H_{t-1}(\boldsymbol{x})}\left(1-f(\boldsymbol{x})h_t(\boldsymbol{x})+\frac{f^2(\boldsymbol{x})h_t^2(\boldsymbol{x})}{2}\right)\right]\\ &= \mathbb{E}_{\boldsymbol{x}\sim\mathcal{D}}\left[e^{-f(\boldsymbol{x})H_{t-1}(\boldsymbol{x})}\left(1-f(\boldsymbol{x})h_t(\boldsymbol{x})+\frac{1}{2}\right)\right].\end{aligned} \tag{8.13}$$

Hence, the ideal classifier is

$$\begin{aligned}h_t(\boldsymbol{x}) &= \underset{h}{\arg\min}\, \ell_{\exp}(H_{t-1}+h \mid \mathcal{D})\\ &= \underset{h}{\arg\min}\, \mathbb{E}_{\boldsymbol{x}\sim\mathcal{D}}\left[e^{-f(\boldsymbol{x})H_{t-1}(\boldsymbol{x})}\left(1-f(\boldsymbol{x})h(\boldsymbol{x})+\frac{1}{2}\right)\right]\\ &= \underset{h}{\arg\max}\, \mathbb{E}_{\boldsymbol{x}\sim\mathcal{D}}\left[e^{-f(\boldsymbol{x})H_{t-1}(\boldsymbol{x})}f(\boldsymbol{x})h(\boldsymbol{x})\right]\\ &= \underset{h}{\arg\max}\, \mathbb{E}_{\boldsymbol{x}\sim\mathcal{D}}\left[\frac{e^{-f(\boldsymbol{x})H_{t-1}(\boldsymbol{x})}}{\mathbb{E}_{\boldsymbol{x}\sim\mathcal{D}}\left[e^{-f(\boldsymbol{x})H_{t-1}(\boldsymbol{x})}\right]}f(\boldsymbol{x})h(\boldsymbol{x})\right],\end{aligned} \tag{8.14}$$

where $\mathbb{E}_{\boldsymbol{x}\sim\mathcal{D}}\left[e^{-f(\boldsymbol{x})H_{t-1}(\boldsymbol{x})}\right]$ is a constant. Let $\mathcal{D}_t$ denote a distribution

$$\mathcal{D}_t(\boldsymbol{x}) = \frac{\mathcal{D}(\boldsymbol{x})e^{-f(\boldsymbol{x})H_{t-1}(\boldsymbol{x})}}{\mathbb{E}_{\boldsymbol{x}\sim\mathcal{D}}\left[e^{-f(\boldsymbol{x})H_{t-1}(\boldsymbol{x})}\right]}. \tag{8.15}$$

According to the definition of mathematical expectation, the ideal classifier is equivalent to

$$\begin{aligned}h_t(\boldsymbol{x}) &= \underset{h}{\arg\max}\, \mathbb{E}_{\boldsymbol{x}\sim\mathcal{D}}\left[\frac{e^{-f(\boldsymbol{x})H_{t-1}(\boldsymbol{x})}}{\mathbb{E}_{\boldsymbol{x}\sim\mathcal{D}}\left[e^{-f(\boldsymbol{x})H_{t-1}(\boldsymbol{x})}\right]}f(\boldsymbol{x})h(\boldsymbol{x})\right]\\ &= \underset{h}{\arg\max}\, \mathbb{E}_{\boldsymbol{x}\sim\mathcal{D}_t}[f(\boldsymbol{x})h(\boldsymbol{x})].\end{aligned} \tag{8.16}$$

Since $f(\boldsymbol{x}), h(\boldsymbol{x}) \in \{-1, +1\}$, we have

$$f(\boldsymbol{x})h(\boldsymbol{x}) = 1 - 2\mathbb{I}(f(\boldsymbol{x}) \neq h(\boldsymbol{x})), \tag{8.17}$$

and the ideal classifier is

$$h_t(\boldsymbol{x}) = \underset{h}{\arg\min}\, \mathbb{E}_{\boldsymbol{x}\sim\mathcal{D}_t}\left[\mathbb{I}(f(\boldsymbol{x}) \neq h(\boldsymbol{x}))\right]. \tag{8.18}$$

From (8.18), we see that the ideal classifier $h_t$ minimizes the classification error under the distribution $\mathcal{D}_t$. Therefore, the weak classifier at round $t$ is trained on the distribution $\mathcal{D}_t$, and its classification error should be less than 0.5 for $\mathcal{D}_t$. This idea is similar to the *residual approximation* to some extent. Considering the relationship between $\mathcal{D}_t$ and $\mathcal{D}_{t+1}$, we have

$$
\begin{aligned}
\mathcal{D}_{t+1}(\boldsymbol{x}) &= \frac{\mathcal{D}_t(\boldsymbol{x})e^{-f(\boldsymbol{x})H_t(\boldsymbol{x})}}{\mathbb{E}_{\boldsymbol{x}\sim\mathcal{D}}\left[e^{-f(\boldsymbol{x})H_t(\boldsymbol{x})}\right]} \\
&= \frac{\mathcal{D}_t(\boldsymbol{x})e^{-f(\boldsymbol{x})H_{t-1}(\boldsymbol{x})}e^{-f(\boldsymbol{x})\alpha_t h_t(\boldsymbol{x})}}{\mathbb{E}_{\boldsymbol{x}\sim\mathcal{D}}\left[e^{-f(\boldsymbol{x})H_t(\boldsymbol{x})}\right]} \\
&= \mathcal{D}_t(\boldsymbol{x}) \cdot e^{-f(\boldsymbol{x})\alpha_t h_t(\boldsymbol{x})} \frac{\mathbb{E}_{\boldsymbol{x}\sim\mathcal{D}}\left[e^{-f(\boldsymbol{x})H_{t-1}(\boldsymbol{x})}\right]}{\mathbb{E}_{\boldsymbol{x}\sim\mathcal{D}}\left[e^{-f(\boldsymbol{x})H_t(\boldsymbol{x})}\right]},
\end{aligned}
\tag{8.19}
$$

which is the update rule of a sample distribution as in line 7 of ◘ Algorithm 8.1.

From (8.11) and (8.19), we can see that the AdaBoost algorithm, as shown in ◘ Algorithm 8.1, can be derived by iteratively optimizing the exponential loss function based on an additive model.

Boosting algorithms require the base learners to learn from specified sample distributions, and this is often accomplished by *re-weighting*; that is, in each round, a new weight is assigned to a training sample according to the new sample distribution. For base learning algorithms that do not accept weighted samples, *re-sampling* can be used; that is, in each round, a new training set is sampled from the new sample distribution. In general, there is not much difference between re-weighting and re-sampling in terms of prediction performance. Note that in each round, there is a sanity check on whether the current base learner satisfies some basic requirements. For example, line 5 of ◘ Algorithm 8.1 checks whether the current base learner is better than random guessing. If the requirements are not met, the current base learner is discarded and the learning process stops. In such cases, the number of rounds could still be far from the pre-specified limit $T$, which may lead to unsatisfactory performance due to the small number of base learners in the ensemble. However, if the re-sampling method is used, there is an option to "restart" to avoid early termination (Kohavi and Wolpert 1996). More specifically, a new training set is sampled according to the sample distribution again after discarding the current disqualified base learner, and then an alternative base learner is trained such that the learning process can continue to finish $T$ rounds.

See Sect. 2.5 for the bias-variance decomposition.

A decision stump is a decision tree with a single layer. See Sect. 4.3.

The size refers to the number of base learners in an ensemble.

From the perspective of bias-variance decomposition, Boosting mainly focuses on reducing bias, and this is why an ensemble of learners with weak generalization ability can be so powerful. To give a more concrete illustration, we use decision stumps as the base learner and run the AdaBoost algorithm on the watermelon data set 3.0$\alpha$ (◘ Table 4.5). The decision boundaries of ensembles of different sizes together with the corresponding base learners are illustrated in ◘ Figure 8.3.

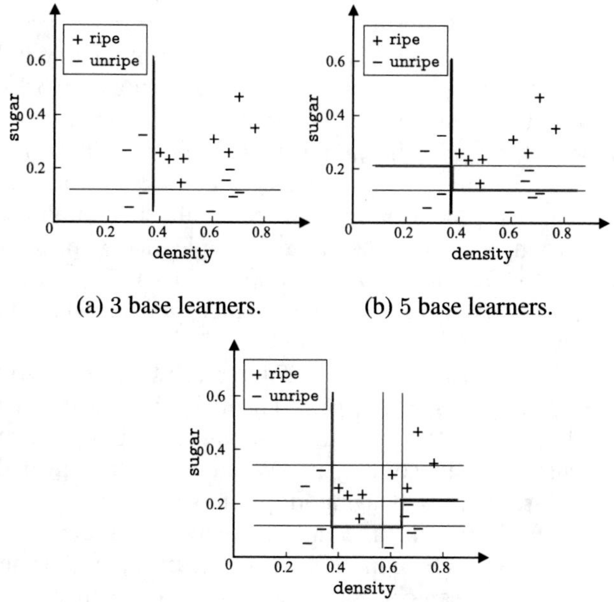


(a) 3 base learners.

(b) 5 base learners.

(c) 11 base learners.

**Fig. 8.3** AdaBoost on the watermelon data set 3.0$\alpha$ with ensemble sizes 3, 5, and 11. The decision boundaries of the ensemble and base learners are shown in red and black, respectively

## 8.3 Bagging and Random Forest

From Sect. 8.1, we know that the generalization ability of an ensemble depends on the independence of base learners. Though strict independence is not possible in practice, we can still make the learners as different as possible. One way of creating different base learners is to partition the original training set into several non-overlapped subsets and use each subset to train a base learner. Because the training subsets are different, the trained base learners are likely to be different as well. However, if the subsets are totally different, then it implies that each subset contains only a small portion of the original training set, possibly leading to poor base learners. Since a good ensemble requires each base learner to be reasonably good, we often allow the subsets to overlap such that each of them contains sufficient samples.

## 8.3.1 Bagging

Bagging stands for Bootstrap AGGregaING.

See Sect. 2.2.3 for bootstrap sampling.

Bagging (Breiman 1996a) is a representative method of parallel ensemble learning based on bootstrap sampling. It works as follows: given a data set with $m$ samples, we randomly pick one sample and copy it to the sampling set; we keep it in the original data set such that it still has a chance to be picked up next time; repeating this process $m$ times gives a data set containing $m$ samples, where some of the original samples may appear more than once while some may never appear. From (2.1), we know that approximately 63.2% of the original samples will appear in the data set.

Applying the above process $T$ times produces $T$ data sets, and each contains $m$ samples. Then, the base learners are trained on these data sets and combined. Such a procedure is the basic workflow of Bagging. When combining the predictions of base learners, Bagging adopts the simple voting method for classification tasks and the simple averaging method for regression tasks. When multiple classes receive the same number of votes, we can choose one at random or further investigate the confidence of votes. The Bagging algorithm is given in ◘ Algorithm 8.2.

8

That is, all base learners are equally weighted through voting or averaging.

**Algorithm 8.2** Bagging

**Input:** Training set: $D = \{(\boldsymbol{x}_1, y_1), (\boldsymbol{x}_2, y_2), \ldots, (\boldsymbol{x}_m, y_m)\}$;
Base learning algorithm $\mathfrak{L}$;
Number of training rounds $T$.

**Process:**

1: **for** $t = 1, 2, \ldots, T$ **do**
2: $h_t = \mathfrak{L}(D, \mathcal{D}_{bs})$.
3: **end for**

**Output:** $H(\boldsymbol{x}) = \underset{y \in \mathcal{Y}}{\arg\max} \sum_{t=1}^{T} \mathbb{I}(h_t(\boldsymbol{x}) = y)$.

$\mathcal{D}_{bs}$ is the distribution of a data set generated by bootstrap.

Suppose that the computational complexity of a base learner is $O(m)$, then the complexity of Bagging is roughly $T(O(m) + O(s))$, where $O(s)$ is the complexity of voting or averaging. Since the complexity $O(s)$ is low and $T$ is a constant that is often not too large, Bagging has the same order of complexity as the base learner, that is, Bagging is an efficient ensemble learning algorithm. Besides, unlike the standard AdaBoost, which only applies to binary classification, Bagging can be applied to multiclass classification and regression without modification.

For AdaBoost, modifications are needed to enable multiclass classification or regression. See (Zhou 2012) for some variants.

The bootstrap sampling brings Bagging another advantage: since each base learner only uses roughly 63.2% of the original training samples for training, the remaining 36.8% samples can be used as a validation set to get an *out-of-bag esti-*

*mate* (Breiman 1996a; Wolpert and Macready 1999) of the generalization ability. To get the out-of-bag estimate, we need to track the training samples used by each base learner. Let $D_t$ denote the set of samples used by the learner $h_t$, and $H^{\mathrm{oob}}(\boldsymbol{x})$ denote the out-of-bag prediction of the sample $\boldsymbol{x}$, that is, considering only the predictions made by base learners that did not use the sample $\boldsymbol{x}$ for training. Then, we have

See Sect. 2.2.3 for out-of-bag estimate.

$$H^{\mathrm{oob}}(\boldsymbol{x}) = \underset{y \in \mathcal{Y}}{\arg\max} \sum_{t=1}^{T} \mathbb{I}(h_t(\boldsymbol{x}) = y) \cdot \mathbb{I}(\boldsymbol{x} \notin D_t), \tag{8.20}$$

and the out-of-bag estimate of the generalization error of Bagging is

$$\epsilon^{\mathrm{oob}} = \frac{1}{|D|} \sum_{(\boldsymbol{x},y) \in D} \mathbb{I}(H^{\mathrm{oob}}(\boldsymbol{x}) \neq y). \tag{8.21}$$

Other than estimating the generalization errors, the out-of-bag samples also have many other uses. For example, when the base learners are decision trees, the out-of-bag samples can be used for pruning branches or estimating the posterior probability of each node, and this is particularly useful when a node contains no training samples. When the base learners are neural networks, the out-of-bag samples can be used to assist early stopping to reduce the risk of overfitting.

From the perspective of bias-variance decomposition, Bagging helps to reduce the variance, and this is particularly useful for unpruned decision trees or neural networks that are unstable to data manipulation. To give a more concrete illustration, we use decision trees based on information gain as base learners to run the Bagging algorithm on the watermelon data set 3.0$\alpha$ (◘ Table 4.5). The decision boundaries of ensembles of different sizes together with the respective base learners are illustrated in ◘ Figure 8.4.

See Sect. 2.5 for the bias-variance decomposition.

### 8.3.2 Random Forest

Random Forest (RF) (Breiman 2001a) is an extension of Bagging, where randomized feature selection is introduced on top of Bagging. Specifically, traditional decision trees select an optimal split feature from the feature set of each node, whereas RF selects from a subset of $k$ features randomly generated from the feature set of the node. The parameter $k$ controls the randomness, where the splitting is the same as in traditional decision trees if $k = d$, and a split feature is randomly

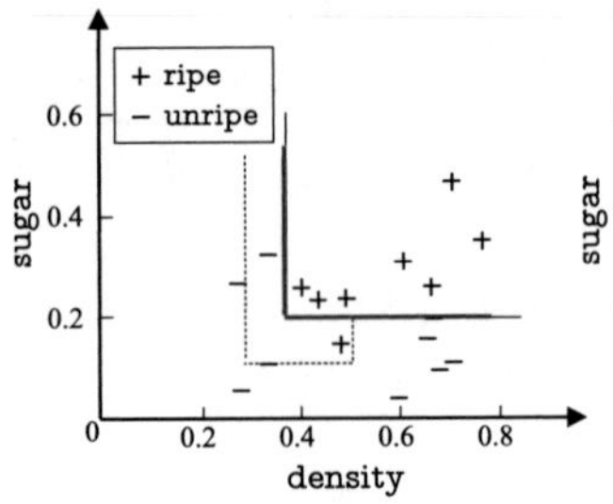


(a) 3 base learners.

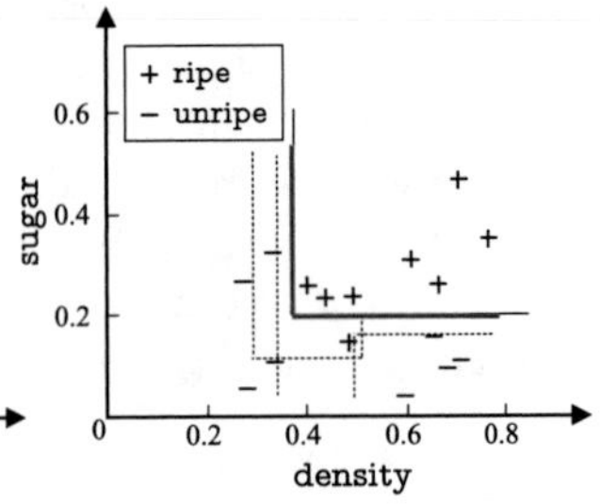


(b) 5 base learners.

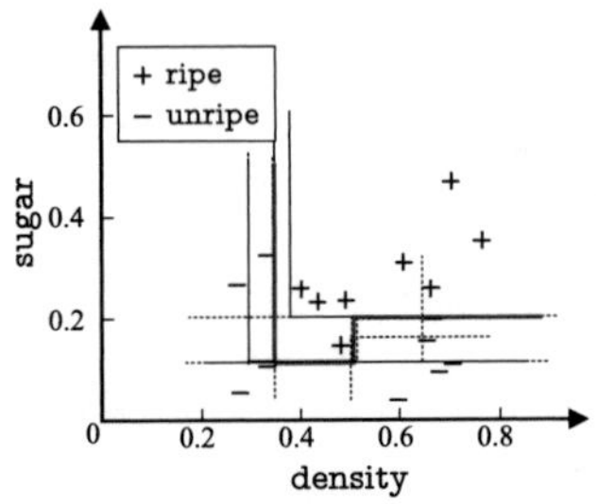


(c) 11 base learners.

**Fig. 8.4** Bagging on the watermelon data set 3.0$\alpha$ with ensemble sizes 3, 5, and 11. The decision boundaries of the ensemble and base learners are shown in red and black, respectively

selected if $k = 1$. Typically, the recommended value of $k$ is $k = \log_2 d$ (Breiman 2001a).

Despite its ease of implementation and low computational cost, RF often shows surprisingly good performance in real-world applications and is honored as a representative of state-of-the-art ensemble learning methods. With a small modification, RF introduces feature-based "diversity" to Bagging by feature manipulation, while Bagging considers sample-based diversity only. RF further enlarges the difference between base learners, which leads to ensembles with better generalization ability.

See Sect. 8.5.3 for sample manipulation and feature manipulation.

The convergence property of RF is comparable to that of Bagging. As illustrated in ◘ Figure 8.5, RF usually starts with a poor performance at the beginning, especially when there is only one base learner in the ensemble since the feature manipulation reduces the performance of each base learner. Eventually, however, RF often converges to a lower generalization error after adding more base learners to the ensemble. It is worth mentioning that it is often more efficient to train an RF than applying Bagging since RF uses "randomized" decision trees that evaluate subsets of features for splitting, whereas

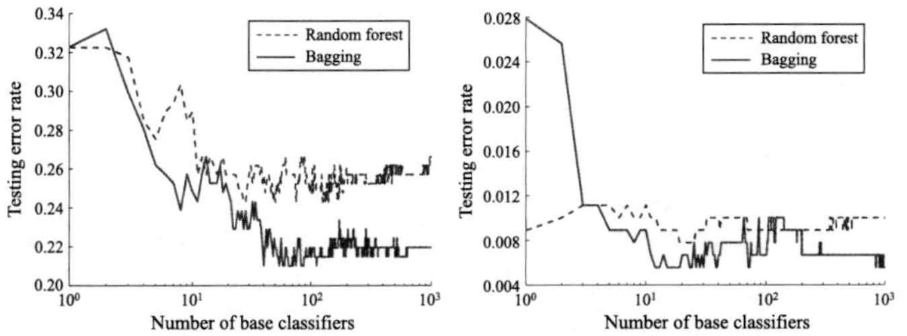


(a) The glass data set. (b) The auto-mpg data set.

**Fig. 8.5** The impact of the ensemble size on RF and Bagging on two UCI data sets

Bagging uses "deterministic" decision trees that evaluate all features for splitting.

## 8.4 Combination Strategies

As illustrated in ◘ Figure 8.6, combining individual learners is beneficial from the following three perspectives (Dietterich 2000):

- Statistical perspective: since the hypothesis space is often large, there are usually multiple hypotheses achieving the same performance. If a single learner is chosen, then the generalization performance solely depends on the quality of this single learner. Combining multiple learners will reduce the risk of incorrectly choosing a single poor learner.
- Computational perspective: learning algorithms are often stuck in local optimum, which may lead to poor generalization performance even when there is abundant training data. If choosing another way by repeating the learning process multiple times, the risk of being stuck in a terrible local

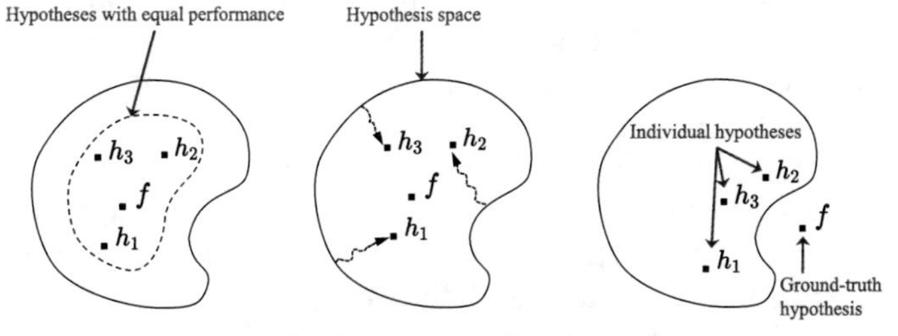


(a) Statistical reason. (b) Computational reason. (c) Representational reason.

**Fig. 8.6** The benefits of combining learners (Dietterich 2000) optimum is reduced, though it is not guaranteed to be the global optimum.

- Representational perspective: sometimes, the ground-truth hypothesis is not represented by any candidates in the hypothesis space of the current learning algorithm. In such cases, it is for sure that a single learner will not find the ground-truth hypothesis. On the other hand, the hypothesis space extends by combining multiple learners, and hence it is more likely to find a better approximation to the ground-truth hypothesis.

Suppose an ensemble contains $T$ individual learners $\{h_1, h_2, \ldots, h_T\}$, where $h_i(\boldsymbol{x})$ is the output of $h_i$ on sample $\boldsymbol{x}$. The rest of this section introduces several typical strategies for combing $h_i$.

### 8.4.1 Averaging

8

Averaging is the most commonly used combination strategy for numerical output $h_i(\boldsymbol{x}) \in \mathbb{R}$. Typical averaging methods include *simple averaging*

$$H(\boldsymbol{x}) = \frac{1}{T}\sum_{i=1}^{T} h_i(\boldsymbol{x}) \tag{8.22}$$

and *weighted averaging*

$$H(\boldsymbol{x}) = \sum_{i=1}^{T} w_i h_i(\boldsymbol{x}), \tag{8.23}$$

where $w_i$ is the weight of individual learner $h_i$, and typically $w_i \geqslant 0, \sum_{i=1}^{T} w_i = 1$. Simple averaging is a special case of weighted averaging when $w_i = 1/T$.

In a study of Stacking regression, (Breiman 1996b) found that the weights must be non-negative to ensure the ensemble performs better than the best single learner. Hence, the non-negative constraint is often applied to the weights of learners.

Weighted averaging has been widely used since the 1950s (Markowitz 1952), and it was first used in ensemble learning in Perrone and Cooper (1993). Weighted averaging plays an important role in ensemble learning since other combination methods can all be viewed as its special cases or variants. Indeed, weighted averaging can be regarded as a fundamental motivation of ensemble learning studies. Given a set of base learners, different ensemble learning methods can be regarded as different ways of assigning the weights.

Typically, weights are learned from training data. However, the learned weights are often unreliable due to data insufficiency or noise. This is particularly true for large ensembles with many base learners because trying to learn too many weights can easily lead to overfitting. In fact, many empirical studies and applications have shown that weighted averaging is not necessarily better than simple averaging (Xu et al. 1992; Ho et al. 1994; Kittler et al. 1998). Generally speaking, the weighted averaging method is a better choice when individual learners have considerable different performance, while the simple averaging method is preferred when individual learners share similar performance.

For example, setting the weights inversely proportional to the estimated errors of individual learners.

### 8.4.2 Voting

For classification, a learner $h_i$ predicts a class label from a set of $N$ class labels $\{c_1, c_2, \ldots, c_N\}$, and a common combination strategy is voting. For ease of discussion, let the $N$-dimensional vector $(h_i^1(\boldsymbol{x}); h_i^2(\boldsymbol{x}); \ldots; h_i^N(\boldsymbol{x}))$ denote the output of $h_i$ on the sample $\boldsymbol{x}$, where $h_i^j(\boldsymbol{x})$ is the output of $h_i$ on $\boldsymbol{x}$ for the class label $c_j$. Then, we can define the following voting methods:

- Majority voting:

$$H(\boldsymbol{x}) = \begin{cases} c_j, & \text{if } \sum_{i=1}^{T} h_i^j(\boldsymbol{x}) > 0.5 \sum_{k=1}^{N} \sum_{i=1}^{T} h_i^k(\boldsymbol{x}); \\ \text{reject}, & \text{otherwise}. \end{cases} \tag{8.24}$$

  That is, output the class label that receives more than half of the votes or refuses to predict if none of the class labels receive more than half of the votes.
- Plurality voting:

$$H(\boldsymbol{x}) = c_{\underset{j}{\arg\max} \sum_{i=1}^{T} h_i^j(\boldsymbol{x})}. \tag{8.25}$$

  That is, output the class label that receives the most votes and randomly select one in case of a tie.
- Weighted voting:

$$H(\boldsymbol{x}) = c_{\underset{j}{\arg\max} \sum_{i=1}^{T} w_i h_i^j(\boldsymbol{x})} \tag{8.26}$$

  That is, a plurality voting with weights assigned to learners, where $w_i$ is the weight of $h_i$, and typically $w_i \geqslant 0$ and $\sum_{i=1}^{T} w_i = 1$ like the constraints on $w_i$ in weighted averaging (8.23).

The standard majority voting (8.24) offers a "reject" option, which is an effective mechanism for tasks requiring reliability (e.g., medical diagnosis). However, if it is compulsory to make a prediction, then plurality voting can be used instead. For tasks that do not allow rejections, both the majority and the plurality voting methods are called majority voting, or just voting.

Equations (8.24)–(8.26) assume no particular value type for the output $h_i^j(\boldsymbol{x})$, but two common value types are

- Class label $h_i^j(\boldsymbol{x}) \in \{0, 1\}$: the output is 1 if $h_i$ predicts the sample $\boldsymbol{x}$ as class $c_j$, and 0 otherwise. The corresponding voting is known as *hard voting*.
- Class probability $h_i^j(\boldsymbol{x}) \in [0, 1]$: an estimate to the posterior probability $P(c_j \mid \boldsymbol{x})$. The corresponding voting is known as *soft voting*.

Different types of $h_i^j(\boldsymbol{x})$ should not be mixed. For some learners, the class labels come with confidence values that can be converted into class probabilities. However, if the confidence values are unnormalized (e.g., the margin values in SVM), then, before using the confidence values as probabilities, we need to apply *calibration* techniques, such as Platt scaling (Platt 2000) and isotonic regression (Zadrozny and Elkan 2001). Interestingly, though the class probabilities are often imprecise, the performance of combining class probabilities usually outperforms that of combining class labels. Note that the class probabilities are not comparable if different types of base learners are used. In such cases, the class probabilities can be converted into class labels before voting, e.g., setting the largest $h_i^j(\boldsymbol{x})$ as 1 and the others as 0.

8

For example, a heterogeneous ensemble consists of different types of base learners.

### 8.4.3 Combining by Learning

When there is abundant training data, a more powerful combination strategy is *combining by learning*: using a meta-learner to combine the individual learners. A representative of such methods is *Stacking* (Wolpert 1992; Breiman 1996b). Here, we call the individual learners as first-level learners and call the learners performing the combination as second-level learners or meta-learners.

Stacking is also a well-known ensemble learning method on its own, and many ensemble learning methods can be viewed as its special cases or variants. We introduce it here since it is also a special combination method.

Stacking starts by training the first-level learners using the original training set and then "generating" a new data set for training the second-level learner. In the new data set, the outputs of the first-level learners are used as the input features, while the labels from the original training set remain unchanged. The Stacking algorithm is given in ◘ Algorithm 8.3, where the first-level learners are assumed heterogeneous.

The first-level learners can also be homogeneous.

**Algorithm 8.3** Stacking

**Input:** Training set $D = \{(\boldsymbol{x}_1, y_1), (\boldsymbol{x}_2, y_2), \ldots, (\boldsymbol{x}_m, y_m)\}$;
First-level learning algorithms $\mathcal{L}_1, \mathcal{L}_2, \ldots, \mathcal{L}_T$;
Second-level learning algorithm $\mathcal{L}$.

**Process:**

**for** $t = 1, 2, \ldots, T$ **do**
  $h_t = \mathcal{L}_t(D)$;
**end for**
$D' = \varnothing$;
**for** $i = 1, 2, \ldots, m$ **do**
  **for** $t = 1, 2, \ldots, T$ **do**
    $z_{it} = h_t(\boldsymbol{x}_i)$;
  **end for**
  $D' = D' \cup ((z_{i1}, z_{i2}, \ldots, z_{iT}), y_i)$;
**end for**
$h' = \mathcal{L}(D')$.

**Output:** $H(\boldsymbol{x}) = h'(h_1(\boldsymbol{x}), h_2(\boldsymbol{x}), \ldots, h_T(\boldsymbol{x}))$.

Generate first-level learner $h_t$ using first-level learning algorithm $\mathcal{L}_t$.

Generate second-level training set.

Generate second-level learner $h'$ using second-level learning algorithm on $\mathcal{D}'$.

When generating the second-level training set from the first-level learners, there would be a high risk of overfitting if the generated training set contains the exact training samples used by the first-level learners. Hence, we often employ cross-validation or leave-one-out methods such that the samples that have not been used for the training of the first-level learners are used for generating the training set for the second-level learner. Taking $k$-fold cross-validation as an example. The original training set is partitioned into $k$ roughly equal-sized partitions $D_1, D_2, \ldots, D_k$. Denote $D_j$ and $\overline{D}_j = D \backslash D_j$ as the testing set and the training set of the $j$th fold, respectively. For $T$ first-level learning algorithms, the first-level learner $h_t^{(j)}$ is obtained by applying the $t$th learning algorithm on the subset $\overline{D}_j$. Let $z_{it} = h_t^{(j)}(\boldsymbol{x}_i)$, then, for each sample $\boldsymbol{x}_i$ in $D_j$, the output from the first-level learners can be written as $\boldsymbol{z}_i = (z_{i1}; z_{i2}; \ldots; z_{iT})$, which is used as the input for the second-level learner with the original label $y_i$. By finishing the entire cross-validation process on the $T$ first-level learners, we obtain a data set $D' = \{(\boldsymbol{z}_i, y_i)\}_{i=1}^m$ for training the second-level learner.

The generalization ability of a Stacking ensemble heavily depends on the representation of the input features for the second-level learner as well as the choice of the second-level learning algorithm. Existing studies show that the Multi-response Linear Regression (MLR) is a good second-level learning algorithm when the first-level learners output class probabilities (Ting and Witten 1999), and the MLR can do an even better job when different sets of features are used (Seewald 2002).

MLR is a classifier based on linear regression. It builds a linear regression model for each class by setting the output of samples belonging to this class as 1 and 0 for the rest. Then, a testing sample is classified as the class with the highest predicted output value.

This is how the StackingC algorithm is implemented in WEKA.

Bayes Model Averaging (BMA) assigns weights to different learners based on posterior probabilities, and it can be regarded as a special implementation of the weighted averaging method. Clarke (2003) compared Stacking and BMA and showed that, in theory, if there is limited noise in the data and the correct data generating model is among the models under consideration, then BMA is guaranteed to be no worse than Stacking. In practice, however, the correct data generating model is by no way guaranteed in the models under consideration, and sometimes even hard to be approximated by the models under consideration. Hence, Stacking usually performs better than BMA in practice since it is more robust than BMA; besides, BMA is also more sensitive to approximation errors.

## 8.5 Diversity

### 8.5.1 Error-Ambiguity Decomposition



As mentioned in Sect. 8.1, to have an ensemble with strong generalization ability, the individual learners should be "accurate and diverse". This section is devoted to providing a brief theoretical analysis of this aspect.

Let us assume that the individual learners $h_1, h_2, \ldots, h_T$ are combined via weighted averaging (8.23) into an ensemble for the regression task $f : \mathbb{R}^d \mapsto \mathbb{R}$. Then, for the sample $\boldsymbol{x}$, the *ambiguity* of the learner $h_i$ is defined as

$$A(h_i|\boldsymbol{x}) = (h_i(\boldsymbol{x}) - H(\boldsymbol{x}))^2, \tag{8.27}$$

and the *ambiguity* of the ensemble is defined as

$$\begin{aligned}\overline{A}(h \mid \boldsymbol{x}) &= \sum_{i=1}^{T} w_i A(h_i \mid \boldsymbol{x}) \\ &= \sum_{i=1}^{T} w_i (h_i(\boldsymbol{x}) - H(\boldsymbol{x}))^2.\end{aligned} \tag{8.28}$$

The ambiguity term represents the degree of disagreement among individual learners on the sample $\boldsymbol{x}$, which reflects the level of diversity in some sense. The squared errors of the individual learner $h_i$ and the ensemble $H$ are, respectively,

$$E(h_i \mid \boldsymbol{x}) = (f(\boldsymbol{x}) - h_i(\boldsymbol{x}))^2, \tag{8.29}$$

$$E(H \mid \boldsymbol{x}) = (f(\boldsymbol{x}) - H(\boldsymbol{x}))^2. \tag{8.30}$$

Let $\overline{E}(h \mid \boldsymbol{x}) = \sum_{i=1}^{T} w_i \cdot E(h_i \mid \boldsymbol{x})$ denote the weighted average error of individual learners, then, we have

$$\begin{aligned}\overline{A}(h \mid \boldsymbol{x}) &= \sum_{i=1}^{T} w_i E(h_i \mid \boldsymbol{x}) - E(H \mid \boldsymbol{x}) \\ &= \overline{E}(h \mid \boldsymbol{x}) - E(H \mid \boldsymbol{x}).\end{aligned} \tag{8.31}$$

Since (8.31) holds for every sample $\boldsymbol{x}$, for all samples we have

$$\sum_{i=1}^{T} w_i \int A(h_i \mid \boldsymbol{x}) p(\boldsymbol{x}) d\boldsymbol{x} = \sum_{i=1}^{T} w_i \int E(h_i \mid \boldsymbol{x}) p(\boldsymbol{x}) d\boldsymbol{x} - \int E(H \mid \boldsymbol{x}) p(\boldsymbol{x}) d\boldsymbol{x}, \tag{8.32}$$

where $p(\boldsymbol{x})$ is the probability density of the sample $\boldsymbol{x}$. Similarly, the generalization error and the ambiguity term of the learner $h_i$ on all samples are, respectively,

Here, we abbreviate $E(h_i)$ and $A(h_i)$ as $E_i$ and $A_i$.

$$E_i = \int E(h_i \mid \boldsymbol{x}) p(\boldsymbol{x}) d\boldsymbol{x}, \tag{8.33}$$

$$A_i = \int A(h_i \mid \boldsymbol{x}) p(\boldsymbol{x}) d\boldsymbol{x}. \tag{8.34}$$

The generalization error of the ensemble is

Here, we abbreviate $E(H)$ as $E$.

$$E = \int E(H \mid \boldsymbol{x}) p(\boldsymbol{x}) d\boldsymbol{x}. \tag{8.35}$$

Let $\overline{E} = \sum_{i=1}^{T} w_i E_i$ denote the weighted average error of individual learners, and $\overline{A} = \sum_{i=1}^{T} w_i A_i$ denote the weighted average ambiguity of individual learners. Then, substituting (8.33)–(8.35) into (8.32) gives

$$E = \overline{E} - \overline{A}. \tag{8.36}$$

The elegant Equation (8.36) clearly shows that the generalization ability of an ensemble depends on the accuracy and diversity of individual learners. The above analysis is known as the *error-ambiguity decomposition* (Krogh and Vedelsby 1995).

Given the above analysis, some readers may propose that we can easily obtain the optimal ensemble by optimizing $\overline{E} - \overline{A}$ directly. Unfortunately, direct optimization of $\overline{E} - \overline{A}$ is hard in practice, not only because both terms are defined in the entire sample space but also because $\overline{A}$ is not a diversity measure that is directly operable; it is only known after we have the ensemble fully constructed. Besides, we should note that the above derivation process is only applicable to regression and is difficult to extend to classification.

### 8.5.2 Diversity Measures

*Diversity measures*, as the name suggests, are for measuring the diversity of individual learners in an ensemble. A typical approach is to measure the pairwise similarity or dissimilarity between learners.

See Sect. 2.3.2 for confusion matrix.

Given a data set $D = \{(\boldsymbol{x}_1, y_1), (\boldsymbol{x}, y_2), \ldots, (\boldsymbol{x}_m, y_m)\}$, the *contingency table* of the classifiers $h_i$ and $h_j$ for binary classification (i.e., $y_i \in \{-1, +1\}$) is

| | $h_i = +1$ | $h_i = -1$ |
|---|---|---|
| $h_j = +1$ | $a$ | $c$ |
| $h_j = -1$ | $b$ | $d$ |

where $a$ is the number of samples predicted as positive by both $h_i$ and $h_j$, and similarly for $b$, $c$, and $d$. With the contingency table, we list some representative diversity measures below as follows:

8

- Disagreement measure:

$$\mathrm{dis}_{ij} = \frac{b+c}{m}, \tag{8.37}$$

where the range of $\mathrm{dis}_{ij}$ is [0, 1], and a larger value indicates a higher diversity.
- Correlation coefficient:

$$\rho_{ij} = \frac{ad - bc}{\sqrt{(a+b)(a+c)(c+d)(b+d)}}, \tag{8.38}$$

where the range of $\rho_{ij}$ is $[-1, 1]$. When $h_i$ and $h_j$ are unrelated, the value is 0. The value is positive when $h_i$ and $h_j$ are positively correlated, and negative when $h_i$ and $h_j$ are negatively correlated.
- $Q$-statistic:

$$Q_{ij} = \frac{ad - bc}{ad + bc}, \tag{8.39}$$

where $Q_{ij}$ has the same sign as the correlation coefficient $\rho_{ij}$, and $|Q_{ij}| \geqslant |\rho_{ij}|$.
- $\kappa$-statistic:

$$\kappa = \frac{p_1 - p_2}{1 - p_2}, \tag{8.40}$$

where $p_1$ is the probability that these two classifiers agree on the prediction, and $p_2$ is the probability of agreement by

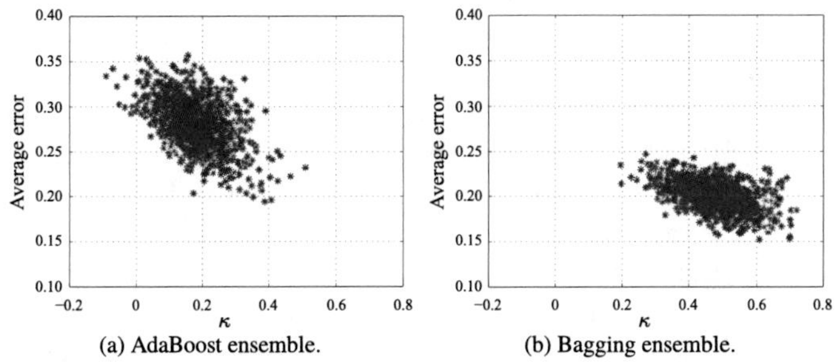


**Fig. 8.7** The $\kappa$-error diagrams on the UCI *tic-tac-toe* data set; every ensemble consists of 50 C4.5 decision trees

chance. Given a data set $D$, these two probabilities can be estimated as

$$p_1 = \frac{a+d}{m}, \tag{8.41}$$

$$p_2 = \frac{(a+b)(a+c)+(c+d)(b+d)}{m^2}. \tag{8.42}$$

If the two classifiers agree on all samples in $D$, then $\kappa = 1$; if the two classifiers agree by chance, then $\kappa = 0$. Normally, $\kappa$ is a non-negative value, but it can be negative in rare cases where the agreement is even less than what is expected by chance.

The aforementioned diversity measures are all pairwise measures that can be easily visualized with 2-D plots. For example, the well-known "$\kappa$-error diagrams", as illustrated in ◘ Figure 8.7, plot each pair of classifiers as a point in a scatter plot, where the x-axis represents their $\kappa$ value, and the y-axis represents their average error. The closer the points to the top of the plot, the lower the accuracy of individual classifiers; the closer the points to the right-hand side of the plot, the less the diversity of individual classifiers.

### 8.5.3 Diversity Generation

Our previous discussions showed that effective ensemble learning relies on diverse individual learners, but how can we enhance diversity? The general idea is to introduce some randomness into the learning process by manipulating the data samples, input features, output representations, and algorithm parameters.

- **Data sample manipulation**: given the original data set, we can manipulate it by generating multiple subsets to train different individual learners. Data sample manipulation is often based on sampling methods such as bootstrap sampling used by Bagging and sequential sampling used by AdaBoost. This approach is widely adopted due to its simplicity and effectiveness. Some base learners, such as decision trees and neural networks, are called *unstable base learners* since they are sensitive to data sample manipulation, that is, a small change to the training samples leads to significant changes to the base learners. Hence, data sampling manipulation is particularly effective for unstable base learners. On the other hand, some base learners, such as linear learners, SVM, naïve Bayes, and $k$-nearest neighbors, are *stable base learners* that are insensitive to data sample manipulation. Hence, other mechanisms are needed, such as input feature manipulation.
- **Input feature manipulation**: the training samples are usually described by a set of features, where different *subspaces* (subsets of features) offer different views of the data, and the individual learners trained on different subspaces will be different. The *Random Subspace* algorithm (Ho 1998) is a well-known ensemble method that relies on input feature manipulation. As illustrated in ■ Algorithm 8.4, the random subspace algorithm extracts multiple subsets of features and uses each subset to train a base learner. For data with many redundant features, applying the random subspace algorithm improves not only the diversity but also the efficiency. Meanwhile, due to the redundant features, the performance of learners trained with a subset of features is often acceptable. Nevertheless, the input feature manipulation is often not a good choice when there is a limited number of features or low feature redundancy.



A subspace usually refers to a lower dimensional feature space mapped from the original higher dimensional feature space. The features describing the lower dimensional space are not necessarily the original features but transformed from the original features. See Chap. 10.

**Algorithm 8.4** Random Subspace

**Input:** Training set $D = \{(\boldsymbol{x}_1, y_1), (\boldsymbol{x}_2, y_2), \ldots, (\boldsymbol{x}_m, y_m)\}$;
Base learning algorithm $\mathcal{L}$;
Number of base learners $T$;
Number of features in subspace $d'$.

**Process:**

**for** $t = 1, 2, \ldots, T$ **do**
  $\mathcal{F}_t = \mathrm{RS}(D, d')$;
  $D_t = \mathrm{Map}_{\mathcal{F}_t}(D)$;
  $h_t = \mathcal{L}(D_t)$.
**end for**

**Output:** $H(\boldsymbol{x}) = \underset{y \in \mathcal{Y}}{\arg\max} \sum_{t=1}^{T} \mathbb{I}\left(h_t\left(\mathrm{Map}_{\mathcal{F}_t}(\boldsymbol{x})\right) = y\right)$.

$d'$ is smaller than the number of original features $d$.

$\mathcal{F}_t$ includes $d'$ randomly selected features, and $D_t$ keeps only the features in $\mathcal{F}_t$.

- **Output representation manipulation**: diversity can also be enhanced by manipulating the output representations. One method is making small changes to the class labels. For example, the Flipping Output method (Breiman 2000) randomly flips the class labels of some samples. Alternatively, we can transform the output representation. For example, the Output Smearing (Breiman 2000) method converts classification outputs into regression outputs before constructing individual learners. Besides, we can also divide the original task into multiple subtasks that can be solved in parallel. For example, the ECOC method (Dietterich and Bakiri 1995) employs error-correcting output codes to divide a multiclass classification task into multiple binary classification tasks to train base learners.

See Sect. 3.5 for ECOC.

- **Algorithm Parameter Manipulation**: base learning algorithms often have parameters, such as the number of hidden neurons and initial connection weights in neural networks. By randomly setting the parameters, we can obtain individual learners that are quite different. For example, the *Negative Correlation* method (Liu and Yao 1999) employs a regularization term to enforce individual neural networks to use different parameters. For algorithms with limited parameters, we can replace some internal components with alternatives for manipulation. For example, using different split feature selection methods in individual decision trees. Note that when a single learner is to be used, we often determine the best parameters by training multiple learners with different parameter settings (e.g., cross-validation), though only one of them is selected. In contrast, ensemble learning utilizes all of these trained learners, and hence the practical computational cost of creating an ensemble is not much higher than creating a single learner.

Different diversity generation mechanisms can be used together. For example, the random forest introduced in Sect. 8.3.2 employs both the data sample manipulation and input feature manipulation. Some methods use more mechanisms at the same time (Zhou 2012).

## 8.6 Further Reading

Zhou (2012) is the main recommended reading about ensemble learning, which provides more detailed discussions on the content covered in this chapter. Kuncheva (2004), Rokach (2010b) are also good references. Schapire and Freund (2012) is a book dedicated to Boosting.

8

Boosting was initially developed in Schapire (1990) to provide a constructive answer to an important theoretical question proposed by Kearns and Valiant (1989): "Is weak learnability equivalent to strong learnability?" The original Boosting algorithm has only theoretical significance. After several years of hard work, Freund and Schapire (1997) proposed AdaBoost. Due to the importance of this work, Yoav Freund and Robert Schapire won the 2003 Gödel Prize, a prestigious award in theoretical computer science. Different ensemble learning methods often have significantly different working mechanisms and theoretical properties. For example, from the bias-variance decomposition point of view, Boosting focuses on reducing the bias while Bagging focuses on reducing the variance. Some attempts, such as MultiBoosting (Webb 2000), have been made to take advantage of both approaches. There already exist considerable theoretical research works on Boosting and Bagging, which can be found in Chaps. 2 and 3 of Zhou (2012).

The derivation given in Sect. 8.2 came from the *statistical view* (Friedman et al. 2000). This school of theory believes that AdaBoost is essentially optimizing exponential loss based on an additive model with an iterative process like Newton's method. It inspires the replace of the iterative method to other optimization methods, leading to AdaBoost variants, such as GradientBoosting (Friedman 2001) and LPBoost (Demiriz et al. 2008). However, there are significant differences between the practical behavior of AdaBoost and that derived from this school of theory (Mease and Wyner 2008), especially it could not explain the fundamentally amazing phenomenon of why AdaBoost does not overfit. Therefore, many researchers argue that the statistical view only explains a learning process that is similar to AdaBoost but not AdaBoost itself. The *margin theory* (Schapire et al. 1998) offers an intuitive explanation of this important phenomenon, but it has been in challenging in the past 15 years until recent studies established final conclusions and shed light on the design of new learning methods. See Zhou (2014) for more information.

A more strict description of this phenomenon is "Why AdaBoost can further improve the generalization ability even after the training error reaches zero?" Indeed, if training continues, overfitting will eventually happen.

In addition to the basic combination methods introduced in this chapter, many other advanced methods exist, including the methods based on Dempster-Shafer theory, Dynamic Classifier Selection, and mixture of experts. This chapter only introduced pairwise diversity measures. Kuncheva and Whitaker (2003), Tang et al. (2006) showed that most existing diversity measures have drawbacks. The understanding of diversity is considered as the holy grail problem in ensemble learning research. More information about combination methods and diversity can be found in Chaps. 4 and 5 of Zhou (2012).

After obtaining an ensemble, trying to eliminate some individuals to get a smaller ensemble is called *ensemble pruning*, which helps reduce the storage and prediction time costs. Early studies mainly focused on pruning in sequential ensemble methods, but it was found that pruning hurts the generalization performance (Rokach 2010a). Zhou et al. (2002) disclosed that the pruning of parallel ensembles can reduce the ensemble size while improving the generalization ability, which opened the door to optimization-based ensemble pruning techniques. More information about this topic can be found in Chap. 6 of Zhou (2012).

Pruning of parallel ensembles is also called *selective ensemble*. However, nowadays, the term selective ensemble is often used as a synonym of ensemble pruning, which is also called *ensemble selection*.

Chapters 7 and 8 in Zhou (2012) discussed the use of ensemble learning in other learning tasks such as clustering, semi-supervised learning, and cost-sensitive learning. Indeed, ensemble learning is now widely used in most learning tasks, and almost all winning solutions in the KDD Cups (a famous data mining competition) employed ensemble learning.

Since an ensemble consists of multiple learners, it is a black box, even if the individual learners have good interpretability. There are attempts to improve the interpretability of ensembles, such as converting an ensemble to a single model or extracting symbolic rules from an ensemble. Research on this topic derived *twice-learning* techniques, such as NeC4.5 (Zhou and Jiang 2004), which can produce a single learner that outperforms the ensemble; later, similar techniques are called knowledge distillation. Besides, visualization techniques are also helpful in improving interpretability. More information on this topic can be found in Chap. 8 of Zhou (2012).

## Exercises

**8.1** Suppose we toss a coin with a probability of $p$ it lands on heads and a probability of $1-p$ it lands on tails. Let $H(n)$ denote the number of heads tossing the coin $n$ times, then the probability of getting at most $k$ heads is

$$P(H(n) \leqslant k) = \sum_{i=0}^{k} \binom{i}{n} p^i (1-p)^{n-i}. \tag{8.43}$$

For $\delta > 0$ and $k = (p-\delta)n$, we have Hoeffding's inequality

$$P(H(n) \leqslant (p-\delta)n) \leqslant e^{-2\delta^2 n}. \tag{8.44}$$

Try to derive (8.3).

**8.2** The exponential loss function is not the only consistent surrogate function for the 0/1 loss function. Considering (8.5), prove that every loss function $\ell(-f(\boldsymbol{x})H(\boldsymbol{x}))$ is a consistent surrogate function for the 0/1 loss function if $H(\boldsymbol{x})$ is monotonically decreasing in the interval $[-\infty, \delta]$ ($\delta > 0$).

8

**8.3** Download or implement the AdaBoost algorithm, and then use unpruned decision trees as base learners to train an AdaBoost ensemble on the watermelon data set 3.0$\alpha$. Compare the results with ◘ Figure 8.4.

The watermelon data set 3.0$\alpha$ is in ◘ Table 4.5.

**8.4** GradientBoosting (Friedman 2001) is a common Boosting algorithm. Try to analyze its difference and commonality to the AdaBoost algorithm.

**8.5** Implement Bagging, and then use decision stumps as base learners to train a Bagging ensemble on the watermelon data set 3.0$\alpha$. Compare the results with ◘ Figure 8.4.

**8.6** Analyze the reasons why Bagging is not effective in improving the performance of naïve Bayes classifiers.

**8.7** Analyze the reasons why it is faster to train a random forest than training a decision tree-based Bagging ensemble.

**8.8** The MultiBoosting algorithm (Webb 2000) employs AdaBoost as the base learner in Bagging, while the Iterative Bagging algorithm (Breiman 2001b) employs Bagging as the base learner in AdaBoost. Discuss the pros and cons of each method.

**8.9** * Design a visualized diversity measure. Use it to evaluate the ensembles obtained in Exercises 8.3 and 8.5, and compare with the $\kappa$-error diagrams.

**8.10** * Design an ensemble learning algorithm that can improve the performance of $k$-nearest neighbors classifiers.

## Break Time

### Short Story: Leo Breiman in His Green Old Age

Leo Breiman (1928–2005) was a great statistician in the twentieth century. At the end of the twentieth century, he publicly stated that the statistics community had made statistics like a branch of abstract mathematics, which divorced from the original intention of statistics. Instead, he believed that "*Statistics existed for the purposes of prediction and explanation and working with data.*" As a statistician, he claimed that his research was more about machine learning since there were more data-related challenges in this area. In fact, Breiman was an outstanding machine learning researcher, who not only developed the CART decision tree, but also made three major contributions to ensemble learning: Bagging, Random Forest, and theoretical arguments about Boosting. Interestingly, all of these were completed after he retired in 1993 from the Department of Statistics, UC Berkeley.



In the early days, Breiman obtained a degree in physics from the California Institute of Technology and decided to study philosophy at Columbia University. However, the head of the Department of Philosophy told Breiman that two of his best Ph.D. students could not find jobs, and hence Breiman changed his mind and turned to study mathematics. After earning his master's and Ph.D. degrees from UC Berkeley, he became a professor at the University of California, Los Angeles (UCLA) teaching probabilities. 7 years later, he was bored of probabilities and resigned from the position. To say goodbye to probabilities, he spent half a year at home to write a book about probabilities and then started a career as a statistical consultant in the industry for 13 years. Then, he came back to the Department of Statistics, UC Berkeley as a professor. Breiman had a variety of life experiences. He spent his sabbatical leave working for UNESCO as an educational statistician in Liberia to teach statistics for out-of-school children. He was also an amateur sculptor, and he even managed business with partners selling ice in Mexico. Random forest, which Breiman thought was the most important research outcome of his life, was developed after he was in his 70s.

## References

Breiman L (1996a) Bagging predictors. Mach Learn 24(2):123–140
Breiman L (1996b) Stacked regressions. Mach Learn 24(1):49–64
Breiman L (2000) Randomizing outputs to increase prediction accuracy. Mach Learn 40(3):113–120
Breiman L (2001a) Random forests. Mach Learn 45(1):5–32
Breiman L (2001b) Using iterated bagging to debias regressions. Mach Learn 45(3):261–277
Clarke B (2003) Comparing Bayes model averaging and stacking when model approximation error cannot be ignored. J Mach Learn Res 4:683–712
Demiriz A, Bennett KP, Shawe-Taylor J (2008) Linear programming Boosting via column generation. Mach Learn 46(1–3):225–254
Dietterich TG (2000) Ensemble methods in machine learning. In: Proceedings of the 1st international workshop on multiple classifier systems (MCS), pp 1–15. Cagliari, Italy
Dietterich TG, Bakiri G (1995) Solving multiclass learning problems via error-correcting output codes. J Artif Intell Res 2:263–286
Freund Y, Schapire RE (1997) A decision-theoretic generalization of on-line learning and an application to boosting. J Comput Syst Sci 55(1):119–139
Friedman J, Hastie T, Tibshirani R (2000) Additive logistic regression: a statistical view of boosting (with discussions). Ann Stat 28(2):337–407
Friedman JH (2001) Greedy function approximation: a gradient Boosting machine. Ann Stat 29(5):1189–1232
Ho TK (1998) The random subspace method for constructing decision forests. IEEE Trans Patt Anal Mach Intell 20(8):832–844
Ho TK, Hull JJ, Srihari SN (1994) Decision combination in multiple classifier systems. IEEE Trans Patt Anal Mach Intell 16(1):66–75
Kearns M, Valiant LG (1989) Cryptographic limitations on learning Boolean formulae and finite automata. In: Proceedings of the 21st annual ACM symposium on theory of computing (STOC), pp 433–444. Seattle, WA
Kittler J, Hatef M, Duin R, Matas J (1998) On combining classifiers. IEEE Trans Patt Anal Mach Intell 20(3):226–239
Kohavi R, Wolpert DH (1996) Bias plus variance decomposition for zero-one loss functions. In: Proceedings of the 13th international conference on machine learning (ICML), pp 275–283. Bari, Italy
Krogh A, Vedelsby J (1995) Neural network ensembles, cross validation and active learning. In: Tesauro G, Touretzky DS, Leen TK (eds) Advances in neural information processing systems 7 (NIPS). MIT Press, Cambridge, MA, pp 231–238
Kuncheva LI (2004) Combining pattern classifiers: methods and algorithms. Wiley, Hoboken, NJ
Kuncheva LI, Whitaker CJ (2003) Measures of diversity in classifier ensembles and their relationship with the ensemble accuracy. Mach Learn 51(2):181–207
Liu Y, Yao X (1999) Ensemble learning via negative correlation. Neural Netw 12(10):1399–1404
Markowitz H (1952) Portfolio selection. J Financ 7(1):77–91
Mease D, Wyner A (2008) Evidence contrary to the statistical view of boosting (with discussions). J Mach Learn Res 9:131–201
Perrone MP, Cooper LN (1993) When networks disagree: ensemble method for neural networks. In: Mammone RJ (ed) Artificial neural networks for speech and vision. Chapman & Hall, New York, NY, pp 126–142

Platt J (2000) Probabilities for SV machines. In: Smola A, Bartlett P, Schölkopf B, Schuurmans D (eds) Advances in large margin classifiers. MIT Press, Cambridge, MA, pp 61–74

Rokach L (2010a) Ensemble-based classifiers. Artif Intell Rev 33(1):1–39

Rokach L (2010b) Pattern classification using ensemble methods. World Scientific, Singapore

Schapire RE (1990) The strength of weak learnability. Mach Learn 5(2):197–227

Schapire RE, Freund Y (2012) Boosting: foundations and algorithms. MIT Press, Cambridge, MA

Schapire RE, Freund Y, Bartlett P, Lee WS (1998) Boosting the margin: a new explanation for the effectiveness of voting methods. Ann Stat 26(5):1651–1686

Seewald AK (2002) How to make stacking better and faster while also taking care of an unknown weakness. In: Proceedings of the 19th international conference on machine learning (ICML), pp 554–561. Sydney, Australia

Tang EK, Suganthan PN, Yao X (2006) An analysis of diversity measures. Mach Learn 65(1):247–271

Ting KM, Witten IH (1999) Issues in stacked generalization. J Artif Intell Res 10:271–289

Webb GI (2000) MultiBoosting: a technique for combining boosting and wagging. Mach Learn 40(2):159–196

Wolpert DH (1992) Stacked generalization. Neural Netw 5(2):241–260

Wolpert DH, Macready WG (1999) An efficient method to estimate Bagging's generalization error. Mach Learn 35(1):41–55

Xu L, Krzyzak A, Suen CY (1992) Methods of combining multiple classifiers and their applications to handwriting recognition. IEEE Trans Syst Man Cybern 22(3):418–435

Zadrozny B, Elkan C (2001) Obtaining calibrated probability estimates from decision trees and naive Bayesian classifiers. In: Proceedings of the 18th international conference on machine learning (ICML), pp 609–616. Williamstown, MA

Zhou Z-H (2012) Ensemble methods: foundations and algorithms. Chapman & Hall/CRC, Boca Raton, FL

Zhou Z-H (2014) Large margin distribution learning. In: Proceedings of the 6th IAPR international workshop on artificial neural networks in pattern recognition (ANNPR), pp 1–11. Montreal, Canada

Zhou Z-H, Jiang Y (2004) NeC4.5: neural ensemble based C4.5. IEEE Trans Knowl Data Eng 16(6):770–773

Zhou Z-H, Wu J, Wei T (2002) Ensembling neural networks: many could be better than all. Artif Intell 137(1–2):239–263

# Clustering

Table of Contents




Z.-H. Zhou, *Machine Learning*,
https://doi.org/10.1007/978-981-15-1967-3_9

## 9.1 Clustering Problem

*Unsupervised learning* aims to discover underlying properties and patterns from unlabeled training samples and lays the foundation for further data analysis. Among various unsupervised learning techniques, the most researched and applied one is *clustering*.

Other unsupervised learning tasks include *density estimation*, *anomaly detection*, etc.

Clustering aims to partition a data set into disjoint subsets, where each subset is called a *cluster*. Through the partitioning, each cluster is potentially corresponding to a concept (category), such as "light green watermelon", "dark green watermelon", "seeded watermelon", "seedless watermelon", or even "locally grown watermelon" and "imported watermelon". Note that clustering algorithms are unaware of such concepts before clustering and are only responsible for creating the clusters. The concept carried by each cluster, however, is interpreted by the user.

For clustering algorithms, a cluster is also called a *class*.

Formally, given a data set $D = \{\boldsymbol{x}_1, \boldsymbol{x}_2, \ldots, \boldsymbol{x}_m\}$ containing $m$ unlabeled samples, where each sample $\boldsymbol{x}_i = (x_{i1}; x_{i2}; \ldots; x_{in})$ is an $n$-dimensional vector. Then, a clustering algorithm partitions the data set $D$ into $k$ disjoint clusters $\{C_l \mid l = 1, 2, \ldots, k\}$, where $C_{l'} \cap_{l' \neq l} C_l = \varnothing$ and $D = \cup_{l=1}^{k} C_l$. Accordingly, we denote $\lambda_j \in \{1, 2, \ldots, k\}$ as the *cluster label* of sample $\boldsymbol{x}_j$ (i.e., $\boldsymbol{x}_j \in C_{\lambda_j}$). Then, the clustering result can be represented as a cluster label vector $\boldsymbol{\lambda} = (\lambda_1; \lambda_2; \ldots; \lambda_m)$ with $m$ elements.

Clustering can also use labeled samples (e.g., Sects. 9.4.2 and 13.6). However, the generated clusters and the class labels are different.

9

Clustering can be used by itself to identify the inherent structure of data, while it can also serve as a pre-processing technique for other learning tasks such as classification. For example, a business may want to classify new users into different "categories", but this may not be easy. In such a case, we can use clustering to group all users into clusters, where each cluster represents a user category. Then, a classification model can be built upon the clusters for classifying the category of new users.

Depending on the learning strategy used, clustering algorithms can be divided into several categories. The representative algorithms of each category will be discussed in the second half of this chapter. Before that, let us first discuss two fundamental problems involved with clustering—performance measure and distance calculation.

## 9.2 Performance Measure

Performance measures for clustering are also called *validity indices*. As classification result is evaluated by performance measures in supervised learning, the clustering result also needs to be evaluated via some validity indices. Besides, once a validity index is selected, we can embed it into the optimization objective of clustering algorithms such that the generated clusters are more aligned to the desired results.

See Sect. 2.3 for performance measures in supervised learning.

So, how does a good clustering look like? Intuitively, we wish "things of a kind come together"; that is, samples in the same cluster should be as similar as possible while samples from different clusters should be as different as possible. In other words, we seek clusters with high *intra-cluster similarity* and low *inter-cluster similarity*.

Roughly speaking, there are two types of clustering validity indices. The first type is *external index*, which compares the clustering result against a *reference model*. The second type is *internal index*, which evaluates the clustering result without using any reference model.

For example, use the clusters provided by domain experts as a reference model.

Given a data set $D = \{\boldsymbol{x}_1, \boldsymbol{x}_2, \ldots, \boldsymbol{x}_m\}$, suppose a clustering algorithm produces the clusters $\mathcal{C} = \{C_1, C_2, \ldots, C_k\}$, and a reference model gives the clusters $\mathcal{C}^* = \{C_1^*, C_2^*, \ldots, C_s^*\}$. Accordingly, let $\boldsymbol{\lambda}$ and $\boldsymbol{\lambda}^*$ denote the clustering label vectors of $\mathcal{C}$ and $\mathcal{C}^*$, respectively. Then, for each pair of samples, we can define the following four terms

Typically, $k \neq s$.

$$a = |SS|, \quad SS = \{(\boldsymbol{x}_i, \boldsymbol{x}_j) \mid \lambda_i = \lambda_j, \lambda_i^* = \lambda_j^*, i < j\}, \tag{9.1}$$

$$b = |SD|, \quad SD = \{(\boldsymbol{x}_i, \boldsymbol{x}_j) \mid \lambda_i = \lambda_j, \lambda_i^* \neq \lambda_j^*, i < j\}, \tag{9.2}$$

$$c = |DS|, \quad DS = \{(\boldsymbol{x}_i, \boldsymbol{x}_j) \mid \lambda_i \neq \lambda_j, \lambda_i^* = \lambda_j^*, i < j\}, \tag{9.3}$$

$$d = |DD|, \quad DD = \{(\boldsymbol{x}_i, \boldsymbol{x}_j) \mid \lambda_i \neq \lambda_j, \lambda_i^* \neq \lambda_j^*, i < j\}, \tag{9.4}$$

where the set $SS$ includes the sample pairs that both samples belong to the same cluster in $\mathcal{C}$ and also belong to the same cluster in $\mathcal{C}^*$; the set $SD$ includes sample pairs that both samples belong to the same cluster in $\mathcal{C}$ but not in $\mathcal{C}^*$; the sets $DS$ and $DD$ can be interpreted similarly. Since each sample pair $(\boldsymbol{x}_i, \boldsymbol{x}_j)$ $(i < j)$ can only appear in one set, we have $a + b + c + d = m(m-1)/2$.

With (9.1)–(9.4), some commonly used external indices can be defined as follows:

- Jaccard Coefficient (JC):

$$\text{JC} = \frac{a}{a+b+c}. \tag{9.5}$$

- Fowlkes and Mallows Index (FMI):

$$\text{FMI} = \sqrt{\frac{a}{a+b} \cdot \frac{a}{a+c}}. \tag{9.6}$$

- Rand Index (RI):

$$\text{RI} = \frac{2(a+d)}{m(m-1)}. \tag{9.7}$$

The above external validity indices take values in the interval [0, 1], and a larger index value indicates better clustering quality.

Internal validity indices evaluate the clustering quality without using a reference model. Given the generated clusters $\mathcal{C} = \{C_1, C_2, \ldots, C_k\}$, we can define the following four terms:

$$\text{avg}(C) = \frac{2}{|C|\,(|C|-1)} \sum_{1 \leqslant i < j \leqslant |C|} \text{dist}(\boldsymbol{x}_i, \boldsymbol{x}_j), \tag{9.8}$$

$$\text{diam}(C) = \max_{1 \leqslant i < j \leqslant |C|} \text{dist}(\boldsymbol{x}_i, \boldsymbol{x}_j), \tag{9.9}$$

$$d_{\min}(C_i, C_j) = \min_{\boldsymbol{x}_i \in C_i, \boldsymbol{x}_j \in C_j} \text{dist}(\boldsymbol{x}_i, \boldsymbol{x}_j), \tag{9.10}$$

$$d_{\text{cen}}(C_i, C_j) = \text{dist}(\boldsymbol{\mu}_i, \boldsymbol{\mu}_j), \tag{9.11}$$

where $\boldsymbol{\mu} = \frac{1}{|C|} \sum_{1 \leqslant i \leqslant |C|} \boldsymbol{x}_i$ denotes the centroid of cluster $C$, and $\text{dist}(\cdot, \cdot)$ measures the distance between two samples. Here, $\text{avg}(C)$ is the average distance between the samples in cluster $C$; $\text{diam}(C)$ is the largest distance between samples in cluster $C$; $d_{\min}(C_i, C_j)$ is the distance between two nearest samples in clusters $C_i$ and $C_j$; and $d_{\text{cen}}(C_i, C_j)$ is the distance between the centroids of clusters $C_i$ and $C_j$.

A larger distance corresponds to a lower similarity. Distance calculations will be discussed in the next section.

With (9.8)–(9.11), some commonly used internal validity indices can be defined as follows:

- Davies–Bouldin Index (DBI):

$$\text{DBI} = \frac{1}{k} \sum_{i=1}^{k} \max_{j \neq i} \left( \frac{\text{avg}(C_i) + \text{avg}(C_j)}{d_{\text{cen}}(C_i, C_j)} \right). \tag{9.12}$$

- Dunn Index (DI):

$$\text{DI} = \min_{1 \leqslant i \leqslant k} \left\{ \min_{j \neq i} \left( \frac{d_{\min}(C_i, C_j)}{\max_{1 \leqslant l \leqslant k} \text{diam}(C_l)} \right) \right\}. \tag{9.13}$$

A smaller value of DBI indicates better clustering quality, while a larger value of DI indicates better clustering quality.

## 9.3 Distance Calculation

Given a function $\text{dist}(\cdot, \cdot)$, if it is a *distance measure*, then it must have the following axioms:

$$\text{Non-negativity: } \text{dist}(\boldsymbol{x}_i, \boldsymbol{x}_j) \geqslant 0; \tag{9.14}$$

$$\text{Identity of indiscernibles: } \text{dist}(\boldsymbol{x}_i, \boldsymbol{x}_j) = 0 \text{ if and only if } \boldsymbol{x}_i = \boldsymbol{x}_j; \tag{9.15}$$

$$\text{Symmetry: } \text{dist}(\boldsymbol{x}_i, \boldsymbol{x}_j) = \text{dist}(\boldsymbol{x}_j, \boldsymbol{x}_i); \tag{9.16}$$

$$\text{Subadditivity: } \text{dist}(\boldsymbol{x}_i, \boldsymbol{x}_j) \leqslant \text{dist}(\boldsymbol{x}_i, \boldsymbol{x}_k) + \text{dist}(\boldsymbol{x}_k, \boldsymbol{x}_j). \tag{9.17}$$

Subadditivity is also known as *triangle inequality*.

Given two samples $\boldsymbol{x}_i = (x_{i1}; x_{i2}; \ldots; x_{in})$ and $\boldsymbol{x}_j = (x_{j1}; x_{j2}; \ldots; x_{jn})$, a commonly used measure is the *Minkowski distance*

$$\text{dist}_{\text{mk}}(\boldsymbol{x}_i, \boldsymbol{x}_j) = \left( \sum_{u=1}^{n} |x_{iu} - x_{ju}|^p \right)^{\frac{1}{p}}. \tag{9.18}$$

Equation (9.18) is the $\text{L}_p$ norm of $\boldsymbol{x}_i - \boldsymbol{x}_j$, that is, $\|\boldsymbol{x}_i - \boldsymbol{x}_j\|_p$.

When $p \geqslant 1$, (9.18) satisfies the distance measure axioms (9.14)–(9.17).

When $p = 2$, the Minkowski distance becomes the *Euclidean distance*

When $p \to \infty$, it becomes the *Chebyshev distance*.

$$\text{dist}_{\text{ed}}(\boldsymbol{x}_i, \boldsymbol{x}_j) = \|\boldsymbol{x}_i - \boldsymbol{x}_j\|_2 = \sqrt{\sum_{u=1}^{n} |x_{iu} - x_{ju}|^2}. \tag{9.19}$$

When $p = 1$, the Minkowski distance becomes the *Manhattan distance*

Also known as the *city block distance*.

$$\text{dist}_{\text{man}}(\boldsymbol{x}_i, \boldsymbol{x}_j) = \|\boldsymbol{x}_i - \boldsymbol{x}_j\|_1 = \sum_{u=1}^{n} |x_{iu} - x_{ju}|. \tag{9.20}$$

We generally divide attributes into *continuous attributes*, which have infinite domains, and *categorical attributes*, which have finite domains. However, for distance calculations, it is more important to consider whether the attributes include *ordinal information*. For example, a categorical attribute with the domain $\{1, 2, 3\}$ is more like a continuous attribute since the distance can be calculated with the attribute values, that is, "1" is closer to "2" than "3". Such attributes are called *ordinal attributes*. In contrast, if the domain of an attribute is discrete like {aircraft, train, ship}, then the distance cannot be

Continuous attributes are also known as *numerical attributes*, and categorical attributes are also known as *nominal attributes*.

directly calculated with the attribute values, and such attributes are *non-ordinal attributes*. Note that the Minkowski distance is only applicable to ordinal attributes.

For non-ordinal attributes, the Value Difference Metric (VDM) (Stanfill and Waltz 1986) can be used instead. Let $m_{u,a}$ denote the number of samples taking value $a$ on the attribute $u$, and $m_{u,a,i}$ denote the number of samples within the $i$th cluster taking value $a$ on the attribute $u$, and $k$ is the number of clusters. Then, the VDM distance between two categorical values $a$ and $b$ of attribute $u$ is

When the class information is given, $k$ is usually set to the number of classes.

$$\mathrm{VDM}_p(a, b) = \sum_{i=1}^{k} \left| \frac{m_{u,a,i}}{m_{u,a}} - \frac{m_{u,b,i}}{m_{u,b}} \right|^p. \tag{9.21}$$

For mixed attribute types, we can combine the Minkowski distance and the VDM. Without loss of generality, we arrange ordinal attributes in front of non-ordinal attributes and let $n_c$ denote the number of ordinal attributes and $n - n_c$ denote the number of non-ordinal attributes. Then, the joint distance measure is

9

$$\mathrm{MinkovDM}_p(\boldsymbol{x}_i, \boldsymbol{x}_j) = \left( \sum_{u=1}^{n_c} |x_{iu} - x_{ju}|^p + \sum_{u=n_c+1}^{n} \mathrm{VDM}_p(x_{iu}, x_{ju}) \right)^{\frac{1}{p}}. \tag{9.22}$$

In cases where different attributes have different importance, we can use *weighted distance*. For example, the weighted Minkowski distance is

$$\mathrm{dist}_{\mathrm{wmk}}(\boldsymbol{x}_i, \boldsymbol{x}_j) = (w_1 \cdot |x_{i1} - x_{j1}|^p + \ldots + w_n \cdot |x_{in} - x_{jn}|^p)^{\frac{1}{p}}, \tag{9.23}$$

where the weights $w_i \geqslant 0$ $(i = 1, 2, \ldots, n)$ represent the importance of attributes, and typically $\sum_{i=1}^{n} w_i = 1$.

We often define *similarity measures* via some kinds of distances, and the larger the distance, the lower the similarity. However, it is worth noting that distances used in defining similarity measures are not required to satisfy all axioms of being distance measures, particularly the subadditivity (9.17). For example, we may seek a similarity measure that considers humans and horses are different, but considers both are similar to centaurs. To make this happen, we need to ensure small distances from humans and horses to centaurs, while ensuring a large distance between humans and horses. Such distances, as illustrated in ◘ Figure 9.1, do not satisfy the subadditivity

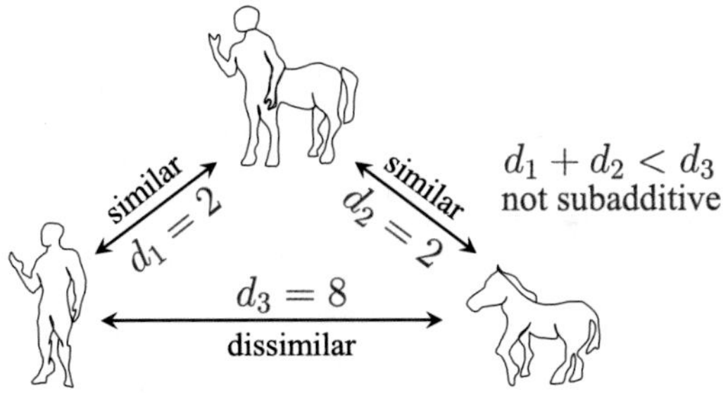


**Fig. 9.1** An example of non-metric distance

Mathematically, setting $d_3 = 3$ gives subadditivity. Semantically, however, $d_3$ should be much larger than $d_1$ and $d_2$.

condition, and hence are called *non-metric distances*. Last but not least, the distance calculation methods introduced in this section are defined in advance, but in some real-world tasks, it is necessary to determine the distance calculation method based on the data samples via *distance metric learning*.

See Sect. 10.6 for distance metric learning.

## 9.4 Prototype Clustering

*Prototype clustering*, also known as *prototype-based clustering*, is a family of clustering algorithms that assumes the clustering structure can be represented by a set of prototypes. Typically, such algorithms start with some initial prototypes, and then iteratively update and optimize the prototypes. Many algorithms have been developed using different prototype representations and optimization methods. The rest of this section discusses several well-known prototype-based clustering algorithms.

A "prototype" refers to a representative data point in the sample space.

### 9.4.1 *k*-Means Clustering

Given a data set $D = \{\boldsymbol{x}_1, \boldsymbol{x}_2, \ldots, \boldsymbol{x}_m\}$, the $k$-means algorithm minimizes the squared error of clusters $\mathcal{C} = \{C_1, C_2, \ldots, C_k\}$:

$$E = \sum_{i=1}^{k} \sum_{\boldsymbol{x} \in C_i} \|\boldsymbol{x} - \boldsymbol{\mu}_i\|_2^2, \tag{9.24}$$

where $\boldsymbol{\mu}_i = \frac{1}{|C_i|} \sum_{\boldsymbol{x} \in C_i} \boldsymbol{x}$ is the mean vector of cluster $C_i$. Intuitively, (9.24) represents the closeness between the mean vector of a cluster and the samples within that cluster, where a smaller $E$ indicates higher intra-cluster similarity.

Nevertheless, minimizing (9.24) is not easy since it requires evaluations of all possible partitions of the data set $D$, which is indeed an NP-hard problem (Aloise et al. 2009). Hence, the $k$-means algorithm takes a greedy strategy and adopts an iterative optimization method to find an approximate solution of (9.24). The algorithm is illustrated in ◻ Algorithm 9.1, where line 1 initializes the mean vectors, and lines 4–8 and lines 9–16 iteratively update the clusters and mean vectors, respectively. When the clusters do not change after one iteration, the current clusters are returned.

We take the watermelon data set 4.0 in ◻ Table 9.1 as an example to demonstrate the $k$-means algorithm. For ease of discussion, let $\boldsymbol{x}_i$ represent the sample with the ID $i$, where $\boldsymbol{x}_i$ is a two-dimensional vector containing the attributes `density` and `sugar`.

The watermelon data set 3.0$\alpha$ in ◻ Table 4.5 is a subset of the watermelon data set 4.0.

The class label of the samples 9–21 is `ripe=false`, and the class label of the rest samples is `ripe=true`. Since we do not use the label information in this section, the labels are omitted in the table.

◻ **Tab. 9.1** The watermelon data set 4.0

| ID | density | sugar | ID | density | sugar | ID | density | sugar |
|---|---|---|---|---|---|---|---|---|
| 1 | 0.697 | 0.460 | 11 | 0.245 | 0.057 | 21 | 0.748 | 0.232 |
| 2 | 0.774 | 0.376 | 12 | 0.343 | 0.099 | 22 | 0.714 | 0.346 |
| 3 | 0.634 | 0.264 | 13 | 0.639 | 0.161 | 23 | 0.483 | 0.312 |
| 4 | 0.608 | 0.318 | 14 | 0.657 | 0.198 | 24 | 0.478 | 0.437 |
| 5 | 0.556 | 0.215 | 15 | 0.360 | 0.370 | 25 | 0.525 | 0.369 |
| 6 | 0.403 | 0.237 | 16 | 0.593 | 0.042 | 26 | 0.751 | 0.489 |
| 7 | 0.481 | 0.149 | 17 | 0.719 | 0.103 | 27 | 0.532 | 0.472 |
| 8 | 0.437 | 0.211 | 18 | 0.359 | 0.188 | 28 | 0.473 | 0.376 |
| 9 | 0.666 | 0.091 | 19 | 0.339 | 0.241 | 29 | 0.725 | 0.445 |
| 10 | 0.243 | 0.267 | 20 | 0.282 | 0.257 | 30 | 0.446 | 0.459 |

Suppose we set $k = 3$, then the algorithm randomly picks up three samples $\boldsymbol{x}_6$, $\boldsymbol{x}_{12}$, and $\boldsymbol{x}_{24}$ as the initial mean vectors, that is,

$$\boldsymbol{\mu}_1 = (0.403; 0.237),\quad \boldsymbol{\mu}_2 = (0.343; 0.099),\quad \boldsymbol{\mu}_3 = (0.478; 0.437).$$

Then, for the sample $\boldsymbol{x}_1 = (0.697; 0.460)$, its distances to the three current mean vectors $\boldsymbol{\mu}_1$, $\boldsymbol{\mu}_2$, and $\boldsymbol{\mu}_3$ are 0.369, 0.506, and 0.220, respectively. Since its distance to $\boldsymbol{\mu}_3$ is the shortest, $\boldsymbol{x}_1$ is assigned to cluster $C_3$. Similarly, we evaluate all samples in the data set and find the following cluster assignments:

**Algorithm 9.1** $k$-Means Clustering

**Input:** Data set $D = \{\boldsymbol{x}_1, \boldsymbol{x}_2, \ldots, \boldsymbol{x}_m\}$;
Number of clusters $k$.

**Process:**

Randomly select $k$ samples as the initial mean vectors $\{\boldsymbol{\mu}_1, \boldsymbol{\mu}_2, \ldots, \boldsymbol{\mu}_k\}$;
**repeat**
  $C_i = \varnothing (1 \leqslant i \leqslant k)$;
  **for** $j = 1, 2, \ldots, m$ **do**
    Compute the distance between sample $\boldsymbol{x}_j$ and each mean vector $\boldsymbol{\mu}_i (1 \leqslant i \leqslant k)$: $d_{ji} = \|\boldsymbol{x}_j - \boldsymbol{\mu}_i\|_2$;
    According to the nearest mean vector, decide the cluster label of $\boldsymbol{x}_j$: $\lambda_j = \arg\min_{i \in \{1,2,\ldots,k\}} d_{ji}$;
    Move $\boldsymbol{x}_j$ to the corresponding cluster: $C_{\lambda_j} = C_{\lambda_j} \cup \{\boldsymbol{x}_j\}$;
  **end for**
  **for** $i = 1, 2, \ldots, k$ **do**
    Compute the updated mean vectors: $\boldsymbol{\mu}'_i = \frac{1}{|C_i|} \sum_{\boldsymbol{x} \in C_i} \boldsymbol{x}$;
    **if** $\boldsymbol{\mu}'_i \neq \boldsymbol{\mu}_i$ **then**
      Update the current mean vector $\boldsymbol{\mu}_i$ to $\boldsymbol{\mu}'_i$;
    **else**
      Leave the current mean vector unchanged.
    **end if**
  **end for**
**until** All mean vectors remain unchanged

**Output:** Clusters $C = \{C_1, C_2, \ldots, C_k\}$ .

In order to avoid long execution time, we often set a maximum number of iterations or a threshold of minimum change to stop the learning early.

$$\begin{aligned} C_1 &= \{\boldsymbol{x}_3, \boldsymbol{x}_5, \boldsymbol{x}_6, \boldsymbol{x}_7, \boldsymbol{x}_8, \boldsymbol{x}_9, \boldsymbol{x}_{10}, \boldsymbol{x}_{13}, \boldsymbol{x}_{14}, \boldsymbol{x}_{17}, \boldsymbol{x}_{18}, \boldsymbol{x}_{19}, \boldsymbol{x}_{20}, \boldsymbol{x}_{23}\}; \\ C_2 &= \{\boldsymbol{x}_{11}, \boldsymbol{x}_{12}, \boldsymbol{x}_{16}\}; \\ C_3 &= \{\boldsymbol{x}_1, \boldsymbol{x}_2, \boldsymbol{x}_4, \boldsymbol{x}_{15}, \boldsymbol{x}_{21}, \boldsymbol{x}_{22}, \boldsymbol{x}_{24}, \boldsymbol{x}_{25}, \boldsymbol{x}_{26}, \boldsymbol{x}_{27}, \boldsymbol{x}_{28}, \boldsymbol{x}_{29}, \boldsymbol{x}_{30}\}. \end{aligned}$$

From $C_1$, $C_2$ and $C_3$, we can calculate the new mean vectors

$$\boldsymbol{\mu}'_1 = (0.493; 0.207), \quad \boldsymbol{\mu}'_2 = (0.394; 0.066), \quad \boldsymbol{\mu}'_3 = (0.602; 0.396).$$

The above process repeats until convergence. For example, as illustrated in ◘ Figure 9.2, the $k$-means algorithm finds the final cluster assignments when the 5th iteration produced the same clusters as the 4th iteration.

### 9.4.2 Learning Vector Quantization

Learning Vector Quantization (LVQ) is another prototype-based clustering algorithm similar to the $k$-means algorithm but requires labeled data samples; that is, the clustering process is assisted by supervised information.

The clustering process can be viewed as identifying *subclasses*, where each subclass corresponds to one cluster.

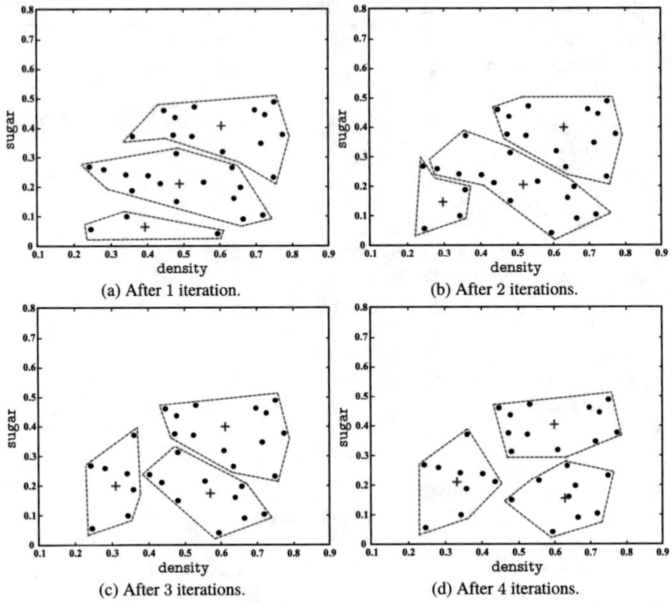


**Fig. 9.2** Results of the $k$-means algorithm on the watermelon data set 4.0 with $k = 3$. The samples and mean vectors are represented by "•" and "+", respectively. The red dashed lines are the boundaries of clusters



Given a data set $D = \{(\boldsymbol{x}_1, y_1), (\boldsymbol{x}, y_2), \ldots, (\boldsymbol{x}_m, y_m)\}$, where each sample $\boldsymbol{x}_j$ is described by an $n$-dimensional feature vector $(x_{j1}; x_{j2}; \ldots; x_{jn})$ and a class label $y_j \in \mathcal{Y}$. LVQ aims to learn a set of $n$-dimensional prototype vectors $\{\boldsymbol{p}_1, \boldsymbol{p}_2, \ldots, \boldsymbol{p}_q\}$, where each prototype vector represents one cluster and its class label is $t_i \in \mathcal{Y}$.

Line 5 of ◘ Algorithm 9.2 takes the "winner-take-all" strategy of competitive learning. SOM can be seen as a clustering algorithm for unlabeled samples, while LVQ can be seen as an extension of SOM that utilizes supervised information. More information about competitive learning and SOM can be found in Sects. 5.5.2 and 5.5.3.

As illustrated in ◘ Algorithm 9.2, the LVQ algorithm starts with an initialization of the prototype vectors in line 1, e.g., randomly selects a sample with class label $t_q$ as the prototype vector for the $q$th cluster. The prototype vectors are iteratively optimized in lines 2–12. In each round, the algorithm randomly selects a labeled sample and finds its nearest prototype vector, and then the prototype vector is updated based on whether the selected sample has the same class label with the prototype vector. In line 12, the current prototype vectors are returned as the final result if the stopping condition is met, e.g., the maximum number of iterations is reached, or there is minor or even no update to the prototype vectors.

**Algorithm 9.2** Learning Vector Quantization

**Input:** Training set $D = \{(\boldsymbol{x}_1, y_1), (\boldsymbol{x}_2, y_2), \ldots, (\boldsymbol{x}_m, y_m)\}$;
Number of prototype vectors $q$;
Initial labels of prototype vectors $\{t_1, t_2, \ldots, t_q\}$;
Learning rate $\eta$.

**Process:**

1: Initialize a set of prototype vectors $\{\boldsymbol{p}_1, \boldsymbol{p}_2, \ldots, \boldsymbol{p}_q\}$;
2: **repeat**
3: Randomly pickup a sample $(\boldsymbol{x}_j, y_j)$ from the data set $D$;
4: Compute the distance between $\boldsymbol{x}_j$ and $\boldsymbol{p}_i (1 \leqslant i \leqslant q)$: $d_{ji} = \|\boldsymbol{x}_j - \boldsymbol{p}_i\|_2$;
5: Find the nearest prototype vector $p_{i^*}$ for $\boldsymbol{x}_j$, where $i^* = \arg\min_{i \in \{1,2,\ldots,q\}} d_{ji}$;
6: **if** $y_j = t_{i^*}$ **then**
7: $\boldsymbol{p}' = \boldsymbol{p}_{i^*} + \eta \cdot (\boldsymbol{x}_j - \boldsymbol{p}_{i^*})$;
8: **else**
9: $\boldsymbol{p}' = \boldsymbol{p}_{i^*} - \eta \cdot (\boldsymbol{x}_j - \boldsymbol{p}_{i^*})$;
10: **end if**
11: Update the prototype vector $\boldsymbol{p}_{i^*}$ to $\boldsymbol{p}'$.
12: **until** The termination condition is met

**Output:** Prototype vectors $\{\boldsymbol{p}_1, \boldsymbol{p}_2, \ldots, \boldsymbol{p}_q\}$.

$\boldsymbol{x}_j$ and $\boldsymbol{p}_{i^*}$ have the same class label.

$\boldsymbol{x}_j$ and $\boldsymbol{p}_{i^*}$ have different class labels.

E.g., the maximum number of iterations is reached.

Lines 6–10, which update the prototype vectors, are the core of LVQ. Intuitively, if the class labels of a sample $\boldsymbol{x}_j$ and its nearest prototype vector $\boldsymbol{p}_{i^*}$ are the same, then $\boldsymbol{p}_{i^*}$ is moved toward $\boldsymbol{x}_j$. As shown in line 7, the prototype vector is updated by

$$\boldsymbol{p}' = \boldsymbol{p}_{i^*} + \eta \cdot (\boldsymbol{x}_j - \boldsymbol{p}_{i^*}), \tag{9.25}$$

and the distance between $\boldsymbol{p}'$ and $\boldsymbol{x}_j$ is

$$\begin{aligned} \|\boldsymbol{p}' - \boldsymbol{x}_j\|_2 &= \|\boldsymbol{p}_{i^*} + \eta \cdot (\boldsymbol{x}_j - \boldsymbol{p}_{i^*}) - \boldsymbol{x}_j\|_2 \\ &= (1 - \eta) \cdot \|\boldsymbol{p}_{i^*} - \boldsymbol{x}_j\|_2 \,. \end{aligned} \tag{9.26}$$

Suppose that the learning rate $\eta \in (0, 1)$, then the prototype vector $\boldsymbol{p}_{i^*}$ becomes closer to $\boldsymbol{x}_j$ after updating to $\boldsymbol{p}'$.

Similarly, if $\boldsymbol{p}_{i^*}$ and $\boldsymbol{x}_j$ have different class labels, then the distance between updated prototype vector and $\boldsymbol{x}_j$ is increased to $(1 + \eta) \cdot \|\boldsymbol{p}_{i^*} - \boldsymbol{x}_j\|_2$, that is, away from $\boldsymbol{x}_j$.

With a set of learned prototype vectors $\{\boldsymbol{p}_1, \boldsymbol{p}_2, \ldots, \boldsymbol{p}_q\}$, clustering assignments can be made for the sample space $\mathcal{X}$. Specifically, each sample $\boldsymbol{x}$ is assigned to the cluster represented by the nearest prototype vector. In other words, a prototype vector $\boldsymbol{p}_i$ defines a region $R_i$, where the distance from any sample in $R_i$ to $\boldsymbol{p}_i$ is not larger than the distance to any other prototype vector $\boldsymbol{p}_{i'} (i' \neq i)$, that is,

If all samples in the region $R_i$ are represented by the prototype vector $\boldsymbol{p}_i$, then we have done a *lossy compression* to the data, which is also called *vector quantization*, hence the name LVQ.

$$R_i = \{\boldsymbol{x} \in \mathcal{X} \mid \|\boldsymbol{x} - \boldsymbol{p}_i\|_2 \leqslant \|\boldsymbol{x} - \boldsymbol{p}_{i'}\|_2 \,, i' \neq i\}. \tag{9.27}$$

Therefore, $\{R_1, R_2, \ldots, R_q\}$ forms a partition of the sample space $\mathcal{X}$, known as the *Voronoi tessellation*.

Next, we take the watermelon data set 4.0 in ◘ Table 9.1 as an example to demonstrate the learning process of LVQ. Let $c_2$ be the class label of the samples with ID 9–21, and $c_1$ be the class label for the rest of the samples. Letting $q = 5$, that is, the goal is to find 5 prototype vectors $\boldsymbol{p}_1$, $\boldsymbol{p}_2$, $\boldsymbol{p}_3$, $\boldsymbol{p}_4$, and $\boldsymbol{p}_5$, and the corresponding class labels are $c_1$, $c_2$, $c_2$, $c_1$, and $c_1$, respectively.

That is, 3 clusters for ripe=true, and 2 clusters for ripe=false.

The algorithm starts by initializing the prototype vector of each cluster to a sample that has the same class label as the pre-defined class label of the cluster. Suppose that the selected samples for these 5 clusters are $\boldsymbol{x}_5$, $\boldsymbol{x}_{12}$, $\boldsymbol{x}_{18}$, $\boldsymbol{x}_{23}$, and $\boldsymbol{x}_{29}$. In the first iteration, if the randomly selected sample is $\boldsymbol{x}_1$, then its distances to the current prototype vectors $\boldsymbol{p}_1$, $\boldsymbol{p}_2$, $\boldsymbol{p}_3$, $\boldsymbol{p}_4$, and $\boldsymbol{p}_5$ are 0.283, 0.506, 0.434, 0.260, and 0.032, respectively. Since $\boldsymbol{p}_5$ is the nearest prototype vector to $\boldsymbol{x}_1$ and they have the same class label $c_1$, LVQ will update $\boldsymbol{p}_5$ to a new prototype vector

9

$$\begin{aligned}
\boldsymbol{p}' &= \boldsymbol{p}_5 + \eta \cdot (\boldsymbol{x}_1 - \boldsymbol{x}_5)\\
&= (0.752; 0.445) + 0.1 \cdot ((0.697; 0.460) - (0.725; 0.445))\\
&= (0.722; 0.447),
\end{aligned}$$

where $\eta = 0.1$ is a pre-specified learning rate. After $\boldsymbol{p}_5$ is updated to $\boldsymbol{p}'$, the learning process continues until convergence. ◘ Figure 9.3 shows the clustering results after different iterations.

### 9.4.3 Mixture-of-Gaussian Clustering

Unlike $k$-means and LVQ, Mixture-of-Gaussian clustering does not use prototype vectors but probabilistic models to represent clustering structures.

Before we discuss the technical details, let us revisit the definition of (multivariate) Gaussian distribution. For a random vector $\boldsymbol{x}$ in an $n$-dimensional sample space $\mathcal{X}$, if $\boldsymbol{x}$ follows a Gaussian distribution, then its probability density function is

Denoted by $\boldsymbol{x} \sim \mathcal{N}(\boldsymbol{\mu}, \boldsymbol{\Sigma})$.

$$p(\boldsymbol{x}) = \frac{1}{(2\pi)^{\frac{n}{2}} \, |\boldsymbol{\Sigma}|^{\frac{1}{2}}} e^{-\frac{1}{2}(\boldsymbol{x}-\boldsymbol{\mu})^{\top}\boldsymbol{\Sigma}^{-1}(\boldsymbol{x}-\boldsymbol{\mu})}, \tag{9.28}$$

$\boldsymbol{\Sigma}$: symmetric positive definite matrix;
$|\boldsymbol{\Sigma}|$: determinant of $\boldsymbol{\Sigma}$;
$\boldsymbol{\Sigma}^{-1}$: inverse of $\boldsymbol{\Sigma}$.

where $\boldsymbol{\mu}$ is an $n$-dimensional mean vector and $\boldsymbol{\Sigma}$ is an $n \times n$ covariance matrix. From (9.28), we can see that a Gaussian distribution is fully determined by its mean vector $\boldsymbol{\mu}$ and covari-

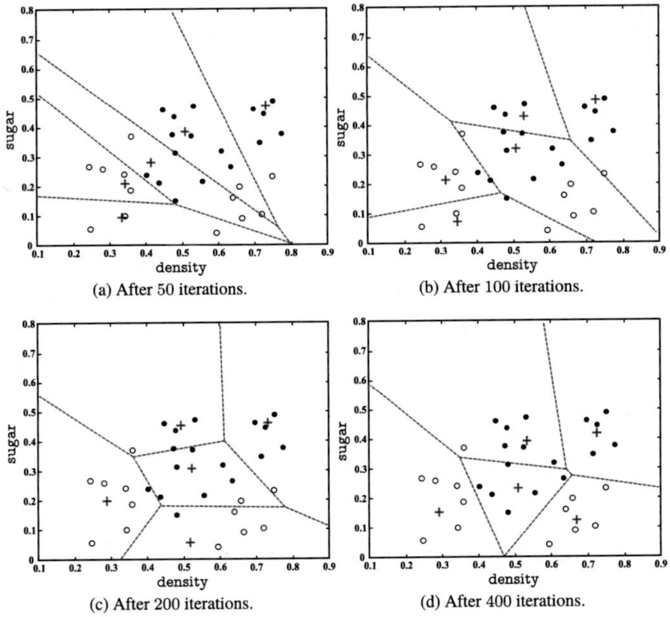


(a) After 50 iterations. (b) After 100 iterations.
(c) After 200 iterations. (d) After 400 iterations.

**Fig. 9.3** Results of the LVQ algorithm after different iterations on the watermelon data set 4.0 with $q = 5$. The symbols "•", "∘", "+" represent, respectively, the class $c_1$ samples, the class $c_2$ samples, and the prototype vectors. The red dashed lines show the Voronoi tessellation

ance matrix $\mathbf{\Sigma}$. To show this dependency more explicitly, we write the probability density function as $p(\boldsymbol{x} \mid \boldsymbol{\mu}, \mathbf{\Sigma})$.

The Mixture-of-Gaussian distribution is defined as

$$p_{\mathcal{M}}(\boldsymbol{x}) = \sum_{i=1}^{k} \alpha_i \cdot p(\boldsymbol{x} \mid \boldsymbol{\mu}_i, \mathbf{\Sigma}_i), \tag{9.29}$$

$p_{\mathcal{M}}(\cdot)$ is also a probability density function where $\int p_{\mathcal{M}}(\boldsymbol{x})d\boldsymbol{x} = 1$.

which consists of $k$ mixture components and each corresponds to a Gaussian distribution. $\boldsymbol{\mu}_i$ and $\mathbf{\Sigma}_i$ are the parameters of the $i$th mixture component, and $\alpha_i > 0$ are the corresponding *mixture coefficients*, where $\sum_{i=1}^{k} \alpha_i = 1$.

Suppose that the samples are generated from a Mixture-of-Gaussian distribution with the following process: select the Gaussian mixture components using the prior distribution defined by $\alpha_1, \alpha_2, \ldots, \alpha_k$, where $\alpha_i$ is the probability of selecting the $i$th mixture component; then, generate samples by sampling from the probability density functions of the selected mixture components.

Let $D = \{\boldsymbol{x}_1, \boldsymbol{x}_2, \ldots, \boldsymbol{x}_m\}$ be a training set generated from the above process, and $z_j \in \{1, 2, \ldots, k\}$ be the random variable of the Gaussian mixture component that generated sample $\boldsymbol{x}_j$, where the values of $z_j$ are unknown. Since the prior probability

$P(z_j = i)$ for $z_j$ corresponds to $\alpha_i (i = 1, 2, \ldots, k)$, the posterior distribution of $z_j$, according to Bayes' theorem, is

$$p_{\mathcal{M}}(z_j = i \mid \boldsymbol{x}_j) = \frac{P(z_j = i) \cdot p_{\mathcal{M}}(\boldsymbol{x}_j \mid z_j = i)}{p_{\mathcal{M}}(\boldsymbol{x}_j)}$$
$$= \frac{\alpha_i \cdot p(\boldsymbol{x}_j \mid \boldsymbol{\mu}_i, \boldsymbol{\Sigma}_i)}{\sum_{l=1}^{k} \alpha_l \cdot p(\boldsymbol{x}_j \mid \boldsymbol{\mu}_l, \boldsymbol{\Sigma}_l)}. \tag{9.30}$$

In other words, $p_{\mathcal{M}}(z_j = i \mid \boldsymbol{x}_j)$ gives the posterior probability that $\boldsymbol{x}_j$ is generated by the $i$th Gaussian mixture component. For ease of discussion, we denote it by $\gamma_{ji}$, where $i = 1, 2, \ldots, k$.

When the Mixture-of-Gaussian distribution is known, the data set $D$ can be divided into $k$ clusters $\mathcal{C} = \{C_1, C_2, \ldots, C_k\}$, and the cluster assignment $\lambda_j$ for each sample $\boldsymbol{x}_j$ is given by

$$\lambda_j = \underset{i \in \{1,2,\ldots,k\}}{\arg\max}\ \gamma_{ji}. \tag{9.31}$$

Hence, from the prototype clustering point of view, the Mixture-of-Gaussian clustering employs probabilistic models (with Gaussian distribution) to represent the prototypes, and the cluster assignments are made by the posterior probabilities of the prototypes.

9

How do we optimize the model parameters $\{(\alpha_i, \boldsymbol{\mu}_i, \boldsymbol{\Sigma}_i) \mid 1 \leqslant i \leqslant k\}$ in (9.29)? One method is to apply the maximum likelihood estimation on the data set $D$, that is, maximizing the (log) likelihood

See Sect. 7.2 for maximum likelihood estimation.

$$LL(D) = \ln\left(\prod_{j=1}^{m} p_{\mathcal{M}}(\boldsymbol{x}_j)\right)$$
$$= \sum_{j=1}^{m} \ln\left(\sum_{i=1}^{k} \alpha_i \cdot p(\boldsymbol{x}_j \mid \boldsymbol{\mu}_i, \boldsymbol{\Sigma}_i)\right). \tag{9.32}$$

The optimization problem is usually solved by the EM algorithm. We give a brief derivation as follows.

See Sect. 7.6 for the EM algorithm.

If the parameters $\{(\alpha_i, \boldsymbol{\mu}_i, \boldsymbol{\Sigma}_i) \mid 1 \leqslant i \leqslant k\}$ maximize (9.32), then, from $\frac{\partial LL(D)}{\partial \boldsymbol{\mu}_i} = 0$, we have

$$\sum_{j=1}^{m} \frac{\alpha_i \cdot p(\boldsymbol{x}_j \mid \boldsymbol{\mu}_i, \boldsymbol{\Sigma}_i)}{\sum_{l=1}^{k} \alpha_l \cdot p(\boldsymbol{x}_j \mid \boldsymbol{\mu}_l, \boldsymbol{\Sigma}_l)} (\boldsymbol{x}_j - \boldsymbol{\mu}_i) = 0. \tag{9.33}$$

From (9.30) and $\gamma_{ji} = p_{\mathcal{M}}(z_j = i \mid \boldsymbol{x}_j)$, we have

$$\boldsymbol{\mu}_i = \frac{\sum_{j=1}^{m} \gamma_{ji} \boldsymbol{x}_j}{\sum_{j=1}^{m} \gamma_{ji}}. \tag{9.34}$$

In other words, the mean of each mixture component can be calculated as a weighted average of the samples, where each sample is weighted by the posterior probability of this sample belonging to the given component. Similarly, from $\frac{\partial LL(D)}{\partial \boldsymbol{\Sigma}_i} = 0$, we have

$$\boldsymbol{\Sigma}_i = \frac{\sum_{j=1}^{m} \gamma_{ji}(\boldsymbol{x}_j - \boldsymbol{\mu}_i)(\boldsymbol{x}_j - \boldsymbol{\mu}_i)^\top}{\sum_{j=1}^{m} \gamma_{ji}}. \tag{9.35}$$

The maximization of $LL(D)$ is subject to the constraints on the mixture coefficients $\alpha_i$: $\alpha_i \geqslant 0$ and $\sum_{i=1}^{k} \alpha_i = 1$. Considering the Lagrange form of $LL(D)$

$$LL(D) + \lambda \left( \sum_{i=1}^{k} \alpha_i - 1 \right), \tag{9.36}$$

where $\lambda$ is the Lagrange multiplier. Setting the derivative of (9.36) with respect to $\alpha_i$ equal to zero gives

$$\sum_{j=1}^{m} \frac{p(\boldsymbol{x}_j \mid \boldsymbol{\mu}_i, \boldsymbol{\Sigma}_i)}{\sum_{l=1}^{k} \alpha_l \cdot p(\boldsymbol{x}_j \mid \boldsymbol{\mu}_l, \boldsymbol{\Sigma}_l)} + \lambda = 0. \tag{9.37}$$

If we multiply both sides of (9.37) by $\alpha_i$ and sum over all mixture components, we obtain $\lambda = -m$. Substituting it into (9.37) gives

$$\alpha_i = \frac{1}{m} \sum_{j=1}^{m} \gamma_{ji}, \tag{9.38}$$

which shows that the mixture coefficient of each Gaussian component is the average posterior probability of each sample belonging to this Gaussian component.

From the above derivation, we have the EM algorithm for the Gaussian mixture model: in each round, use the current parameters to calculate the posterior probability $\gamma_{ji}$ that each sample is belonging to each Gaussian component (E-step), and then update the parameters $\{(\alpha_i, \boldsymbol{\mu}_i, \boldsymbol{\Sigma}_i) \mid 1 \leqslant i \leqslant k\}$ based on (9.34), (9.35), and (9.38) (M-step).

The Mixture-of-Gaussian clustering algorithm is given in ◘ Algorithm 9.3. The algorithm starts by initializing the parameters in line 1. Then, in lines 2–12, the parameters are iteratively updated using the EM algorithm. When the maximum number of iterations is reached or the log-likelihood function $LL(D)$ is no longer increasing (or the increment is small), the EM algorithm stops, and the cluster assignments are made in lines 14–17.

---

**Algorithm 9.3** Mixture-of-Gaussian Clustering

**Input:** Data set $D = \{x_1, x_2, \ldots, x_m\}$;
  Number of Gaussian mixture components $k$.
**Process:**
1: Initialize the parameters $\{(\alpha_i, \mu_i, \Sigma_i) \mid 1 \leqslant i \leqslant k\}$ of the Mixture-of-Gaussian distribution;
2: **repeat**

The E-step of the EM algorithm.

3:   **for** $j = 1, 2, \ldots, m$ **do**
4:     According to (9.30), compute the posterior probabilities that $x_j$ is generated by each Gaussian mixture component, i.e., $\gamma_{ji} = p_{\mathcal{M}}(z_j = i \mid x_j)(1 \leqslant i \leqslant k)$;
5:   **end for**

The M-step of the EM algorithm.

6:   **for** $i = 1, 2, \ldots, k$ **do**
7:     Compute the updated mean vector: $\mu'_i = \frac{\sum_{j=1}^{m} \gamma_{ji} x_j}{\sum_{j=1}^{m} \gamma_{ji}}$;
8:     Compute the updated covariance matrix: $\Sigma'_i = \frac{\sum_{j=1}^{m} \gamma_{ji}(x_j - \mu'_i)(x_j - \mu'_i)^{\top}}{\sum_{j=1}^{m} \gamma_{ji}}$;
9:     Compute the updated mixture coefficients: $\alpha'_i = \frac{1}{m}\sum_{j=1}^{m} \gamma_{ji}$;
10:   **end for**
11:   Update the model parameters $\{(\alpha_i, \mu_i, \Sigma_i) \mid 1 \leqslant i \leqslant k\}$ to $\{(\alpha'_i, \mu'_i, \Sigma'_i) \mid 1 \leqslant i \leqslant k\}$;

E.g., the maximum number of iterations is reached.

12: **until** The termination condition is met
13: $C_i = \varnothing (1 \leqslant i \leqslant k)$;
14: **for** $j = 1, 2, \ldots, m$ **do**
15:   Determine the cluster label $\lambda_j$ of $x_j$ according to (9.31);
16:   Move $x_j$ to the corresponding cluster: $C_{\lambda_j} = C_{\lambda_j} \cup \{x_j\}$.
17: **end for**
**Output:** Clusters $\mathcal{C} = \{C_1, C_2, \ldots, C_k\}$.

---

We take the watermelon data set 4.0 in ◘ Table 9.1 as an example to give a more concrete demonstration. Suppose that the number of Gaussian mixture components is $k = 3$, and the algorithm starts with the following parameter initialization: $\alpha_1 = \alpha_2 = \alpha_3 = \frac{1}{3}$; $\mu_1 = x_6$; $\mu_2 = x_{22}$; $\mu_3 = x_{27}$; $\Sigma_1 = \Sigma_2 = \Sigma_3 = \begin{pmatrix} 0.1 & 0.0 \\ 0.0 & 0.1 \end{pmatrix}$.

In the first iteration, the algorithm computes the posterior probabilities of samples given that they are generated by each mixture component. Taking $x_1$ as an example, the posterior probabilities computed by (9.30) are $\gamma_{11} = 0.219$, $\gamma_{12} = 0.404$, and $\gamma_{13} = 0.377$. After computing the posterior probabilities of all samples with respect to all mixture components, we obtain the following updated model parameters:

$$\alpha'_1 = 0.361, \quad \alpha'_2 = 0.323, \quad \alpha'_3 = 0.316$$
$$\mu'_1 = (0.491; 0.251), \quad \mu'_2 = (0.571; 0.281), \quad \mu'_3 = (0.534; 0.295)$$
$$\Sigma'_1 = \begin{pmatrix} 0.025 & 0.004 \\ 0.004 & 0.016 \end{pmatrix}, \quad \Sigma'_2 = \begin{pmatrix} 0.023 & 0.004 \\ 0.004 & 0.017 \end{pmatrix}, \quad \Sigma'_3 = \begin{pmatrix} 0.024 & 0.005 \\ 0.005 & 0.016 \end{pmatrix}$$

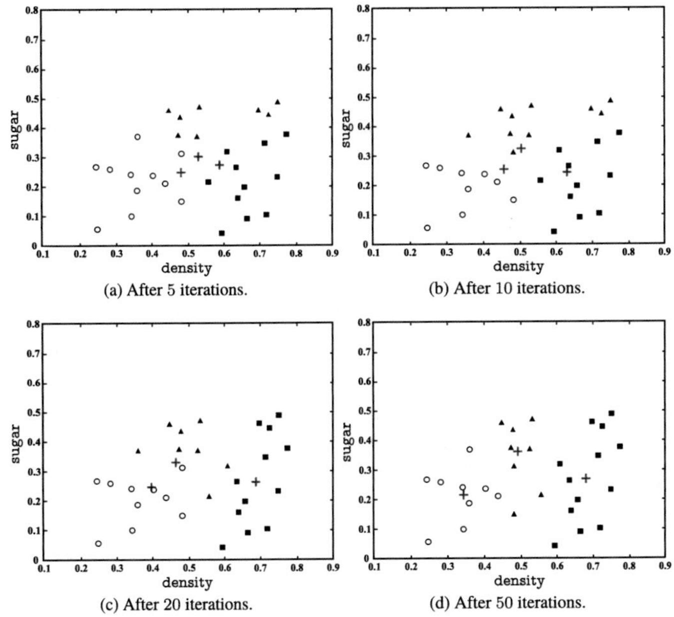


(a) After 5 iterations. (b) After 10 iterations. (c) After 20 iterations. (d) After 50 iterations.

**Fig. 9.4** Results of the Mixture-of-Gaussian algorithm after different iterations on the watermelon data set 4.0 with $k = 3$. The symbols "○", "■", "▲" represent the samples of cluster $C_1$, $C_2$, and $C_3$, respectively. The mean vectors of Gaussian mixture components are denoted by "+"

The above updating process repeats until convergence. ◘ Figure 9.4 shows the clustering results after different iterations.

## 9.5 Density Clustering

*Density clustering*, also known as *density-based clustering*, is a family of clustering algorithms that assumes the clustering structure can be determined by the densities of sample distributions. Typically, density clustering algorithms evaluate the connectivity between samples from the density perspective and expand the clusters by adding connectable samples.

DBSCAN (Ester et al. 1996) is a representative density clustering algorithm, which characterizes the density of sample distributions by a pair of "neighborhood" parameters ($\epsilon$, $MinPts$). Given a data set $D = \{\boldsymbol{x}_1, \boldsymbol{x}_2, \ldots, \boldsymbol{x}_m\}$, we define the following concepts:

DBSCAN stands for Density-Based Spatial Clustering of Applications with Noise.

- $\epsilon$-neighborhood: for $\boldsymbol{x}_j \in D$, its $\epsilon$-neighborhood includes all samples in $D$ that have a distance to $\boldsymbol{x}_j$ no larger than $\epsilon$, that is, $N_\epsilon(\boldsymbol{x}_j) = \{\boldsymbol{x}_i \in D \mid \text{dist}(\boldsymbol{x}_i, \boldsymbol{x}_j) \leqslant \epsilon\}$.
- Core object: $\boldsymbol{x}_j$ is a core object if its $\epsilon$-neighborhood includes at least $MinPts$ samples, that is, $|N_\epsilon(\boldsymbol{x}_j)| \geqslant MinPts$.

In the rest of this chapter, unless otherwise stated, the distance function $\text{dist}(\cdot, \cdot)$ is assumed to be the Euclidean distance.

In general, directly density-reachable is not symmetric.

Density-reachable is subadditive but not symmetric.

Density-connected is symmetric.

- Directly density-reachable: $\boldsymbol{x}_j$ is said to be directly density-reachable by $\boldsymbol{x}_i$ if $\boldsymbol{x}_i$ is a core object and $\boldsymbol{x}_j$ is in the $\epsilon$-neighborhood of $\boldsymbol{x}_i$.
- Density-reachable: $\boldsymbol{x}_j$ is said to be density-reachable by $\boldsymbol{x}_i$ if there exists a sequence of samples $p_1, p_2, \ldots, p_n$, where $p_1 = \boldsymbol{x}_i$, $p_n = \boldsymbol{x}_j$, and $p_{i+1}$ is directly density-reachable by $p_i$.
- Density-connected: $\boldsymbol{x}_i$ and $\boldsymbol{x}_j$ are said to be density-connected if there exists $\boldsymbol{x}_k$ such that both $\boldsymbol{x}_i$ and $\boldsymbol{x}_j$ are density-reachable by $\boldsymbol{x}_k$.

The above concepts are illustrated in ◘ Figure 9.5.

With these concepts, the DBSCAN algorithm defines a cluster as follows: the largest set of density-connected samples derived by density-reachable relationships. Formally, given the neighborhood parameters ($\epsilon$, *MinPts*), a cluster $C \subseteq D$ is a nonempty subset with the following properties:

The samples in $D$ that do not belong to any clusters are considered as noisy or anomaly samples.

$$\text{Connectivity: } \boldsymbol{x}_i \in C, \boldsymbol{x}_j \in C \Rightarrow \boldsymbol{x}_i \text{ and } \boldsymbol{x}_j \text{ are density-connected;} \tag{9.39}$$

$$\text{Maximality: } \boldsymbol{x}_i \in C, \boldsymbol{x}_j \text{ is density-reachable by } \boldsymbol{x}_i \Rightarrow \boldsymbol{x}_j \in C. \tag{9.40}$$

Then, how do we find all clusters satisfying the above properties? Actually, if $\boldsymbol{x}$ is a core object and let $X = \{\boldsymbol{x}' \in D \mid \boldsymbol{x}' \text{ is density-reachable by } \boldsymbol{x}\}$ denote the set of samples density-reachable by $\boldsymbol{x}$, then it can be proved that $X$ is a cluster that satisfies both the connectivity and the maximality. Following this observation, the DBSCAN algorithm generates clusters by expanding from core objects, as illustrated in ◘ Algorithm 9.4. More specifically, lines 1–7 find all core objects based on the given neighborhood parameters ($\epsilon$, *MinPts*); lines 10–24 generate a cluster by randomly selecting one of the core objects as a *seed* and expanding from it to include all density-reachable samples; lines 10–24 repeat until all core objects have been selected.

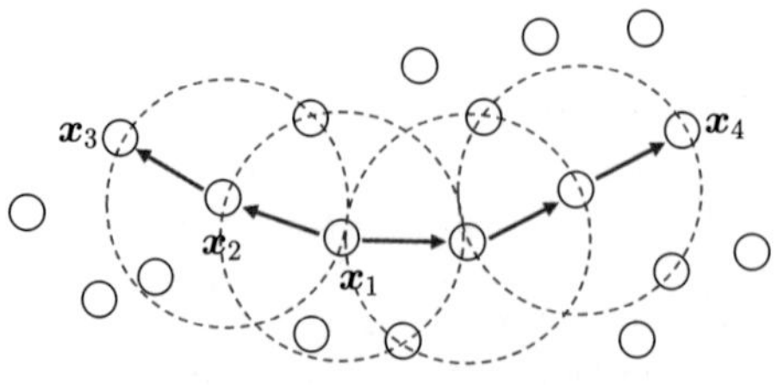


**Fig. 9.5** The basic concepts of DBSCAN (*MinPts* = 3): the dashed circles show the $\epsilon$-neighborhood; $\boldsymbol{x}_1$ is a core object; $\boldsymbol{x}_2$ is directly density-reachable by $\boldsymbol{x}_1$; $\boldsymbol{x}_3$ is density-reachable by $\boldsymbol{x}_1$; $\boldsymbol{x}_3$ and $\boldsymbol{x}_4$ are density-connected

---

**Algorithm 9.4 DBSCAN**

**Input:** Data set $D = \{x_1, x_2, \ldots, x_m\}$;
Neighborhood parameters $(\epsilon, MinPts)$.

**Process:**
1: Initialize the set of core objects: $\Omega = \varnothing$;
2: **for** $j = 1, 2, \ldots, m$ **do**
3:    Determine the $\epsilon$-neighborhood $N_\epsilon(x_j)$ of sample $x_j$;
4:    **if** $\left| N_\epsilon(x_j) \right| \geqslant MinPts$ **then**
5:       Add sample $x_j$ to the set of core objects: $\Omega = \Omega \cup \{x_j\}$;
6:    **end if**
7: **end for**
8: Initialize the number of clusters: $k = 0$;
9: Initialize the set of unprocessed samples: $\Gamma = D$;
10: **while** $\Omega \neq \varnothing$ **do**
11:    Keep a copy of the current unprocessed data set: $\Gamma_{old} = \Gamma$;
12:    Randomly select a core object $o \in \Omega$, and initialize the queue $Q = \langle o \rangle$;
13:    $\Gamma = \Gamma \backslash \{o\}$;
14:    **while** $Q \neq \varnothing$ **do**
15:       Dequeue the first sample $q$ from $Q$;
16:       **if** $|N_\epsilon(q)| \geqslant MinPts$ **then**
17:          Letting $\Delta = N_\epsilon(q) \cap \Gamma$;
18:          Enqueue the samples in $\Delta$ into $Q$;
19:          $\Gamma = \Gamma \backslash \Delta$;
20:       **end if**
21:    **end while**
22:    $k = k + 1$, generate cluster $C_k = \Gamma_{old} \backslash \Gamma$;
23:    $\Omega = \Omega \backslash C_k$.
24: **end while**

**Output:** Clusters $\mathcal{C} = \{C_1, C_2, \ldots, C_k\}$.

---

We take the watermelon data set 4.0 in ◼ Table 9.1 as an example to give a more concrete demonstration. Suppose that the neighborhood parameters are $(\epsilon = 0.11, MinPts = 5)$. We start by finding the $\epsilon$-neighborhood for every sample so that we can identify the set of core objects: $\Omega = \{x_3, x_5, x_6, x_8, x_9, x_{13}, x_{14}, x_{18}, x_{19}, x_{24}, x_{25}, x_{28}, x_{29}\}$. Then, we randomly select a core object from $\Omega$ as a seed and expand from it to include all density-reachable samples. These samples form a cluster. Without loss of generality, suppose the core object $x_8$ is selected as the seed, then the first generated cluster is

$$C_1 = \{\boldsymbol{x}_6, \boldsymbol{x}_7, \boldsymbol{x}_8, \boldsymbol{x}_{10}, \boldsymbol{x}_{12}, \boldsymbol{x}_{18}, \boldsymbol{x}_{19}, \boldsymbol{x}_{20}, \boldsymbol{x}_{23}\}.$$

After that, DBSCAN removes all core objects in $C_1$ from $\Omega$, that is, $\Omega = \Omega \backslash C_1 = \{\boldsymbol{x}_3, \boldsymbol{x}_5, \boldsymbol{x}_9, \boldsymbol{x}_{13}, \boldsymbol{x}_{14}, \boldsymbol{x}_{24}, \boldsymbol{x}_{25}, \boldsymbol{x}_{28}, \boldsymbol{x}_{29}\}$. Then, the next cluster is generated by randomly selecting another core object from the updated $\Omega$ as seed. The process repeats until there is no more element in $\Omega$. ◘ Figure 9.6 shows the clusters generated in different rounds. In addition to $C_1$, the other three generated clusters are

$$C_2 = \{\boldsymbol{x}_3, \boldsymbol{x}_4, \boldsymbol{x}_5, \boldsymbol{x}_9, \boldsymbol{x}_{13}, \boldsymbol{x}_{14}, \boldsymbol{x}_{16}, \boldsymbol{x}_{17}, \boldsymbol{x}_{21}\};$$
$$C_3 = \{\boldsymbol{x}_1, \boldsymbol{x}_2, \boldsymbol{x}_{22}, \boldsymbol{x}_{26}, \boldsymbol{x}_{29}\};$$
$$C_4 = \{\boldsymbol{x}_{24}, \boldsymbol{x}_{25}, \boldsymbol{x}_{27}, \boldsymbol{x}_{28}, \boldsymbol{x}_{30}\}.$$



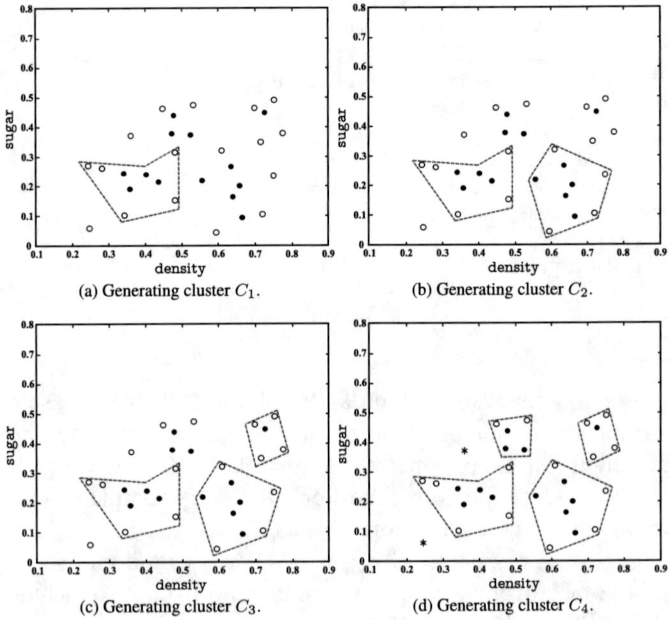


(a) Generating cluster $C_1$. (b) Generating cluster $C_2$.

(c) Generating cluster $C_3$. (d) Generating cluster $C_4$.

**Fig. 9.6** Results of the DBSCAN algorithm with $\epsilon = 0.11$ and $MinPts = 5$. The symbols "•", "∘", "∗" represent the core objects, the non-core objects, and the noisy samples, respectively. The red dashed lines show the clusters

## 9.6 Hierarchical Clustering

*Hierarchical clustering* aims to create a tree-like clustering structure by dividing a data set at different layers. The hierarchy of clusters can be formed by taking either a bottom-up strategy or a top-down strategy.

AGNES is a representative hierarchical clustering algorithm that takes the bottom-up strategy. The algorithm starts by considering each sample in the data set as an initial cluster. Then, in each round, two nearest clusters are merged as a new cluster, and this process repeats until the number of clusters meets the pre-specified value. Here, the key is how to measure the distance between clusters. Since each cluster is a set of data points, we need to define a distance measure about sets. For example, given clusters $C_i$ and $C_j$, we can define the following distances:

AGNES stands for *A Gglomerative NESting*.

*Hausdorff distance* is often used for computing the distance between sets. See Exercise 9.2.

$$\text{Minimum distance: } d_{\min}(C_i, C_j) = \min_{\boldsymbol{x}\in C_i, \boldsymbol{z}\in C_j} \operatorname{dist}(\boldsymbol{x}, \boldsymbol{z}), \quad (9.41)$$

$$\text{Maximum distance: } d_{\max}(C_i, C_j) = \max_{\boldsymbol{x}\in C_i, \boldsymbol{z}\in C_j} \operatorname{dist}(\boldsymbol{x}, \boldsymbol{z}), \quad (9.42)$$

$$\text{Average distance: } d_{\mathrm{avg}}(C_i, C_j) = \frac{1}{|C_i|\,|C_j|} \sum_{\boldsymbol{x}\in C_i} \sum_{\boldsymbol{z}\in C_j} \operatorname{dist}(\boldsymbol{x}, \boldsymbol{z}). \quad (9.43)$$

The minimum distance between two clusters is determined by their two nearest samples; the maximum distance is determined by the two farthest samples from the clusters; the average distance is determined by all samples in both clusters. When the cluster distances are measured by $d_{\min}$, $d_{\max}$, or $d_{\mathrm{avg}}$, the corresponding AGNES algorithms are called *single-linkage*, *complete-linkage*, or *average-linkage*, respectively.

**Algorithm 9.5** AGNES

**Input:** Data set $D = \{\boldsymbol{x}_1, \boldsymbol{x}_2, \ldots, \boldsymbol{x}_m\}$;
Cluster distance metric function $d$;
Number of clusters $k$.

Typically denoted by $d_{\min}$, $d_{\max}$, or $d_{\text{avg}}$;

**Process:**

1: **for** $j = 1, 2, \ldots, m$ **do**
2: $\quad C_j = \{\boldsymbol{x}_j\}$;
3: **end for**
4: **for** $i = 1, 2, \ldots, m$ **do**
5: $\quad$ **for** $j = i + 1, \ldots, m$ **do**
6: $\quad\quad M(i, j) = d(C_i, C_j)$;
7: $\quad\quad M(j, i) = M(i, j)$;
8: $\quad$ **end for**
9: **end for**
10: Set the current number of clusters: $q = m$;
11: **while** $q > k$ **do**
12: $\quad$ Find two clusters $C_{i^*}$ and $C_{j^*}$ that have the shortest distance;
13: $\quad$ Merge $C_{i^*}$ and $C_{j^*}$: $C_{i^*} = C_{i^*} \cup C_{j^*}$;
14: $\quad$ **for** $j = j^* + 1, j^* + 2, \ldots, q$ **do**
15: $\quad\quad$ Change the index of $C_j$ to $C_{j-1}$;
16: $\quad$ **end for**
17: $\quad$ Delete the $j^*$th row and $j^*$th column of the distance matrix $M$;
18: $\quad$ **for** $j = 1, 2, \ldots, q - 1$ **do**
19: $\quad\quad M(i^*, j) = d(C_{i^*}, C_j)$;
20: $\quad\quad M(j, i^*) = M(i^*, j)$;
21: $\quad$ **end for**
22: $\quad q = q - 1$.
23: **end while**

**Output:** Clusters $\mathcal{C} = \{C_1, C_2, \ldots, C_k\}$.

Initialize single-sample clusters. (lines 1–3)

Initialize the cluster distance matrix. (lines 4–9)

$i^* < j^*$. (line 12)



The pseudocode of AGNES is given in ◘ Algorithm 9.5. In lines 1–9, the algorithm starts by initializing the distance matrix using the initial single-sample clusters. Then, in lines 11–23, the algorithm finds and merges the two clusters with the shortest distance, and then update the distance matrix accordingly. This process repeats until the number of clusters meets the pre-specified value.

We take the watermelon data set 4.0 in ◘ Table 9.1 as an example to demonstrate AGNES. Suppose the algorithm repeats until all samples appear in one cluster (i.e., $k = 1$). Then, we obtain a *dendrogram*, as illustrated in ◘ Figure 9.7, where each row links a set of clusters.

Cutting at a specified row of the dendrogram yields the corresponding clusters. For example, if we cut along the dashed line in ◘ Figure 9.7, we obtain the following 7 clusters:

$C_1 = \{\boldsymbol{x}_1, \boldsymbol{x}_{26}, \boldsymbol{x}_{29}\}$; $\quad C_2 = \{\boldsymbol{x}_2, \boldsymbol{x}_3, \boldsymbol{x}_4, \boldsymbol{x}_{21}, \boldsymbol{x}_{22}\}$;

$C_3 = \{\boldsymbol{x}_{23}, \boldsymbol{x}_{24}, \boldsymbol{x}_{25}, \boldsymbol{x}_{27}, \boldsymbol{x}_{28}, \boldsymbol{x}_{30}\}$; $\quad C_4 = \{\boldsymbol{x}_5, \boldsymbol{x}_7\}$;

$C_5 = \{\boldsymbol{x}_9, \boldsymbol{x}_{13}, \boldsymbol{x}_{14}, \boldsymbol{x}_{16}, \boldsymbol{x}_{17}\}$; $\quad C_6 = \{\boldsymbol{x}_6, \boldsymbol{x}_8, \boldsymbol{x}_{10}, \boldsymbol{x}_{15}, \boldsymbol{x}_{18}, \boldsymbol{x}_{19}, \boldsymbol{x}_{20}\}$;

$C_7 = \{\boldsymbol{x}_{11}, \boldsymbol{x}_{12}\}$.

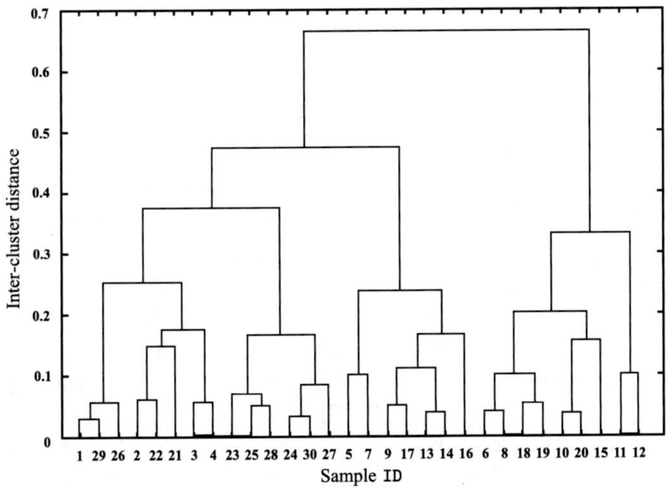


**Fig. 9.7** The dendrogram (using $d_{\max}$) generated by the AGNES algorithm on the watermelon data set 4.0. The x-axis shows the *ID* of samples and the y-axis shows the distances between clusters

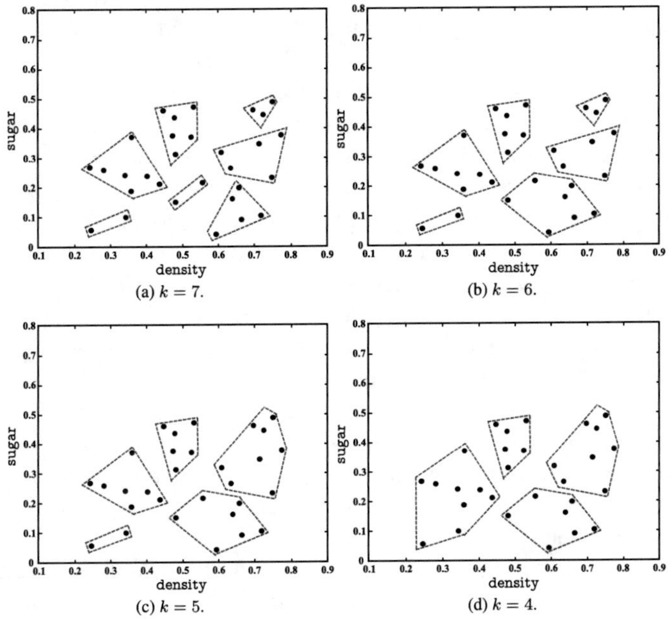


**Fig. 9.8** Results of the AGNES algorithm (using $d_{\max}$) on the watermelon data set 4.0 with different number of clusters ($k = 7, 6, 5, 4$). The symbol "•" represents samples, and the red dashed lines show the clusters

The higher the cutting level, the fewer clusters are produced. ◘ Figure 9.8 shows the clusters obtained at different cutting points of ◘ Figure 9.7.

## 9.7 Further Reading

Clustering is an area of machine learning in which new algorithms emerge the most in amounts and the fastest in time. One reason behind is that there is no objective criterion for judging the quality of a clustering result. In fact, given a data set, it is always possible to propose a new algorithm based on some criteria that have not yet been considered by existing algorithms (Estivill-Castro 2002). Compared to other machine learning areas, clustering is an area that has not been systematically developed yet, and hence the well-known textbook (Mitchell 1997) did not even have a dedicated chapter on clustering. Nevertheless, we still include this chapter to introduce some representative clustering algorithms since clustering techniques are very important in real-world applications. More information about clustering can be found in dedicated books and survey articles, such as Jain (2009), Jain and Dubes (1988), Jain et al. (1999), Xu and Wunsch II (2005).

For example, the same set of fruits can be clustered by size, color, or even origin.



In addition to the validity indices introduced in Sect. 9.2, other common performance measures include $F$-statistic, *mutual information*, *average silhouette width* (Rousseeuw 1987), etc. More information can be found in Halkidi et al. (2001), Jain and Dubes (1988), Maulik and Bandyopadhyay (2002).

Distance calculation is the core of many learning problems. Minkowski distance describes a general form of distance calculation. Besides, inner product distance and cosine similarity are also commonly used; see Deza and Deza (2009) for more details. MinkovDM is given in Zhou and Yu (2005). Nonmetric distances (Jacobs et al. 2000; Tan et al. 2009) are often used in applications involving complex semantics, such as pattern recognition and image retrieval. Distance metric learning can be directly embedded into the learning process of clustering (Xing et al. 2003).

See Sect. 10.6 for distance metric learning.

$k$-means can be seen as a special case of Mixture-of-Gaussian clustering when the mixture components have equal covariances and each sample is assigned to only one mixture component. $k$-means has been reinvented many times by researchers from different fields, such as Steinhaus in 1956, Lloyd in 1957, and MacQueen in 1967 (Jain and Dubes 1988; Jain 2009). There are many variants of $k$-means, such as $k$-medoids (Kaufman and Rousseeuw 1987) whose prototype vectors are always training samples, $k$-modes (Huang 1998) which supports categorical features, Fuzzy $C$-means (FCM) (Bezdek

Convex clustering structures refer to clustering structures with an ellipsoidal shape.

Bregman distances, also known as Bregman divergences, are a family of distances that do not have subadditivity nor symmetry.

1981) which is a *soft clustering* algorithm that allows samples belong to different prototypes simultaneously. It is worth pointing out that $k$-means and its variants only perform well on convex clustering structures. Recent studies show that adopting Bregman distances can significantly improve the performance on more clustering structures (Banerjee et al. 2005). Introducing kernel tricks to $k$-means gives the kernel $k$-means (Schölkopf et al. 1998). Dhillon et al. (2004) showed that kernel $k$-means is closely related to spectral clustering (von Luxburg 2007), which can be seen as applying $k$-means after dimensionality reduction by Laplacian Eigenmap. The number of clusters $k$ is typically specified by users. Though there are some heuristic methods for determining $k$ automatically (Pelleg and Moore 2000, Tibshirani et al. 2001), the common practice is still to select the best one by trying different $k$ values.

See Chap. 10 for dimensionality reduction.

Each round of the standard LVQ only updates the nearest prototype vector of the current sample. Improved versions, such as LVQ2 and LVQ3 (Kohonen 2001), allow updating multiple prototype vectors in parallel resulting in much faster convergence speed. McLachlan and Peel (2000) introduced Mixture-of-Gaussian clustering in detail, and more information about the underlying EM optimization can be found in Bilmes (1998), Jain and Dubes (1988).

In addition to DBSCAN (Ester et al. 1996), there are other density-based clustering methods based on different density representations of sample distributions, such as OPTICS (Ankerst et al. 1999) and DENCLUE (Hinneburg and Keim 1998). AGNES (Kaufman and Rousseeuw 1990) takes a bottom-up strategy to generate hierarchical clustering structures. By contrast, DIANA (Kaufman and Rousseeuw 1990) takes a top-down strategy. Both AGNES and DIANA do not allow backtrace the clusters that have been merged or split, which has been improved in other hierarchical clustering methods such as BIRCH (Zhang et al. 1996) and ROCK (Guha et al. 1999).

Clustering ensembles combine multiple clustering learners to reduce the negative influence caused by random factors in the clustering process and the incompatibility between the assumptions and the actual clustering structures. More information can be found in Chap. 7 of (Zhou 2012).

Clustering methods and distance calculations are often used for anomaly detection (Hodge and Austin 2004; Chandola et al. 2009). For example, we can consider anomaly samples as those far away from all cluster centers or in regions with extremely low densities. Recent studies have proposed *isolation*-based methods for efficient anomaly detection (Liu et al. 2012).

## Exercises

**9.1** Prove (1) when $p \geqslant 1$, Minkowski distance satisfies the four axioms of distance measures; (2) when $0 \leqslant p < 1$, Minkowski distance does not satisfy subadditivity but satisfies non-negativity, identity of indiscernibles, and symmetry; (3) when $p$ approaches positive infinity, Minkowski distance equals to the maximum absolute distance of respective components, that is,

$$\lim_{p \to +\infty} \left( \sum_{u=1}^{n} \left| x_{iu} - x_{ju} \right|^p \right)^{\frac{1}{p}} = \max_u \left| x_{iu} - x_{ju} \right| .$$

**9.2** For sets $X$ and $Z$ from the same sample space, their distance can be calculated using the *Hausdorff distance* as

$$\mathrm{dist}_H(X, Z) = \max\left(\mathrm{dist}_h(X, Z), \mathrm{dist}_h(Z, X)\right), \tag{9.44}$$

where

$$\mathrm{dist}_h(X, Z) = \max_{\boldsymbol{x} \in X} \min_{\boldsymbol{z} \in Z} \|\boldsymbol{x} - \boldsymbol{z}\|_2 . \tag{9.45}$$

Prove that the Hausdorff distance satisfies the four axioms of distance measures.

**9.3** Discuss whether we can use the $k$-means algorithm to find the optimal solution that minimizes (9.24).

**9.4** Implement and run the $k$-means algorithm on the watermelon data set 4.0 with three different $k$ values and three different initial centroids. Discuss what kind of initial centroids can lead to good results.

The watermelon data set 4.0 is in ◘ Table 9.1 in Sect. 9.4.1.

**9.5** Based on the definition of DBSCAN, $X$ is the set of samples that are density-reachable by a core object $\boldsymbol{x}$. Prove that $X$ satisfies both connectivity (9.39) and maximality (9.40).

**9.6** Analyze the difference between using minimum distance and maximum distance in the AGNES algorithm.

**9.7** We say a clustering is convex clustering if every cluster has a convex hull (i.e., a convex polyhedron that contains all samples in the cluster) and the convex hulls do not intersect. Analyze which clustering algorithms introduced in this chapter can only produce convex clusters and which clustering algorithms can produce non-convex clusters.

**9.8** Design a new clustering performance measure, and compare it to other measures introduced in Sect. 9.2.

**9.9** * Design a non-metric distance that can be used for mixture types of features.

**9.10** * Design an improved $k$-means algorithm that can automatically determine the number of clusters; implement and run it on the watermelon data set 4.0.

## Break Time

### Short Story: Manhattan Distance and Hermann Minkowski

Manhattan distance, also known as *Taxicab geometry*, is a word invented by the distinguished German mathematician Hermann Minkowski (1864–1909). The Manhattan distance between two points is the sum of the absolute differences of their Cartesian coordinates, and this is exactly the shortest travel distance between two locations in a city with grid street plan. For example, in Manhattan, the travel distance from the corner of Fifth Avenue and the 33rd street to the corner of Third Avenue and the 23rd street is $(5 - 3) + (33 - 23) = 12$ blocks.

Aleksota is now Kaunas, Lithuania.
Königsberg, now Kaliningrad, Russia, is the origin of the famous "Seven Bridges" problem.

Minkowski was born in a Jewish family in Aleksota, Russian. At the age of 8, he moved with his family to Königsberg to escape the persecution of Jews in Russia. In Königsberg, he lived quite close to David Hilbert, who became a great mathematician later. Minkowski was a child prodigy who read Shakespeare, Schiller, and Goethe, and was able to recite almost the entire "Faust" by memory. At the age of 8, he attended gymnasium and then finished the 8-year course in five and a half years. At the age of 17, he established the theory of quadratic forms, for which he was awarded the "Grand Prix des Sciences Mathématiques" by the French Academy of Sciences. On September 21, 1908, Minkowski delivered the famous talk "Space and Time", in which he proposed four-dimensional geometry, clearing the road for the general theory of relativity. Unfortunately, three months later he died suddenly of appendicitis.

Four-dimensional geometry is also called *Minkowski space*.

In 1896, Minkowski was a teacher at ETH Zurich, and Albert Einstein was one of his students. Max Born, the winner of Nobel Prize, said that he found the "armory" of relativity from Minkowski's works on mathematics. After Minkowski passed away, his friend Hilbert collected his unpublished work and published the book *Gesammelte Abhandlungen von Hermann Minkowski* in 1911. Minkowski's brother, Oskar Minkowski, was the "father of insulin"; his nephew, Rudolph Minkowski, was a well-known German-American astronomer.

## References

Aloise D, Deshpande A, Hansen P, Popat P (2009) NP-hardness of Euclidean sum-of-squares clustering. Mach Learn 75(2):245–248

Ankerst M, Breunig M, Kriegel H-P, Sander J (1999) OPTICS: ordering points to identify the clustering structure. In: Proceedings of the ACM SIGMOD international conference on management of data (SIGMOD), Philadelphia, PA, pp 49–60

Banerjee A, Merugu S, Dhillon I, Ghosh J (2005) Clustering with bregman divergences. J Mach Learn Res 6:1705–1749

Bezdek JC (1981) Pattern recognition with fuzzy objective function algorithms. Plenum Press, New York, NY

Bilmes JA (1998) A gentle tutorial of the EM algorithm and its applications to parameter estimation for Gaussian mixture and hidden Markov models. Technical Report TR-97-021, Department of Electrical Engineering and Computer Science, University of California at Berkeley, Berkeley, CA

Chandola V, Banerjee A, umar V (2009) Anomaly detection: a survey. ACM Comput Surv 41(3):Article 15

Deza M, Deza E (2009) Encyclopedia of Distances. Springer, Berlin

Dhillon IS, Guan Y, Kulis B (2004) Kernel $k$-means: Spectral clustering and normalized cuts. In: Proceedings of the 10th ACM SIGKDD international conference on knowledge discovery and data mining (KDD), Seattle, WA, pp 551–556

Ester M, Kriegel HP, Sander J, Xu X (1996) A density-based algorithm for discovering clusters in large spatial databases. In: Proceedings of the 2nd international conference on knowledge discovery and data mining (KDD), Portland, OR, pp 226–231

Estivill-Castro V (2002) Why so many clustering algorithms—a position paper. SIGKDD Explor 1(4):65–75

Guha S, Rastogi R, Shim K (1999) ROCK: a robust clustering algorithm for categorical attributes. In: Proceedings of the 15th international conference on data engineering (ICDE), Sydney, Australia, pp 512–521

Halkidi M, Batistakis Y, Vazirgiannis M (2001) On clustering validation techniques. J Intell Inf Syst 27(2–3):107–145

Hinneburg A, Keim DA (1998) An efficient approach to clustering in large multimedia databases with noise. In: Proceedings of the 4th international conference on knowledge discovery and data mining (KDD), New York, NY, pp 58–65

Hodge VJ, Austin J (2004) A survey of outlier detection methodologies. Artif Intell Rev 22(2):85–126

Huang Z (1998) Extensions to the $k$-means algorithm for clustering large data sets with categorical values. Data Min Knowl Discov 2(3):283–304

Jacobs DW, Weinshall D, Gdalyahu Y (2000) Classification with non-metric distances: image retrieval and class representation. IEEE Trans Patt Anal Mach Intell 6(22):583–600

Jain AK (2009) Data clustering: 50 years beyond $k$-means. Patt Recogn Lett 371(8):651–666

Jain AK, Dubes RC (1988) Algorithms for clustering data. Prentice Hall, Upper Saddle River, NJ

Jain AK, Murty MN, Flynn PJ (1999) Data clustering: a review. ACM Comput Surv 3(31):264–323

Kaufman L, Rousseeuw PJ (1987) Clustering by means of medoids. In: Dodge Y (ed) Statistical data analysis based on the $L_1$-Norm and related methods. Elsevier, Amsterdam, Netherlands, pp 405–416

Kaufman L, Rousseeuw PJ (1990) Finding groups in data: an introduction to cluster analysis. Wiley, New York, NY
Kohonen T (2001) Self-organizing maps, 3rd edn. Springer, Berlin
Liu FT, Ting KM, Zhou Z-H (2012) Isolation-based anomaly detection. ACM Trans Knowl Discov Data 6(1):Article 3
Maulik U, Bandyopadhyay S (2002) Performance evaluation of some clustering algorithms and validity indices. IEEE Trans Patt Anal Mach Intell 24(12):1650–1654
McLachlan G, Peel D (2000) Finite mixture models. Wiley, New York, NY
Mitchell T (1997) Machine learning. McGraw Hill, New York, NY
Pelleg D, Moore A (2000) X-means: extending *k*-means with efficient estimation of the number of clusters. In: Proceedings of the 17th international conference on machine learning (ICML), Stanford, CA, pp 727–734
Rousseeuw PJ (1987) Silhouettes: a graphical aid to the interpretation and validation of cluster analysis. J Comput Appl Math 20:53–65
Schölkopf B, Smola A, Müller K-R (1998) Nonlinear component analysis as a kernel eigenvalue problem. Neural Comput 10(5):1299–1319
Stanfill C, Waltz D (1986) Toward memory-based reasoning. Commun ACM 29(2):1213–1228
Tan X, Chen S, Zhou Z-H, Liu J (2009) Face recognition under occlusions and variant expressions with partial similarity. IEEE Trans Inf Forensics Secur 2(4):217–230
Tibshirani R, Walther G, Hastie T (2001) Estimating the number of clusters in a data set via the gap statistic. J R Stat Soc -Ser B 63(2):411–423
von Luxburg U (2007) A tutorial on spectral clustering. Stat Comput 17(4):395–416
Xing EP, Ng AY, Jordan MI, Russell S (2003) Distance metric learning, with application to clustering with side-information. In: Becker S, Thrun S, Obermayer K (eds) Advances in neural information processing systems 15 (NIPS). MIT Press, Cambridge, MA, pp 505–512
Xu R, Wunsch D II (2005) Survey of clustering algorithms. IEEE Trans Neural Netw 3(16):645–678
Zhang T, Ramakrishnan R, Livny M (1996) BIRCH: an efficient data clustering method for very large databases. In: Proceedings of the ACM SIGMOD international conference on management of data (SIGMOD), Montreal, Canada, pp 103–114
Zhou Z-H (2012) Ensemble methods: foundations and algorithms. Chapman & Hall/CRC, Boca Raton, FL
Zhou Z-H, Yu Y (2005) Ensembling local learners through multimodal perturbation. IEEE Trans Syst Man Cybern -Part B: Cybern 35(4):725–735

# Dimensionality Reduction and Metric Learning

Table of Contents




Z.-H. Zhou, *Machine Learning*,
https://doi.org/10.1007/978-981-15-1967-3_10

## 10.1 k-Nearest Neighbor Learning

$k$-Nearest Neighbor ($k$NN) is a commonly used supervised learning method with a simple mechanism: given a testing sample, find the $k$ nearest training samples based on some distance metric, and then use these $k$ "neighbors" to make predictions. Typically, for classification problems, *voting* can be used to predict the testing sample as the most frequent class label in the $k$ neighbors; for regression problems, *averaging* can be used to predict the testing sample as the average of the $k$ real-valued outputs. Besides, the samples can be weighted by the distances in the way that a closer sample is assigned a higher weight.

The principle of $k$NN agrees with the proverb that "One takes the behavior of one's company."

See Sect. 8.4.

Compared to other methods that were introduced earlier, $k$-Nearest Neighbor has something unique: it does not have an explicit training process! In fact, it is a representative of *lazy learning*, which simply stores the samples in the training phase and does nothing until the testing samples are received. In contrast, *eager learning* refers to the methods that learn from samples in the training phase.

Figure 10.1 provides an illustration of the $k$NN classifier. We can see that the parameter $k$ plays an important role since different $k$ values may lead to very different classification results. In addition, different distance calculations may also lead to significantly different "neighborhood", and consequently, different classification results.



We start our discussion with One-Nearest Neighbor Classifier (1NN, i.e., $k = 1$) for binary classification problems. Here, we assume the distance calculations are "appropriate" such that we can identify $k$ appropriate neighbors.

Given a testing sample $\boldsymbol{x}$, suppose its nearest neighbor is $\boldsymbol{z}$, then the misclassification rate of 1NN is the probability that $\boldsymbol{x}$ and $\boldsymbol{z}$ have different class labels, that is,

$$P(err) = 1 - \sum_{c \in \mathcal{Y}} P(c \mid \boldsymbol{x}) P(c \mid \boldsymbol{z}). \tag{10.1}$$

Let us assume that the samples are *i.i.d.*, and we can always find a sample within an arbitrarily small positive range $\delta$ for any $\boldsymbol{x}$. In other words, for any testing sample, the training sample $\boldsymbol{z}$ in (10.1) can always be found within an arbitrarily small range. Let $c^* = \arg\max_{c \in \mathcal{Y}} P(c \mid \boldsymbol{x})$ denote the result of the Bayes optimal classifier. Then, we have

See Sect. 7.1 for the Bayes optimal classifier.

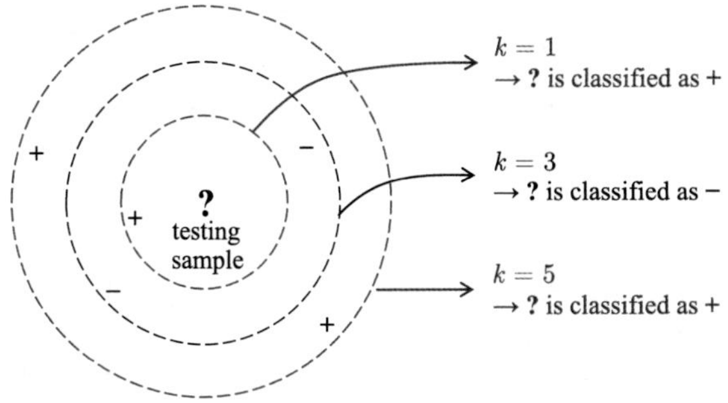


**Fig. 10.1** The $k$NN classifier. The dashed lines are equidistant lines; the testing sample is classified as positive when $k = 1$ or $k = 5$ and is classified as negative when $k = 3$

$$
\begin{aligned}
P(err) &= 1 - \sum_{c \in \mathcal{Y}} P(c \mid \boldsymbol{x}) P(c \mid \boldsymbol{z}) \\
&\simeq 1 - \sum_{c \in \mathcal{Y}} P^2(c \mid \boldsymbol{x}) \\
&\leqslant 1 - P^2(c^* \mid \boldsymbol{x}) \\
&= (1 + P(c^* \mid \boldsymbol{x}))(1 - P(c^* \mid \boldsymbol{x})) \\
&\leqslant 2 \times (1 - P(c^* \mid \boldsymbol{x})).
\end{aligned} \tag{10.2}
$$

From (10.2), we can make a somewhat surprising conclusion: though the 1NN classifier is simple, its generalization error is at most twice the error of the Bayes optimal classifier!

Here, we only provide a brief discussion for beginners. See Cover and Hart (1967) for a more detailed analysis.

## 10.2 Low-Dimensional Embedding

Our discussions in the previous section rely on an important assumption: we can always find a sample within an arbitrarily small positive range $\delta$ for any $\boldsymbol{x}$, which means the sampling must be sufficiently dense, that is, *dense sampling*. However, this assumption is impractical. For example, when $\delta = 0.001$ and there is only one feature, for (10.2) to hold, we need 1000 samples uniformly distributed within the normalized value range of the given feature. However, this is only the case when the dimension is 1. If the dimension increases, the situation will change significantly. For instance, given 20 features, we would need $(10^3)^{20} = 10^{60}$ samples to satisfy the dense sampling requirement. Since the number of features can easily go beyond thousands in practice, it is generally impractical to fulfill the dense sampling requirement. Besides, distance calculations, as required by many learning methods, become difficult in the

For reference: the number of elementary particles in the observable universe is about $10^{80}$ (each dust particle contains billions of elementary particles).

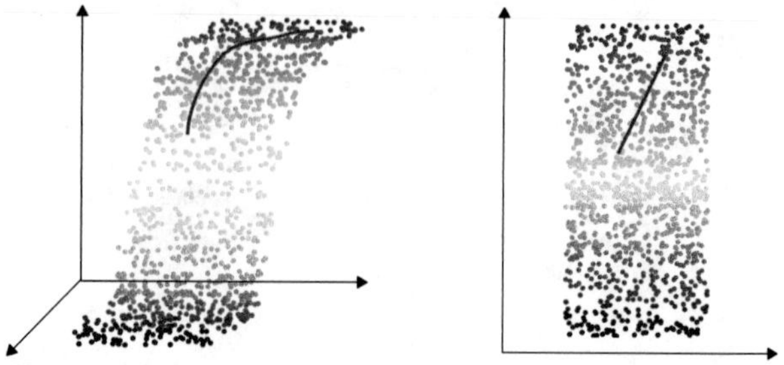

(a) Observation in 3-dimensional space. (b) Observation in 2-dimensional space.

**Fig. 10.2** An illustration of low-dimensional embedding

high-dimensional space. For example, even the inner product is no longer easy in high-dimensional cases.

In fact, high dimensionality leads to many issues, such as data sparsity and the difficulty of distance calculation, that are faced by all machine learning methods. This phenomenon is known as the *curse of dimensionality* (Bellman 1957).

Another important approach is feature selection. See Chap. 11.

A general approach to alleviating the curse of dimensionality is *dimensionality reduction*, which aims to convert the original high-dimensional feature space into a low-dimensional *subspace* via mathematical transformations. The sampling is much denser in the subspace, and the distance calculations also become easier. We may wonder why it is possible to perform dimensionality reduction while keeping the necessary information for further learning? It is because, in many cases, although the observed or collected samples are high dimensional, the useful information for the learning tasks could be just a low-dimensional distribution; that is, there is a low-dimensional *embedding* in the high-dimensional space. ◘ Figure 10.2 provides an intuitive example, in which learning becomes easier in the subspace of the low-dimensional embedding.

Multiple Dimensional Scaling (MDS) (Cox and Cox 2001) is a classic dimensionality reduction technique that projects a data set from the original space into a lower dimensional space while preserving the distances between samples. ◘ Figure 10.2 provides an illustration of MDS.

We briefly discuss how MDS works. Let $\mathbf{D} \in \mathbb{R}^{m\times m}$ be a distance matrix of $m$ samples in the original space, where the distance between the samples $\boldsymbol{x}_i$ and $\boldsymbol{x}_j$ is the element $dist_{ij}$ on the $i$th row and $j$th column. The task is to obtain the sample representation $\mathbf{Z} \in \mathbb{R}^{d'\times m}$ in the $d'$-dimensional space, where $d' \leqslant d$, and the distance between any two samples is the same in both spaces, i.e., $\|z_i - z_j\| = dist_{ij}$.

Let $\mathbf{B} = \mathbf{Z}^{\mathrm{T}}\mathbf{Z} \in \mathbb{R}^{m \times m}$ be the inner product matrix of the samples in the dimension-reduced space, where $b_{ij} = z_i^{\mathrm{T}} z_j$. Then, we have

$$\begin{aligned} dist_{ij}^2 &= \|z_i\|^2 + \|z_j\|^2 - 2z_i^{\mathrm{T}} z_j \\ &= b_{ii} + b_{jj} - 2b_{ij}. \end{aligned} \tag{10.3}$$

For ease of discussion, suppose the dimension-reduced samples are zero-centered, that is, $\sum_{i=1}^{m} z_i = \mathbf{0}$. We notice that both the row-wise and column-wise sums of $\mathbf{B}$ are 0, that is, $\sum_{i=1}^{m} b_{ij} = \sum_{j=1}^{m} b_{ij} = 0$. Then, we can derive

$\mathbf{0} \in \mathbb{R}^{d'}$ is an all-zero vector of length $d'$.

$$\sum_{i=1}^{m} dist_{ij}^2 = \mathrm{tr}(\mathbf{B}) + mb_{jj}, \tag{10.4}$$

$$\sum_{j=1}^{m} dist_{ij}^2 = \mathrm{tr}(\mathbf{B}) + mb_{ii}, \tag{10.5}$$

$$\sum_{i=1}^{m}\sum_{j=1}^{m} dist_{ij}^2 = 2m\,\mathrm{tr}(\mathbf{B}), \tag{10.6}$$

where $\mathrm{tr}(\cdot)$ is the trace of a matrix, and $\mathrm{tr}(\mathbf{B}) = \sum_{i=1}^{m} \|z_i\|^2$. Let

$$dist_{i\cdot}^2 = \frac{1}{m}\sum_{j=1}^{m} dist_{ij}^2, \tag{10.7}$$

$$dist_{\cdot j}^2 = \frac{1}{m}\sum_{i=1}^{m} dist_{ij}^2, \tag{10.8}$$

$$dist_{\cdot\cdot}^2 = \frac{1}{m^2}\sum_{i=1}^{m}\sum_{j=1}^{m} dist_{ij}^2, \tag{10.9}$$

then, from (10.3)–(10.9), we have

$$b_{ij} = -\frac{1}{2}(dist_{ij}^2 - dist_{i\cdot}^2 - dist_{\cdot j}^2 + dist_{\cdot\cdot}^2), \tag{10.10}$$

which makes it possible to calculate the inner product matrix $\mathbf{B}$ from the distance matrix $\mathbf{D}$ that is not changed by dimensionality reduction.

By eigenvalue decomposition, we have $\mathbf{B} = \mathbf{V}\mathbf{\Lambda}\mathbf{V}^{\mathrm{T}}$, where $\mathbf{\Lambda} = \mathrm{diag}(\lambda_1, \lambda_2, \ldots, \lambda_d)$ is the diagonal matrix consisting of the eigenvalues $\lambda_1 \geqslant \lambda_2 \geqslant \ldots \geqslant \lambda_d$, and $\mathbf{V}$ is the matrix of eigenvectors. Suppose that there are $d^*$ non-zero eigenvalues that form the diagonal matrix $\mathbf{\Lambda}_* = \mathrm{diag}(\lambda_1, \lambda_2, \ldots, \lambda_{d*})$, and let $\mathbf{V}_*$ denote the corresponding eigenvector matrix. Then, $\mathbf{Z}$ can be expressed as

$$\mathbf{Z} = \mathbf{\Lambda}_*^{1/2}\mathbf{V}_*^{\mathrm{T}} \in \mathbb{R}^{d^* \times m}. \tag{10.11}$$

In practice, however, we only require the dimension-reduced distances to be close to the original distances rather than the same. In such cases, we can take the $d' \ll d$ largest eigenvalues to make a diagonal matrix $\tilde{\mathbf{\Lambda}} = \mathrm{diag}(\lambda_1, \lambda_2, \ldots, \lambda_{d'})$, and denote $\tilde{\mathbf{V}}$ as the corresponding eigenvector matrix. Then, $\mathbf{Z}$ can be expressed as

$$\mathbf{Z} = \tilde{\mathbf{\Lambda}}^{1/2}\tilde{\mathbf{V}}^{\mathrm{T}} \in \mathbb{R}^{d' \times m}. \tag{10.12}$$

The pseudocode of MDS is given in ◘ Algorithm 10.1.

---

**Algorithm 10.1** Multiple Dimensional Scaling

**Input:** Distance matrix $\mathbf{D} \in \mathbb{R}^{m \times m}$, where $dist_{ij}$ is the distance from $\boldsymbol{x}_i$ to $\boldsymbol{x}_j$;
Dimension $d'$ of the low-dimensional space.

**Process:**
1: Compute $dist_{i\cdot}^2$, $dist_{\cdot j}^2$, and $dist_{\cdot\cdot}^2$ according to (10.7)–(10.9);
2: Compute inner product matrix $\mathbf{B}$ according to (10.10);
3: Perform eigenvalue decomposition on matrix $\mathbf{B}$;
4: Take the $d'$ largest eigenvalues to make a diagonal matrix $\tilde{\mathbf{\Lambda}}$, and $\tilde{\mathbf{V}}$ is the corresponding eigenvector matrix.

**Output:** The matrix $\tilde{\mathbf{V}}\tilde{\mathbf{\Lambda}}^{1/2} \in \mathbb{R}^{m \times d'}$, where each row gives the coordinates of the sample in the low-dimensional space.

---

In general, the simplest method to obtain a lower dimensional subspace is applying linear transformations to the original high-dimensional space. Given the samples $\mathbf{X} = (\boldsymbol{x}_1, \boldsymbol{x}_2, \ldots, \boldsymbol{x}_m) \in \mathbb{R}^{d \times m}$ in a $d$-dimensional space, the transformed samples in the $d'$-dimensional space are

$$\mathbf{Z} = \mathbf{W}^{\mathrm{T}}\mathbf{X}, \tag{10.13}$$

where $d' \leqslant d$, $\mathbf{W} \in \mathbb{R}^{d \times d'}$ is the transformation matrix, and $\mathbf{Z} \in \mathbb{R}^{d' \times m}$ is the sample representations in the new space.

Typically, $d' \ll d$.

The transformation matrix $\mathbf{W}$ can be seen as $d'$ basis vectors with a dimension of $d$, and $z_i = \mathbf{W}^{\mathrm{T}}\boldsymbol{x}_i$ is a $d'$-dimensional feature vector obtained by multiplying the original feature vector $\boldsymbol{x}_i$ by each of the $d'$ basis vectors. In other words, $z_i$ is the coordinate vector of $\boldsymbol{x}_i$ in the new coordinate system $\{\mathbf{w}_1, \mathbf{w}_2, \ldots, \mathbf{w}_{d'}\}$. When $\mathbf{w}_i$ and $\mathbf{w}_j (i \neq j)$ are orthogonal, the new coordinate system is an orthogonal coordinate system, and $\mathbf{W}$ is an orthogonal transformation matrix. We notice that the features in the new space are linear combinations of the features in the original space.

The methods that perform dimensionality reduction by linear transformations are called linear dimensionality reduction methods. Such methods follow the basic form of (10.13), though different methods may require the lower dimensional subspace to satisfy different conditions, that is, adding some constraints to $\mathbf{W}$. The next section will introduce a prevalent dimensionality reduction method that requires the lower dimensional subspace to have the maximum variance.

Typically, we evaluate the effectiveness of a dimensionality reduction by comparing the performance of the learner before and after the dimensionality reduction, where the performance improvement indicates the effectiveness. When the dimensionality is reduced to 2 or 3, we can apply visualization techniques to inspect dimensionality reduction effectiveness.

## 10.3 Principal Component Analysis

Principal Component Analysis (PCA) is one of the most commonly used dimensionality reduction methods. Before introducing the technical details, let us consider the following question: for the samples in an orthogonal feature space, how can we use a hyperplane (i.e., a line in two-dimensional space generalized to high-dimensional space) to represent the samples? Intuitively, if such a hyperplane exists, then it perhaps needs to have the following properties:

- Minimum reconstruction error: the samples should have short distances to this hyperplane;
- Maximum variance: the projections of samples onto the hyperplane should stay away from each other.

Interestingly, the above two properties lead to two equivalent derivations of PCA. First, let us derive PCA by minimizing the reconstruction error.

Suppose the samples are zero-centered (i.e., $\sum_i \boldsymbol{x}_i = \mathbf{0}$). Let $\{\mathbf{w}_1, \mathbf{w}_2, \ldots, \mathbf{w}_d\}$ denote the new coordinate system after projection, where $\mathbf{w}_i$ is an orthonormal basis vector, that is, $\|\mathbf{w}_i\|_2 = 1$ and $\mathbf{w}_i^{\mathrm{T}}\mathbf{w}_j = 0 (i \neq j)$. If some of the coordinates are removed from the new coordinate system (i.e., the dimension is reduced to $d' < d$), then the projection of the sample $\boldsymbol{x}_i$ in the lower dimensional coordinate system is $\boldsymbol{z}_i = (z_{i1}; z_{i2}; \ldots; z_{id'})$, where $z_{ij} = \mathbf{w}_j^{\mathrm{T}}\boldsymbol{x}_i$ is the coordinate of $\boldsymbol{x}_i$ in the $j$th dimension of the lower dimensional coordinate system. If we reconstruct $\boldsymbol{x}_i$ from $\boldsymbol{z}_i$, then we have $\hat{\boldsymbol{x}}_i = \sum_{j=1}^{d'} z_{ij}\mathbf{w}_j$.

For the entire training set, the total distance between the original samples $\boldsymbol{x}_i$ and the projection-reconstructed samples

$\hat{\boldsymbol{x}}_i$ is

const is a constant.

$$\sum_{i=1}^{m}\left\|\sum_{j=1}^{d'} z_{ij}\mathbf{w}_j - \boldsymbol{x}_i\right\|_2^2 = \sum_{i=1}^{m} \boldsymbol{z}_i^{\mathrm{T}}\boldsymbol{z}_i - 2\sum_{i=1}^{m} \boldsymbol{z}_i^{\mathrm{T}}\mathbf{W}^{\mathrm{T}}\boldsymbol{x}_i + \text{const}$$
$$\propto -\operatorname{tr}\left(\mathbf{W}^{\mathrm{T}}\left(\sum_{i=1}^{m}\boldsymbol{x}_i\boldsymbol{x}_i^{\mathrm{T}}\right)\mathbf{W}\right), \tag{10.14}$$

where $\mathbf{W} = (\mathbf{w}_1, \mathbf{w}_2, \ldots, \mathbf{w}_d)$. According to the property of minimum reconstruction error, (10.14) should be minimized. Since $\mathbf{w}_j$ is the orthonormal basis vector and $\sum_i \boldsymbol{x}_i\boldsymbol{x}_i^{\mathrm{T}}$ is the covariance matrix, we have

Strictly speaking, the covariance matrix is $\frac{1}{m-1}\sum_{i=1}^{m}\boldsymbol{x}_i\boldsymbol{x}_i^{\mathrm{T}}$, but the constant term makes no difference here.

$$\min_{\mathbf{W}} \; -\operatorname{tr}(\mathbf{W}^{\mathrm{T}}\mathbf{X}\mathbf{X}^{\mathrm{T}}\mathbf{W})$$
$$\text{s.t. } \mathbf{W}^{\mathrm{T}}\mathbf{W} = \mathbf{I}, \tag{10.15}$$

which is the optimization objective of PCA.

Another interpretation of PCA is from the maximum variance perspective. If we wish the projections of samples to stay away from each other, then the variance of the projected samples should be maximized, as illustrated in ◘ Figure 10.3.

The covariance matrix of the projected samples is $\sum_i \mathbf{W}^{\mathrm{T}}\boldsymbol{x}_i\boldsymbol{x}_i^{\mathrm{T}}\mathbf{W}$, where $\mathbf{W}^{\mathrm{T}}\boldsymbol{x}_i$ is the projection of $\boldsymbol{x}_i$ on the hyperplane in the new space. Then, we can write the optimization objective as

$$\max_{\mathbf{W}} \; \operatorname{tr}(\mathbf{W}^{\mathrm{T}}\mathbf{X}\mathbf{X}^{\mathrm{T}}\mathbf{W})$$
$$\text{s.t. } \mathbf{W}^{\mathrm{T}}\mathbf{W} = \mathbf{I}. \tag{10.16}$$

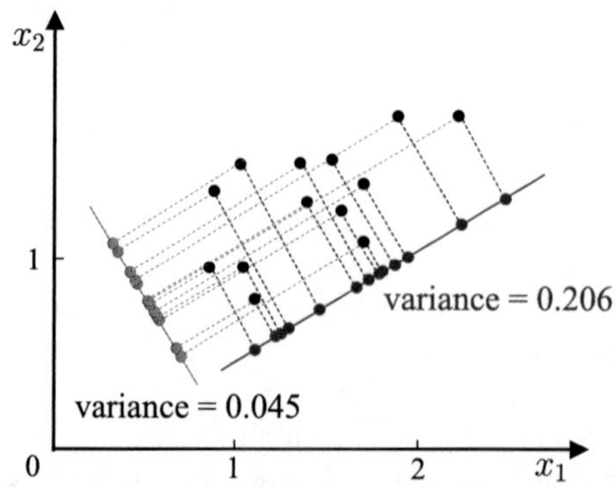


**Fig. 10.3** To make the projections of the samples away from each other as far as possible (the red lines), we need to maximize the variance of the projected samples

We notice that (10.15) and (10.16) are equivalent.

Applying Language multipliers to (10.15) or (10.16) gives

$$\mathbf{X}\mathbf{X}^{\mathrm{T}}\mathbf{w}_i = \lambda_i \mathbf{w}_i. \tag{10.17}$$

To obtain the solution of PCA, we perform eigenvalue decomposition on the covariance matrix $\mathbf{X}\mathbf{X}^{\mathrm{T}}$ and obtain the sorted eigenvalues: $\lambda_1 \geqslant \lambda_2 \geqslant \ldots \geqslant \lambda_d$. Then, we construct the solution $\mathbf{W}^* = (\mathbf{w}_1, \mathbf{w}_2, \ldots, \mathbf{w}_{d'})$ using the eigenvectors of the first $d'$ eigenvalues. The pseudocode of PCA is given in ◘ Algorithm 10.2.

In practice, we often do singular value decomposition on $\mathbf{X}$ instead of eigenvalue decomposition on $\mathbf{X}\mathbf{X}^{\mathrm{T}}$, and both decompositions lead to the same $\mathbf{W}$.

PCA can also be seen as incrementally selecting the direction that maximizes the covariance, that is, find the eigenvalues of the covariance matrix $\sum_i \boldsymbol{x}_i \boldsymbol{x}_i^{\mathrm{T}}$ and take the eigenvector $\mathbf{w}_1$ corresponding to the largest eigenvalue; then find the eigenvalues of $\sum_i \boldsymbol{x}_i \boldsymbol{x}_i^{\mathrm{T}} - \lambda_1 \mathbf{w}_1 \mathbf{w}_1^{\mathrm{T}}$ and then take the eigenvector $\mathbf{w}_2$ corresponding to the largest eigenvalue; ... Since the components of $\mathbf{W}$ are orthogonal and

$$\sum_{i=1}^{m} \boldsymbol{x}_i \boldsymbol{x}_i^{\mathrm{T}} = \sum_{j=1}^{d} \lambda_j \mathbf{w}_j \mathbf{w}_j^{\mathrm{T}},$$

it can be proved that the incremental approach is equivalent to the approach of selecting the first $d'$ eigenvalues at one time.

**Algorithm 10.2** Principal Component Analysis

**Input:** Data set $D = \{\boldsymbol{x}_1, \boldsymbol{x}_2, \ldots, \boldsymbol{x}_m\}$;
Dimension $d'$ of the lower dimensional space.

**Process:**

1: Center all samples: $\boldsymbol{x}_i \leftarrow \boldsymbol{x}_i - \frac{1}{m}\sum_{i=1}^{m} \boldsymbol{x}_i$;
2: Compute the covariance matrix $\mathbf{X}\mathbf{X}^{\mathrm{T}}$ of samples;
3: Perform eigenvalue decomposition on the covariance matrix $\mathbf{X}\mathbf{X}^{\mathrm{T}}$;
4: Take the eigenvectors $\mathbf{w}_1, \mathbf{w}_2, \ldots, \mathbf{w}_{d'}$ corresponding to the $d'$ largest eigenvalues.

**Output:** The projection matrix $\mathbf{W}^* = (\mathbf{w}_1, \mathbf{w}_2, \ldots, \mathbf{w}_{d'})$.

Typically, the dimension $d'$ of the lower dimensional space is specified by the user or selected by doing cross-validation with different $d'$ values, that is, comparing the performance of a $k$NN classifier (or other learners with low computational cost) in the dimension-reduced spaces produced by different $d'$ values. For PCA, a threshold $t$ (e.g., $t = 95\%$) can also be set from the reconstruction perspective to find the minimum $d'$ subject to

$$\frac{\sum_{i=1}^{d'} \lambda_i}{\sum_{i=1}^{d} \lambda_i} \geqslant t. \tag{10.18}$$

With just $\mathbf{W}^*$ and the mean vector of samples, PCA can project new samples to the lower dimensional space by applying simple vector subtraction and matrix-vector multiplication operations. The lower dimensional space and the original high-dimensional space are different since the eigenvectors corresponding to the $d-d'$ smallest eigenvalues are discarded. Information loss is an unavoidable consequence of dimensionality reduction, but it is often necessary: on the one hand, reducing the dimensionality can lead to relatively denser sampling although the number of samples remains the same, which is an important motivation of performing dimensionality reduction; on the other hand, the eigenvectors of the smallest eigenval-

The mean vector is used to zero-center the new samples via vector subtraction.

ues often relate to the noises, that is, discarding them helps to reduce the noise.

## 10.4 Kernelized PCA

Linear dimensionality reduction methods transform a high-dimensional space into a low-dimensional space via a linear mapping. In practice, however, non-linear mappings are often needed to find the proper low-dimensional embedding. ◘ Figure 10.4 shows an example of embedding data points to an S-shaped surface in a three-dimensional space, where the data points are sampled from a squared region of a two-dimensional space. If we apply linear dimensionality reduction methods to the three-dimensional space, we will lose the original low-dimensional structure. We call the original low-dimensional space, from which the data points are sampled, as the *intrinsic* low-dimensional space.

A general approach to non-linear dimensionality reduction is to *kernelize* linear dimensionality reduction methods via kernel tricks. Next, we give a demonstration with the representative Kernelized PCA (KPCA) (Schölkopf et al. 1998).

See Sect. 6.6 for kernel methods.



Suppose that we project data from the high-dimensional feature space to a hyperplane spanned by $\mathbf{W} = (\mathbf{w}_1, \mathbf{w}_2, \ldots, \mathbf{w}_d)$. Then, according to (10.17), we have the following for $\mathbf{w}_j$:

$$\left(\sum_{i=1}^{m} z_i z_i^{\mathrm{T}}\right) \mathbf{w}_j = \lambda_j \mathbf{w}_j, \tag{10.19}$$

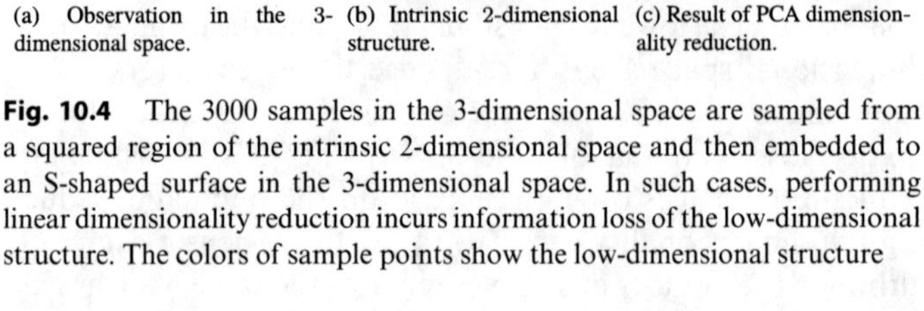

(a) Observation in the 3-dimensional space. (b) Intrinsic 2-dimensional structure. (c) Result of PCA dimensionality reduction.

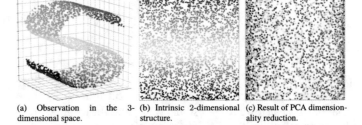

**Fig. 10.4** The 3000 samples in the 3-dimensional space are sampled from a squared region of the intrinsic 2-dimensional space and then embedded to an S-shaped surface in the 3-dimensional space. In such cases, performing linear dimensionality reduction incurs information loss of the low-dimensional structure. The colors of sample points show the low-dimensional structure

where $z_i$ is the image of $\boldsymbol{x}_i$ in the high-dimensional feature space. Then, we have

$$\begin{aligned}\mathbf{w}_j &= \frac{1}{\lambda_j}\left(\sum_{i=1}^{m} z_i z_i^{\mathrm{T}}\right)\mathbf{w}_j = \sum_{i=1}^{m} z_i \frac{z_i^{\mathrm{T}}\mathbf{w}_j}{\lambda_j} \\ &= \sum_{i=1}^{m} z_i \alpha_i^j, \end{aligned} \tag{10.20}$$

where $\alpha_i^j = \frac{1}{\lambda_j} z_i^{\mathrm{T}}\mathbf{w}_j$ is the $j$th component of $\boldsymbol{\alpha}_i$. Suppose that $z_i$ is obtained by mapping the original sample $\boldsymbol{x}_i$ via $\phi$, that is, $z_i = \phi(\boldsymbol{x}_i), i = 1, 2, \ldots, m$. If the explicit form of $\phi$ is known, then we can use it to map the samples to the high-dimensional feature space, and then apply PCA. Equation (10.19) can be rewritten as

$$\left(\sum_{i=1}^{m} \phi(\boldsymbol{x}_i)\phi(\boldsymbol{x}_i)^{\mathrm{T}}\right)\mathbf{w}_j = \lambda_j \mathbf{w}_j, \tag{10.21}$$

and (10.20) can be rewritten as

$$\mathbf{w}_j = \sum_{i=1}^{m} \phi(\boldsymbol{x}_i)\alpha_i^j. \tag{10.22}$$

Since the exact form of $\phi$ is generally unknown, we introduce the kernel function

$$\kappa(\boldsymbol{x}_i, \boldsymbol{x}_j) = \phi(\boldsymbol{x}_i)^{\mathrm{T}}\phi(\boldsymbol{x}_j). \tag{10.23}$$

Substituting (10.22) and (10.23) into (10.21) gives

$$\mathbf{K}\boldsymbol{\alpha}^j = \lambda_j \boldsymbol{\alpha}^j, \tag{10.24}$$

where $\mathbf{K}$ is the kernel matrix corresponding to $\kappa$, $(\mathbf{K})_{ij} = \kappa(\boldsymbol{x}_i, \boldsymbol{x}_j)$, and $\boldsymbol{\alpha}^j = (\alpha_1^j; \alpha_2^j; \ldots; \alpha_m^j)$. We notice that (10.24) is an eigenvalue decomposition problem, and hence we simply take the eigenvectors corresponding to the $d'$ largest eigenvalues in $\mathbf{K}$.

For a new sample $\boldsymbol{x}$, its projected coordinate in the $j$th ($j = 1, 2, \ldots, d'$) dimension is

$$\begin{aligned} z_j &= \mathbf{w}_j^{\mathrm{T}}\phi(\boldsymbol{x}) = \sum_{i=1}^{m} \alpha_i^j \phi(\boldsymbol{x}_i)^{\mathrm{T}}\phi(\boldsymbol{x}) \\ &= \sum_{i=1}^{m} \alpha_i^j \kappa(\boldsymbol{x}_i, \boldsymbol{x}), \end{aligned} \tag{10.25}$$

where $\alpha_i$ has been normalized. From (10.25), we see that KPCA is computationally expensive since it sums over all samples to compute the projected coordinate.

## 10.5 Manifold Learning

*Manifold learning* is a dimensionality reduction approach that utilizes some concepts of topological manifolds. A *manifold* is a topological space that is locally homeomorphic to the Euclidean space. It locally meets the properties of European space, which means we can use Euclidean distance, and this fact motivates a new way of dimensionality reduction. If a low-dimensional manifold is embedded in the high-dimensional space, then the samples, which seem to be very complex in the high-dimensional space, locally have the same properties of samples in the Euclidean space. Therefore, we can establish dimensionality reduction mappings locally, and then extend to the entire space. When the dimensionality is reduced to two or three, we can naturally visualize the data, and hence manifold learning is also useful for visualization purposes. The rest of this section introduces two representative manifold learning methods.



### 10.5.1 Isometric Mapping

Isometric Mapping (Isomap) (Tenenbaum et al. 2000) is motivated by the fact that the straight-line Euclidean distances measured in a high-dimensional space can be misleading since the straight lines in the high-dimensional space may not exist in the low-dimensional manifold embedding. As illustrated in ◘ Figure 10.5a, the distance between two points in the low-dimensional manifold embedding is the *geodesic distance*. Imagine an insect crawling from one point to another, and it has to crawl on the surface. Then, the red curve in ◘ Figure 10.5a is the shortest path, that is, the geodesic on the S-shaped surface. The geodesic distance is the intrinsic distance between the two points. In such cases, measuring the straight-line distances in the high-dimensional space is inappropriate.

Then, how can we compute the geodesic distances? Recall that a manifold is locally homeomorphic to the Euclidean space, which means we can find neighboring points of each point based on the Euclidean distance, and hence a neighborhood graph can be constructed. In the neighborhood graph, only neighboring points are connected, and the geodesic distance is measured by finding the shortest path between two points in the neighborhood graph. From ◘ Figure 10.5b, we

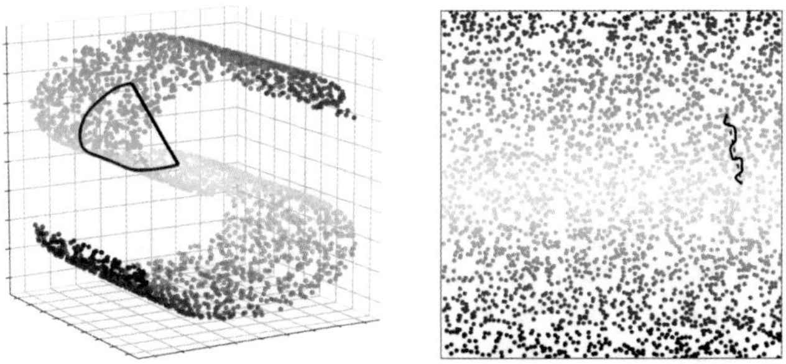

(a) The geodesic distance and the straight-line distance.

(b) The geodesic distance and the neighborhood distance.

**Fig. 10.5** We cannot use the straight-line distance in the high-dimensional space to compute the geodesic distance (red) in the low-dimensional manifold embedding, but an approximation is possible with the neighborhood distance

can see that the shortest path in the neighborhood graph provides a good approximation to the geodesic distance in the low-dimensional manifold embedding.

To find the shortest path between two points in a neighborhood graph, we can use the well-known Dijkstra algorithm or the Floyd algorithm. With the distance between any two points, we can then use the MDS method introduced in Sect. 10.2 to obtain the coordinates of samples in the low-dimensional space. The pseudocode of Isomap is given in ◘ Algorithm 10.3.

These two algorithms were proposed by the 1972 Turing Award winner E. W. Dijkstra and the 1978 Turing Award winner R. Floyd, respectively.

---

**Algorithm 10.3** Isometric Mapping

**Input:** Data set $D = \{\boldsymbol{x}_1, \boldsymbol{x}_2, \ldots, \boldsymbol{x}_m\}$;
Nearest neighbor parameter $k$;
Dimension $d'$ of the low-dimensional space.

**Process:**

1: **for** $i = 1, 2, \ldots, m$ **do**
2: Find the $k$ nearest neighbors of $\boldsymbol{x}_i$;
3: Set the distances between $\boldsymbol{x}_i$ and the $k$ nearest neighbors to the Euclidean distance, and set the distances from $\boldsymbol{x}_i$ to other sample points as positive infinity;
4: **end for**
5: Compute the distance $\mathrm{dist}(\boldsymbol{x}_i, \boldsymbol{x}_j)$ between every pair sample points using a shortest path algorithm;
6: Use $\mathrm{dist}(\boldsymbol{x}_i, \boldsymbol{x}_j)$ as input for the MDS algorithm;
7: **return** The output of the MDS algorithm.

**Output:** The low-dimensional projections $Z = \{z_1, z_2, \ldots, z_m\}$ of the data set $D$.

---

See Sect. 10.2 for MDS.

Isomap only gives the low-dimensional coordinates of training samples, but how can we project new samples to the low-dimensional space? A general approach is to train a regression model using the high-dimensional coordinates of training samples as the input and the corresponding low-dimensional coordinates as the output. Then, use the trained regression model to predict the low-dimensional coordinates of new samples. Such a method looks ad hoc, but currently there seems to be no better solution.

There are two general approaches to constructing the neighborhood graph. The first approach is to specify the number of neighbors. For example, using the $k$ nearest neighbors measured by Euclidean distance, which gives the $k$-nearest neighbor graph. The other approach is to specify a distance threshold $\epsilon$ to consider all points with a distance smaller than $\epsilon$ as neighbors, and the generated graph is called a $\epsilon$-nearest neighbor graph. However, both approaches have the same limitation: if the specified neighborhood range, either $k$ or $\epsilon$, is too large, then "short circuit" may happen, in which some distant points are incorrectly considered close to each other; on the other hand, if the specified neighborhood range is too small, then "open circuit" may happen, in which some regions become disconnected from each other. Either case misleads the consequent calculation of the shortest path.

### 10.5.2 Locally Linear Embedding



Unlike Isomap, which maintains the distances between samples, Locally Linear Embedding (LLE) (Roweis and Saul 2000) aims to keep the linear relationships between neighboring samples. As illustrated in ◘ Figure 10.6, suppose the coordinates of a sample point $\boldsymbol{x}_i$ can be reconstructed via a linear combination of the coordinates of its neighboring samples $\boldsymbol{x}_j$, $\boldsymbol{x}_k$, and $\boldsymbol{x}_l$, that is,

$$\boldsymbol{x}_i = w_{ij}\boldsymbol{x}_j + w_{ik}\boldsymbol{x}_k + w_{il}\boldsymbol{x}_l. \tag{10.26}$$

LLE aims to keep the relationship of (10.26) in the low-dimensional space.

LLE starts by identifying the neighborhood indices set $Q_i$ for sample $\boldsymbol{x}_i$, and then find the linear reconstruction weights $\mathbf{w}_i$ of samples in $Q_i$:

$$\begin{aligned} \min_{\mathbf{w}_1,\mathbf{w}_2,\ldots,\mathbf{w}_m} \quad & \sum_{i=1}^{m} \left\| \boldsymbol{x}_i - \sum_{j\in Q_i} w_{ij}\boldsymbol{x}_j \right\|_2^2 \\ \text{s.t.} \quad & \sum_{j\in Q_i} w_{ij} = 1, \end{aligned} \tag{10.27}$$

where $\boldsymbol{x}_i$ and $\boldsymbol{x}_j$ are known. Letting $C_{jk} = (\boldsymbol{x}_i - \boldsymbol{x}_j)^{\mathrm{T}}(\boldsymbol{x}_i - \boldsymbol{x}_k)$, then $w_{ij}$ has a closed form solution

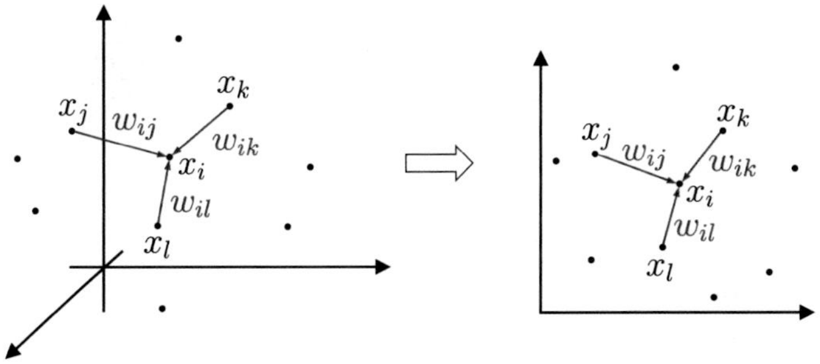


**Fig. 10.6** The reconstruction relationship of samples in the high-dimensional space are preserved in the low-dimensional space

$$w_{ij} = \frac{\sum_{k \in Q_i} C_{jk}^{-1}}{\sum_{l,s \in Q_i} C_{ls}^{-1}}. \tag{10.28}$$

Since LLE preserves $\mathbf{w}_i$ in the low-dimensional space, the low-dimensional coordinates $z_i$ of $\boldsymbol{x}_i$ can be obtained by

$$\min_{z_1,z_2,\ldots,z_m} \sum_{i=1}^{m} \left\| z_i - \sum_{j \in Q_i} w_{ij} z_j \right\|_2^2 \tag{10.29}$$

We notice that (10.27) and (10.29) have the same form of optimization objectives. The only difference is that (10.27) optimizes $\mathbf{w}_i$, whereas (10.29) optimizes $z_i$, that is, the low-dimensional coordinates of $\boldsymbol{x}_i$.

Letting $\mathbf{Z} = (z_1, z_2, \ldots, z_m) \in \mathbb{R}^{d' \times m}$, $(\mathbf{W})_{ij} = w_{ij}$, and

$$\mathbf{M} = (\mathbf{I} - \mathbf{W})^{\mathrm{T}}(\mathbf{I} - \mathbf{W}), \tag{10.30}$$

then (10.29) can be rewritten as

$$\begin{aligned} &\min_{\mathbf{Z}}\ \mathrm{tr}(\mathbf{Z}\mathbf{M}\mathbf{Z}^{\mathrm{T}}) \\ &\text{s.t. } \mathbf{Z}\mathbf{Z}^{\mathrm{T}} = \mathbf{I}. \end{aligned} \tag{10.31}$$

We can solve (10.31) by eigenvalue decomposition: $\mathbf{Z}^{\mathrm{T}}$ is the matrix consisting of the eigenvectors corresponding to the $d'$ smallest eigenvalues of $\mathbf{M}$.

The pseudocode of LLE is given in ■ Algorithm 10.4. From line 4, we see that $\boldsymbol{x}_i$ and $z_i$ are not impacted by any changes to a non-neighbor sample $\boldsymbol{x}_j$. This idea of restricting the impact of changes in a local region is adopted in many other places as well.

**Algorithm 10.4** Locally Linear Embedding

**Input:** Data set $D = \{\boldsymbol{x}_1, \boldsymbol{x}_2, \ldots, \boldsymbol{x}_m\}$;
Nearest neighbor parameter $k$;
Dimension $d'$ of the low-dimensional space.

**Process:**

```
1: for i = 1, 2, ..., m do
2:     Find the k nearest neighbors of x_i;
3:     Compute w_ij using (10.27) for j ∈ Q_i;
4:     Set w_ij = 0 for j ∉ Q_i;
5: end for
6: Obtain M using (10.30);
7: Perform eigenvalue decomposition on M;
8: return the eigenvectors corresponding to the d' smallest eigenvalues
   of M.
```

**Output:** The low-dimensional projections $Z = \{z_1, z_2, \ldots, z_m\}$ of the data set $D$.



## 10.6 Metric Learning

In machine learning, the main purpose of dimensionality reduction is to find a lower dimensional space, in which the learning performance is better than that in the original high-dimensional space. Since each space corresponds to a distance metric defined on the sample features, the searching of an appropriate space is indeed searching for an appropriate distance metric. Then, why not "learn" the appropriate distance metric directly? This question motivated *metric learning*, also known as *distance metric learning*.

To learn a distance metric, we must express it in a learnable form. We have seen many distance metrics in Sect. 9.3, but all of them have a fixed form without adjustable parameters that can be improved by learning from data. Therefore, we need an extension first.

For two $d$-dimensional samples $\boldsymbol{x}_i$ and $\boldsymbol{x}_j$, their squared Euclidean distance can be written as

The Euclidean distance is squared for the convenience of subsequent calculations.

$$\mathrm{dist}^2_{\mathrm{ed}}(\boldsymbol{x}_i, \boldsymbol{x}_j) = \|\boldsymbol{x}_i - \boldsymbol{x}_j\|_2^2 = dist^2_{ij,1} + dist^2_{ij,2} + \ldots + dist^2_{ij,d}, \tag{10.32}$$

where $dist_{ij,k}$ is the distance between $\boldsymbol{x}_i$ and $\boldsymbol{x}_j$ on the $k$th dimension. Suppose different features have different importance, then we can introduce the feature weights $\mathbf{w}$ as

$$\begin{aligned}\mathrm{dist}^2_{\mathrm{wed}}(\boldsymbol{x}_i, \boldsymbol{x}_j) &= \|\boldsymbol{x}_i - \boldsymbol{x}_j\|_2^2 \\ &= w_1 \cdot dist^2_{ij,1} + w_2 \cdot dist^2_{ij,2} + \ldots + w_d \cdot dist^2_{ij,d} \\ &= (\boldsymbol{x}_i - \boldsymbol{x}_j)^{\mathrm{T}} \mathbf{W} (\boldsymbol{x}_i - \boldsymbol{x}_j),\end{aligned} \tag{10.33}$$

where $w_i \geqslant 0$, $\mathbf{W} = \mathrm{diag}(\mathbf{w})$ is a diagonal matrix, and $(\mathbf{W})_{ii} = w_i$.

The adjustable weights $\mathbf{W}$ are the parameters to be learned from data. We notice that the non-diagonal elements of $\mathbf{W}$ are zero, which means the axes are orthogonal (i.e., the features are independent). In practice, however, this is often not true. For example, weight and volume of watermelons are positively correlated, that is, the axes are not orthogonal. Hence, by replacing $\mathbf{W}$ in (10.33) with a symmetric positive semidefinite matrix $\mathbf{M}$, we have the *Mahalanobis distance*

$$\mathrm{dist}^2_{\mathrm{mah}}(\boldsymbol{x}_i, \boldsymbol{x}_j) = (\boldsymbol{x}_i - \boldsymbol{x}_j)^{\mathrm{T}}\mathbf{M}(\boldsymbol{x}_i - \boldsymbol{x}_j) = \left\|\boldsymbol{x}_i - \boldsymbol{x}_j\right\|^2_{\mathbf{M}}, \tag{10.34}$$

Mahalanobis distance is named after Indian mathematician P. C. Mahalanobis. In the standard Mahalanobis distance, $\mathbf{M}$ is the inverse of the covariance matrix (i.e., $\mathbf{M} = \boldsymbol{\Sigma}^{-1}$). However, the value of $\mathbf{M}$ is made more flexible in metric learning.

where $\mathbf{M}$ is also called the *metric matrix*, and metric learning is about learning $\mathbf{M}$. In order to ensure the distances are non-negative and symmetric, $\mathbf{M}$ must be a positive semidefinite matrix, that is, there must be a matrix $\mathbf{P}$ such that $\mathbf{M} = \mathbf{P}\mathbf{P}^{\mathrm{T}}$.

To learn $\mathbf{M}$, we need to set an objective. Suppose we wish to improve the performance of the nearest neighbor classifier, then $\mathbf{M}$ can be directly embedded into the classifier's performance measure, and the optimal $\mathbf{M}$ is obtained by optimizing the performance measure. We take Neighborhood Component Analysis (NCA) (Goldberger et al. 2005) as an example to discuss the process.

Nearest neighbor classifiers usually employ the voting method for classification, that is, each neighboring sample has 1 vote, and each non-neighboring sample has 0 vote. Instead of using binary voting, we can also use the probabilistic voting method, in which the probability of a sample $\boldsymbol{x}_j$ influences the classification of a sample $\boldsymbol{x}_i$ is given by

$$p_{ij} = \frac{\exp\left(-\left\|\boldsymbol{x}_i - \boldsymbol{x}_j\right\|^2_{\mathbf{M}}\right)}{\sum_l \exp\left(-\left\|\boldsymbol{x}_i - \boldsymbol{x}_l\right\|^2_{\mathbf{M}}\right)}, \tag{10.35}$$

where $p_{ij}$ is maximized when $i = j$. The impact of $\boldsymbol{x}_j$ on $\boldsymbol{x}_i$ decreases as their distance increases. If the objective is to maximize the LOO accuracy, then the accuracy of $\boldsymbol{x}_i$ is the probability that other samples can correctly classify it:

See Sect. 2.2.2 for LOO.

$$p_i = \sum_{j \in \Omega_i} p_{ij}, \tag{10.36}$$

where $\Omega_i$ is the index set of samples with the same class label as $\boldsymbol{x}_i$. Then, the LOO accuracy for the entire data set is

$$\sum_{i=1}^{m} p_i = \sum_{i=1}^{m} \sum_{j \in \Omega_i} p_{ij}. \tag{10.37}$$

Substituting (10.35) into (10.37) and considering $\mathbf{M} = \mathbf{P}\mathbf{P}^{\mathrm{T}}$, we have the optimization objective of NCA:

$$\min_{\mathbf{P}} \ 1 - \sum_{i=1}^{m} \sum_{j \in \Omega_i} \frac{\exp\left(-\left\|\mathbf{P}^{\mathrm{T}}\boldsymbol{x}_i - \mathbf{P}^{\mathrm{T}}\boldsymbol{x}_j\right\|_2^2\right)}{\sum_l \exp\left(-\left\|\mathbf{P}^{\mathrm{T}}\boldsymbol{x}_i - \mathbf{P}^{\mathrm{T}}\boldsymbol{x}_l\right\|_2^2\right)}. \tag{10.38}$$

It can be solved using Stochastic Gradient Descent (Goldberger et al. 2005).

Solving (10.38) gives the distance metric matrix $\mathbf{M}$ that maximizes the LOO accuracy of the nearest neighbor classifier.

In addition to the supervised learning objectives, such as minimizing the error rate, we can also introduce domain knowledge into the optimization objective of metric learning. For example, if we already know that some samples are similar or dissimilar, we can define a *must-link* constraints set $\mathcal{M}$ and a *cannot-link* constraints set $\mathcal{C}$, where $(\boldsymbol{x}_i, \boldsymbol{x}_j) \in \mathcal{M}$ indicates $\boldsymbol{x}_i$ and $\boldsymbol{x}_j$ are similar, and $(\boldsymbol{x}_i, \boldsymbol{x}_k) \in \mathcal{C}$ indicates $\boldsymbol{x}_i$ and $\boldsymbol{x}_k$ are dissimilar. Since we wish similar samples to have smaller distances and dissimilar samples to have larger distances, we can obtain the metric matrix $\mathbf{M}$ by solving the following convex optimization problem (Xing et al. 2003):

$$\begin{aligned} \min_{\mathbf{M}} \quad & \sum_{(\boldsymbol{x}_i, \boldsymbol{x}_j) \in \mathcal{M}} \|\boldsymbol{x}_i - \boldsymbol{x}_j\|_{\mathbf{M}}^2 \\ \text{s.t.} \quad & \sum_{(\boldsymbol{x}_i, \boldsymbol{x}_k) \in \mathcal{C}} \|\boldsymbol{x}_i - \boldsymbol{x}_k\|_{\mathbf{M}} \geqslant 1, \\ & \mathbf{M} \succeq 0, \end{aligned} \tag{10.39}$$

where the constraint $\mathbf{M} \succeq 0$ ensures that $\mathbf{M}$ is positive semidefinite. Equation (10.39) minimizes the total distance between similar samples while keeping the individual distances between dissimilar samples no less than 1.

Metric learning methods generally do not require $\mathbf{M}$ to be low-rank.

rank($\mathbf{M}$) is the rank of $\mathbf{M}$.

Different metric learning methods obtain "good" positive semidefinite symmetric distance metric matrices $\mathbf{M}$ based on different goals. When $\mathbf{M}$ is a low-rank matrix, by eigenvalue decomposition of $\mathbf{M}$, we can always obtain a set of rank($\mathbf{M}$) orthogonal basis vectors, where rank($\mathbf{M}$) is less than the original number of features $d$. Hence, the learning outcome of metric learning can derive a dimension-reduced matrix $\mathbf{P} \in \mathbb{R}^{d \times \mathrm{rank}(\mathbf{M})}$ that can be used for dimensionality reduction.

## 10.7 Further Reading

Representatives of lazy learning include $k$-nearest neighbor learners and lazy decision trees (Friedman et al. 1996). Naïve Bayes classifiers can be used for both lazy learning and eager learning. See Aha (1997) for more information about lazy learning.

PCA is a representative unsupervised linear dimensionality reduction method. A representative supervised linear dimensionality reduction method is LDA (Fisher 1936) (see Sect. 3.4), and its kernelized version is KLDA (Baudat and Anouar 2000) (see Sect. 6.6). Maximizing the correlation between two variable sets gives Canonical Correlation Analysis (CCA) (Hotelling 1936), and its kernelized version is KCCA (Harden et al. 2004), which has numerous applications in multi-view learning. Some pattern recognition studies found that a direct dimensionality reduction on the object matrix (e.g., an image) often leads to better performance than dimensionality reduction on a reshaped vector (e.g., converting an image to a vector). Following this observation, many methods were proposed including 2DPCA (Yang et al. 2004), 2DLDA (Ye et al. 2005), $(2D)^2$PCA (Zhang and Zhou 2005), and also tensor-based methods (Kolda and Bader 2009).

See Sect. 13.5 for multi-view learning.

Other than Isomap and LLE, popular manifold learning methods include Laplacian Eigenmaps (LE) (Belkin and Niyogi 2003), Local Tangent Space Alignment (LTSA) (Zhang and Zha 2004), etc. Locality Preserving Projections (LPP) (He and Niyogi 2004) is a linear dimensionality reduction method based on LE. For supervised learning, the low-dimensional space adjusted by the class information is often superior to the intrinsic low-dimensional space (Geng and Zhou 2005). It is worth noting that neighborhood preservation in manifold learning requires a dense sample, and this is indeed a major barrier in high-dimensional situations. Hence, in practice, the dimensionality reduction performance of manifold learning is often not as good as expected. Nevertheless, the idea of neighborhood preservation has significantly influenced other branches of machine learning research, such as the well-known manifold assumption and the manifold regularization in semi-supervised learning (Belkin et al. 2006). Yan et al. (2007) provided a unified framework for dimensionality reduction from a graph embedding point of view.

See Chap. 13.

The must-link and cannot-link constraints have already been used in semi-supervised clustering (Wagstaff et al. 2001) before they were used in metric learning. In metric learning, such constraints are applied to all samples at the same time (Xing et al. 2003), and hence the corresponding methods are called global metric learning methods. Some attempts

See Sect. 13.6 for semi-supervised clustering.

have also been made to incorporate local constraints (e.g., local triplet constraints), leading to local distance metric learning methods (Weinberger and Saul 2009). There are even attempts to find an appropriate distance metric for every sample (Frome et al. 2007, Zhan et al. 2009). In terms of learning and optimization, different metric learning methods usually adopt different optimization techniques. For example, Yang et al. (2006) converted the metric learning problem to a binary classification problem using a discriminant analysis method under a probabilistic framework. Davis et al. (2007) converted the metric learning problem to a Bregman optimization problem, which enables online learning, under the framework of information theory.

## Exercises

**10.1** Implement the $k$-nearest neighbors algorithm and compare its decision boundary to the decision boundary of decision trees on the watermelon data set 3.0$\alpha$.

The watermelon data set 3.0$\alpha$ is in ◘ Table 4.5.

**10.2** Let *err* and *err** be, respectively, the expected error rates of the nearest neighbors classifier and the Bayes optimal classifier. Prove

$$err^* \leqslant err \leqslant err^* \left(2 - \frac{|\mathcal{Y}|}{|\mathcal{Y}-1|} \times err^*\right). \tag{10.40}$$

**10.3** *Centering* should be done before dimensionality reduction, and a typical method is to convert the covariance matrix $\mathbf{XX}^{\mathrm{T}}$ to $\mathbf{XHH}^{\mathrm{T}}\mathbf{X}^{\mathrm{T}}$, where $\mathbf{H} = \mathbf{I} - \frac{1}{m}\mathbf{11}^{\mathrm{T}}$. Discuss the effects of the centering process.

**10.4** In practice, we often perform singular value decomposition of the centered sample matrix $\mathbf{X}$ instead of eigenvalue decomposition of the covariance matrix $\mathbf{XX}^{\mathrm{T}}$. Discuss the reasons of taking this approach.

**10.5** We often require the projection matrix in dimensionality reduction to be orthogonal. Discuss the pros and cons of using orthogonal and non-orthogonal projection matrices.

E.g., princomp function in MATLAB.

**10.6** Use the PCA function in any software packages of your choice to perform dimensionality reduction on the Yale face data set, and investigate the images corresponding to the first 20 feature vectors.

The Yale face data set is available at ▶ http://vision.ucsd.edu/content/yale-face-database.

**10.7** Analyze the connections between kernelized linear dimensionality reduction and manifold learning, and discuss their pros and cons.

**10.8** * The short circuit and open circuit in $k$-nearest neighbor graphs and $\epsilon$-nearest neighbor graphs are troublesome for Isomap. Design a method to alleviate the problem.

**10.9** * Design a method to find the low-dimensional coordinates of new samples after performing LLE dimensionality reduction.

**10.10** Discuss how to ensure the distance metric produced by metric learning satisfies the four axioms of distance measures.

See Sect. 9.3 for the axioms of distance measures.

## Break Time

### Short Story: Principal Component Analysis and Karl Pearson

Principal Component Analysis (PCA) is, by far, the most commonly used dimensionality reduction method. It has many other names, such as Singular Value Decomposition (SVD) of the scatter matrix in linear algebra, factor analysis in statistics, Karhünen-Loève transform in signal processing, Hotelling transform in image analysis, Latent Semantic Analysis (LSA) in text analysis, Proper Orthogonal Decomposition (POD) in mechanical engineering, Empirical Orthogonal Function (EOF) in meteorology, Experimental Modal Analysis (EMA) in structural dynamics, and the Schmidt–Mirsky theorem in psychometrics.

Karl Pearson (1857–1936) invented PCA in 1901. Pearson was known to be a "walking encyclopedia", who was a statistician, applied mathematician, philosopher, historian, folklorist, theologian, anthropologist, linguist, social activist, education reformer, and writer. After graduated from King's College, Cambridge in 1879, Pearson visited many universities, including Heidelberg University and the University of Berlin. In 1884, he was appointed to a professor of applied mathematics at University College, London (UCL), and became a fellow of the Royal Society at the age of 39. In 1892, he published the book *The Grammar of Science*, which offered some inspiration to Einstein's theory of relativity. Pearson has made tremendous contributions to statistics, such as correlation coefficients, standard deviation, the chi-square test, and the method of moments. Pearson laid the foundation of the hypothesis testing theory and the statistical decision theory, and he is often considered as the "father of statistics".

Pearson started his research on statistics under the influence of two biologists F. Galton and W. Welton, who worked on quantitative analysis of the theory of evolution. Pearson together with Galton and Welton founded the prestigious journal *Biometrika* in 1901, and Pearson was the editor-in-chief for the rest of his life. His only son Egon Pearson is also a prominent statistician, and the Neyman–Pearson lemma is named after him. Egon succeeded his father as a professor of statistics at UCL and as the editor-in-chief of the journal *Biometrika*. Later, Egon became the president of the Royal Statistical Society.

Galton is the cousin of C. Darwin and the inventor of *eugenics*.

## References

Aha D (ed) (1997) Lazy learning. Kluwer, Norwell

Baudat G, Anouar F (2000) Generalized discriminant analysis using a kernel approach. Neural Comput 12(10):2385–2404

Belkin M, Niyogi P (2003) Laplacian eigenmaps for dimensionality reduction and data representation. Neural Comput 15(6):1373–1396

Belkin M, Niyogi P, Sindhwani V (2006) Manifold regularization: a geometric framework for learning from labeled and unlabeled examples. J Mach Learn Res 7:2399–2434

Bellman RE (1957) Dynamic programming. Princeton University Press, Princeton

Cover TM, Hart PE (1967) Nearest neighbor pattern classification. IEEE Trans Inf Theory 13(1):21–27

Cox TF, Cox MA (2001) Multidimensional scaling. Chapman & Hall/CRC, London

Davis JV, Kulis B, Jain P, Sra S, Dhillon IS (2007) Information-theoretic metric learning. In: Proceedings of the 24th international conference on machine learning (ICML), Corvalis, OR, pp 209–216

Fisher RA (1936) The use of multiple measurements in taxonomic problems. Ann Eugen 7(2):179–188

Friedman JH, Kohavi R, Yun Y (1996) Lazy decision trees. In: Proceedings of the 13th national conference on artificial intelligence (AAAI), Portland, OR, pp 717–724

Frome A, Singer Y, Malik J (2007) Image retrieval and classification using local distance functions. In: Scholkopf B, Platt JC, Hoffman T (eds) Advances in neural information processing systems 19 (NIPS). MIT Press, Cambridge, pp 417–424

Geng X, Zhan D-C, Zhou Z-H (2005) Supervised nonlinear dimensionality reduction for visualization and classification. IEEE Trans Syst Man Cybern-Part B: Cybern 35(6):1098–1107

Goldberger J, Hinton GE, Roweis ST, Salakhutdinov RR (2005) Neighbourhood components analysis. In: Saul LK, Weiss Y, Bottou L (eds) Advances in neural information processing systems 17 (NIPS). MIT Press, Cambridge, pp 513–520

Harden DR, Szedmak S, Shawe-Taylor J (2004) Canonical correlation analysis: an overview with application to learning methods. Neural Comput 16(12):2639–2664

He X, Niyogi P (2004) Locality preserving projections. In: Thrun S, Saul LK, Scholkopf B (eds) Advances in neural information processing systems 16 (NIPS). MIT Press, Cambridge, pp 153–160

Hotelling H (1936) Relations between two sets of variates. Biometrika 28(3–4):321–377

Kolda TG, Bader BW (2009) Tensor decompositions and applications. SIAM Rev 51(3):455–500

Roweis ST, Saul LK (2000) Nonlinear dimensionality reduction by locally linear embedding. Science 290(5500):2323–2326

Schölkopf B, Smola A, Müller K-R (1998) Nonlinear component analysis as a kernel eigenvalue problem. Neural Comput 10(5):1299–1319

Tenenbaum JB, de Silva V, Langford JC (2000) A global geometric framework for nonlinear dimensionality reduction. Science 290(5500):2319–2323

Wagstaff K, Cardie C, Rogers S, Schrödl S (2001) Constrained $k$-means clustering with background knowledge. In: Proceedings of the 18th international conference on machine learning (ICML), Williamstown, MA, pp 577–584

Weinberger KQ, Saul LK (2009) Distance metric learning for large margin nearest neighbor classification. J Mach Learn Res 10:207–244

Xing EP, Ng AY, Jordan MI, Russell S (2003) Distance metric learning, with application to clustering with side-information. In: Becker S, Thrun S, Obermayer K (eds) Advances in neural information processing systems 15 (NIPS). MIT Press, Cambridge, MA, pp 505–512

Yan S, Xu D, Zhang B, Zhang H-J (2007) Graph embedding and extensions: a general framework for dimensionality reduction. IEEE Trans Pattern Anal Mach Intell 29(1):40–51

Yang J, Zhang D, Frangi AF, Yang J-Y (2004) Two-dimensional PCA: A new approach to appearance-based face representation and recognition. IEEE Trans Pattern Anal Mach Intell 26(1):131–137

Yang L, Jin R, Sukthankar R, Liu Y (2006) An efficient algorithm for local distance metric learning. In: Proceedings of the 21st national conference on artificial intelligence (AAAI), Boston, MA, pp 543–548

Ye J, Janardan R, Li Q (2005) Two-dimensional linear discriminant analysis. In: Saul LK, Weiss Y, Bottou L (eds) Advances in neural information processing systems 17 (NIPS). MIT Press, Cambridge, pp 1569–1576

Zhan D-C, Li Y-F, Zhou Z-H (2009) Learning instance specific distances using metric propagation. In: Proceedings of the 26th international conference on machine learning (ICML), Montreal, Canada, pp 1225–1232

Zhang D, Zhou Z-H (2005) $(2D)^2$PCA: 2-directional 2-dimensional PCA for efficient face representation and recognition. Neurocomputing 69(1–3):224–231

Zhang Z, Zha H (2004) Principal manifolds and nonlinear dimension reduction via local tangent space alignment. SIAM J Sci Comput 26(1):313–338

# Feature Selection and Sparse Learning

Table of Contents




Z.-H. Zhou, *Machine Learning*,
https://doi.org/10.1007/978-981-15-1967-3_11

## 11.1 Subset Search and Evaluation

Watermelons can be described by many attributes, such as color, root, sound, texture, and surface, but experienced people can determine the ripeness with only the root and sound information. In other words, not all attributes are equally important for the learning task. In machine learning, attributes are also called *features*. Features that are useful for the current learning task are called *relevant features*, and those useless ones are called *irrelevant features*. The process of selecting relevant features from a given feature set is called *feature selection*.

Feature selection is an important step of *data preprocessing* that often needs to be done before training the learners. The reasons for feature selection are twofold. Firstly, the curse of dimensionality is a common issue in practical learning problems due to the large number of features. If we can identify the relevant features, then the subsequent learning process will deal with a much lower dimensionality, and hence the curse of dimensionality is alleviated. From this point of view, feature selection shares a similar motivation with dimensionality reduction, as discussed in Chap. 10. Secondly, eliminating irrelevant features often reduces the difficulty of learning because the learner is more likely to discover the truth without being distracted by irrelevant information.

It is worth noting that the feature selection process must not discard important features. Otherwise, the performance of subsequent learning will be hurt by information loss. Furthermore, for the same data set, the important features are often different for different learning tasks, and hence when we say some features are "irrelevant", we refer to a specific learning task. There is another category of features called *redundant features*, whose information can be derived from other features. For example, considering a cubic object, if the features base length and base width are known, then base area is a redundant feature since it can be calculated from base length and base width. Since redundant features usually provide no additional information, removing them can reduce the workload of the learning process. However, in some cases, the redundant features can make the learning easier. For example, suppose the task is to estimate the volume of a cubic, then it will become easier if using the redundant feature base area. More specifically, if a redundant feature happens to be an *intermediate concept* of the learning task, then it is a helpful redundant feature. For ease of discussion, we assume all data sets in this chapter contain no redundant feature, and all essential information is available in the given features.

When there is no prior domain knowledge, we need to evaluate all possible feature subsets to select the one that contains all essential information. However, this is computationally impractical due to the combinatorial explosion. A more practical method is to generate a *candidate subset* of features and then evaluate its quality, and then, based on the evaluation result, generate the next candidate subset. This process continues until we cannot find any better candidate subsets. There are two questions in this process: how to generate the next candidate subset of features based on the current evaluation result and how to evaluate the quality of a candidate subset of features.

The first question leads to the *subset search* problem. Given a feature set $\{a_1, a_2, \ldots, a_d\}$, we can consider every feature as a candidate subset, and then evaluate these $d$ single-feature subsets. Suppose $\{a_2\}$ is optimal, then it is the selected set in the first round. After that, we add one more feature to the selected set of the previous round to generate $d - 1$ two-feature candidate subsets. Among them, suppose the optimum is $\{a_2, a_4\}$, and it is better than $\{a_2\}$, then it is the selected set of the current round. The candidate subset generation process stops when the optimal $(k + 1)$-feature subset in the $(k + 1)$th round is not better than the selected set of the previous round, and the previously selected $k$-feature subset is returned as the outcome of feature selection. Such an incremental approach of adding relevant features is called *forward search*. Similarly, if we start with a complete feature set and gradually remove irrelevant features, then it is called *backward search*. Moreover, we can combine the forward and backward approaches into *bidirectional search* to gradually select relevant features (once selected, these features are always kept in the subsequent rounds) and remove irrelevant features.

Also known as the *generate and search* of subset.

All of the above are greedy approaches since they only consider the optimal set in the current round. For example, suppose $a_5$ is better than $a_6$ in the 3rd round, then we have the selected set $\{a_2, a_4, a_5\}$. However, it is possible that, in the 4th round, $\{a_2, a_4, a_6, a_8\}$ is better than any $\{a_2, a_4, a_5, a_i\}$. Unfortunately, such problems are unavoidable unless we do an exhaustive search.

The second question leads to the *subset evaluation* problem. Given a data set $D$, where the proportion of the $i$th class in $D$ is $p_i$ $(i = 1, 2, \ldots, |\mathcal{Y}|)$. For ease of discussion, let us assume all features are discrete. Then, given a feature subset $A$, the feature values split $D$ into $V$ subsets $\{D^1, D^2, \ldots, D^V\}$, where each subset includes the samples taking the same values on $A$. After that, we can compute the information gain of the feature subset $A$:

Suppose that each feature has $v$ possible values, then $V = v^{|A|}$ is potentially very large. Hence, in practice, we often reuse the evaluation results from the previous round of subset search as a start point.

$$\mathrm{Gain}(A) = \mathrm{Ent}(D) - \sum_{v=1}^{V} \frac{|D^v|}{|D|} \mathrm{Ent}(D^v), \tag{11.1}$$

where the information entropy Ent($D$) is defined as

See Sect. 4.2.1 for information entropy.

$$\mathrm{Ent}(D) = -\sum_{k=1}^{|\mathcal{Y}|} p_k \log_2 p_k. \tag{11.2}$$

The larger the information gain Gain($A$) is, the more useful information the feature subset $A$ contains for classification. Therefore, we can use the training set $D$ to evaluate every candidate feature subset.

More generally, the feature subset $A$ partitions the data set $D$, where each partition corresponds to a value assignment of $A$. Since the label information $Y$ gives the ground truth partitions of $D$, we can evaluate $A$ by checking the difference between these two partitions. The smaller the difference, the better the subset $A$. Information gain is only one of the options to measure the difference, and any methods that can measure the difference between two partitions can be used here.

Many *diversity measures*, after some modifications, can be used for evaluating feature subsets, e.g., disagreement measure and correlation coefficient. See Sect. 8.5.2.

Putting a feature subset search method and a subset evaluation method together gives a feature selection method. When we combine forward search with information entropy, it looks similar to decision trees. Actually, decision trees can be used for feature selection as well, where the set of splitting features is the selected feature subset. Other feature selection methods essentially combine, either explicitly or implicitly, one or more subset search and subset evaluation methods, though not that apparent as decision tree-based feature selection.

Commonly used feature selection methods can be roughly classified into three categories, namely *filter*, *wrapper*, and *embedding* methods.



## 11.2 Filter Methods

Filter methods select features without considering the subsequent learners, and hence they act like "filters" before the training process.

Relief (Relevant Features) (Kira and Rendell 1992) is a well-known filter feature selection method, which incorporates a relevance statistic to measure feature importance. This statistic is a vector in which each component corresponds to the importance of an original feature, and thus the importance of a feature subset is determined by the sum of the corresponding components in the vector. To select a feature subset, we select features with a component greater than a user-specified threshold $\tau$. Alternatively, we can specify the number of features $k$, and then select the features corresponding to the $k$ largest components.

The key of Relief is to determine the relevance statistics for a given training set $\{(\boldsymbol{x}_1, y_1), (\boldsymbol{x}_2, y_2), \ldots, (\boldsymbol{x}_m, y_m)\}$. Relief starts by finding two nearest neighbors $\boldsymbol{x}_{i,nh}$ and $\boldsymbol{x}_{i,nm}$ for each sample $\boldsymbol{x}_i$, where the first one, known as a *near-hit*, has the same class label as $\boldsymbol{x}_i$, while the second one, known as a *near-miss*, has a different class label. Then, the relevance statistics component of feature $j$ is

$$\delta^j = \sum_i -\text{diff}(x_i^j, x_{i,\text{nh}}^j)^2 + \text{diff}(x_i^j, x_{i,\text{nm}}^j)^2, \tag{11.3}$$

where $x_a^j$ is the value of feature $j$ of sample $\boldsymbol{x}_a$. $\text{diff}(x_a^j, x_b^j)$ depends on the type of feature $j$: when $j$ is discrete, $\text{diff}(x_a^j, x_b^j) = 0$ if $x_a^j = x_b^j$ and 1 otherwise; when $j$ is continuous, $\text{diff}(x_a^j, x_b^j) = \left|x_a^j - x_b^j\right|$ given that $x_a^j$ and $x_b^j$ are normalized to $[0, 1]$.

From (11.3), we see that, for feature $j$, if the distance from $\boldsymbol{x}_i$ to its near-hit $\boldsymbol{x}_{i,\text{nh}}$ is shorter than the distance to its near-miss $\boldsymbol{x}_{i,\text{nm}}$, then feature $j$ is said to be useful for distinguishing different classes, and hence the relevance statistics component of feature $j$ is increased; otherwise, if the distance from $\boldsymbol{x}_i$ to its near-hit $\boldsymbol{x}_{i,\text{nh}}$ is greater than the distance to its near-miss $\boldsymbol{x}_{i,\text{nm}}$, then feature $j$ is considered as useless for distinguishing different classes, and hence the relevance statistics component of feature $j$ is decreased. By averaging the results calculated on all samples, we obtain the relevance statistics components of all features, where a higher value indicates a better discriminative ability.

The calculation of relevance statistics in Relief is implicitly related to the idea of metric learning. See Sect. 10.6 for metric learning.

The index $i$ in (11.3) indicates samples used for averaging. In practice, Relief only needs to average over a sampled subset rather than the entire data set (Kira and Rendell 1992). The computational complexity of Relief is linear to the number of sampling and the number of the original features, and hence it is a highly efficient filter feature selection method.

Relief was originally designed for binary classification problems, and its variant Relief-F (Kononenko 1994) can handle multiclass classification problems. Given a data set $D$ with $|\mathcal{Y}|$ classes. For a sample $\boldsymbol{x}_i$ of class $k$ ($k \in \{1, 2, \ldots, |\mathcal{Y}|\}$), Relief-F starts by finding the near-hit $\boldsymbol{x}_{i,nh}$ from class $k$ samples, and then find a near-miss from each class other than $k$, denoted by $\boldsymbol{x}_{i,l,nm}$ ($l = 1, 2, \ldots, |\mathcal{Y}|\,; l \neq k$). Accordingly, the relevance statistics component of feature $j$ becomes

$$\delta^j = \sum_i -\text{diff}\left(x_i^j, x_{i,\text{nh}}^j\right)^2 + \sum_{l \neq k}\left(p_l \times \text{diff}\left(x_i^j, x_{i,l,\text{nm}}^j\right)^2\right), \tag{11.4}$$

where $p_l$ is the proportion of class $l$ samples in the data set $D$.

## 11.3 Wrapper Methods

Unlike filter methods, which do not consider the subsequent learners, wrapper methods directly use the performance of subsequent learners as the evaluation metric for feature subsets. In other words, wrapper methods aim to find the most useful feature subset "tailored" for the given learner.

Generally speaking, wrapper methods are usually better than filter methods in terms of the learner's final performance since the feature selection is optimized for the given learner. However, wrapper methods are often much more computationally expensive since they train the learner multiple times during the feature selection.

Las Vegas Wrapper (LVW) (Liu and Setiono 1996) is a typical wrapper method. It searches feature subsets using a randomized strategy under the framework of the Las Vegas method, and the subsets are evaluated based on the final classification error. Pseudocode of LVW is given in ◘ Algorithm 11.1.

The Las Vegas method and the Monte Carlo method are two randomized methods named after gambling cities. Their main difference is that, when there is a time constraint, the Las Vegas method will give a satisfactory solution or give no solution, whereas the Monte Carlo method always gives a solution, though not necessarily satisfy the requirements. Both methods can give a satisfactory solution when there is no time constraint.

Line 8 of ◘ Algorithm 11.1 estimates the error of learner $\mathfrak{L}$ using cross-validation on the feature subset $A'$. If the error of $A'$ is smaller than the error of the current feature subset $A^*$, or their errors are comparable but the size of $A'$ is smaller, then $A'$ is set as the new optimal subset.

It is worth noting that each randomly generated subset is evaluated by training the learner one more time, which is computationally expensive. Hence, LVW introduces a parameter $T$ to limit the number of iterations. However, when the number of original features is large (i.e., $|A|$ is large), LVW may run for a long time if we set $T$ to a large number. In other words, LVW may not produce a solution if there is a constraint on the running time.

**Algorithm 11.1** Las Vegas Wrapper

**Input:** Data set $D$;
Feature set $A$;
Learning algorithm $\mathfrak{L}$;
Parameter $T$ of the stopping condition.

**Process:**

$E = \infty$;
$d = |A|$;
$A^* = A$;
$t = 0$;
**while** $t < T$ **do**
  Generate random feature subset $A'$;
  $d' = |A'|$;
  $E' = \text{CrossValidation}(\mathfrak{L}(D^{A'}))$;
  **if** $(E' < E) \vee ((E' = E) \wedge (d' < d))$ **then**
    $t = 0$;
    $E = E'$;
    $d = d'$;
    $A^* = A'$;
  **else**
    $t = t + 1$.
  **end if**
**end while**

**Output:** Feature subset $A^*$.

Initialization.

Use cross-validation to estimate the error of the learner on $A'$.

Stop if there is no update in $T$ consecutive rounds.

## 11.4 Embedded Methods and $L_1$ Regularization

The feature selection process and the learner training process are clearly separated in both filter methods and wrapper methods. By contrast, embedded methods unify the feature selection process and the learner training process into a joint optimization process, that is, the features are automatically selected during the training.

Given a data set $D = \{(\boldsymbol{x}_1, y_1), (\boldsymbol{x}_2, y_2), \ldots, (\boldsymbol{x}_m, y_m)\}$, where $\boldsymbol{x} \in \mathbb{R}^d$ and $y \in \mathbb{R}$. Taking a simple linear regression model as an example, suppose the squared error is used as the loss function, then the optimization objective is

$$\min_{\mathbf{w}} \sum_{i=1}^{m} (y_i - \mathbf{w}^\top \boldsymbol{x}_i)^2. \tag{11.5}$$

Equation (11.5) can easily overfit the data when there is a large number of features but a small number of samples. To alleviate overfitting, we can introduce a regularization term to (11.5). If we use $L_2$ regularization, then we have

See Sect. 6.4 for regularization.

$$\min_{\mathbf{w}} \sum_{i=1}^{m} (y_i - \mathbf{w}^\top \boldsymbol{x}_i)^2 + \lambda \|\mathbf{w}\|_2^2 . \tag{11.6}$$

In 1943, A. Tikhonov published ridge regression in the journal *Proceedings of the USSR Academy of Sciences*. To acknowledge its author, ridge regression is also called *Tikhonov regression* and $L_2$ regularization is also called *Tikhonov reguarlization*.

where $\lambda > 0$ is the regularization parameter. Equation (11.6) is called *ridge regression* (Tikhonov and Arsenin 1977), which significantly reduces the risk of overfitting by introducing $L_2$ regularization.

We can also replace $L_2$ norm with $L_p$ norm. For example, when $p = 1$ (i.e., $L_1$ norm), we have

$$\min_{\mathbf{w}} \sum_{i=1}^{m} (y_i - \mathbf{w}^\top \boldsymbol{x}_i)^2 + \lambda \|\mathbf{w}\|_1 , \tag{11.7}$$

which is known as Least Absolute Shrinkage and Selection Operator (LASSO) (Tibshirani 1996).

Though both $L_1$ and $L_2$ regularization can help reduce the risk of overfitting, the former enjoys an extra benefit: $L_1$ norm is more likely to result in a *sparse* solution than $L_2$ norm, that is, fewer non-zero components in $\mathbf{w}$.

It seems that $L_0$ norm is a natural choice for inducing sparsity on $\mathbf{w}$ (i.e., minimize the number of non-zero components in $\mathbf{w}$). However, $L_0$ norm is discontinuous, which makes the optimization difficult, and hence it is often approximated by $L_1$ norm.

To see the difference between $L_1$ and $L_2$ norms, let us look at an intuitive example. Suppose $\boldsymbol{x}$ has only two features, then, for both (11.6) and (11.7), the solution $\mathbf{w}$ has two components $w_1$ and $w_2$. Using these two components as two axes, we plot the *contours* of the first terms in (11.6) and (11.7), that is, connecting the points with an equal squared error in the space of $(w_1, w_2)$. Then, we plot the contours for $L_1$ norm and $L_2$ norm, respectively, that is, connecting the points with equal $L_1$ norm and connecting the points with equal $L_2$ norm. The plotted contours are illustrated in ◘ Figure 11.1. The solutions of (11.6) and (11.7) need to make a trade-off between the squared error term and the regularization term. In other words, the solutions lie on the intersections between the contours of the squared error term and the regularization term. From ◘ Figure 11.1, we can see that, when using $L_1$ norm, the intersections often lie on the axes, that is, either $w_1$ or $w_2$ is 0. By contrast, when using $L_2$ norm, the intersections often lie in a quadrant, that is, neither $w_1$ nor $w_2$ is 0. In other words, $L_1$ norm is more likely to result in a sparse solution than $L_2$ norm.

That is, those features corresponding to non-zero components of $\mathbf{w}$

A sparse solution $\mathbf{w}$ implies that only the original features corresponding to non-zero components of $\mathbf{w}$ are included in the final model. Hence, a model trained with $L_1$ regularization is likely to use only some of the original features. In other words, the learning method based on $L_1$ regularization is an embedded feature selection method that unifies the process of feature selection and the process of training.

The $L_1$ regularization problem can be solved by Proximal Gradient Descent (PGD) (Combettes and Wajs 2005). More specifically, let $\nabla$ denote the differential operator, and we consider the optimization objective

$$\min_{\boldsymbol{x}} \ f(\boldsymbol{x}) + \lambda \|\boldsymbol{x}\|_1 . \tag{11.8}$$

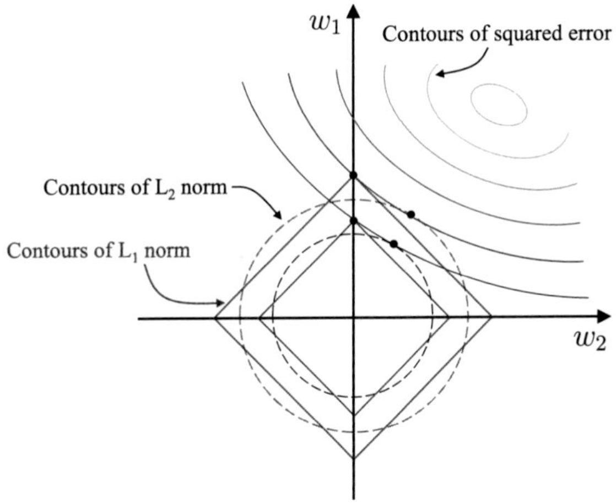


**Fig. 11.1** It is easier to obtain sparse solutions using $L_1$ regularization than using $L_2$ regularization

If $f(\boldsymbol{x})$ is differentiable and $\nabla f$ satisfies the $L$-Lipschitz condition, that is, there exists a constant $L > 0$ such that

$$\left\|\nabla f(\boldsymbol{x}') - \nabla f(\boldsymbol{x})\right\|_2 \leqslant L \left\|\boldsymbol{x}' - \boldsymbol{x}\right\|_2 \quad (\forall \boldsymbol{x}, \boldsymbol{x}'). \tag{11.9}$$

Then, in the vicinity of $\boldsymbol{x}_k$, $f(\boldsymbol{x})$ can be approximated by the second-order Taylor expansion as

$$\begin{aligned}\hat{f}(\boldsymbol{x}) &\simeq f(\boldsymbol{x}_k) + \langle \nabla f(\boldsymbol{x}_k), \boldsymbol{x} - \boldsymbol{x}_k \rangle + \frac{L}{2} \|\boldsymbol{x} - \boldsymbol{x}_k\|^2 \\ &= \frac{L}{2} \left\| \boldsymbol{x} - \left( \boldsymbol{x}_k - \frac{1}{L} \nabla f(\boldsymbol{x}_k) \right) \right\|_2^2 + \text{const},\end{aligned} \tag{11.10}$$

where $\langle \cdot, \cdot \rangle$ is the inner product and const is a constant independent of $\boldsymbol{x}$. The minimum of (11.10) is obtained at $\boldsymbol{x}_{k+1}$:

$$\boldsymbol{x}_{k+1} = \boldsymbol{x}_k - \frac{1}{L} \nabla f(\boldsymbol{x}_k). \tag{11.11}$$

Therefore, if we minimize $f(\boldsymbol{x})$ with gradient descent method, then each step of gradient descent iteration is equivalent to minimizing the quadratic function $\hat{f}(\boldsymbol{x})$. Similarly, extending this idea to (11.8) gives

$$\boldsymbol{x}_{k+1} = \underset{\boldsymbol{x}}{\arg\min} \frac{L}{2} \left\| \boldsymbol{x} - \left( \boldsymbol{x}_k - \frac{1}{L} \nabla f(\boldsymbol{x}_k) \right) \right\|_2^2 + \lambda \|\boldsymbol{x}\|_1, \tag{11.12}$$

that is, each gradient descent iteration takes the minimization of $L_1$ norm into account.

For (11.12), we can first calculate $z = \boldsymbol{x}_k - \frac{1}{L}\nabla f(\boldsymbol{x}_k)$, and then solve

$$\boldsymbol{x}_{k+1} = \arg\min_{\boldsymbol{x}} \frac{L}{2} \|\boldsymbol{x} - z\|_2^2 + \lambda \|\boldsymbol{x}\|_1 . \tag{11.13}$$

Let $x^i$ denote the $i$th component of $\boldsymbol{x}$. Then, by expanding the components of $\boldsymbol{x}$ in (11.13), we find out that there is no term in the form of $x^i x^j$ $(i \neq j)$, which means that the components of $\boldsymbol{x}$ are not interfering with each other. Hence, (11.13) has a closed-form solution

Exercise 11.8.

$$x_{k+1}^i = \begin{cases} z^i - \lambda/L, & \lambda/L < z^i; \\ 0, & \left|z^i\right| \leqslant \lambda/L; \\ z^i + \lambda/L, & z^i < -\lambda/L, \end{cases} \tag{11.14}$$

where $x_{k+1}^i$ and $z^i$ are the $i$th component of $\boldsymbol{x}_{k+1}$ and $z$, respectively. Therefore, PGD can quickly solve LASSO or any other models that are based on $\text{L}_1$ norm.

## 11.5 Sparse Representation and Dictionary Learning

We can consider the data set $D$ as a matrix, in which each row corresponds to a sample, and each column corresponds to a feature. Feature selection assumes that the features are *sparse*, which means that many columns of the matrix are irrelevant to the learning problem. By identifying and removing these columns, we can obtain a smaller matrix that is easier to learn. It also reduces the computation and storage overheads and improves the interpretability of the learned model.

With the input factors reduced, the "input-output" relations within the learned model become more clear.

Now, let us consider another kind of sparsity: there are many zero elements in the matrix $D$, but the zero elements do not lie in the whole rows or columns. Such data is quite common in practice. For example, in document classification problems, each document is a sample, and each word within is a feature with the frequency as its value. In other words, each row in the matrix $D$ corresponds to a document, and each column corresponds to a word. The intersection of a row and a column is the frequency of the word in the document. How many columns are we talking about here? Taking English as an example, according to the Oxford English Dictionary, there are 171, 476 words, which correspond to 171, 476 columns in the matrix. Even if we consider only the most important words, as in the Oxford 3000 list, there are still 3000 columns. However, since most of these words do not appear in a single typical document, every row in the matrix contains a large number of zero elements, and the zero elements of different documents often appear in quite different columns.

The above sparse representation can benefit some learning tasks. For example, an important reason that the linear support vector machine performs well on text classification tasks is that the sparse representation of word frequencies makes most problems linearly separable. Meanwhile, sparse representations will lead to less storage overheads since there are efficient methods to store the sparse matrix.

See Sects. 6.3 and 12.4.

Is it possible to convert a dense data set $D$ (i.e., a typical non-sparse data set) into a sparse representation to take its advantages? It is worth noting that the sparse representation we are looking for is "appropriately sparse" rather than "overly sparse". For example, using the Oxford 3000 list may give a useful representation that is appropriately sparse, whereas using the Oxford English Dictionary may produce a less useful and overly sparse representation.

Of course, in real-world learning problems, such as image classification, there is no Oxford 3000 list available, and hence we need to learn such a *dictionary*. The dictionary can convert samples from a dense representation to a sparse representation that leads to an easier learning problem as well as a simpler model. The process of constructing an appropriate dictionary is called *dictionary learning*, also known as *sparse coding*. However, the two names have slightly different emphases: dictionary learning emphasizes the process of learning the dictionary, while sparse coding emphasizes the process of converting samples to sparse representations. Since both are done in the same process of optimization, we do not distinguish them in this book, and both are called dictionary learning.

Dictionary is also called a *codebook*.

Dictionary learning is also known as *codebook learning*.

Given a data set $\{\boldsymbol{x}_1, \boldsymbol{x}_2, \ldots, \boldsymbol{x}_m\}$, the simplest form of dictionary learning is

$$\min_{\mathbf{B},\boldsymbol{\alpha}_i} \sum_{i=1}^{m} \|\boldsymbol{x}_i - \mathbf{B}\boldsymbol{\alpha}_i\|_2^2 + \lambda \sum_{i=1}^{m} \|\boldsymbol{\alpha}_i\|_1 , \tag{11.15}$$

where $\mathbf{B} \in \mathbb{R}^{d\times k}$ is the dictionary matrix, $k$ is the vocabulary size, which is usually specified by the user, and $\boldsymbol{\alpha}_i \in \mathbb{R}^k$ is the sparse representation of sample $\boldsymbol{x}_i \in \mathbb{R}^d$. In (11.15), the first term seeks for $\boldsymbol{\alpha}_i$ that can reconstruct $\boldsymbol{x}_i$, and the second term seeks for $\boldsymbol{\alpha}_i$ that is sparse.

Compared to LASSO, (11.15) is more complicated since it not only needs to learn $\boldsymbol{\alpha}_i$, which is similar to $\mathbf{w}$ in (11.7), but also the dictionary matrix $\mathbf{B}$. However, as inspired by LASSO, we can solve (11.15) with alternating optimization.

In the first step of alternating optimization, we fix the dictionary $\mathbf{B}$ to optimize $\boldsymbol{\alpha}_i$. By expanding (11.15) as the components of $\boldsymbol{\alpha}_i$, we find out that there is no cross-term like $\alpha_i^u\alpha_i^v$ $(u \neq v)$, and hence, similar to LASSO, we can find $\boldsymbol{\alpha}_i$ for each sample $\boldsymbol{x}_i$:

$$\min_{\boldsymbol{\alpha}_i} \ \|\boldsymbol{x}_i - \mathbf{B}\boldsymbol{\alpha}_i\|_2^2 + \lambda \|\boldsymbol{\alpha}_i\|_1 . \tag{11.16}$$

In the second step, we initialize with $\boldsymbol{\alpha}_i$ to optimize the dictionary $\mathbf{B}$, that is, (11.15) becomes

$$\min_{\mathbf{B}} \ \|\mathbf{X} - \mathbf{B}\mathbf{A}\|_F^2 , \tag{11.17}$$

where $\mathbf{X} = (\boldsymbol{x}_1, \boldsymbol{x}_2, \ldots, \boldsymbol{x}_m) \in \mathbb{R}^{d\times m}$, $\mathbf{A} = (\boldsymbol{\alpha}_1, \boldsymbol{\alpha}_2, \ldots, \boldsymbol{\alpha}_m) \in \mathbb{R}^{k\times m}$, and $\|\cdot\|_F$ is the Frobenius norm. There are different methods to solve (11.17), and a frequently used one is K-SVD (Aharon et al. 2006), which takes a column-wise updating strategy. Let $\boldsymbol{b}_i$ denote the $i$th column of the dictionary matrix $\mathbf{B}$, and $\boldsymbol{\alpha}^i$ denote the $i$th row of the sparse representation matrix $\mathbf{A}$, then (11.17) can be rewritten as

$$\begin{aligned}
\min_{\mathbf{B}} \ \|\mathbf{X} - \mathbf{B}\mathbf{A}\|_F^2 &= \min_{\boldsymbol{b}_i} \ \left\|\mathbf{X} - \sum_{j=1}^{k} \boldsymbol{b}_j\boldsymbol{\alpha}^j\right\|_F^2 \\
&= \min_{\boldsymbol{b}_i} \ \left\|\left(\mathbf{X} - \sum_{j\neq i} \boldsymbol{b}_j\boldsymbol{\alpha}^j\right) - \boldsymbol{b}_i\boldsymbol{\alpha}^i\right\|_F^2 \\
&= \min_{\boldsymbol{b}_i} \ \left\|\mathbf{E}_i - \boldsymbol{b}_i\boldsymbol{\alpha}^i\right\|_F^2 .
\end{aligned} \tag{11.18}$$

When updating the $i$th column of the dictionary matrix, $\mathbf{E}_i = \mathbf{X} - \sum_{j\neq i} \boldsymbol{b}_j\boldsymbol{\alpha}^j$ is fixed since all other columns are fixed. Hence, (11.18) is, theoretically, minimized by finding the orthogonal vectors corresponding to the largest singular values of $\mathbf{E}_i$. However, since the singular value decomposition of $\mathbf{E}_i$ updates both $\boldsymbol{b}_i$ and $\boldsymbol{\alpha}^i$, it may break the sparsity of $\mathbf{A}$. To avoid this, K-SVD takes some special treatments for $\mathbf{E}_i$ and $\boldsymbol{\alpha}^i$: before the singular value decomposition, we keep only the non-zero elements of $\boldsymbol{\alpha}^i$, and, for $\mathbf{E}_i$, we keep only the product terms of $\boldsymbol{b}_i$ and the non-zero elements of $\boldsymbol{\alpha}^i$; by doing so, the sparsity obtained in the first step is preserved.

After initializing the dictionary matrix $\mathbf{B}$, repeating the above two steps gives the final dictionary $\mathbf{B}$ and a sparse representation $\boldsymbol{\alpha}_i$ for each sample $\boldsymbol{x}_i$. The above dictionary learning process has a user-specified parameter $k$ that controls the vocabulary size, which relates to the sparseness.

## 11.6 Compressed Sensing

In real-world applications, we often need to recover the full information from the partially available information. Taking telecommunication as an example, we convert analog signals into digital signals. According to the Nyquist sampling theorem, if the sampling rate is more than twice the analog

The Nyquist sampling theorem provides a sufficient but not necessary condition for signal recovery.

signal's highest frequency, then the sampled digital signal contains all information in the analog signal. That is, we can fully reconstruct the original analog signals from the converted digital signals. However, the sampled digital signal is often compressed for ease of transmission and storage, which may lead to information loss. Besides, there could be more information loss during the transmission, e.g., packet loss. In such cases, the information is not received in full, then can we still fully reconstruct the original signal? One solution to such problems is *compressed sensing* (Donoho 2006; Candès et al. 2006).

Also known as *compressive sensing*.

Suppose there is a discrete signal $\boldsymbol{x}$ with a length $m$, then, by sampling at a rate much lower than the sampling rate required by the Nyquist sampling theorem, we obtain a sampled signal $\boldsymbol{y}$ with a length $n$ ($n \ll m$):

$\boldsymbol{y}$ is also called a *measurement value*.

$$\boldsymbol{y} = \boldsymbol{\Phi x}, \tag{11.19}$$

where $\boldsymbol{\Phi} \in \mathbb{R}^{n \times m}$ is the measurement matrix of signal $\boldsymbol{x}$, which determines the sampling rate and how the sampled signal is constructed.

The measurement value $\boldsymbol{y}$ can be easily calculated with the discrete signal $\boldsymbol{x}$ and the measurement matrix $\boldsymbol{\Phi}$. However, can a receiver recover the original signal $\boldsymbol{x}$ from the received measurement value and measurement matrix?

In general, the answer is "no". Since $n \ll m$, (11.19), which consists of $\boldsymbol{y}$, $\boldsymbol{x}$, and $\boldsymbol{\Phi}$, must be an under-determined equation whose numerical solution is difficult to find.

Suppose there is a linear transformation $\boldsymbol{\Psi} \in \mathbb{R}^{m \times m}$ that can represent $\boldsymbol{x}$ as $\boldsymbol{\Psi s}$, then $\boldsymbol{y}$ can be expressed by

We assume $\boldsymbol{x}$ is not sparse.

$$\boldsymbol{y} = \boldsymbol{\Phi \Psi s} = \mathbf{A}\boldsymbol{s}, \tag{11.20}$$

where $\mathbf{A} = \boldsymbol{\Phi\Psi} \in \mathbb{R}^{n \times m}$. If we can recover $\boldsymbol{s}$ from $\boldsymbol{y}$, then the signal $\boldsymbol{x}$ can be recovered by $\boldsymbol{x} = \boldsymbol{\Psi s}$.

At first glance, (11.20) is not helpful since the inverse problem (i.e., recovering the signal $\boldsymbol{s}$) is still under-determined. Interestingly, however, the inverse problem can be easily solved if $\boldsymbol{s}$ is sparse! It is because the sparsity significantly reduces the impacts of unknown factors. When $\boldsymbol{s}$ is sparse, $\boldsymbol{\Psi}$ in (11.20) is called a sparse basis, and $\mathbf{A}$ acts like a dictionary that converts the signal to a sparse representation.

In many applications, we can obtain sparse $\boldsymbol{s}$. For example, the digital signals of image or audio are usually non-sparse in the time domain, but they can be converted to sparse signals in the frequency domain using mathematical transforms, such as Fourier transform, cosine transform, and wavelet transform.

Unlike feature selection and sparse representation, compressed sensing focuses on how to recover the original signal from the partially observed samples via the inherent sparsity of the signal. It is commonly agreed that compressed sensing has two stages: the *sensing and measurement* stage and the *reconstruction and recovery* stage. The sensing and measurement stage focuses on obtaining a sparse representation of the original signal via methods such as Fourier transform, wavelet transform, as well as dictionary learning and sparse coding that have been introduced in Sect. 11.5. Many of these techniques have already been developed in other fields, such as signal processing, before compressed sensing was proposed. The reconstruction and recovery stage focuses on reconstructing the original signal from the partial observations by exploiting the sparsity property. This stage is the heart of compressed sensing, and we usually refer to this part when talking about compressed sensing.

The theories behind compressed sensing are a bit complex. Hence, we only briefly introduce the Restricted Isometry Property (RIP) (Candès 2008).

For a $n \times m$ ($n \ll m$) matrix $\mathbf{A}$, we say $\mathbf{A}$ satisfies the $k$-restricted isometry property ($k$-RIP) if there exists a constant $\delta_k \in (0, 1)$ such that

$$(1-\delta_k)\,\|\boldsymbol{s}\|_2^2 \leqslant \|\mathbf{A}_k \boldsymbol{s}\|_2^2 \leqslant (1+\delta_k)\,\|\boldsymbol{s}\|_2^2 \tag{11.21}$$

holds for every vector $\boldsymbol{s}$ and submatrix $\mathbf{A}_k \in \mathbb{R}^{n\times k}$ of $\mathbf{A}$. In such cases, the sparse signal $\boldsymbol{s}$ can be almost fully reconstructed from $\boldsymbol{y}$, and subsequently $\boldsymbol{x}$:

11

$$\begin{aligned} &\min_{\boldsymbol{s}} \ \|\boldsymbol{s}\|_0 \\ &\text{s.t.}\ \ \boldsymbol{y} = \mathbf{A}\boldsymbol{s}. \end{aligned} \tag{11.22}$$

However, since (11.22) involves the minimization of $\mathrm{L}_0$ norm, it is an NP-hard problem. Fortunately, under certain conditions, minimizing $\mathrm{L}_0$ norm shares the same solution with minimizing $\mathrm{L}_1$ norm (Candès et al. 2006). Hence, in practice, we solve

$$\begin{aligned} &\min_{\boldsymbol{s}} \ \|\boldsymbol{s}\|_1 \\ &\text{s.t.}\ \ \boldsymbol{y} = \mathbf{A}\boldsymbol{s}, \end{aligned} \tag{11.23}$$

which means the compressed sensing problem is solved as an $\mathrm{L}_1$ norm minimization problem. For example, (11.23) can be converted to a form that is equivalent to LASSO, which can then be solved by proximal gradient descent, that is, using Basis Pursuit De-Noising (Chen et al. 1998).

The techniques for reconstructing full information from partial information have important uses in many real-world applications. For example, by collecting the ratings provided by readers, an online book store can infer the preferences of readers, which can then be used for personalized recommendations. However, the preference information is only partially available since none of the readers has read all books, and none of the books has been read by all users. ◘ Table 11.1 gives an example of ratings provided by four readers. The ratings are processed into a 1–5 scale, from the least to the most preferred. Because readers only rate the books they read, the table contains many unknown values.

This is a typical *collaborative filtering* problem.

Considering the ratings in ◘ Table 11.1 as a partially available signal, is it possible to recover the original signal via the mechanism of compressed sensing?

Recall that a pre-requisite of recovering an under-sampled signal via compressed sensing is that the signal has a sparse representation. Then, do the readers' preference data have a sparse representation? The answer is "yes". In general, the ratings depend on many factors such as genre, author, and bookbinding; for ease of discussion, suppose the ratings in ◘ Table 11.1 are only about the genre. In our example, `Gulliver's Travels` and `Robinson Crusoe` are novel books; `The Story of Mankind` and `A Study of History` are history books; `Gitanjali` is a poem book. Typically, books with the same genre are likely to have the same readers. With this assumption, books can be grouped by genre, where the number of groups is far less than the number of books. In other words, the signal in ◘ Table 11.1 is sparse in terms of genre. Hence, the mechanism of compressed sensing should be applicable to the rating recovery problem.

**◘ Tab. 11.1** The book ratings by some readers

| | Gulliver's Travels | Robinson Crusoe | The Story of Mankind | A Study of History | Gitanjali |
|---|---|---|---|---|---|
| Emma | 5 | ? | ? | 3 | 2 |
| Michael | ? | 5 | 3 | ? | 5 |
| John | 5 | 3 | ? | ? | ? |
| Sarah | 3 | ? | 5 | 4 | ? |

The above problem can be solved with the *matrix completion* technique (Candès and Recht 2009) as follows:

Also known as *low-rank matrix recovery*.

$$\begin{aligned} &\min_{\mathbf{X}} \ \operatorname{rank}(\mathbf{X}) \\ &\text{s.t.} \ (\mathbf{X})_{ij} = (\mathbf{A})_{ij}, \ (i,j) \in \Omega, \end{aligned} \tag{11.24}$$

where $\mathbf{X}$ is the sparse signal to be recovered; $\text{rank}(\mathbf{X})$ is the rank of the matrix $\mathbf{X}$; $\mathbf{A}$ is the observed signal, e.g., the rating matrix in ◘ Table 11.1; $\Omega$ is the set of subscript index pairs $(i, j)$ of the non-missing (i.e., not "?") elements $(\mathbf{A})_{ij}$ in $\mathbf{A}$. The constraint in (11.24) states that the elements $(\mathbf{X})_{ij}$ in the recovered matrix $\mathbf{X}$ must have the same values as the corresponding elements in the observed matrix $\mathbf{A}$.

Similar to (11.22), (11.24) is an NP-hard problem. The convex hull of $\text{rank}(\mathbf{X})$ on the set $\{\mathbf{X} \in \mathbb{R}^{m \times n} : \|\mathbf{X}\|_F^2 \leqslant 1\}$ is the *nuclear norm* of $\mathbf{X}$:

Nuclear norm is also known as *trace norm*.

$$\|\mathbf{X}\|_* = \sum_{j=1}^{\min\{m,n\}} \sigma_j(\mathbf{X}), \tag{11.25}$$

where $\sigma_j(\mathbf{X})$ is the singular value of $\mathbf{X}$. Equation (11.25) states that the nuclear norm of a matrix is the sum of its singular values. Therefore, we can find an approximate solution of (11.24) by minimizing the nuclear norm of the matrix:

$$\begin{aligned} \min_{\mathbf{X}} \quad & \|\mathbf{X}\|_* \\ \text{s.t.} \quad & (\mathbf{X})_{ij} = (\mathbf{A})_{ij}, \ (i,j) \in \Omega. \end{aligned} \tag{11.26}$$

Since (11.26) is a convex optimization problem, it can be solved with Semidefinite Programming (SDP). Theoretical studies show that, under certain conditions, if $\text{rank}(\mathbf{A}) = r$ and $n \ll m$, then $\mathbf{A}$ can be perfectly recovered from $O(mr \log^2 m)$ observed elements (Recht 2011).

See Appendix B.3 for SDP.

11

## 11.7 Further Reading

Feature selection is one of the earliest research areas of machine learning. Early studies were mainly based on the "feature generation and search"–"evaluation" process. Many search techniques from artificial intelligence were introduced to the process of subset generation and search, such as the branch and bound method (Narendra and Fukunaga 1977) and the floating search method (Pudil et al. 1994). In terms of subset evaluation, many criteria were borrowed from information theory, such as information entropy and Akaike Information Criterion (AIC) (Akaike 1974). Blum and Langley (1997) discussed different subset evaluation criteria, while (Forman 2003) conducted extensive experimental comparisons.

In the early days, most feature selection studies focused on filter methods. Later on, wrapper methods emerged (Kohavi and John 1997), followed by embedded methods (Weston and Tipping 2003). The construction of decision trees can also be regarded as a kind of feature selection, and hence, from this point of view, embedded methods can be traced back to ID3 (Quinlan 1986). A considerable amount of literature has conducted empirical comparisons on different feature selection methods (Yang and Pederson 1997; Jain and Zongker 1997). More information on feature selection can be found in Guyon and Elisseeff (2003), Liu et al. (2010) and books dedicated to this subject Liu and Motoda (1998, 2007).

Least Angle RegresSion (LARS) (Efron et al. 2004) is an embedded feature selection method for linear regression that selects the feature with the highest correlation to the residual in each round. LASSO (Tibshirani 1996) can be implemented by slightly modifying LARS. LASSO was extended to other variants, such as Group LASSO (Yuan and Lin 2006) that considers feature group structures, and Fused LASSO (Tibshirani et al. 2005) that considers feature sequence structures. Since LASSO methods are not strictly convex, there may exist multiple solutions, and this problem can be solved with Elastic Net (Zou and Hastie 2005).

For dictionary learning and sparse coding (Aharon et al. 2006), we can manipulate the sparsity by changing the dictionary size, as well as the dictionary "structures". For example, we can assume the dictionary has *group structures*, that is, the variables in the same group are either all zero or all non-zero. Such a property is known as *group sparsity* and is utilized by group sparse coding (Bengio et al. 2009). Sparse coding and block sparse coding have many applications in image feature extraction, and more information can be found in Mairal et al. (2008), Wang et al. (2010).

For example, if some words express a concept, these words form a group. When this concept is not present in a document, all variables corresponding to this group are zero.

Compressed sensing (Donoho 2006; Candès et al. 2006) motivated robust principal component analysis (Candès et al. 2011) and matrix completion-based collaborative filtering (Recht et al. 2010). Baraniuk (2007) provided a brief introduction to compressed sensing. After converting $L_0$ norm into $L_1$ norm, we can find the solutions using basis pursuit de-noising of LASSO, as well as other techniques such as *Basis Pursuit* (Chen et al. 1998) and *Matching Pursuit* (Mallat and Zhang 1993). Liu and Ye (2009) proposed a projection-based method for efficient sparse learning and provided SLEP, a software package for sparse learning (► https://github.com/divelab/slep/).

## Exercises

The watermelon data set 3.0 is in ▪ Table 4.3.

**11.1** Implement and run the Relief algorithm on the watermelon data set 3.0.

**11.2** Write down the pseudocode of the Relief-F algorithm.

**11.3** The Relief algorithm evaluates the importance of each feature. Design an improved version such that the importance of each feature pair can be evaluated.

**11.4** Design an improved version of LVW such that the algorithm is guaranteed to produce a solution even if there is a time constraint.

**11.5** With the assistance of ▪ Figure 11.1, use examples to describe the situations when $L_1$ regularization cannot produce sparse solutions.

**11.6** Discuss the relationships between ridge regression and support vector machines.

**11.7** Discuss the difficulties of solving $L_0$ norm regularization directly.

**11.8** Derive the closed-form solution (11.14) for the $L_1$ norm minimization problem.

**11.9** Discuss the difference and commonality between the uses of sparsity in dictionary learning and compressed sensing.

**11.10** * Improve (11.15) such that the learned dictionary has group sparsity.

## Break Time

### Short story: Monte Carlo method and Stanisaw Ulam

Stanisł aw Ulam (1909–1984) is a well-known Polish mathematician, who has made significant contributions to ergodic theory, number theory, set theory, etc. The Ulam sequence is named after him.

Ulam was born in Lemberg, Austria-Hungary. He studied at the Lwów Polytechnic Institute, where he obtained his Ph.D. degree in 1933. Two years later, in 1935, John von Neumann invited Ulam to visit the Institute for Advanced Study in Princeton. In 1940, Ulam became an assistant professor at the University of Wisconsin–Madison, and a citizen of the United States in the next year. In 1943, Ulam joined the Manhattan Project, in which he, together with Edward Teller, developed the Teller–Ulam design, which became the basis for most thermonuclear weapons.

Lemberg is now Lviv, Ukraine. The city was temporarily conquered by Austria-Hungary between 1867 and 1918 and returned to Poland after World War I. In 1939, the city became part of the former Soviet Union, and later on, it became part of the independent nation of Ukraine in 1991.

Both John von Neumann and Edward Teller were born in Hungary.

Electronic Numerical Integrator and Computer (ENIAC), one of the earliest electronic general-purpose digital computers, was used in the Manhattan Project shortly after its invention. Ulam was keenly aware of the possibility of using computers to estimate probabilistic variables via hundreds of simulations. John von Neumann immediately recognized the importance of this idea and supported Ulam. In 1947, Ulam implemented the proposed statistical method and successfully applied it to the calculations of the nuclear chain reaction. Because Ulam often mentioned his uncle, Michał Ulam, "who just had to go to Monte Carlo" to gamble, Nicolas Metropolis dubbed the statistical method "Monte Carlo method".

The Metropolis–Hasting algorithm is a representative Monte Carlo method named after Metropolis.

## References

Aharon M, Elad M, Bruckstein A (2006) K-SVD: an algorithm for designing overcomplete dictionaries for sparse representation. IEEE Trans Image Process 54(11):4311–4322

Akaike H (1974) A new look at the statistical model identification. IEEE Trans Autom Control 19(6):716–723

Baraniuk RG (2007) Compressive sensing. IEEE Signal Process Mag 24(4):118–121

Bengio S, Pereira F, Singer Y, Strelow D (2009) Group sparse coding. In: Bengio Y, Schuurmans D, Lafferty JD, Williams CKI, Culotta A (eds) Advances in neural information processing systems 22 (NIPS). MIT Press, Cambridge, pp 82–89

Blum A, Langley P (1997) Selection of relevant features and examples in machine learning. Artif Intell 97(1–2):245–271

Candès EJ (2008) The restricted isometry property and its implications for compressed sensing. Comptes Rendus Math 346(9–10):589–592

Candès EJ, Recht B (2009) Exact matrix completion via convex optimization. Found Comput Math 9(6):717–772

Candès EJ, Romberg J, Tao T (2006) Robust uncertainty principles: exact signal reconstruction from highly incomplete frequency information. IEEE Trans Inf Theory 52(2):489–509

Candès EJ, Li X, Ma Y, Wright J (2011) Robust principal component analysis? J ACM 58(3). Article 11

Chen SS, Donoho DL, Saunders MA (1998) Atomic decomposition by basis pursuit. SIAM J Sci Comput 20(1):33–61

Combettes PL, Wajs VR (2005) Signal recovery by proximal forward-backward splitting. Multiscale Model Simul 4(4):1168–1200

Donoho DL (2006) Compressed sensing. IEEE Trans Inf Theory 52(4):1289–1306

Efron B, Hastie T, Johnstone I, Tibshirani R (2004) Least angle regression. Ann Stat 32(2):407–499

Forman G (2003) An extensive empirical study of feature selection metrics for text classification. J Mach Learn Res 3:1289–1305

Guyon I, Elisseeff A (2003) An introduction to variable and feature selection. J Mach Learn Res 3:1157–1182

Jain A, Zongker D (1997) Feature selection: evaluation, application, and small sample performance. IEEE Trans Pattern Anal Mach Intell 19(2):153–158

Kira K, Rendell LA (1992) The feature selection problem: Traditional methods and a new algorithm. I: Proceedings of the 10th national conference on artificial intelligence (AAAI), San Jose, CA, pp 129–134

Kohavi R, John GH (1997) Wrappers for feature subset selection. Artificial Intelligence 97(1–2):273–324

Kononenko I (1994) Estimating attributes: analysis and extensions of RELIEF. In: Proceedings of the 7th European conference on machine learning (ECML), Catania, Italy, pp 171–182

Liu H, Motoda H (1998) Feature selection for knowledge discovery and data mining. Kluwer, Boston

Liu H, Motoda H (2007) Computational methods of feature selection. Chapman & Hall/CRC, Boca Raton

Liu H, Motoda H, Setiono R, Zhao Z (2010) Feature selection: an ever evolving frontier in data mining. In: Proceedings of the 4th workshop on feature selection in data mining (FSDM), Hyderabad, India, pp 4–13

Liu H, Setiono R (1996) Feature selection and classification — a probabilistic wrapper approach. In: Proceedings of the 9th international conference on industrial and engineering applications of artificial intelligence and expert systems (IEA/AIE), Fukuoka, Japan, pp 419–424
Liu J, Ye J (2009) Efficient Euclidean projections in linear time. In: Proceedings of the 26th international conference on machine learning (ICML), Montreal, Canada, pp 657–664
Mairal J, Elad M, Sapiro G (2008) Sparse representation for color image restoration. IEEE Trans Image Process 17(1):53–69
Mallat SG, Zhang ZF (1993) Matching pursuits with time-frequency dictionaries. IEEE Trans Signal Process 41(12):3397–3415
Narendra PM, Fukunaga K (1977) A branch and bound algorithm for feature subset selection. IEEE Trans Comput C-26(9):917–922
Pudil P, Novovičová J, Kittler J (1994) Floating search methods in feature selection. Pattern Recognit Lett 15(11):1119–1125
Quinlan JR (1986) Induction of decision trees. Mach Learn 1(1):81–106
Recht B (2011) A simpler approach to matrix completion. J Mach Learn Res 12:3413–3430
Recht B, Fazel M, Parrilo P (2010) Guaranteed minimum-rank solutions of linear matrix equations via nuclear norm minimization. SIAM Rev 52(3):471–501
Tibshirani R (1996) Regression shrinkage and selection via the LASSO. J R Stat Soc: Ser B 58(1):267–288
Tibshirani R, Saunders M, Rosset S, Zhu J, Knight K (2005) Sparsity and smoothness via the fused LASSO. J R Stat Soc: Ser B 67(1):91–108
Tikhonov AN, Arsenin VY (1977) Solutions of ill-posed problems. Winston, Washington, DC
Wang J, Yang J, Yu K, Lv F, Huang T, Gong Y (2010) Locality-constrained linear coding for image classification. In: Proceedings of the IEEE computer society conference on computer vision and pattern recognition (CVPR), San Francisco, CA, pp 3360–3367
Weston J, Elisseeff A, Schölkopf B, Tipping M (2003) Use of the zero norm with linear models and kernel methods. J Mach Learn Res 3:1439–1461
Yang Y, Pederson JO (1997) A comparative study on feature selection in text categorization. In: Proceedings of the 14th international conference on machine learning (ICML), Nashville, TN, pp 412–420
Yuan M, Lin Y (2006) Model selection and estimation in regression with grouped variables. J R Stat Soc-Ser B 68(1):49–67
Zou H, Hastie T (2005) Regularization and variable selection via the elastic net. J R Stat Soc-Ser B 67(2):301–320

# 12

# Computational Learning Theory

Table of Contents




Z.-H. Zhou, *Machine Learning*,
https://doi.org/10.1007/978-981-15-1967-3_12

## 12.1 Basic Knowledge

As the name suggests, *computational learning theory* is about "learning" by "computation" and is the theoretical foundation of machine learning. It aims to analyze the difficulties of learning problems, provides theoretical guarantees for learning algorithms, and guides the algorithm design based on theoretical analysis.

Given a data set $D = \{(\boldsymbol{x}_1, y_1), (\boldsymbol{x}_2, y_2), \ldots, (\boldsymbol{x}_m, y_m)\}$, where $\boldsymbol{x}_i \in \mathcal{X}$. In this chapter, we focus on binary classification problems (i.e., $y_i \in \mathcal{Y} = \{-1, +1\}$) unless otherwise stated. Suppose there is an underlying unknown distribution $\mathcal{D}$ over all samples in $\mathcal{X}$, and all samples in $D$ are drawn independently from the distribution $\mathcal{D}$, that is, *i.i.d.* samples.

Let $h$ be a mapping from $\mathcal{X}$ to $\mathcal{Y}$, and its generalization error is

$$E(h; \mathcal{D}) = P_{\boldsymbol{x}\sim\mathcal{D}}(h(\boldsymbol{x}) \neq y). \tag{12.1}$$

The empirical error of $h$ over $D$ is

$$\widehat{E}(h; D) = \frac{1}{m}\sum_{i=1}^{m}\mathbb{I}(h(\boldsymbol{x}_i) \neq y_i). \tag{12.2}$$

Since $D$ contains *i.i.d.* samples drawn from $\mathcal{D}$, the expectation of the empirical error of $h$ equals to the generalization error. When it is clear from the context, we abbreviate $E(h; \mathcal{D})$ and $\widehat{E}(h; D)$ as $E(h)$ and $\widehat{E}(h)$, respectively. The maximum error we can tolerate for a learned model, also known as the *error parameter*, is an upper bound of $E(h)$, denoted by $\epsilon$, where $E(h) \leqslant \epsilon$.

The rest of this chapter studies the gap between the empirical error and the generalization error. A mapping $h$ is said to be consistent with $D$ if the empirical error of $h$ on the data set $D$ is 0. For any two mappings $h_1, h_2 \in \mathcal{X} \rightarrow \mathcal{Y}$, their difference can be measured by the *disagreement*

$$d(h_1, h_2) = P_{\boldsymbol{x}\sim\mathcal{D}}(h_1(\boldsymbol{x}) \neq h_2(\boldsymbol{x})). \tag{12.3}$$

For ease of reference, we list a few frequently used inequalities below

- **Jensen's inequality**: for every convex function $f(x)$, we have

$$f(\mathbb{E}(x)) \leqslant \mathbb{E}(f(x)). \tag{12.4}$$

- **Hoeffding's inequality** (Hoeffding 1963): if $x_1, x_2, \ldots, x_m$ are $m$ independent random variables with $0 \leqslant x_i \leqslant 1$, then, for any $\epsilon > 0$, we have

$$P\left(\frac{1}{m}\sum_{i=1}^{m} x_i - \frac{1}{m}\sum_{i=1}^{m} \mathbb{E}(x_i) \geqslant \epsilon\right) \leqslant \exp(-2m\epsilon^2), \tag{12.5}$$

$$P\left(\left|\frac{1}{m}\sum_{i=1}^{m} x_i - \frac{1}{m}\sum_{i=1}^{m} \mathbb{E}(x_i)\right| \geqslant \epsilon\right) \leqslant 2\exp(-2m\epsilon^2). \tag{12.6}$$

- **McDiarmid's inequality** (McDiarmid 1989): if $x_1, x_2, \ldots, x_m$ are $m$ independent random variables, and for any $1 \leq i \leq m$, the function $f$ satisfies

$$\sup_{x_1,\ldots,x_m,x_i'} \left|f(x_1,\ldots,x_m) - f(x_1,\ldots,x_{i-1},x_i',x_{i+1},\ldots,x_m)\right| \leqslant c_i,$$

then, for any $\epsilon > 0$, we have

$$P(f(x_1,\ldots,x_m) - \mathbb{E}(f(x_1,\ldots,x_m)) \geqslant \epsilon) \leqslant \exp\left(\frac{-2\epsilon^2}{\sum_i c_i^2}\right), \tag{12.7}$$

$$P(|f(x_1,\ldots,x_m) - \mathbb{E}(f(x_1,\ldots,x_m))| \geqslant \epsilon) \leqslant 2\exp\left(\frac{-2\epsilon^2}{\sum_i c_i^2}\right). \tag{12.8}$$

## 12.2 PAC Learning

Probably Approximately Correct (PAC) learning theory (Valiant 1984) is one of the most fundamental components of computational learning theory.

Let $c$ denote a *concept*, which provides a mapping from the sample space $\mathcal{X}$ to the label space $\mathcal{Y}$, and $c$ determines the ground-truth label $y$ of the sample $\boldsymbol{x}$. A concept $c$ is said to be a target concept if $c(\boldsymbol{x}) = y$ holds for every sample $(\boldsymbol{x}, y)$. The set of all target concepts that we wish to learn is called a *concept class*, denoted by $\mathcal{C}$.

The set of all possible concepts for a given learning algorithm $\mathfrak{L}$ is called a *hypothesis space*, denoted by $\mathcal{H}$. Since the ground-truth concept class is unknown to learning algorithms, $\mathcal{H}$ and $\mathcal{C}$ are usually different. A learning algorithm constructs $\mathcal{H}$ by collecting all concepts that are believed to be the target concepts. Since it is unknown whether the collected concepts are ground-truth target concepts, $h \in \mathcal{H}$ is referred to as a *hypoth-*

The hypothesis space of a learning algorithm $\mathfrak{L}$ is different from the hypothesis space of the learning problem as discussed in Sect. 1.3.

*esis*, which provides a mapping from the sample space $\mathcal{X}$ to the label space $\mathcal{Y}$.

If $c \in \mathcal{H}$, then $\mathcal{H}$ contains a hypothesis that can correctly classify all instances, and such a learning problem is said to be *separable* or *consistent* with respect to the learning algorithm $\mathfrak{L}$. If $c \notin \mathcal{H}$, then $\mathcal{H}$ does not contain any hypothesis that can correctly classify all instances, and such a learning problem is said to be *non-separable* or *inconsistent* with respect to the learning algorithm $\mathfrak{L}$.

Given a training set $D$, we wish the learning algorithm $\mathfrak{L}$ can learn a hypothesis $h$ that is close to the target concept $c$. Readers may wonder why not learn the exact target concept $c$? The reason is that the machine learning process is subject to many factors. For example, since the training set $D$ usually contains finite samples, there often exist many *equivalent hypotheses* that cannot be distinguished by learning algorithms on $D$. Also, there exists some randomness when sampling $D$ from $\mathcal{D}$, and hence the hypotheses learned from different equal-sized training sets could be different. Therefore, instead of learning the exact target concept $c$, we wish to learn a hypothesis $h$ with an error bounded by a given value with high confidence, that is, a hypothesis that is probably approximately correct (i.e., PAC). Let $1-\delta$ denote the confidence, and we have the formal definition as follows:

See Sect. 1.4.

In general, the fewer the training samples, the higher the randomness.

**Definition 12.1** (*PAC Identify*) A learning algorithm $\mathfrak{L}$ is said to PAC identify the concept class $\mathcal{C}$ from the hypothesis space $\mathcal{H}$ if, for any $c \in \mathcal{C}$ and distribution $\mathcal{D}$, and $\epsilon, \delta \in (0, 1)$, the learning algorithm $\mathfrak{L}$ outputs a hypothesis $h \in \mathcal{H}$ satisfying

$$P(E(h) \leqslant \epsilon) \geqslant 1 - \delta. \tag{12.9}$$

Such a learning algorithm $\mathfrak{L}$ has a probability of at least $1-\delta$ of learning an approximation of the target concept $c$ with an error of at most $\epsilon$. Following Definition 12.1, we can further define the following:

**Definition 12.2** (*PAC Learnable*) A target concept class $\mathcal{C}$ is said to be PAC learnable with respect to the hypothesis space $\mathcal{H}$ if there exists a learning algorithm $\mathfrak{L}$ such that, for any $\epsilon, \delta \in (0, 1)$ and distribution $\mathcal{D}$, the learning algorithm $\mathfrak{L}$ can PAC identify the concept class $\mathcal{C}$ from the hypothesis space $\mathcal{H}$ for any $m \geqslant \mathrm{poly}(1/\epsilon, 1/\delta, \mathrm{size}(\boldsymbol{x}), \mathrm{size}(c))$, where $\mathrm{poly}(\cdot,\cdot,\cdot,\cdot)$ is a polynomial function and $m$ is the number of *i.i.d.* training samples drawn from the distribution $\mathcal{D}$.

The sample size $m$ is related to the error $\epsilon$, the confidence $1-\delta$, the complexity of data $\mathrm{size}(\boldsymbol{x})$, and the complexity of target concept $\mathrm{size}(c)$.

For learning algorithms, it is necessary to consider the running time complexity. Hence, we further define:

**Definition 12.3** (*PAC Learning Algorithm*) A concept class $\mathcal{C}$ is said to be efficiently PAC learnable by its PAC learning algorithm $\mathfrak{L}$ if $\mathcal{C}$ is PAC learnable by $\mathfrak{L}$ within a polynomial time $\text{poly}(1/\epsilon, 1/\delta, \text{size}(\boldsymbol{x}), \text{size}(c))$.

Suppose the learning algorithm $\mathfrak{L}$ processes each sample with a constant time, then the running time complexity is equivalent to the sample complexity, and we could focus only on the sample complexity:

**Definition 12.4** (*Sample Complexity*) The sample complexity of a PAC learning algorithm $\mathfrak{L}$ is the smallest sample size $m \geqslant \text{poly}(1/\epsilon, 1/\delta, \text{size}(\boldsymbol{x}), \text{size}(c))$ required by $\mathfrak{L}$.

PAC learning provides a formal framework for describing the learning ability of learning algorithms, and many important questions can be discussed theoretically under this framework. For example, what are the requirements for learning a good model for a given learning problem? What are the conditions for an algorithm to learn effectively? How many training samples are required to learn a good model?

A hypothesis space $\mathcal{H}$ includes all possible output hypotheses of a learning algorithm $\mathfrak{L}$, and a key element of PAC learning is the complexity of $\mathcal{H}$. If the hypothesis space is the same as the concept class (i.e., $\mathcal{H} = \mathcal{C}$), then $\mathcal{C}$ is said to be *properly PAC learnable* with respect to $\mathcal{H}$. Intuitively, it means the ability of the learning algorithm properly matches the learning problem. However, it is impractical to assume that $\mathcal{H} = \mathcal{C}$ since we do not know the concept class for real problems, let alone some learning algorithm $\mathfrak{L}$ with $\mathcal{H}$ is exact $\mathcal{C}$. Therefore, it is more realistic to study the cases when the hypothesis space and the concept class are different (i.e., $\mathcal{H} \neq \mathcal{C}$). In general, a larger $\mathcal{H}$ is more likely to contain the target concept we are looking for, though the larger hypothesis space also makes it more difficult to find the target concept. $\mathcal{H}$ is called a *finite hypothesis space* if $|\mathcal{H}|$ is finite, and an *infinite hypothesis space* otherwise.

## 12.3 Finite Hypothesis Space

### 12.3.1 Separable Case

In separable cases, the target concept $c$ is in the hypothesis space $\mathcal{H}$ (i.e., $c \in \mathcal{H}$). Then, given a training set $D$ with size $m$, how can we find a hypothesis from $\mathcal{H}$ satisfying the constraint of a given error parameter?

It is natural to come up with the following learning strategy. Since the labels of the samples in $D$ are assigned by the target concept $c \in \mathcal{H}$, any hypotheses that misclassify any samples in $D$ must not be the target concept $c$. Hence, we simply eliminate all hypotheses that are inconsistent with $D$ and keep the rest. When the training set $D$ is sufficiently large, we can keep eliminating inconsistent hypotheses from $\mathcal{H}$ until there is only one hypothesis left, which must be the target concept $c$. In practice, however, since the training data is usually limited, we may end up with more than one hypothesis that is consistent with $D$, and we cannot distinguish them without additional information.

Given that the training data is limited, how many samples do we need to learn a good approximation of the target concept $c$? For PAC learning, we say a training set $D$ is sufficient for a learning algorithm $\mathfrak{L}$ if $\mathfrak{L}$ can find an $\epsilon$-approximation of the target concept with a probability of at least $1-\delta$.

We first estimate the probability of having a hypothesis that performs perfectly on the training set but still with a generalization error greater than $\epsilon$. Suppose the generalization error of a hypothesis $h$ is greater than $\epsilon$, then, for any *i.i.d.* sample $(\boldsymbol{x}, y)$ drawn from the distribution $\mathcal{D}$, we have

12

$$\begin{aligned} P(h(\boldsymbol{x})=y) &= 1-P(h(\boldsymbol{x}) \neq y) \\ &= 1-E(h) \\ &< 1-\epsilon. \end{aligned} \tag{12.10}$$

Since $D$ contains $m$ samples independently drawn from $\mathcal{D}$, the probability that $h$ and $D$ are consistent is given by

$$\begin{aligned} P((h(\boldsymbol{x}_1)=y_1) \wedge \ldots \wedge (h(\boldsymbol{x}_m)=y_m)) &= (1-P(h(\boldsymbol{x}) \neq y))^m \\ &< (1-\epsilon)^m. \end{aligned} \tag{12.11}$$

Though we do not know which hypothesis $h \in \mathcal{H}$ will be the output by the learning algorithm $\mathfrak{L}$, we only need to ensure that the total probability of having any hypotheses that are consistent with $D$ and have generalization errors greater than $\epsilon$ is not greater than $\delta$. That is, ensuring the total probability

$$P(h \in \mathcal{H} : E(h) > \epsilon \wedge \widehat{E}(h) = 0) < |\mathcal{H}|\,(1-\epsilon)^m$$
$$< |\mathcal{H}|\,e^{-m\epsilon} \qquad (12.12)$$

is not greater than $\delta$, that is,

$$|\mathcal{H}|\,e^{-m\epsilon} \leqslant \delta. \qquad (12.13)$$

Hence, we have

$$m \geqslant \frac{1}{\epsilon}(\ln|\mathcal{H}| + \ln\frac{1}{\delta}), \qquad (12.14)$$

which shows that every finite hypothesis space $\mathcal{H}$ is PAC learnable, and the required sample size is given by (12.14). As the number of samples increases, the generalization error of the output hypothesis $h$ converges toward 0 at a convergence rate of $O(\frac{1}{m})$.

### 12.3.2 Non-separable Case

For difficult learning problems, the target concept $c$ is usually not in the hypothesis space $\mathcal{H}$. Suppose $\widehat{E}(h) \neq 0$ for any $h \in \mathcal{H}$, that is, every hypothesis in $\mathcal{H}$ misclassifies at least one training example, then, from Hoeffding's inequality, we have:

**Lemma 12.1** *Let $D$ be a training set containing $m$ samples independently drawn from a distribution $\mathcal{D}$. Then, for any $h \in \mathcal{H}$ and $0 < \epsilon < 1$, we have*

$$P(\widehat{E}(h) - E(h) \geqslant \epsilon) \leqslant \exp(-2m\epsilon^2), \qquad (12.15)$$
$$P(E(h) - \widehat{E}(h) \geqslant \epsilon) \leqslant \exp(-2m\epsilon^2), \qquad (12.16)$$
$$P\left(\left|E(h) - \widehat{E}(h)\right| \geqslant \epsilon\right) \leqslant 2\exp(-2m\epsilon^2). \qquad (12.17)$$

**Corollary 12.1** *Let $D$ be a training set containing $m$ samples independently drawn from a distribution $\mathcal{D}$. Then, for any $h \in \mathcal{H}$ and $0 < \epsilon < 1$, the following holds with a probability of at least $1-\delta$:*

$$\widehat{E}(h) - \sqrt{\frac{\ln(2/\delta)}{2m}} \leqslant E(h) \leqslant \widehat{E}(h) + \sqrt{\frac{\ln(2/\delta)}{2m}}. \qquad (12.18)$$

Corollary 12.1 shows that, for a large $m$, the empirical error of $h$ is a good approximation to its generalization error. For finite hypothesis spaces, we have

**Theorem 12.1** *Let $\mathcal{H}$ be a finite hypothesis space. Then, for any $h \in \mathcal{H}$ and $0 < \delta < 1$, we have*

$$P\left(\left|E(h)-\widehat{E}(h)\right| \leqslant \sqrt{\frac{\ln|\mathcal{H}|+\ln(2/\delta)}{2m}}\right) \geqslant 1-\delta. \quad (12.19)$$

**Proof** Let $h_1, h_2, \ldots, h_{|\mathcal{H}|}$ denote the hypotheses in $\mathcal{H}$, and we have

$$\begin{aligned}&P(\exists h \in \mathcal{H}: \left|E(h)-\widehat{E}(h)\right| > \epsilon)\\&= P\left(\left(\left|E_{h_1}-\widehat{E}_{h_1}\right| > \epsilon\right) \vee \ldots \vee \left(\left|E_{h_{|\mathcal{H}|}}-\widehat{E}_{h_{|\mathcal{H}|}}\right| > \epsilon\right)\right)\\&\leqslant \sum_{h\in\mathcal{H}} P(\left|E(h)-\widehat{E}(h)\right| > \epsilon).\end{aligned}$$

From (12.17), we have

$$\sum_{h\in\mathcal{H}} P(\left|E(h)-\widehat{E}(h)\right| > \epsilon) \leqslant 2\,|\mathcal{H}| \exp(-2m\epsilon^2),$$

which proves (12.19) by letting $\delta = 2\,|\mathcal{H}| \exp(-2m\epsilon^2)$. □

That is to find the best hypothesis in $\mathcal{H}$.

A learning algorithm $\mathfrak{L}$ cannot learn an $\epsilon$-approximation of the target concept $c$ if $c \notin \mathcal{H}$. However, for a given hypothesis space $\mathcal{H}$, the hypothesis $h \in \mathcal{H}$ with the smallest generalization error is still a reasonably good target. In other words, instead of targeting at $c$, we find an $\epsilon$-approximation of $h$, i.e., $\arg\min_{h\in\mathcal{H}} E(h)$. This approach generalizes PAC learning to *agnostic learning* in which $c \notin \mathcal{H}$. Accordingly, we define

**Definition 12.5** (*Agnostic PAC learnable*) A hypothesis space $\mathcal{H}$ is said to be agnostic PAC learnable if there exists a learning algorithm $\mathfrak{L}$ such that, for any $\epsilon, \delta \in (0, 1)$ and distribution $\mathcal{D}$, the learning algorithm $\mathfrak{L}$ outputs a hypothesis $h \in \mathcal{H}$ satisfying

$$P(E(h) - \min_{h'\in\mathcal{H}} E(h') \leqslant \epsilon) \geqslant 1-\delta, \quad (12.20)$$

for any $m \geqslant \text{poly}(1/\epsilon, 1/\delta, \text{size}(\boldsymbol{x}), \text{size}(c))$, where $m$ is the number of *i.i.d.* training samples drawn from the distribution $\mathcal{D}$.

Similar to PAC learnable, a hypothesis space $\mathcal{H}$ is said to be efficiently agnostic PAC learnable by its agnostic PAC learning algorithm $\mathfrak{L}$ if $\mathcal{H}$ is agnostic PAC learnable by $\mathfrak{L}$ within a polynomial time $\text{poly}(1/\epsilon, 1/\delta, \text{size}(\boldsymbol{x}), \text{size}(c))$. The sample complexity of the learning algorithm $\mathfrak{L}$ is the smallest sample size $m$ satisfying the above requirements.

## 12.4 VC Dimension

Hypothesis spaces in real-world applications are usually infinite, such as all intervals in the real domain and all hyperplanes in the $\mathbb{R}^d$ space. To study the learnability of such cases, we need to measure the complexity of hypothesis spaces. A general approach is to consider the *Vapnik–Chervonenkis dimension* (VC dimension) (Vapnik and Chervonenkis 1971). We first introduce three concepts: *growth function*, *dichotomy*, and *shattering*.

Given a hypothesis space $\mathcal{H}$ and a set of instances $D = \{\boldsymbol{x}_1, \boldsymbol{x}_2, \ldots, \boldsymbol{x}_m\}$, where each hypothesis $h \in \mathcal{H}$ can label every instance in $D$. The labeling result is denoted by

$$h|_D = \{(h(\boldsymbol{x}_1), h(\boldsymbol{x}_2), \ldots, h(\boldsymbol{x}_m))\},$$

any element of which is called a *dichotomy*. The number of dichotomies generated by the hypotheses in $\mathcal{H}$ over $D$ increases as $m$ increases.

For example, in binary classification problems, there are at most 4 dichotomies given 2 instances, and 8 dichotomies given 3 instances.

**Definition 12.6** For $m \in \mathbb{N}$ , the growth function $\Pi_{\mathcal{H}}(m)$ of a hypothesis space $\mathcal{H}$ is defined as

$\mathbb{N}$ is the natural number domain.

$$\Pi_{\mathcal{H}}(m) = \max_{\{\boldsymbol{x}_1,\ldots,\boldsymbol{x}_m\}\subseteq\mathcal{X}} |\{(h(\boldsymbol{x}_1), \ldots, h(\boldsymbol{x}_m)) \mid h \in \mathcal{H}\}| . \tag{12.21}$$

The growth function $\Pi_{\mathcal{H}}(m)$ gives the largest number of dichotomies that the hypothesis space $\mathcal{H}$ can generate over $m$ instances. The more dichotomies, the more representation power, that is, the better adaptability to learning problems. The growth function describes the representation power of a hypothesis space $\mathcal{H}$, which also reflects the complexity of the hypothesis space. We can now use a growth function to present the relationship between the empirical error and the generalization error:

**Theorem 12.2** *For any* $m \in \mathbb{N}$, $0 < \epsilon < 1$, *and* $h \in \mathcal{H}$, *we have*

$$P\left(\left|E(h) - \widehat{E}(h)\right| > \epsilon\right) \leqslant 4\Pi_{\mathcal{H}}(2m)\exp\left(-\frac{m\epsilon^2}{8}\right). \tag{12.22}$$

The proof can be found in Vapnik and Chervonenkis (1971).

Different hypotheses in $\mathcal{H}$ may generate identical or different dichotomies over $D$. The number of dichotomies could be finite even for an infinite hypothesis space $\mathcal{H}$; for example, there are at most $2^m$ dichotomies over $m$ instances. We say that a hypothesis space $\mathcal{H}$ can *shatter* a data set $D$ if $\mathcal{H}$ can generate all possible dichotomies of $D$, that is, $\Pi_{\mathcal{H}}(m) = 2^m$.

We can now formally define the VC dimension as follows:

**Definition 12.7** The VC dimension of a hypothesis space $\mathcal{H}$ is the size of the largest instance set $D$ shattered by $\mathcal{H}$:

$$\mathbf{VC}(\mathcal{H}) = \max\{m : \Pi_{\mathcal{H}}(m) = 2^m\}. \tag{12.23}$$

$\mathrm{VC}(\mathcal{H}) = d$ says that there exists an instance set $D$ of size $d$ that can be shattered by $\mathcal{H}$. However, it does not mean every instance set $D$ of size $d$ can be shattered by $\mathcal{H}$. Some readers may have recognized that the definition of the VC dimension does not involve the underlying data distribution $\mathcal{D}$! In other words, the VC dimension of a hypothesis space $\mathcal{H}$ can be calculated even if the data distribution is unknown.

In general, we can calculate the VC dimension of $\mathcal{H}$ as follows: the VC dimension of $\mathcal{H}$ is $d$ if there exists an instance set of size $d$ shattered by $\mathcal{H}$ while there is no instance set of size $d + 1$ shattered by $\mathcal{H}$. We illustrate the calculation of the VC dimension with the following two examples:

12

**Example 12.1** (*Interval* $[a, b]$ *in the real domain*) Let $\mathcal{H} = \{h_{[a,b]} : a, b \in \mathbb{R}, a \leqslant b\}$ denote the set of all closed intervals in the real domain $\mathcal{X} = \mathbb{R}$. For every $x \in \mathcal{X}$, we have $h_{[a,b]}(x) = +1$ if $x \in [a, b]$; otherwise, $h_{[a,b]}(x) = -1$. Letting $x_1 = 0.5$ and $x_2 = 1.5$, then, $\{x_1, x_2\}$ is shattered by the hypotheses $\{h_{[0,1]}, h_{[0,2]}, h_{[1,2]}, h_{[2,3]}\}$ from $\mathcal{H}$, hence the VC dimension of $\mathcal{H}$ is at least 2. However, there is no hypothesis $h_{[a,b]} \in \mathcal{H}$ that can generate the dichotomy $\{(x_3, +), (x_4, -), (x_5, +)\}$ for a data set containing any 3 instances $\{x_3, x_4, x_5\}$, where $x_3 < x_4 < x_5$. Hence, the VC dimension of $\mathcal{H}$ is 2.

**Example 12.2** (*Linear separators in the* 2*-dimensional real plane*) Let $\mathcal{H}$ denote the set of all linear separators in the 2-dimensional real plane $\mathcal{X} = \mathbb{R}^2$. From ◘ Figure 12.1 we see that there exists a data set of size 3 shattered by $\mathcal{H}$, whereas there is no instance set of size 4 shattered by $\mathcal{H}$. Hence, the VC dimension of the hypothesis space $\mathcal{H}$ of all linear separators in the 2-dimensional real plane is 3.

From Definition 12.7, we see the following relationship between the VC dimension and the growth function (Sauer 1972):

**Lemma 12.2** *If the VC dimension of a hypothesis space $\mathcal{H}$ is $d$, then, for any $m \in \mathbb{N}$, we have*

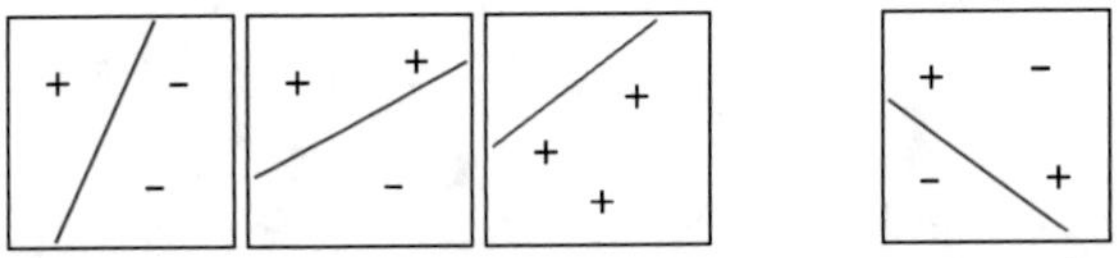

All of the $2^3$=8 dichotomies can be made by linear separators

(a) 3 instances.

At least one of the $2^4$=16 dichotomies cannot be made by linear separators

(b) 4 instances.

**Fig. 12.1** The VC dimension of the hypothesis space of all linear separators in the 2-dimensional real plane is 3

$$\Pi_{\mathcal{H}}(m) \leqslant \sum_{i=0}^{d} \binom{m}{i}. \tag{12.24}$$

Also known as *Sauer's Lemma*.

**Proof** We will proceed by induction. The theorem holds when $m = 1$, and $d = 0$ or $d = 1$. Hypothesizing that the theorem holds for $(m-1, d-1)$ and $(m-1, d)$. Letting $D = \{\boldsymbol{x}_1, \boldsymbol{x}_2, \ldots, \boldsymbol{x}_m\}$ and $D' = \{\boldsymbol{x}_1, \boldsymbol{x}_2, \ldots, \boldsymbol{x}_{m-1}\}$, we have

$$\mathcal{H}_{|D} = \{(h(\boldsymbol{x}_1), h(\boldsymbol{x}_2), \ldots, h(\boldsymbol{x}_m)) \mid h \in \mathcal{H}\},$$
$$\mathcal{H}_{|D'} = \{(h(\boldsymbol{x}_1), h(\boldsymbol{x}_2), \ldots, h(\boldsymbol{x}_{m-1})) \mid h \in \mathcal{H}\}.$$

Since every hypothesis $h \in \mathcal{H}$ classifies $\boldsymbol{x}_m$ as either $+1$ or $-1$, every sequence appeared in $\mathcal{H}_{|D'}$ will appear in $\mathcal{H}_{|D}$ once or twice. Let $\mathcal{H}_{D'|D}$ denote the set of sequences from $\mathcal{H}_{|D'}$ that appear twice in $\mathcal{H}_{|D}$, that is,

$$\mathcal{H}_{D'|D} = \{(y_1, y_2, \ldots, y_{m-1}) \in \mathcal{H}_{|D'} \mid \exists h, h' \in \mathcal{H},$$
$$(h(\boldsymbol{x}_i) = h'(\boldsymbol{x}_i) = y_i) \wedge (h(\boldsymbol{x}_m) \neq h'(\boldsymbol{x}_m)),\ 1 \leqslant i \leqslant m-1\}.$$

Since the sequences in $\mathcal{H}_{D'|D}$ appear twice in $\mathcal{H}_{|D}$ but once in $\mathcal{H}_{|D'}$, we have

$$\left|\mathcal{H}_{|D}\right| = \left|\mathcal{H}_{|D'}\right| + \left|\mathcal{H}_{D'|D}\right|. \tag{12.25}$$

For the data set $D'$ of size $m-1$, we have, from the induction assumption,

$$\left|\mathcal{H}_{|D'}\right| \leqslant \Pi_{\mathcal{H}}(m-1) \leqslant \sum_{i=0}^{d} \binom{m-1}{i}. \tag{12.26}$$

Let $Q$ denote the set of instances shattered by $\mathcal{H}_{D'|D}$. From the definition of $\mathcal{H}_{D'|D}$, we know that $\mathcal{H}_{|D}$ can shatter $Q \cup \{\boldsymbol{x}_m\}$. Since the VC dimension of $\mathcal{H}$ is $d$, the largest possible VC dimension of $\mathcal{H}_{D'|D}$ is $d-1$. Therefore, we have

$$\left|\mathcal{H}_{D'|D}\right| \leqslant \Pi_{\mathcal{H}}(m-1) \leqslant \sum_{i=0}^{d-1}\binom{m-1}{i}. \tag{12.27}$$

From (12.25)–(12.27), we have

$\binom{m-1}{-1} = 0.$

$$\begin{aligned}
\left|\mathcal{H}_{|D}\right| &\leqslant \sum_{i=0}^{d}\binom{m-1}{i} + \sum_{i=0}^{d-1}\binom{m-1}{i} \\
&= \sum_{i=0}^{d}\left(\binom{m-1}{i} + \binom{m-1}{i-1}\right) \\
&= \sum_{i=0}^{d}\binom{m}{i}.
\end{aligned}$$

From the arbitrariness of data set $D$, Lemma 12.2 follows.

□

From Lemma 12.2, we can calculate the upper bound of the growth function:

**Corollary 12.2** *If the VC dimension of a hypothesis space $\mathcal{H}$ is $d$, then, for any integer $m \geqslant d$, we have*

$e$ is Euler's number.

$$\Pi_{\mathcal{H}}(m) \leqslant \left(\frac{e \cdot m}{d}\right)^d. \tag{12.28}$$

12

**Proof**

$m \geqslant d$.

$$\begin{aligned}
\Pi_{\mathcal{H}}(m) &\leqslant \sum_{i=0}^{d}\binom{m}{i} \\
&\leqslant \sum_{i=0}^{d}\binom{m}{i}\left(\frac{m}{d}\right)^{d-i} \\
&= \left(\frac{m}{d}\right)^d \sum_{i=0}^{d}\binom{m}{i}\left(\frac{d}{m}\right)^i \\
&\leqslant \left(\frac{m}{d}\right)^d \sum_{i=0}^{m}\binom{m}{i}\left(\frac{d}{m}\right)^i \\
&= \left(\frac{m}{d}\right)^d \left(1 + \frac{d}{m}\right)^m \\
&\leqslant \left(\frac{e \cdot m}{d}\right)^d.
\end{aligned}$$

□

From Corollary 12.2 and Theorem 12.2, we have the generalization error bound in terms of the VC dimension, also known as the VC bound:

**Theorem 12.3** *If the VC dimension of a hypothesis space $\mathcal{H}$ is $d$, then, for any $m > d$, $\delta \in (0, 1)$, and $h \in \mathcal{H}$, we have*

$$P\left(\left|E(h) - \widehat{E}(h)\right| \leqslant \sqrt{\frac{8d \ln \frac{2em}{d} + 8 \ln \frac{4}{\delta}}{m}}\right) \geqslant 1 - \delta. \tag{12.29}$$

**Proof** Setting $4\Pi_{\mathcal{H}}(2m) \exp(-\frac{m\epsilon^2}{8}) \leqslant 4(\frac{2em}{d})^d \exp(-\frac{m\epsilon^2}{8}) = \delta$, we have

$$\epsilon = \sqrt{\frac{8d \ln \frac{2em}{d} + 8 \ln \frac{4}{\delta}}{m}},$$

which completes the proof by substituting the above equation into Theorem 12.2. □

From Theorem 12.3, the generalization error bound in (12.29) is dependent only on the sample size $m$ and converges toward 0 at a convergence rate of $O(\frac{1}{\sqrt{m}})$. Since the VC bound is independent of the data distribution $\mathcal{D}$ and the data set $D$, it is *distribution-free* and *data-independent*.

Let $h$ denote the hypothesis output by a learning algorithm $\mathfrak{L}$. Then, we say $\mathfrak{L}$ satisfies the Empirical Risk Minimization (ERM) principle if

$$\widehat{E}(h) = \min_{h' \in \mathcal{H}} \widehat{E}(h'). \tag{12.30}$$

Then, we have the following theorem:

**Theorem 12.4** *Every hypothesis space $\mathcal{H}$ with a finite VC dimension is (agnostic) PAC learnable.*

**Proof** Suppose $\mathfrak{L}$ is a learning algorithm satisfying the ERM principle, and $h$ is the hypothesis output by $\mathfrak{L}$. Let $g$ be the hypothesis with the smallest generalization error in $\mathcal{H}$, that is,

$$E(g) = \min_{h \in \mathcal{H}} E(h). \tag{12.31}$$

Letting

$$\delta' = \frac{\delta}{2},$$

$$\sqrt{\frac{(\ln 2/\delta')}{2m}} = \frac{\epsilon}{2}. \tag{12.32}$$

From Corollary 12.1, the following holds with a probability of at least $1 - \delta/2$:

$$\widehat{E}(g) - \frac{\epsilon}{2} \leqslant E(g) \leqslant \widehat{E}(g) + \frac{\epsilon}{2}. \tag{12.33}$$

Setting

$$\sqrt{\frac{8d\ln\frac{2em}{d} + 8\ln\frac{4}{\delta'}}{m}} = \frac{\epsilon}{2}, \tag{12.34}$$

then, from Theorem 12.3, we have

$$P\left(E(h) - \widehat{E}(h) \leqslant \frac{\epsilon}{2}\right) \geqslant 1 - \frac{\delta}{2}. \tag{12.35}$$

Hence, the following holds with a probability of at least $1 - \delta$:

$$\begin{aligned} E(h) - E(g) &\leqslant \widehat{E}(h) + \frac{\epsilon}{2} - \left(\widehat{E}(g) - \frac{\epsilon}{2}\right) \\ &= \widehat{E}(h) - \widehat{E}(g) + \epsilon \\ &\leqslant \epsilon. \end{aligned}$$

We can solve $m$ from (12.32) and (12.34). Then, from the arbitrariness of $\mathcal{H}$, we have Theorem 12.4. □

12

## 12.5 Rademacher Complexity

From Sect. 12.4, we see that the VC bound is distribution-free and data-independent (i.e., it is valid for any data distribution), which makes the analysis of generalization error bound "universal". However, since it does not take the data set into account, the VC bound is generally loose, especially for "poor" data distributions that are far from the typical situation in learning problems.

*Rademacher complexity* presents another characterization of the complexity of the hypothesis space, and the difference from the VC dimension lies in consideration of data distribution in some sense.

*Rademacher complexity* is named after the German mathematician H. Rademacher (1892–1969).

Given a data set $D = \{(\boldsymbol{x}_1, y_1), (\boldsymbol{x}_2, y_2), \ldots, (\boldsymbol{x}_m, y_m)\}$, the empirical error of a hypothesis $h$ is given by

$$
\begin{aligned}
\widehat{E}(h) &= \frac{1}{m}\sum_{i=1}^{m}\mathbb{I}(h(\boldsymbol{x}_i) \neq y_i) \\
&= \frac{1}{m}\sum_{i=1}^{m}\frac{1-y_ih(\boldsymbol{x}_i)}{2} \\
&= \frac{1}{2} - \frac{1}{2m}\sum_{i=1}^{m}y_ih(\boldsymbol{x}_i),
\end{aligned} \tag{12.36}
$$

where $\frac{1}{m}\sum_{i=1}^{m} y_ih(\boldsymbol{x}_i)$ represents the consistency between the predicted values $h(\boldsymbol{x}_i)$ and the ground-truth labels $y_i$. It takes the maximum value 1 if $h(\boldsymbol{x}_i) = y_i$ for all $i \in \{1, 2, \ldots, m\}$. In other words, the hypothesis with the smallest empirical error is

$$
\arg\max_{h\in\mathcal{H}} \frac{1}{m}\sum_{i=1}^{m} y_ih(\boldsymbol{x}_i). \tag{12.37}
$$

In practice, however, the data set may have been corrupted by some noises, that is, the label $y_i$ of sample $(\boldsymbol{x}_i, y_i)$ is affected by some random factors and is no longer the ground-truth label of $\boldsymbol{x}_i$. In such cases, sometimes it is better to select a hypothesis that has considered the influence of random noises, rather than the best hypothesis over the training set.

We introduce the Rademacher random variable $\sigma_i$, which takes value $+1$ or $-1$ with an equal probability of 0.5. With $\sigma_i$, we rewrite (12.37) as

$$
\sup_{h\in\mathcal{H}} \frac{1}{m}\sum_{i=1}^{m} \sigma_ih(\boldsymbol{x}_i). \tag{12.38}
$$

It is likely that we cannot find the maximum value since $\mathcal{H}$ is infinite. Hence, we replace the maximum by the supremum.

We consider all hypotheses in $\mathcal{H}$ and take the expectation over (12.38) as

$$
\mathbb{E}_{\boldsymbol{\sigma}}\left[\sup_{h\in\mathcal{H}} \frac{1}{m}\sum_{i=1}^{m} \sigma_ih(\boldsymbol{x}_i)\right], \tag{12.39}
$$

where $\boldsymbol{\sigma} = \{\sigma_1, \sigma_2, \ldots, \sigma_m\}$. Equation (12.39) takes value in $[0, 1]$ and expresses the representation power of the hypothesis $\mathcal{H}$. For example, (12.39) equals to 0 when $|\mathcal{H}| = 1$, that is, there is only one hypothesis in $\mathcal{H}$; (12.39) equals to 1 when $|\mathcal{H}| = 2^m$ and $\mathcal{H}$ shatters $D$, that is, for any $\boldsymbol{\sigma}$, there exists a hypothesis such that $h(\boldsymbol{x}_i) = \sigma_i$ $(i = 1, 2, \ldots, m)$.

Let $\mathcal{F} : \mathcal{Z} \rightarrow \mathbb{R}$ be a real-valued function space, and $Z = \{\boldsymbol{z}_1, \boldsymbol{z}_2, \ldots, \boldsymbol{z}_m\}$ be a set of *i.i.d.* instances, where $\boldsymbol{z}_i \in \mathcal{Z}$. By replacing $\mathcal{X}$ and $\mathcal{H}$ in (12.39) with $\mathcal{Z}$ and $\mathcal{F}$, respectively, we have

**Definition 12.8** The empirical Rademacher complexity of a function space $\mathcal{F}$ with respect to $Z$ is defined as

$$\widehat{R}_Z(\mathcal{F}) = \mathbb{E}_\sigma\left[\sup_{f\in\mathcal{F}} \frac{1}{m}\sum_{i=1}^{m}\sigma_i f(z_i)\right]. \tag{12.40}$$

The empirical Rademacher complexity measures the correlation between the function space $\mathcal{F}$ and the random noise in the data set $Z$. To analyze the correlation between $\mathcal{F}$ and $\mathcal{D}$ over $\mathcal{Z}$, we can take the expectation over the data set $Z$ with $m$ *i.i.d.* samples drawn from $\mathcal{D}$:

**Definition 12.9** The Rademacher complexity of a function space $\mathcal{F}$ with respect to a distribution $\mathcal{D}$ over $\mathcal{Z}$ is defined as

$$R_m(\mathcal{F}) = \mathbb{E}_{Z\subseteq\mathcal{Z}:|Z|=m}\left[\widehat{R}_Z(\mathcal{F})\right]. \tag{12.41}$$

Based on the Rademacher complexity, we can define the generalization error bound of function space $\mathcal{F}$ (Mohri et al. 2012):

**Theorem 12.5** *Let $\mathcal{F} : \mathcal{Z} \to [0, 1]$ be a real-valued function space, and $Z = \{z_1, z_2, \ldots, z_m\}$ be a set of i.i.d. samples drawn from $\mathcal{D}$ over $\mathcal{Z}$. Then, for any $\delta \in (0, 1)$ and $f \in \mathcal{F}$, the following holds with a probability of at least $1 - \delta$:*

$$\mathbb{E}[f(z)] \leqslant \frac{1}{m}\sum_{i=1}^{m} f(z_i) + 2R_m(\mathcal{F}) + \sqrt{\frac{\ln(1/\delta)}{2m}}, \tag{12.42}$$

$$\mathbb{E}[f(z)] \leqslant \frac{1}{m}\sum_{i=1}^{m} f(z_i) + 2\widehat{R}_Z(\mathcal{F}) + 3\sqrt{\frac{\ln(2/\delta)}{2m}}. \tag{12.43}$$

**Proof** Letting

$$\widehat{E}_Z(f) = \frac{1}{m}\sum_{i=1}^{m} f(z_i),$$

$$\Phi(Z) = \sup_{f\in\mathcal{F}} \mathbb{E}[f] - \widehat{E}_Z(f),$$

and let $Z'$ be another data set that is the same as $Z$ except for one instance. Suppose that $z_m \in Z$ and $z'_m \in Z'$ are the two different instances. Then, we have

12

$$\begin{aligned}\Phi(Z') - \Phi(Z) &= \left(\sup_{f\in\mathcal{F}} \mathbb{E}[f] - \widehat{E}_{Z'}(f)\right) - \left(\sup_{f\in\mathcal{F}} \mathbb{E}[f] - \widehat{E}_{Z}(f)\right) \\ &\leqslant \sup_{f\in\mathcal{F}} \widehat{E}_{Z}(f) - \widehat{E}_{Z'}(f) \\ &= \sup_{f\in\mathcal{F}} \frac{f(z_m) - f(z'_m)}{m} \\ &\leqslant \frac{1}{m}.\end{aligned}$$

Similarly, we have

$$\Phi(Z) - \Phi(Z') \leqslant \frac{1}{m},$$

$$\left|\Phi(Z) - \Phi(Z')\right| \leqslant \frac{1}{m}.$$

According to McDiarmid's inequality (12.7), for any $\delta \in (0, 1)$, the following holds with a probability of at least $1 - \delta$:

$$\Phi(Z) \leqslant \mathbb{E}_Z[\Phi(Z)] + \sqrt{\frac{\ln(1/\delta)}{2m}}, \tag{12.44}$$

where the upper bound of $\mathbb{E}_Z[\Phi(Z)]$ is given by

$$\begin{aligned}\mathbb{E}_Z[\Phi(Z)] &= \mathbb{E}_Z\left[\sup_{f\in\mathcal{F}} \mathbb{E}[f] - \widehat{E}_Z(f)\right] \\ &= \mathbb{E}_Z\left[\sup_{f\in\mathcal{F}} \mathbb{E}_{Z'}\left[\widehat{E}_{Z'}(f) - \widehat{E}_Z(f)\right]\right] \\ &\leqslant \mathbb{E}_{Z,Z'}\left[\sup_{f\in\mathcal{F}} \widehat{E}_{Z'}(f) - \widehat{E}_Z(f)\right] \\ &= \mathbb{E}_{Z,Z'}\left[\sup_{f\in\mathcal{F}} \frac{1}{m}\sum_{i=1}^{m}(f(z'_i) - f(z_i))\right] \\ &= \mathbb{E}_{\sigma,Z,Z'}\left[\sup_{f\in\mathcal{F}} \frac{1}{m}\sum_{i=1}^{m}\sigma_i(f(z'_i) - f(z_i))\right] \\ &\leqslant \mathbb{E}_{\sigma,Z'}\left[\sup_{f\in\mathcal{F}} \frac{1}{m}\sum_{i=1}^{m}\sigma_i f(z'_i)\right] + \mathbb{E}_{\sigma,Z}\left[\sup_{f\in\mathcal{F}} \frac{1}{m}\sum_{i=1}^{m}-\sigma_i f(z_i)\right] \\ &= 2\mathbb{E}_{\sigma,Z}\left[\sup_{f\in\mathcal{F}} \frac{1}{m}\sum_{i=1}^{m}\sigma_i f(z_i)\right] \\ &= 2R_m(\mathcal{F}).\end{aligned}$$

Using Jensen's inequality (12.4) and the convexity of the supremum function.

$\sigma_i$ and $-\sigma_i$ follow the same distribution.

The above gives the proof of (12.42). From Definition 12.9, we know that changing one instance in $Z$ will change the value of $\widehat{R}_Z(\mathcal{F})$ at most $1/m$. According to McDiarmid's inequality (12.7), the following holds with a probability of at least $1 - \delta/2$:

$$R_m(\mathcal{F}) \leqslant \widehat{R}_Z(\mathcal{F}) + \sqrt{\frac{\ln(2/\delta)}{2m}}. \tag{12.45}$$

Then, from (12.44), the following holds with a probability of at least $1 - \delta/2$:

$$\Phi(Z) \leqslant \mathbb{E}_Z[\Phi(Z)] + \sqrt{\frac{\ln(2/\delta)}{2m}}.$$

Hence, the following holds with a probability of at least $1 - \delta$:

$$\Phi(Z) \leqslant 2\widehat{R}_Z(\mathcal{F}) + 3\sqrt{\frac{\ln(2/\delta)}{2m}}. \tag{12.46}$$

The above gives the proof of (12.43). □

Since $\mathcal{F}$ in Theorem 12.5 is a real-valued function over the interval [0, 1], Theorem 12.5 is applicable to regression problems only. For binary classification problems, we have the following theorem:

**Theorem 12.6** *Let $\mathcal{H} : \mathcal{X} \rightarrow \{-1, +1\}$ be a hypothesis space and $D = \{\boldsymbol{x}_1, \boldsymbol{x}_2, \ldots, \boldsymbol{x}_m\}$ be a set of i.i.d. instances drawn from $\mathcal{D}$ over $\mathcal{X}$. Then, for any $\delta \in (0, 1)$ and $h \in \mathcal{H}$, the following holds with a probability of at least $1 - \delta$:*

$$E(h) \leqslant \widehat{E}(h) + R_m(\mathcal{H}) + \sqrt{\frac{\ln(1/\delta)}{2m}}, \tag{12.47}$$

$$E(h) \leqslant \widehat{E}(h) + \widehat{R}_D(\mathcal{H}) + 3\sqrt{\frac{\ln(2/\delta)}{2m}}. \tag{12.48}$$

**Proof** Let $\mathcal{H}$ be a hypothesis space of binary classification problems. By letting $\mathcal{Z} = \mathcal{X} \times \{-1, +1\}$, $h \in \mathcal{H}$ can be transformed to

$$f_h(\boldsymbol{z}) = f_h(\boldsymbol{x}, y) = \mathbb{I}(h(\boldsymbol{x}) \neq y), \tag{12.49}$$

which transforms the hypothesis space $\mathcal{H}$ with an output domain of $\{-1, +1\}$ to a function space $\mathcal{F}_{\mathcal{H}} = \{f_h : h \in \mathcal{H}\}$ with an output domain of [0, 1]. From Definition 12.8, we have

$$\begin{aligned}
\widehat{R}_Z(\mathcal{F}_{\mathcal{H}}) &= \mathbb{E}_{\boldsymbol{\sigma}}\left[\sup_{f_h\in\mathcal{F}_{\mathcal{H}}}\frac{1}{m}\sum_{i=1}^{m}\sigma_i f_h(\boldsymbol{x}_i, y_i)\right]\\
&= \mathbb{E}_{\boldsymbol{\sigma}}\left[\sup_{h\in\mathcal{H}}\frac{1}{m}\sum_{i=1}^{m}\sigma_i \mathbb{I}(h(\boldsymbol{x}_i)\neq y_i)\right]\\
&= \mathbb{E}_{\boldsymbol{\sigma}}\left[\sup_{h\in\mathcal{H}}\frac{1}{m}\sum_{i=1}^{m}\sigma_i \frac{1-y_i h(\boldsymbol{x}_i)}{2}\right]\\
&= \frac{1}{2}\mathbb{E}_{\boldsymbol{\sigma}}\left[\frac{1}{m}\sum_{i=1}^{m}\sigma_i + \sup_{h\in\mathcal{H}}\frac{1}{m}\sum_{i=1}^{m}(-y_i\sigma_i h(\boldsymbol{x}_i))\right]\\
&= \frac{1}{2}\mathbb{E}_{\boldsymbol{\sigma}}\left[\sup_{h\in\mathcal{H}}\frac{1}{m}\sum_{i=1}^{m}(-y_i\sigma_i h(\boldsymbol{x}_i))\right]\\
&= \frac{1}{2}\mathbb{E}_{\boldsymbol{\sigma}}\left[\sup_{h\in\mathcal{H}}\frac{1}{m}\sum_{i=1}^{m}(\sigma_i h(\boldsymbol{x}_i))\right]\\
&= \frac{1}{2}\widehat{R}_D(\mathcal{H}). \qquad (12.50)
\end{aligned}$$

$-y_i\sigma_i$ and $\sigma_i$ follow the same distribution.

Taking the expectation of (12.50) gives

$$R_m(\mathcal{F}_{\mathcal{H}}) = \frac{1}{2}R_m(\mathcal{H}). \qquad (12.51)$$

From (12.50), (12.51), and Theorem 12.5, we have Theorem 12.6 proved. □

Theorem 12.6 gives the generalization error bound based on the Rademacher complexity, also known as the Rademacher bound. In comparison with Theorem 12.3, the VC bound is distribution-free and data-independent, whereas the Rademacher bound depends on the distribution $\mathcal{D}$ in (12.47) and the data set $D$ in (12.48). In other words, the Rademacher bound depends on the data distribution of the specific learning problem. The Rademacher bound is generally tighter than the VC bound since it is "tailored" for the specific learning problem.

For the Rademacher complexity and the growth function, we have

**Theorem 12.7** *The Rademacher complexity $R_m(\mathcal{H})$ and the growth function $\Pi_{\mathcal{H}}(m)$ of a hypothesis space $\mathcal{H}$ satisfy*

See Mohri et al. (2012) for proof.

$$R_m(\mathcal{H}) \leqslant \sqrt{\frac{2\ln \Pi_{\mathcal{H}}(m)}{m}}. \qquad (12.52)$$

From (12.47), (12.52), and Corollary 12.2, we have

$$E(h) \leqslant \widehat{E}(h) + \sqrt{\frac{2d \ln \frac{em}{d}}{m}} + \sqrt{\frac{\ln(1/\delta)}{2m}}. \tag{12.53}$$

In other words, we can derive the VC bound from the Rademacher complexity and the growth function.

## 12.6 Stability

The generalization error bound, based on either the VC dimension or Rademacher complexity, is independent of the specific learning algorithm. Hence, the analysis applies to all learning algorithms and enables us to study the nature of learning problems without considering specific design of learning algorithms. However, if we wish the analysis to be algorithm-dependent, then we need to take a different approach, and one direction is stability analysis.

As the name suggests, the *stability* of an algorithm concerns about whether a minor change of the input will cause a significant change in the output. The input of learning algorithms is a data set, so we need to define the changes on data sets.

Given a data set $D = \{z_1 = (\boldsymbol{x}_1, y_1), z_2 = (\boldsymbol{x}_2, y_2), \ldots, z_m = (\boldsymbol{x}_m, y_m)\}$, where $\boldsymbol{x}_i \in \mathcal{X}$ are *i.i.d.* instances drawn from distribution $\mathcal{D}$ and $y_i \in \{-1, +1\}$. Let $\mathcal{H} : \mathcal{X} \to \{-1, +1\}$ be a hypothesis space, and $\mathfrak{L}_D \in \mathcal{H}$ be the hypothesis learned by a learning algorithm $\mathfrak{L}$ on the training set $D$. Then, we consider the following changes on $D$:

- Let $D^{\backslash i}$ denote the set $D$ with the $i$th sample $z_i$ excluded, that is,

$$D^{\backslash i} = \{z_1, z_2, \ldots, z_{i-1}, z_{i+1}, \ldots, z_m\},$$

- Let $D^i$ denote the set $D$ with the $i$th sample $z_i$ replaced with $z'_i$, that is,

$$D^i = \{z_1, z_2, \ldots, z_{i-1}, z'_i, z_{i+1}, \ldots, z_m\},$$

  where $z'_i = (\boldsymbol{x}'_i, y'_i)$, and $\boldsymbol{x}'_i$ follows distribution $\mathcal{D}$ and is independent of $D$.

A loss function $\ell(\mathfrak{L}_D(\boldsymbol{x}), y) : \mathcal{Y} \times \mathcal{Y} \to \mathbb{R}^+$, abbreviated as $\ell(\mathfrak{L}_D, z)$, characterizes the difference between the predicted label $\mathfrak{L}_D(\boldsymbol{x})$ and the ground-truth label $y$. We now introduce several types of loss with respect to the hypothesis $\mathfrak{L}_D$ as follows:

- Generalization loss:

$$\ell(\mathfrak{L}, D) = \mathbb{E}_{\boldsymbol{x} \in \mathcal{X}, z=(\boldsymbol{x}, y)} \left[\ell(\mathfrak{L}_D, z)\right]; \tag{12.54}$$

12

- Empirical loss:

$$\widehat{\ell}(\mathfrak{L}, D) = \frac{1}{m}\sum_{i=1}^{m} \ell(\mathfrak{L}_D, z_i); \tag{12.55}$$

- Leave-one-out loss:

$$\ell_{\mathrm{loo}}(\mathfrak{L}, D) = \frac{1}{m}\sum_{i=1}^{m} \ell(\mathfrak{L}_{D^{\setminus i}}, z_i). \tag{12.56}$$

We define the *uniform stability* as follows:

**Definition 12.10** A learning algorithm $\mathfrak{L}$ is said to satisfy the $\beta$-uniform stability with respect to loss function $\ell$ if, for any $\boldsymbol{x} \in \mathcal{X}$ and $z = (\boldsymbol{x}, y)$, $\mathfrak{L}$ satisfies

$$\left|\ell(\mathfrak{L}_D, z) - \ell(\mathfrak{L}_{D^{\setminus i}}, z)\right| \leqslant \beta,\ i = 1, 2, \ldots, m. \tag{12.57}$$

If a learning algorithm $\mathfrak{L}$ satisfies the $\beta$-uniform stability with respect to a loss function $\ell$, then

$$\begin{aligned}&\left|\ell(\mathfrak{L}_D, z) - \ell(\mathfrak{L}_{D^i}, z)\right| \\ &\leqslant \left|\ell(\mathfrak{L}_D, z) - \ell(\mathfrak{L}_{D^{\setminus i}}, z)\right| + \left|\ell(\mathfrak{L}_{D^i}, z) - \ell(\mathfrak{L}_{D^{\setminus i}}, z)\right| \\ &\leqslant 2\beta,\end{aligned}$$

which means that the stability of excluding an instance implies the stability of replacing an instance.

If the loss function $\ell$ is bounded as $0 \leqslant \ell(\mathfrak{L}_D, z) \leqslant M$ for all $D$ and $z = (\boldsymbol{x}, y)$, then, we have [Bousquet and Elisseeff (2002)]

**Theorem 12.8** *Given a data set $D$ with $m$ i.i.d. instances drawn from the distribution $\mathcal{D}$. If a learning algorithm $\mathfrak{L}$ satisfies the $\beta$-uniform stability with respect to a loss function $\ell$ upper bounded by $M$, then, for any $m \geqslant 1$ and $\delta \in (0, 1)$, the following holds with a probability of at least $1 - \delta$:*

See Bousquet and Elisseeff (2002) for proof.

$$\ell(\mathfrak{L}, \mathcal{D}) \leqslant \widehat{\ell}(\mathfrak{L}, D) + 2\beta + (4m\beta + M)\sqrt{\frac{\ln(1/\delta)}{2m}}, \tag{12.58}$$

$$\ell(\mathfrak{L}, \mathcal{D}) \leqslant \ell_{\mathrm{loo}}(\mathfrak{L}, D) + \beta + (4m\beta + M)\sqrt{\frac{\ln(1/\delta)}{2m}}. \tag{12.59}$$

Theorem 12.8 shows the generalization error bound of the learning algorithm $\mathfrak{L}$ derived from the stability analysis. From (12.58), we see the convergence rate between the empirical error and the generalization error is $\beta\sqrt{m}$. When $\beta = O(\frac{1}{m})$, the convergence rate becomes $O(\frac{1}{\sqrt{m}})$, which is consistent with those of VC bound and Rademacher bound in comparisons with Theorems 12.3 and 12.6.

The stability analysis of learning algorithm focuses on $|\widehat{\ell}(\mathfrak{L}, D) - \ell(\mathfrak{L}, D)|$, whereas the complexity analysis of the hypothesis space considers $\sup_{h\in\mathcal{H}}|\widehat{E}(h)-E(h)|$. In other words, the stability analysis does not necessarily consider every hypothesis in $\mathcal{H}$, but only analyzes the generalization error bound of the output hypothesis $\mathfrak{L}_D$ based on the properties (stability) of $\mathfrak{L}$. So, what is the relationship between stability and learnability?

To ensure the generalization ability of a stable learning algorithm $\mathfrak{L}$, we must assume $\beta\sqrt{m} \to 0$, that is, the empirical loss converges to the generalization loss; otherwise, learnability can hardly be discussed. For ease of computation, letting $\beta = \frac{1}{m}$ and substituting into (12.58), we have

$$\ell(\mathfrak{L}, \mathcal{D}) \leqslant \widehat{\ell}(\mathfrak{L}, D) + \frac{2}{m} + (4+M)\sqrt{\frac{\ln(1/\delta)}{2m}}. \tag{12.60}$$

Given a loss function $\ell$, a learning algorithm $\mathfrak{L}$ is an ERM learning algorithm satisfying the ERM principle if its output hypothesis minimizes the empirical loss. We have the following theorem on stability and learnability:

Minimizing empirical error and minimizing empirical loss are sometimes different since there exist some poor loss functions $\ell$ such that minimizing the loss does not minimize the empirical error. For ease of discussion, this chapter assumes that minimizing the loss always minimizes the empirical error.

12

**Theorem 12.9** *If an ERM learning algorithm $\mathfrak{L}$ is stable, then the hypothesis space $\mathcal{H}$ is learnable.*

**Proof** Let $g$ be the hypothesis with the minimum generalization loss in $\mathcal{H}$, that is,

$$\ell(g, \mathcal{D}) = \min_{h\in\mathcal{H}} \ell(h, \mathcal{D}).$$

Letting

$$\epsilon' = \frac{\epsilon}{2},$$
$$\frac{\delta}{2} = 2\exp\left(-2m(\epsilon')^2\right),$$

then, from Hoeffding's inequality (12.6), the following holds with a probability of at least $1-\delta/2$ when $m \geqslant \frac{2}{\epsilon^2}\ln\frac{4}{\delta}$:

$$\left|\ell(g, \mathcal{D}) - \widehat{\ell}(g, D)\right| \leqslant \frac{\epsilon}{2}.$$

For (12.60), by setting

$$\frac{2}{m} + (4 + M)\sqrt{\frac{\ln(2/\delta)}{2m}} = \frac{\epsilon}{2},$$

we have $m = O(\frac{1}{\epsilon^2} \ln \frac{1}{\delta})$. Hence, the following holds with a probability of at least $1 - \delta/2$:

$$\ell(\mathfrak{L}, \mathcal{D}) \leqslant \widehat{\ell}(\mathfrak{L}, D) + \frac{\epsilon}{2}.$$

Therefore, the following holds with a probability of at least $1 - \delta$:

$$\begin{aligned}\ell(\mathfrak{L}, \mathcal{D}) - \ell(g, \mathcal{D}) &\leqslant \widehat{\ell}(\mathfrak{L}, D) + \frac{\epsilon}{2} - \left(\widehat{\ell}(g, D) - \frac{\epsilon}{2}\right) \\ &\leqslant \widehat{\ell}(\mathfrak{L}, D) - \widehat{\ell}(g, D) + \epsilon \\ &\leqslant \epsilon,\end{aligned}$$

which proves Theorem 12.9. □

Readers may wonder, why we can derive the learnability of a hypothesis space from the stability of a learning algorithm. Learning algorithm and hypothesis space are very different things. However, it is worth noting that stability is not irrelevant to hypothesis space as they are indeed connected by a loss function $\ell$ according to the definition of stability.

## 12.7 Further Reading

Valiant (1984) proposed PAC learning, which motivated a branch of machine learning research known as *Computational Learning Theory*. A good introductory textbook on this topic is Kearns and Vazirani (1994). The most important academic conference in this field is the Conference on Learning Theory (COLT).

Vapnik and Chervonenkis (1971) proposed the VC dimension, which makes it possible to study the complexity of infinite hypothesis spaces. Sauer's Lemma is named after Sauer (1972), while the same result was also derived in Vapnik and Chervonenkis (1971), Shelah (1972), respectively. This chapter mainly focuses on binary classification problems, and as for multiclass classification problems, the VC dimension can be extended to the Natarajan dimension (Natarajan 1989; Ben-David et al. 1995).

The VC dimension is named after the surnames of the two authors.

Rademacher complexity was introduced to machine learning by Koltchinskii and Panchenko (2000) and received more attention after Bartlett and Mendelson (2002). Bartlett et al. (2002) proposed the local Rademacher complexity, which can derive a tighter generalization error bound for noisy data.

Bousquet and Elisseeff (2002) introduced the stability analysis of machine learning algorithms, and motivated many studies on the relationship between stability and learnability. For example, Mukherjee et al. (2006), Shalev-Shwartz et al. (2010) showed the equivalence of ERM stability and ERM learnability. Since not all learning algorithms satisfy the ERM principle, Shalev-Shwartz et al. (2010) further studied the relationship between stability and learnability with respect to Asymptotical Empirical Risk Minimization (AERM).

This chapter mainly focuses on deterministic learning problems, that is, there is a deterministic label $y$ for each sample $\boldsymbol{x}$. Though most supervised learning problems are deterministic, there are also stochastic learning problems in which the label of an instance does not firmly belong to a single class but is decided by a posterior probability function conditioned on feature values. See Devroye et al. (1996) for more discussions on the generalization error bound in stochastic learning problems.

## Exercises

**12.1** Prove Jensen's inequality (12.4).

**12.2** Prove Lemma 12.1.

**12.3** Prove Corollary 12.1.

Hint: letting $\delta = 2e^{-2m\epsilon^2}$.

**12.4** Prove that the hypothesis space consisting of all linear hyperplanes in $\mathbb{R}^d$ has a VC dimension of $d + 1$.

**12.5** Calculate the VC dimension of the hypothesis space of decision stumps.

**12.6** Prove that the VC dimension of the hypothesis space of decision tree classifiers can be infinite.

**12.7** Prove that the VC dimension of the hypothesis space of $k$-nearest neighbors classifiers can be infinite.

**12.8** Prove that the Rademacher complexity of the constant function $c$ is 0.

**12.9** Given function spaces $\mathcal{F}_1$ and $\mathcal{F}_2$, prove that $R_m(\mathcal{F}_1 + \mathcal{F}_2) \leqslant R_m(\mathcal{F}_1) + R_m(\mathcal{F}_2)$, where $R_m(\cdot)$ is the Rademacher complexity.

**12.10** * Considering Theorem 12.8, discuss the rationality of estimating an algorithm's generalization ability via cross-validation.

## Break Time

**Short Story: Leslie G. Valiant—The Father of Computational Learning Theory**

Theoretical Computer Science (TCS) is an intersection of computer science and mathematics that focuses on mathematical topics of computing. A famous TCS problem is the "P versus NP problem".

Computational learning theory, as a subfield of machine learning, is the intersection of machine learning and TCS. If we are talking about computational learning theory, we have to talk about the British computer scientist Leslie G. Valiant (1949–). Valiant studied at King's College, Cambridge, Imperial College London, and the University of Warwick, where he earned his Ph.D. degree in 1974. Before he became a professor at Harvard University in 1982, he taught at Carnegie Mellon University, the University of Leeds, and the University of Edinburgh. In 1984, *Communications of the ACM* published Valiant's paper titled "A theory of the learnable", in which PAC learning theory was proposed and laid the foundations of computational learning theory. In 2010, Valiant received the Turing Award for his seminal contributions to PAC learning theory, the complexity of enumeration and of algebraic computation, and the theory of parallel and distributed computing. The ACM Turing Award committee pointed out that Valiant's paper published in 1984 created a new research area known as computational learning theory that puts machine learning on a sound mathematical footing. *ACM Computing News* also published an article titled "ACM Turing Award Goes to Innovator in Machine Learning" to emphasize the contributions of this first Turing Award recipient from machine learning.

## References

Bartlett PL, Mendelson S (2002) Rademacher and Gaussian complexities: risk bounds and structural results. J Mach Learn Res 3:463–482
Bartlett PL, Bousquet O, Mendelson S (2002). Localized rademacher complexities. Sydney, Australia, pp 44–58
Ben-David S, Cesa-Bianchi N, Haussler D, Long PM (1995) Characterizations of learnability for classes of $\{0, \ldots, n\}$-valued functions. J Comput Syst Sci 50(1):74–86
Bousquet O, Elisseeff A (2002) Stability and generalization. J Mach Learn Res 2:499–526
Devroye L, Gyorfi L, Lugosi G (eds) (1996) A probabilistic theory of pattern recognition. Springer, New York
Hoeffding W (1963) Probability inequalities for sums of bounded random variables. J Am Stat Assoc 58(301):13–30
Kearns MJ, Vazirani UV (1994) An introduction to computational learning theory. MIT Press, Cambridge
Koltchinskii V, Panchenko D (2000) Rademacher processes and bounding the risk of function learning. In: Gine E, Mason DM, Wellner JA (eds) High dimensional probability II. Birkhäuser Boston, Cambridge, pp 443–457
McDiarmid C (1989) On the method of bounded differences. Surv Comb 141(1):148–188
Mohri M, Rostamizadeh A, Talwalkar A (2012) Foundations of machine learning. MIT Press, Cambridge
Mukherjee S, Niyogi P, Poggio T, Rifkin RM (2006) Learning theory: stability is sufficient for generalization and necessary and sufficient for consistency of empirical risk minimization. Adv Comput Math 25(1–3):161–193
Natarajan BK (1989) On learning sets and functions. Mach Learn 4(1):67–97
Sauer N (1972) On the density of families of sets. J Comb Theory - Ser A 13(1):145–147
Shalev-Shwartz S, Shamir O, Srebro N, Sridharan K (2010) Learnability, stability and uniform convergence. J Mach Learn Res 11:2635–2670
Shelah S (1972) A combinatorial problem; stability and order for models and theories in infinitary languages. Pac J Math 41(1):247–261
Valiant LG (1984) A theory of the learnable. Commun ACM 27(11):1134–1142
Vapnik VN, Chervonenkis A (1971) On the uniform convergence of relative frequencies of events to their probabilities. Theory Probab Its Appl 16(2):264–280

# Semi-Supervised Learning

Table of Contents




Z.-H. Zhou, *Machine Learning*,
https://doi.org/10.1007/978-981-15-1967-3_13

## 13.1 Unlabeled Samples

We come to the watermelon field during the harvest season, and the ground is covered with many watermelons. The melon farmer brings a handful of melons and says that they are all ripe melons, and then points at a few melons in the ground and says that these are not ripe, and they would take a few more days to grow up. Based on this information, can we build a model to determine which melons in the field are ripe for picking? For sure, we can use the ripe and unripe watermelons told by the farmers as positive and negative samples to train a classifier. However, is it too few to use only a handful of melons as training samples? Can we use all the watermelons in the field as well?

Formally, we have a data set $D_l = \{(\boldsymbol{x}_1, y_1), (\boldsymbol{x}_2, y_2), \ldots, (\boldsymbol{x}_l, y_l)\}$ with $l$ labeled samples, where the labels, ripe or unripe, are known. Besides, we have another data set $D_u = \{\boldsymbol{x}_{l+1}, -\boldsymbol{x}_{l+2}, \ldots, \boldsymbol{x}_{l+u}\}$ containing $u$ unlabeled samples without labels, where $l \ll u$. If we use traditional supervised learning methods, then we can only use $D_l$ and have to discard all information in $D_u$. In this case, the generalization ability of a learned model may not be satisfactory when $D_l$ is small. So, can $D_u$ be used in the process of building the model?

A straightforward approach is to label all samples in $D_u$ before learning, that is, we ask the farmers to check all watermelons in the field. Other than this labor-intensive approach, is there a "cheaper" way?

One alternative is that we can use $D_l$ to train a model first, take this model to pick a watermelon in the field, ask the farmer whether it is ripe, then add this newly obtained labeled sample to $D_l$ to retrain a model, and then pick a melon again .... Since we only select the most useful watermelons to improve the performance of the model, the labeling cost is greatly reduced. This kind of learning paradigm is called *active learning*, and its goal is to use as few *query* as possible to get the best performance, that is, to ask the farmer as few times as possible.

For example, we can train an SVM model with $D_l$ and select the unlabeled sample that is closest to the classification hyperplane for query.

Active learning introduces additional expert knowledge and interactions to transform some unlabeled samples into labeled samples, but can we use unlabeled samples to improve the generalization performance without interacting with experts? Though it sounds unrealistic, the answer is "yes"!

Although the unlabeled samples do not contain label information, they contain information about the data distribution,

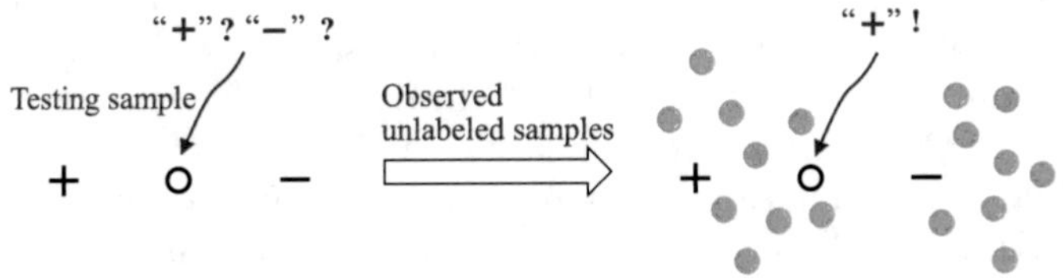


**Fig. 13.1** An illustration of utilizing unlabeled samples. The gray dots on the right-hand side are unlabeled samples

which can benefit the modeling if the samples are independently drawn from the same data distribution as the labeled samples. ◘ Figure 13.1 gives an intuitive illustration: when the testing sample lies between a positive sample and a negative sample, we can hardly tell which class it belongs to; however, if we can observe the unlabeled samples, we are confident that it belongs to the positive class.

When a learner can automatically improve its performance by learning from unlabeled samples without external interactions, it is called *semi-supervised learning*. Semi-supervised learning is highly demanded by real-world applications, since in practice we can easily collect a large number of unlabeled samples, but "labeling" them is costly. For example, in computer-aided medical image analysis, we can obtain a large number of medical images, whereas it is impractical to ask medical experts to mark all lesions. The phenomenon of "limited labeled data, abundant unlabeled data" is even more common on the Internet. For example, when building Web page recommendation models, it is generally required to request users to label Web pages that they are interested in. However, users are generally unwilling to spend time labeling Web pages, and therefore, only a few labeled Web pages are available. Nevertheless, there are countless Web pages on the Internet that can be used as unlabeled data. As we will see later, semi-supervised learning provides a solution for utilizing the "cheap" unlabeled samples.

To utilize unlabeled samples, we must make some assumptions to connect the underlying data distribution disclosed by the unlabeled samples to the labels. A common choice is the *clustering assumption*, which assumes that the data contains clustering structures, and the samples within the same cluster belong to the same class. ◘ Figure 13.1 shows an example of the use of the clustering assumption in exploiting unlabeled samples, where the samples to be predicted and the positive samples are brought together by the unlabeled samples, leading to the conclusion that the testing sample is more likely to be classified as positive, compared with the far separated negative samples. Another commonly used assumption is the *manifold*

The concept of *manifold* is the basis of manifold learning. See Sect. 10.5.

As the clustering assumption considers the class labels, it is mostly for classification problems.

*assumption*, which assumes that the samples are distributed on a manifold structure where neighboring samples have similar output values. On the one hand, the degree of "neighboring" is usually described by "similarity", and therefore the manifold assumption can be seen as an extension of the clustering assumption. It is worth noting that the manifold assumption has no limit on the output value, so it can be used for more types of learning tasks. In fact, no matter we use the clustering assumption or manifold assumption, it essentially assumes that "similar samples have similar outputs".

Semi-supervised learning can be further categorized into *pure semi-supervised learning* and *transductive learning*, where the former assumes the unlabeled samples are not the test samples, and the latter assumes the unlabeled samples are exactly the test samples on which the generalization performance is optimized. In other words, pure semi-supervised learning takes an *open-world assumption*, that is, the model we wish to learn can be applied to any unobserved samples. By contrast, transductive learning takes a *closed-world assumption*, that is, the model only predicts the observed unlabeled samples. ◘ Figure 13.2 illustrates the difference between active learning, pure semi-supervised learning, and transductive learning. Pure semi-supervised learning and transductive learning are often jointly called semi-supervised learning, which is the term used in the rest of the book unless otherwise stated.

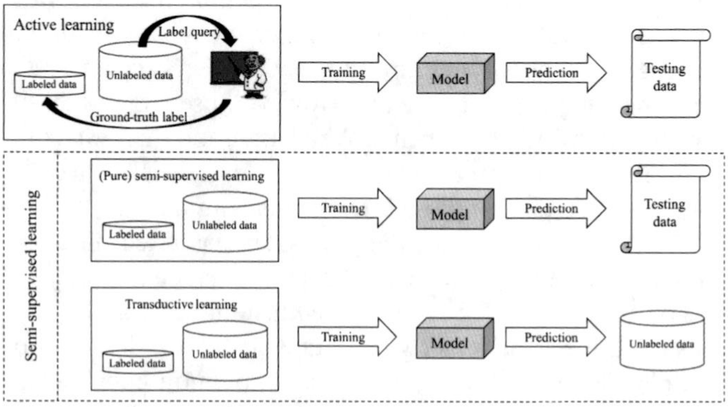


**Fig. 13.2** Active learning, (pure) semi-supervised learning, and transductive learning

13

## 13.2 Generative Methods

*Generative methods*, which are directly based on certain generative models, assume that all samples, labeled or unlabeled, are "generated" from the same underlying model. This assumption enables us to connect the unlabeled samples to the learning objective, where the labels of unlabeled samples are treated as missing parameters that can be estimated with maximum likelihood estimation using the EM algorithm. The main difference among generative methods lies in the underlying assumptions made by their generative models, that is, different assumptions of the underlying model lead to different generative methods.

See Sect. 7.6 for the EM algorithm.

Given a sample $\boldsymbol{x}$ with its ground-truth class label $y \in \mathcal{Y}$, where $\mathcal{Y} = \{1, 2, \ldots, N\}$ is the set of all possible classes. We assume that the sample is generated by a Gaussian mixture model in which each Gaussian mixture component corresponds to a class. In other words, the sample is generated based on the following probability density function:

This assumption implies the one-to-one relationship between mixture components and classes.

$$p(\boldsymbol{x}) = \sum_{i=1}^{N} \alpha_i \cdot p(\boldsymbol{x} \mid \boldsymbol{\mu}_i, \boldsymbol{\Sigma}_i), \tag{13.1}$$

where $\alpha_i \geqslant 0$ is the mixture coefficient and $\sum_{i=1}^{N} \alpha_i = 1$; $p(\boldsymbol{x} \mid \boldsymbol{\mu}_i, \boldsymbol{\Sigma}_i)$ is the probability of the sample $\boldsymbol{x}$ belonging to the $i$th Gaussian mixture component; $\boldsymbol{\mu}_i$ and $\boldsymbol{\Sigma}_i$ are the parameters of the $i$th Gaussian mixture component.

See Sect. 9.4 for Gaussian mixture models.

Let $f(\boldsymbol{x}) \in \mathcal{Y}$ denote the predicted label of $\boldsymbol{x}$, and $\Theta \in \{1, 2, \ldots, N\}$ denote the Gaussian mixture components that $\boldsymbol{x}$ belongs to. Maximizing the posterior probability, we have

$$\begin{aligned} f(\boldsymbol{x}) &= \underset{j \in \mathcal{Y}}{\arg\max}\, p(y = j \mid \boldsymbol{x}) \\ &= \underset{j \in \mathcal{Y}}{\arg\max} \sum_{i=1}^{N} p(y = j, \Theta = i \mid \boldsymbol{x}) \\ &= \underset{j \in \mathcal{Y}}{\arg\max} \sum_{i=1}^{N} p(y = j \mid \Theta = i, \boldsymbol{x}) \cdot p(\Theta = i \mid \boldsymbol{x}), \end{aligned} \tag{13.2}$$

where

$$p(\Theta = i \mid \boldsymbol{x}) = \frac{\alpha_i \cdot p(\boldsymbol{x} \mid \boldsymbol{\mu}_i, \boldsymbol{\Sigma}_i)}{\sum_{i=1}^{N} \alpha_i \cdot p(\boldsymbol{x} \mid \boldsymbol{\mu}_i, \boldsymbol{\Sigma}_i)} \tag{13.3}$$

is the posterior probability that $\boldsymbol{x}$ being generated by the $i$th Gaussian mixture component, where $p(y = j \mid \Theta = i, \boldsymbol{x})$ is the probability of $\boldsymbol{x}$ belonging to class $j$ and is generated by the $i$th Gaussian mixture component. Since we assume that each class corresponds to one Gaussian mixture component, $p(y = j \mid \Theta = i, \boldsymbol{x})$ only depends on the Gaussian mixture component $\Theta$ that $\boldsymbol{x}$ belongs to, which means, we can simply rewrite it as $p(y = j \mid \Theta = i)$ instead. Without loss of generality, we can assume that the $i$th class corresponds to the $i$th Gaussian mixture component, that is, $p(y = j \mid \Theta = i) = 1$ if and only if $i = j$, and $p(y = j \mid \Theta = i) = 0$ otherwise.

We notice that estimating $p(y = j \mid \Theta = i, \boldsymbol{x})$ in (13.2) requires the label of $\boldsymbol{x}$, and hence we can only use the labeled samples. However, $p(\Theta = i \mid \boldsymbol{x})$ does not involve the label, and hence it can be expected to have a more accurate estimation with the help of a large amount of unlabeled samples, that is, the estimation of (13.2) could be more accurate with unlabeled samples. This shows how the unlabeled samples can improve classification performance.

Semi-supervised learning usually assumes that the number of unlabeled samples is much larger than that of labeled samples, though this assumption is not necessary.

Given a labeled data set $D_l = \{(\boldsymbol{x}_1, y_1), (\boldsymbol{x}_2, y_2), \ldots, (\boldsymbol{x}_l, y_l)\}$ and an unlabeled data set $D_u = \{\boldsymbol{x}_{l+1}, \boldsymbol{x}_{l+2}, \ldots, \boldsymbol{x}_{l+u}\}$, where $l \ll u$ and $l + u = m$. Assuming that all samples are *i.i.d.* and are generated by the same Gaussian mixture model. Then, we use maximum likelihood estimation to estimate the parameters $\{(\alpha_i, \boldsymbol{\mu}_i, \boldsymbol{\Sigma}_i) \mid 1 \leqslant i \leqslant N\}$ of the Gaussian mixture model, and the log-likelihood of $D_l \cup D_u$ is formed as

$$LL(D_l \cup D_u) = \sum_{(\boldsymbol{x}_j, y_j) \in D_l} \ln \left( \sum_{i=1}^{N} \alpha_i \cdot p(\boldsymbol{x}_j \mid \boldsymbol{\mu}_i, \boldsymbol{\Sigma}_i) \cdot p(y_i \mid \Theta = i, \boldsymbol{x}_j) \right) + \sum_{\boldsymbol{x}_j \in D_u} \ln \left( \sum_{i=1}^{N} \alpha_i \cdot p(\boldsymbol{x}_j \mid \boldsymbol{\mu}_i, \boldsymbol{\Sigma}_i) \right). \tag{13.4}$$

There are two terms in (13.4): a supervised term based on labeled data set $D_l$ and an unsupervised term based on unlabeled data set $D_u$. The parameter estimation of Gaussian mixture model can be done using the EM algorithm with the following iterative update rules:

See Sect. 9.4 for the EM algorithm on Mixture-of-Gaussian clustering.

Parameters could be initialized by the labeled samples.

- E-step: compute the probabilities of an unlabeled sample $\boldsymbol{x}_j$ belonging to each Gaussian mixture component based on the current model parameters:

$$\gamma_{ji} = \frac{\alpha_i \cdot p(\boldsymbol{x}_j \mid \boldsymbol{\mu}_i, \boldsymbol{\Sigma}_i)}{\sum_{i=1}^{N} \alpha_i \cdot p(\boldsymbol{x}_j \mid \boldsymbol{\mu}_i, \boldsymbol{\Sigma}_i)}; \tag{13.5}$$

- M-step: update the model parameters based on $\gamma_{ji}$, where $l_i$ is the number of labeled samples belonging to the $i$th class:

$$\boldsymbol{\mu}_i = \frac{1}{\sum_{\boldsymbol{x}_j \in D_u} \gamma_{ji} + l_i} \left( \sum_{\boldsymbol{x}_j \in D_u} \gamma_{ji} \boldsymbol{x}_j + \sum_{(\boldsymbol{x}_j, y_j) \in D_l \wedge y_j = i} \boldsymbol{x}_j \right), \tag{13.6}$$

$$\boldsymbol{\Sigma}_i = \frac{1}{\sum_{\boldsymbol{x}_j \in D_u} \gamma_{ji} + l_i} \left( \sum_{\boldsymbol{x}_j \in D_u} \gamma_{ji} (\boldsymbol{x}_j - \boldsymbol{\mu}_i)(\boldsymbol{x}_j - \boldsymbol{\mu}_i)^\top + \sum_{(\boldsymbol{x}_j, y_j) \in D_l \wedge y_j = i} (\boldsymbol{x}_j - \boldsymbol{\mu}_i)(\boldsymbol{x}_j - \boldsymbol{\mu}_i)^\top \right), \tag{13.7}$$

$$\alpha_i = \frac{1}{m} \left( \sum_{\boldsymbol{x}_j \in D_u} \gamma_{ji} + l_i \right). \tag{13.8}$$

By repeating the above process until convergence, we obtain the model parameters for classifying new samples using (13.3) and (13.2).

Other variants of generative methods can be derived by replacing the Gaussian mixture model by other models in the above process, such as mixture of experts models (Miller and Uyar 1997) and naïve Bayes models (Nigam et al. 2000). Generative methods are simple and easy to implement and can outperform other semi-supervised learning methods when the labeled samples are extremely limited. However, such methods heavily rely on accurate model assumptions, that is, the assumed generative model must match the ground-truth data distribution; otherwise, incorporating the unlabeled samples can even reduce the generalization performance (Cozman and Cohen 2002). Unfortunately, it is often difficult to make an accurate model assumption without sufficiently reliable domain knowledge.

## 13.3 Semi-Supervised SVM

Semi-Supervised Support Vector Machine (S3VM) is the extension of SVM for semi-supervised learning. Compared to the standard SVM, which aims to find the separating hyperplane with the maximum margin, S3VM aims to find a separating hyperplane that not only separates the labeled samples but also lies in a low-density region of all samples. As illustrated in ◘ Figure 13.3, the assumption here is *low-density separation*, which is an extension of the clustering assumption under linear separating hyperplanes.

See Chap. 6 for SVM.

A well-known S3VM is Transductive Support Vector Machine (TSVM) (Joachims 1999), which is designed for binary classification problems. TSVM considers all possible label

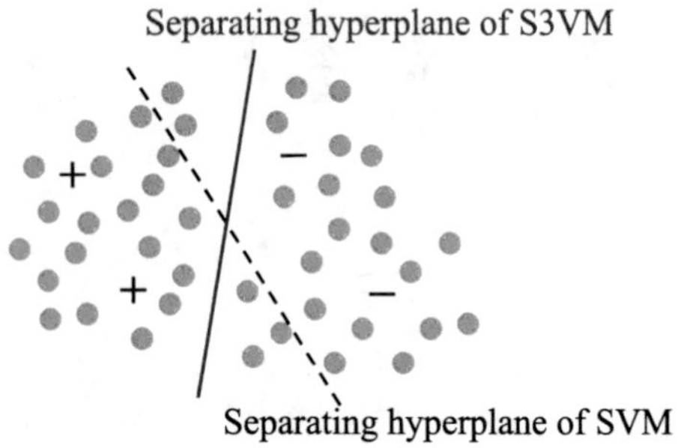


**Fig. 13.3** S3VM and low-density separation ("+" and "−" are, respectively, the labeled positive samples and the labeled negative samples; gray dots are the unlabeled samples)

assignments of unlabeled samples, that is, it temporarily treats each unlabeled sample as a positive or negative sample during optimization. By examining unlabeled samples with all possible label assignments, TSVM aims to find a separating hyperplane that maximizes the margin for both labeled samples and unlabeled samples with label assignment. Once the separating hyperplane is determined, the final label assignment for the unlabeled sample is its prediction.

Formally, given a labeled data set $D_l = \{(\boldsymbol{x}_1, y_1), (\boldsymbol{x}_2, y_2), \ldots, (\boldsymbol{x}_l, y_l)\}$ and an unlabeled data set $D_u = \{\boldsymbol{x}_{l+1}, \boldsymbol{x}_{l+2}, \ldots, \boldsymbol{x}_{l+u}\}$, where $y_i \in \{-1, +1\}$, $l \ll u$, and $l + u = m$. The learning objective of TSVM is to predict the labels $\hat{\mathbf{y}} = (\hat{y}_{l+1}, \hat{y}_{l+2}, \ldots, \hat{y}_{l+u})$, $\hat{y}_i \in \{-1, +1\}$, for all samples in $D_u$ such that

13

$$\begin{aligned}
\min_{\mathbf{w}, b, \hat{\mathbf{y}}, \boldsymbol{\xi}} \quad & \frac{1}{2} \|\mathbf{w}\|_2^2 + C_l \sum_{i=1}^{l} \xi_i + C_u \sum_{i=l+1}^{m} \xi_i \\
\text{s.t.} \quad & y_i(\mathbf{w}^\top \boldsymbol{x}_i + b) \geqslant 1 - \xi_i, \ i = 1, 2, \ldots, l, \\
& \hat{y}_i(\mathbf{w}^\top \boldsymbol{x}_i + b) \geqslant 1 - \xi_i, \ i = l+1, l+2, \ldots, m, \\
& \xi_i \geqslant 0, \ i = 1, 2, \ldots, m,
\end{aligned} \tag{13.9}$$

where $(\mathbf{w}, b)$ determines the separating hyperplane; $\boldsymbol{\xi}$ is the slack vector, where $\xi_i$ $(i = 1, 2, \ldots, l)$ corresponds to the loss of each labeled sample, and $\xi_i$ $(i = l+1, l+2, \ldots, m)$ corresponds to the loss of each unlabeled sample; $C_l$ and $C_u$ are user-specified trade-off parameters to balance the importance of model complexity, labeled samples, and unlabeled samples.

Trying out all possible label assignments of unlabeled samples is an exhaustive process that is only feasible for a small number of unlabeled samples. In practice, more efficient optimization strategies are usually needed.

TSVM takes an iterative local search strategy to find an approximate solution of (13.9). To be specific, it first learns an SVM from the labeled samples by ignoring any terms or constraints in (13.9) that involving $C_u$ or $\hat{\boldsymbol{y}}$. Then, the learned SVM produces the label assignment by predicting all unlabeled samples, where the predictions are called *pseudo-labels*. Substituting the obtained $\hat{\boldsymbol{y}}$ into (13.9) gives a standard SVM learning problem, from which we obtain an updated separating hyperplane and an updated slack vector. Since the pseudo-labels may not be accurate yet, we set $C_u$ to a value smaller than $C_l$ such that the labeled samples have larger weights than the unlabeled samples. After that, TSVM finds two unlabeled samples that have different label assignments and are likely to be incorrectly labeled. Then, by swapping the labels of these two samples, we can obtain the updated separating hyperplane and slack vector from (13.9) and repeat this process. After each iteration, we increase the influence of unlabeled samples by increasing $C_u$ gradually, and the iteration repeats until $C_u = C_l$. Once converged, the obtained SVM not only gives labels of unlabeled samples but can also predict unobserved samples. The pseudocode of TSVM is given in ◘ Algorithm 13.1.

**Algorithm 13.1** Transductive support vector machine

**Input:** Labeled data set $D_l = \{(\boldsymbol{x}_1, y_1), (\boldsymbol{x}_2, y_2), \ldots, (\boldsymbol{x}_l, y_l)\}$;
Unlabeled data set $D_u = \{\boldsymbol{x}_{l+1}, \boldsymbol{x}_{l+2}, \ldots, \boldsymbol{x}_{l+u}\}$;
Trade-off parameters $C_l$ and $C_u$.

**Process:**

1: Train a SVM$_l$ using $D_l$;
2: Use SVM$_l$ to predict the samples in $D_u$ to get $\hat{\boldsymbol{y}} = (\hat{y}_{l+1}, \hat{y}_{l+2}, \ldots, \hat{y}_{l+u})$;
3: Initialize $C_u \ll C_l$;
4: **while** $C_u < C_l$ **do**
5: Solve (13.9) using $D_l$, $D_u$, $\hat{\boldsymbol{y}}$, $C_l$, $C_u$ to get $(\mathbf{w}, b)$ and $\boldsymbol{\xi}$;
6: **while** $\exists\{i, j \mid (\hat{y}_i\hat{y}_j < 0) \wedge (\xi_i > 0) \wedge (\xi_j > 0) \wedge (\xi_i + \xi_j > 2)\}$ **do**
7: $\hat{y}_i = -\hat{y}_i$;
8: $\hat{y}_j = -\hat{y}_j$;
9: Solve (13.9) again using $D_l$, $D_u$, $\hat{\boldsymbol{y}}$, $C_l$, $C_u$ to get $(\mathbf{w}, b)$ and $\boldsymbol{\xi}$;
10: **end while**
11: $C_u = \min\{2C_u, C_l\}$.
12: **end while**

**Output:** Predictions of the unlabeled samples: $\hat{\boldsymbol{y}} = (\hat{y}_{l+1}, \hat{y}_{l+2}, \ldots, \hat{y}_{l+u})$.

$\hat{\boldsymbol{y}}$ is known at this point.

Adjust $\hat{y}_i$ and $\hat{y}_j$.

Increase the influence of unlabeled samples.

The learning process of TSVM may suffer from the class imbalance problem, in which the number of samples in one class is far more than the number of samples in the other class. To alleviate this problem, we can slightly adjust ◘ Algorithm 13.1. Specifically, we split the term $C_u$ in the optimization objective to $C_u^+$ and $C_u^-$, which correspond to unla-

See Sect. 3.6 for class imbalance problem and the motivation of (13.10).

beled samples with positive and negative pseudo-labels, respectively. Then, TSVM is initialized with

$$C_u^+ = \frac{u_-}{u_+} C_u^-, \tag{13.10}$$

where $u_+$ and $u_-$ are the number of unlabeled samples with positive and negative pseudo-labels, respectively.

In lines 6–10 of ■ Algorithm 13.1, the label assignments $\hat{y}_i$ and $\hat{y}_j$ of two unlabeled samples $\boldsymbol{x}_i$ and $\boldsymbol{x}_j$ are likely to be incorrect if $\hat{y}_i$ and $\hat{y}_j$ are different and the corresponding slack variables satisfy $\xi_i + \xi_j > 2$; in such a case, the values of $\hat{y}_i$ and $\hat{y}_j$ need to be swapped, and (13.9) is optimized again such that the objective value of (13.9) decreases after each iteration.

See Joachims (1999) for the proof of convergence

The iterative updating process is an expensive large-scale optimization problem because the algorithm iterates over each pair of unlabeled samples that are likely to be incorrectly labeled. Hence, one research focus of S3VM is the design of efficient optimization strategies, such as LDS (Chapelle and Zien 2005), which uses gradient descent on graph kernel functions, and meanS3VM (Li et al. 2009), which uses label mean estimation.

## 13.4 Graph-Based Semi-Supervised Learning

Training data sets can often be encoded into a graph structure, in which each node corresponds to a sample, and an edge connects two nodes if their corresponding samples are highly similar (or correlated). The *strength* of an edge indicates the degree of similarity (or correlation). Imagine that the nodes of labeled samples are colored, and the nodes of unlabeled samples are uncolored. Then, semi-supervised learning is like the process of spreading the "color" over the graph. Since a graph corresponds to a matrix, we can derive and analyze semi-supervised learning algorithms via matrix operations.



Given a labeled data set $D_l = \{(\boldsymbol{x}_1, y_1), (\boldsymbol{x}_2, y_2), \ldots, (\boldsymbol{x}_l, y_l)\}$ and an unlabeled data set $D_u = \{\boldsymbol{x}_{l+1}, \boldsymbol{x}_{l+2}, \ldots, \boldsymbol{x}_{l+u}\}$, where $l \ll u$ and $l + u = m$. We start by constructing a graph $G = (V, E)$ from $D_l \cup D_u$, where $V = \{\boldsymbol{x}_1, \ldots, \boldsymbol{x}_l, \boldsymbol{x}_{l+1}, \ldots, \boldsymbol{x}_{l+u}\}$ is the node set and $E$ is the edge set; the edge set $E$ can be represented as an *affinity matrix* which is often defined based on the Gaussian function:

$$(\mathbf{W})_{ij} = \begin{cases} \exp\left(\frac{-\|\boldsymbol{x}_i - \boldsymbol{x}_j\|_2^2}{2\sigma^2}\right), & \text{if } i \neq j; \\ 0, & \text{otherwise}, \end{cases} \tag{13.11}$$

where $i, j \in \{1, 2, \ldots, m\}$, and $\sigma > 0$ is a user-specified bandwidth parameter of Gaussian function.

From graph $G = (V, E)$, we aim to learn a real-valued function $f : V \to \mathbb{R}$ with a classification rule $y_i = \text{sign}(f(\boldsymbol{x}_i))$, where $y_i \in \{-1, +1\}$. Intuitively, similar samples share similar labels, and hence we can define an *energy function* (Zhu et al. 2003) with respect to $f$:

Minimizing the energy function gives the optimal result.

$$\begin{aligned} E(f) &= \frac{1}{2}\sum_{i=1}^{m}\sum_{j=1}^{m}(\mathbf{W})_{ij}(f(\boldsymbol{x}_i) - f(\boldsymbol{x}_j))^2 \\ &= \frac{1}{2}\left(\sum_{i=1}^{m} d_i f^2(\boldsymbol{x}_i) + \sum_{j=1}^{m} d_j f^2(\boldsymbol{x}_j) - 2\sum_{i=1}^{m}\sum_{j=1}^{m}(\mathbf{W})_{ij} f(\boldsymbol{x}_i) f(\boldsymbol{x}_j)\right) \\ &= \sum_{i=1}^{m} d_i f^2(\boldsymbol{x}_i) - \sum_{i=1}^{m}\sum_{j=1}^{m}(\mathbf{W})_{ij} f(\boldsymbol{x}_i) f(\boldsymbol{x}_j) \\ &= \boldsymbol{f}^\top(\mathbf{D} - \mathbf{W})\boldsymbol{f}, \end{aligned} \tag{13.12}$$

where $\boldsymbol{f} = (\boldsymbol{f}_l; \boldsymbol{f}_u)$, and $\boldsymbol{f}_l = (f(\boldsymbol{x}_1); f(\boldsymbol{x}_2); \ldots; f(\boldsymbol{x}_l))$ and $\boldsymbol{f}_u = (f(\boldsymbol{x}_{l+1}); f(\boldsymbol{x}_{l+2}); \ldots; f(\boldsymbol{x}_{l+u}))$ are the predictions of function $f$ on the labeled and unlabeled samples, respectively; $\mathbf{D} = \text{diag}(d_1, d_2, \ldots, d_{l+u})$ is a diagonal matrix in which the diagonal element $d_i = \sum_{j=1}^{l+u}(\mathbf{W})_{ij}$ is the sum of the $i$th row of matrix $\mathbf{W}$.

Since $\mathbf{W}$ is a symmetric matrix, $d_i$ is also the sum of the $i$th column.

A function $f$ with the minimal energy should satisfy $f(\boldsymbol{x}_i) = y_i$ $(i = 1, 2, \ldots, l)$ for labeled samples and satisfy $\boldsymbol{\Delta} \boldsymbol{f} = \mathbf{0}$ for unlabeled samples, where $\boldsymbol{\Delta} = \mathbf{D} - \mathbf{W}$ is a *Laplacian matrix*. For $\mathbf{W}$ and $\mathbf{D}$, we partition the matrix by the $l$th row and the $l$th column and group the elements into four blocks: $\mathbf{W} = \begin{bmatrix} \mathbf{W}_{ll} & \mathbf{W}_{lu} \\ \mathbf{W}_{ul} & \mathbf{W}_{uu} \end{bmatrix}$ and $\mathbf{D} = \begin{bmatrix} \mathbf{D}_{ll} & \mathbf{0}_{lu} \\ \mathbf{0}_{ul} & \mathbf{D}_{uu} \end{bmatrix}$. Then, (13.12) can be rewritten as

$$E(f) = (\boldsymbol{f}_l^\top \ \boldsymbol{f}_u^\top)\left(\begin{bmatrix} \mathbf{D}_{ll} & \mathbf{0}_{lu} \\ \mathbf{0}_{ul} & \mathbf{D}_{uu} \end{bmatrix} - \begin{bmatrix} \mathbf{W}_{ll} & \mathbf{W}_{lu} \\ \mathbf{W}_{ul} & \mathbf{W}_{uu} \end{bmatrix}\right)\begin{bmatrix} \boldsymbol{f}_l \\ \boldsymbol{f}_u \end{bmatrix} \tag{13.13}$$

$$= \boldsymbol{f}_l^\top(\mathbf{D}_{ll} - \mathbf{W}_{ll})\boldsymbol{f}_l - 2\boldsymbol{f}_u^\top \mathbf{W}_{ul}\boldsymbol{f}_l + \boldsymbol{f}_u^\top(\mathbf{D}_{uu} - \mathbf{W}_{uu})\boldsymbol{f}_u. \tag{13.14}$$

By setting $\frac{\partial E(f)}{\partial \boldsymbol{f}_u} = \mathbf{0}$, we have

$$\boldsymbol{f}_u = (\mathbf{D}_{uu} - \mathbf{W}_{uu})^{-1}\mathbf{W}_{ul}\boldsymbol{f}_l. \tag{13.15}$$

Let

$$\mathbf{P} = \mathbf{D}^{-1}\mathbf{W} = \begin{bmatrix} \mathbf{D}_{ll}^{-1} & \mathbf{0}_{lu} \\ \mathbf{0}_{ul} & \mathbf{D}_{uu}^{-1} \end{bmatrix} \begin{bmatrix} \mathbf{W}_{ll} & \mathbf{W}_{lu} \\ \mathbf{W}_{ul} & \mathbf{W}_{uu} \end{bmatrix}$$
$$= \begin{bmatrix} \mathbf{D}_{ll}^{-1}\mathbf{W}_{ll} & \mathbf{D}_{ll}^{-1}\mathbf{W}_{lu} \\ \mathbf{D}_{uu}^{-1}\mathbf{W}_{ul} & \mathbf{D}_{uu}^{-1}\mathbf{W}_{uu} \end{bmatrix}, \tag{13.16}$$

we have $\mathbf{P}_{uu} = \mathbf{D}_{uu}^{-1}\mathbf{W}_{uu}$ and $\mathbf{P}_{ul} = \mathbf{D}_{uu}^{-1}\mathbf{W}_{ul}$. Then, (13.15) can be rewritten as

$$\begin{aligned} \boldsymbol{f}_u &= (\mathbf{D}_{uu}(\mathbf{I} - \mathbf{D}_{uu}^{-1}\mathbf{W}_{uu}))^{-1}\mathbf{W}_{ul}\boldsymbol{f}_l \\ &= (\mathbf{I} - \mathbf{D}_{uu}^{-1}\mathbf{W}_{uu})^{-1}\mathbf{D}_{uu}^{-1}\mathbf{W}_{ul}\boldsymbol{f}_l \\ &= (\mathbf{I} - \mathbf{P}_{uu})^{-1}\mathbf{P}_{ul}\boldsymbol{f}_l. \end{aligned} \tag{13.17}$$

By using the label information of $D_l$ as $\boldsymbol{f}_l = (y_1; y_2; \ldots; y_l)$ and substituting it into (13.17), we have $\boldsymbol{f}_u$, which is then used to predict the unlabeled samples.

The above process is a *label propagation* method for binary classification. Next, we introduce another label propagation method (Zhou et al. 2004) that is applicable to multiclass classification.

Suppose $y_i \in \mathcal{Y}$. We start by constructing a graph $G = (V, E)$ from $D_l \cup D_u$, where $V = \{\boldsymbol{x}_1, \ldots, \boldsymbol{x}_l, \ldots, \boldsymbol{x}_{l+u}\}$ is the node set and $E$ is the edge set. The weight matrix $\mathbf{W}$ of the edge set $E$ is the same as (13.11). $\mathbf{D} = \mathrm{diag}(d_1, d_2, \ldots, d_{l+u})$ is a diagonal matrix in which the diagonal element $d_i = \sum_{j=1}^{l+u}(\mathbf{W})_{ij}$ is the sum of the $i$th row of $\mathbf{W}$. We define a $(l+u) \times |\mathcal{Y}|$ non-negative label matrix $\mathbf{F} = (\mathbf{F}_1^{\top}, \mathbf{F}_2^{\top}, \ldots, \mathbf{F}_{l+u}^{\top})^{\top}$, where the $i$th row $\mathbf{F}_i = ((\mathbf{F})_{i1}, (\mathbf{F})_{i2}, \ldots, (\mathbf{F})_{i|\mathcal{Y}|})$ is the label vector of sample $\boldsymbol{x}_i$, and the classification rule is $y_i = \arg\max_{1 \leqslant j \leqslant |\mathcal{Y}|}(\mathbf{F})_{ij}$.

For $i = 1, 2, \ldots, m$ and $j = 1, 2, \ldots, |\mathcal{Y}|$, we initialize $\mathbf{F}$ as

13

$$\mathbf{F}(0) = (\mathbf{Y})_{ij} = \begin{cases} 1, & \text{if } (1 \leqslant i \leqslant l) \wedge (y_i = j); \\ 0, & \text{otherwise}, \end{cases} \tag{13.18}$$

where the first $l$ rows of $\mathbf{Y}$ refer to the label vectors of the $l$ labeled samples.

We construct a label propagation matrix $\mathbf{S} = \mathbf{D}^{-\frac{1}{2}}\mathbf{W}\mathbf{D}^{-\frac{1}{2}}$ based on $\mathbf{W}$, where $\mathbf{D}^{-\frac{1}{2}} = \mathrm{diag}\left(\frac{1}{\sqrt{d_1}}, \frac{1}{\sqrt{d_2}}, \ldots, \frac{1}{\sqrt{d_{l+u}}}\right)$. Then, we have the iterative formula

$$\mathbf{F}(t+1) = \alpha\mathbf{S}\mathbf{F}(t) + (1-\alpha)\mathbf{Y}, \tag{13.19}$$

where $\alpha \in (0, 1)$ is a user-specified parameter for balancing the importance of the label propagation term $\mathbf{S}\mathbf{F}(t)$ and the initialization term $\mathbf{Y}$. Iterating (13.19) until convergence, we have

$$\mathbf{F}^* = \lim_{t\to\infty} \mathbf{F}(t) = (1-\alpha)(\mathbf{I}-\alpha\mathbf{S})^{-1}\mathbf{Y}, \tag{13.20}$$

which can be used to obtain the labels $(\hat{y}_{l+1}, \hat{y}_{l+2}, \ldots, \hat{y}_{l+u})$ of unlabeled samples. The pseudocode of the algorithm is given in ▪ Algorithm 13.2.

---

**Algorithm 13.2** Iterative label propagation

**Input:** Labeled data set $D_l = \{(\boldsymbol{x}_1, y_1), (\boldsymbol{x}_2, y_2), \ldots, (\boldsymbol{x}_l, y_l)\}$;
Unlabeled data set $D_u = \{\boldsymbol{x}_{l+1}, \boldsymbol{x}_{l+2}, \ldots, \boldsymbol{x}_{l+u}\}$;
Graph construction parameter $\sigma$;
Trade-off parameter $\alpha$.

**Process:**
1: Use (13.11) and parameter $\sigma$ to obtain $\mathbf{W}$;
2: Use $\mathbf{W}$ to construct the label propagation matrix $\mathbf{S} = \mathbf{D}^{-\frac{1}{2}}\mathbf{W}\mathbf{D}^{-\frac{1}{2}}$;
3: Initialize $\mathbf{F}(0)$ according to (13.18);
4: $t = 0$;
5: **repeat**
6: $\quad\mathbf{F}(t+1) = \alpha\mathbf{S}\mathbf{F}(t) + (1-\alpha)\mathbf{Y}$;
7: $\quad t = t+1$;
8: **until** Converged to $\mathbf{F}^*$
9: **for** $i = l+1, l+2, \ldots, l+u$ **do**
10: $\quad\hat{y}_i = \arg\max_{1\leqslant j\leqslant|\mathcal{Y}|}(\mathbf{F}^*)_{ij}$.
11: **end for**

**Output:** Predictions of the unlabeled samples: $\hat{\boldsymbol{y}} = (\hat{y}_{l+1}, \hat{y}_{l+2}, \ldots, \hat{y}_{l+u})$.

---

Note that ▪ Algorithm 13.2 corresponds to the following regularization framework:

$$\min_{\mathbf{F}} \frac{1}{2}\left(\sum_{i,j=1}^{l+u}(\mathbf{W})_{ij}\left\|\frac{1}{\sqrt{d_i}}\mathbf{F}_i - \frac{1}{\sqrt{d_j}}\mathbf{F}_j\right\|^2\right) + \mu\sum_{i=1}^{l}\|\mathbf{F}_i - \mathbf{Y}_i\|^2, \tag{13.21}$$

where $\mu > 0$ is the regularization parameter. Considering that the labeled samples are limited while there are many unlabeled samples, we can introduce the $\mathrm{L}_2$ regularization term $\mu\sum_{i=l+1}^{l+u}\|\mathbf{F}_i\|^2$ of unlabeled samples to alleviate the overfitting problem. When $\mu = \frac{1-\alpha}{\alpha}$, the optimal solution of (13.21) equals to the converged solution $\mathbf{F}^*$ of ▪ Algorithm 13.2. See Sect. 11.4.

The second term of (13.21) forces the predicted labels of labeled samples to be similar to the ground-truth labels, and the first term forces nearby samples to have similar labels, that is, the basic semi-supervised learning assumption that we have exploited for (13.12). The only difference is that (13.21) works for discrete class labels, whereas (13.12) considers continuous output.

The concepts in graph-based semi-supervised learning methods are well-defined, and the algorithms' properties can be easily analyzed from the matrix operations. However, such methods also have some deficiencies. Firstly, the memory cost can be very high since the calculations involve matrics of size $O(m^2)$ for a sample size of $O(m)$, that is, it is difficult to apply such algorithms to large-scale problems. Secondly, the graphs are constructed from the training samples, making it challenging to identify the positions of new samples. Hence, with new samples arrived, we need to re-construct the graph and redo the label propagation, or introduce an extra prediction mechanism, such as joining $D_l$ and $D_u$ (labeled by label propagation) to train a classifier (e.g., SVM) for predicting new samples.

## 13.5 Disagreement-Based Methods

Unlike the previously introduced methods that utilize unlabeled samples via a single learner, *disagreement-based methods* train multiple learners and the *disagreement* between learners is crucial for the exploitation of unlabeled data.

Disagreement is also called diversity.

*Co-training* (Blum and Mitchell 1998) is a representative disagreement-based method. Since it was originally designed for *multi-view* data, it is also a representative method for *multi-view learning*. Before introducing co-training, let us first take a look at the multi-view data.

In many real-world applications, a data object can have multiple *attribute sets*, where each attribute set provides a *view* of the data object. For example, a movie can have different attribute sets such as the attribute set about the visual images, the attribute set about the audio, the attribute set about the subtitles, and the attribute set about the online reviews. For ease of discussion, we consider only the visual and audio attribute sets. Then, a movie clip can be represented as a sample $(\langle \boldsymbol{x}^1, \boldsymbol{x}^2\rangle, y)$, where $\boldsymbol{x}^i$ is the feature vector of the sample in view $i$. Let $\boldsymbol{x}^1$ denote the feature vector in the visual view, and $\boldsymbol{x}^2$ denote the feature vector in the audio view. Let $y$ denote the label, indicating the genre of movies, such as action and romance. Then, data in the form of $(\langle \boldsymbol{x}^1, \boldsymbol{x}^2\rangle, y)$ is called multi-view data.

Suppose that different views are *compatible*, that is, they share the same output space $\mathcal{Y}$. Let $\mathcal{Y}^1$ denote the label space for visual information, and $\mathcal{Y}^2$ denote the label space for audio information. Then, we have $\mathcal{Y} = \mathcal{Y}^1 = \mathcal{Y}^2$. For example, both label spaces must be {action, romance} rather than $\mathcal{Y}^1$ = {action, romance} and $\mathcal{Y}^2$ = {literary, horror}. Under the above assumption, there are many benefits of explicitly considering multiple views. For example, it is hard to tell the genre of a movie when we only see two people looking at each other visually. However, it is likely to be a romance movie if we also hear the phrase "I love you" from the audio. On the other hand, if both the visual and the audio information suggest action, then putting them together, it is very likely to be an action movie. Under the assumption of *compatibility*, the above example shows that the *complementarity* between different views is useful for constructing learners.

Co-training can exploit the *compatibility* and *complementarity* of multi-view data. Suppose that the data has two sufficient and conditionally independent views, where "sufficient" means that every view contains enough information to produce an optimal learner, and "conditionally independent" means that two views are independent given the class labels. Under this setting, there is a simple approach to utilize unlabeled samples. We first use the labeled samples to train a classifier on each view and then use each classifier to predict the pseudo-labels of the unlabeled samples that the classifier is most confident with. After that, we supply the unlabeled samples with pseudo-labels to the other classifiers as labeled samples for retraining .... Such a "learning from each other and progressing together" process repeats until convergence, that is, no more changes in both classifiers or the predefined maximum number of iterations is reached. The pseudocode of the algorithm is given in ◘ Algorithm 13.3. The computational cost would be very high if we consider the classification confidence of all unlabeled samples in each round. Therefore, Blum and Mitchell (1998) created a pool of unlabeled samples that are drawn and replenished in each round. The estimation of classification confidence depends on the base learning algorithm $\mathfrak{L}$. For example, it can be obtained from the posterior probability of naïve Bayes classifier or the margin in SVM.

Though co-training looks simple, theoretical studies (Blum and Mitchell 1998) showed that the generalization performance of weak classifiers can be boosted to arbitrary high using unlabeled samples given that the views are sufficient and conditionally independent. In practice, however, since the views rarely hold conditional independence, the improvement in performance is often not that significant as in theory. Notably, the performance of weak learners can still be considerably improved even with weaker conditions (Zhou and Li 2010).

See Chap. 8 for weak classifiers.

For example, the visual and audio of a movie are not conditionally independent.

Co-training was originally designed for multi-view data, but some variants have been developed for single-view data. In these variants, different learners are realized by different algorithms (Goldman and Zhou 2000), different data sampling (Zhou and Li 2005b), or even different parameter settings (Zhou and Li 2005a). The performance can still be improved in these variants by utilizing unlabeled samples. Later on, theoretical studies revealed that multi-view data is unnecessary

**Algorithm 13.3** Co-training

The superscript of $\boldsymbol{x}_i$ indicates the two views rather than ordinal relationships, that is, $\langle \boldsymbol{x}_1^1, \boldsymbol{x}_1^2 \rangle$ and $\langle \boldsymbol{x}_1^2, \boldsymbol{x}_1^1 \rangle$ refer to the same sample.
Let $p, n \ll s$.

**Input:** Labeled data set $D_l = \{(\langle \boldsymbol{x}_1^1, \boldsymbol{x}_1^2 \rangle, y_1), \ldots, (\langle \boldsymbol{x}_l^1, \boldsymbol{x}_l^2 \rangle, y_l)\}$;
Unlabeled data set $D_u = \{\langle \boldsymbol{x}_{l+1}^1, \boldsymbol{x}_{l+2}^2 \rangle, \ldots, \langle \boldsymbol{x}_{l+u}^1, \boldsymbol{x}_{l+u}^2 \rangle\}$;
Buffer pool size $s$;
Number of positive samples in each round $p$;
Number of negative samples in each round $n$;
Base learning algorithm $\mathfrak{L}$;
Number of learning rounds $T$.

**Process:**

Construct buffer pool $D_s$ by randomly selecting $s$ samples from $D_u$;
$D_u = D_u \backslash D_s$;
Initialize the labeled training set for each view.
**for** $j = 1, 2$ **do**
  $D_l^j = \{(\boldsymbol{x}_i^j, y_i) \mid (\langle \boldsymbol{x}_i^j, \boldsymbol{x}_i^{3-j} \rangle, y_i) \in D_l\}$;
**end for**
**for** $t = 1, 2, \ldots, T$ **do**
  **for** $j = 1, 2$ **do**
    Train $h_j$ on the view $j$ using the labeled samples.
    $h_j \leftarrow \mathfrak{L}(D_l^j)$;
    Check the classification confidences of $h_j$ on $D_s^j = \{\boldsymbol{x}_i^j \mid \langle \boldsymbol{x}_i^j, \boldsymbol{x}_i^{3-j} \rangle \in D_s\}$, and select $p$ samples $D_p \subset D_s$ with the highest confidence of being positive and $n$ samples $D_n \subset D_s$ with the highest confidence of being negative;
    Use $D_p^j$ to generate pseudo-labeled positive samples $\tilde{D}_p^{3-j} = \{(\boldsymbol{x}_i^{3-j}, +1) \mid \boldsymbol{x}_i^j \in D_p^j\}$;
    Use $D_n^j$ to generate pseudo-labeled negative samples $\tilde{D}_n^{3-j} = \{(\boldsymbol{x}_i^{3-j}, -1) \mid \boldsymbol{x}_i^j \in D_n^j\}$;
    $D_s = D_s \backslash (D_p \cup D_n)$;
  **end for**
  **if** $h_1$ and $h_2$ are unchanged **then**
    **break**
  **else**
    **for** $j = 1, 2$ **do**
      Extend the labeled data set.
      $D_l^j = D_l^j \cup \left(\tilde{D}_p^j \cup \tilde{D}_n^j\right)$;
    **end for**
    Randomly select $2p + 2n$ samples from $D_u$ and add them to $D_s$.
  **end if**
**end for**

**Output:** Classifiers $h_1$ and $h_2$.



for such algorithms; instead, as long as the weak learners have a significant disagreement (or difference), exchanging pseudo-labels can improve generalization performance (Zhou and Li 2010). In other words, different views, different algorithms, different data sampling, and different parameter settings are just possible ways for generating differences, not necessary conditions.

Hence, such methods are called *disagreement-based methods*.

By choosing the suitable base learners, disagreement-based methods are less affected by model assumptions, non-convexity of loss function, and data scale problems. These simple and effective methods have a solid theoretical foundation and a broad scope of applications. To use them, we need to train multiple learners with significant differences and acceptable performance, but when there are very limited labeled samples, especially when the data does not have multiple views, it is not easy to do without ingenious design.

## 13.6 Semi-Supervised Clustering

Clustering is a typical unsupervised learning problem in which we do not rely on supervision information. In practice, however, we often have access to some additional supervision information, which can be utilized to obtain better clustering results via *semi-supervised clustering*.

Typically, there are two kinds of supervision information in clustering tasks. The first kind is *must-link* and *cannot-link* constraints, where must-link means the samples must belong to the same cluster, and cannot-link means the samples must not belong to the same cluster. The second kind of supervision information refers to a small number of labeled samples.

See Sect. 10.6.

Constrained $k$-means (Wagstaff et al. 2001) is a representative semi-supervised clustering algorithm that utilizes supervision information in the form of must-link and cannot-link constraints. Given a data set $D = \{\boldsymbol{x}_1, \boldsymbol{x}_2, \ldots, \boldsymbol{x}_m\}$, a must-link set $\mathcal{M}$, and a cannot-link set $\mathcal{C}$, where $(\boldsymbol{x}_i, \boldsymbol{x}_j) \in \mathcal{M}$ means $\boldsymbol{x}_i$ and $\boldsymbol{x}_j$ must be in the same cluster and $(\boldsymbol{x}_i, \boldsymbol{x}_j) \in \mathcal{C}$ means $\boldsymbol{x}_i$ and $\boldsymbol{x}_j$ must not be in the same cluster. Constrained $k$-means, which is an extension of $k$-means, ensures that the constraints in $\mathcal{M}$ and $\mathcal{C}$ are satisfied during the clustering process; otherwise it will return an error when the constraints are violated. The pseudocode of constrained $k$-means is given in ◘ Algorithm 13.4. Taking the watermelon data set 4.0 as an example, suppose that must-link constraints exist for the sample pairs $\boldsymbol{x}_4$ and $\boldsymbol{x}_{25}$, $\boldsymbol{x}_{12}$ and $\boldsymbol{x}_{20}$, $\boldsymbol{x}_{14}$ and $\boldsymbol{x}_{17}$, and cannot-link constraints exist for the sample pairs $\boldsymbol{x}_2$ and $\boldsymbol{x}_{21}$, $\boldsymbol{x}_{13}$ and $\boldsymbol{x}_{23}$, $\boldsymbol{x}_{19}$ and $\boldsymbol{x}_{23}$, that is,

See Sect. 9.4.1 for $k$-means.

The watermelon data set 4.0 is in ◘ Table 9.1 in Sect. 9.4.1.

$$\mathcal{M} = \{(\boldsymbol{x}_4, \boldsymbol{x}_{25}), (\boldsymbol{x}_{25}, \boldsymbol{x}_4), (\boldsymbol{x}_{12}, \boldsymbol{x}_{20}), (\boldsymbol{x}_{20}, \boldsymbol{x}_{12}), (\boldsymbol{x}_{14}, \boldsymbol{x}_{17}), (\boldsymbol{x}_{17}, \boldsymbol{x}_{14})\},$$
$$\mathcal{C} = \{(\boldsymbol{x}_2, \boldsymbol{x}_{21}), (\boldsymbol{x}_{21}, \boldsymbol{x}_2), (\boldsymbol{x}_{13}, \boldsymbol{x}_{23}), (\boldsymbol{x}_{23}, \boldsymbol{x}_{13}), (\boldsymbol{x}_{19}, \boldsymbol{x}_{23}), (\boldsymbol{x}_{23}, \boldsymbol{x}_{19})\}.$$

We set $k = 3$ as the number of clusters and randomly pick $\boldsymbol{x}_6$, $\boldsymbol{x}_{12}$, and $\boldsymbol{x}_{27}$ as the initial mean vectors. Under this setting, ◘ Figure 13.4 shows the results of the constrained $k$-means clustering after different iterations. Since the mean vectors do not change after the 5th iteration (compared to the 4th iteration), we have the final clustering result

---

**Algorithm 13.4** Constrained $k$-means clustering

---

**Input:** Data set $D = \{x_1, x_2, \ldots, x_m\}$;
   Must-link constraints set $\mathcal{M}$;
   Cannot-link constraints set $\mathcal{C}$;
   Number of clusters $k$.

**Process:**

1: Randomly select $k$ samples as the initial mean vectors $\{\mu_1, \mu_2, \ldots, \mu_k\}$
2: **repeat**

Initialize $k$ empty clusters.

3:    $C_j = \varnothing (1 \leqslant j \leqslant k)$
4:    **for** $i = 1, 2, \ldots, m$ **do**
5:       Compute the distance between sample $x_i$ and each mean vector $\mu_j (1 \leqslant j \leqslant k)$: $d_{ij} = \|x_i - \mu_j\|_2$;
6:       $\mathcal{K} = \{1, 2, \ldots, k\}$;
7:       is_merged=false;
8:       **while** ¬is_merged **do**
9:          Use $\mathcal{K}$ to find the nearest cluster of $x_i$: $r = \arg\min_{j \in \mathcal{K}} d_{ij}$;
10:          Check if putting $x_i$ into $C_r$ violates any constraint in $\mathcal{M}$ or $\mathcal{C}$;
11:          **if** ¬is_violated **then**
12:             $C_r = C_r \cup \{x_i\}$;
13:             is_merged=true
14:          **else**
15:             $\mathcal{K} = \mathcal{K} \setminus \{r\}$;
16:             **if** $\mathcal{K} = \varnothing$ **then**
17:                **break** and return an error message
18:             **end if**
19:          **end if**
20:       **end while**
21:    **end for**
22:    **for** $j = 1, 2, \ldots, k$ **do**

Update the mean vectors.

23:       $\mu_j = \frac{1}{|C_j|} \sum_{x \in C_j} x$;
24:    **end for**
25: **until** All mean vectors remain unchanged

**Output:** Clusters $\mathcal{C} = \{C_1, C_2, \ldots, C_k\}$

---

$$C_1 = \{x_3, x_5, x_7, x_9, x_{13}, x_{14}, x_{16}, x_{17}, x_{21}\};$$
$$C_2 = \{x_6, x_8, x_{10}, x_{11}, x_{12}, x_{15}, x_{18}, x_{19}, x_{20}\};$$
$$C_3 = \{x_1, x_2, x_4, x_{22}, x_{23}, x_{24}, x_{25}, x_{26}, x_{27}, x_{28}, x_{29}, x_{30}\}.$$

Supervision information also exists in the form of a small number of labeled samples. Given a data set $D = \{x_1, x_2, \ldots, x_m\}$ and a small set of labeled samples $S = \bigcup_{j=1}^{k} S_j \subset D$, where $S_j \neq \varnothing$ is the set of samples in the $j$th cluster. It is straightforward to use such supervision information: we can use the labeled samples as the *seeds* to initialize the $k$ centroids of the $k$-means algorithm, and these seeds are retained in their

*The labels here refer to cluster labels rather than class labels.*

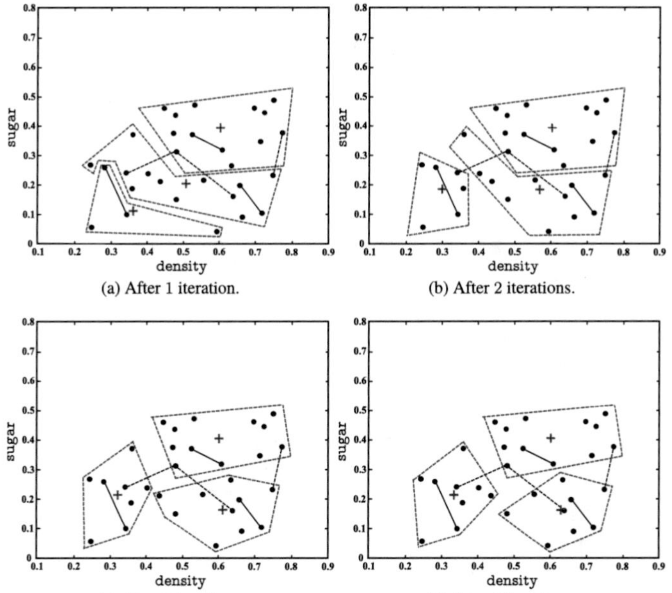


(a) After 1 iteration. (b) After 2 iterations.

(c) After 3 iterations. (d) After 4 iterations.

**Fig. 13.4** Results of the constrained $k$-means algorithm after different iterations on the watermelon data set 4.0 with $k = 3$. The symbols "$\bullet$" and "$+$" represent, respectively, the samples and the mean vectors. The must-link constraints and the cannot-link constraints are, respectively, represented by the solid lines and the dashed lines. The red dashed lines show the clusters

clusters during the learning process. Such a procedure gives the Constrained Seed $k$-means (Basu et al. 2002), as illustrated in ◘ Algorithm 13.5.

Taking the watermelon data set 4.0 as an example, suppose that the following samples are labeled and set as the seeds:

$$S_1 = \{\boldsymbol{x}_4, \boldsymbol{x}_{25}\},\ S_2 = \{\boldsymbol{x}_{12}, \boldsymbol{x}_{20}\},\ S_3 = \{\boldsymbol{x}_{14}, \boldsymbol{x}_{17}\},$$

which are used to initialize the mean vectors. Under this setting, ◘ Figure 13.5 shows the results of the constrained seed $k$-means clustering after different iterations. As the mean vectors do not change after the 4th iteration (compared to the 3rd iteration), we have the final clustering result

$$C_1 = \{\boldsymbol{x}_1, \boldsymbol{x}_2, \boldsymbol{x}_4, \boldsymbol{x}_{22}, \boldsymbol{x}_{23}, \boldsymbol{x}_{24}, \boldsymbol{x}_{25}, \boldsymbol{x}_{26}, \boldsymbol{x}_{27}, \boldsymbol{x}_{28}, \boldsymbol{x}_{29}, \boldsymbol{x}_{30}\};$$
$$C_2 = \{\boldsymbol{x}_6, \boldsymbol{x}_7, \boldsymbol{x}_8, \boldsymbol{x}_{10}, \boldsymbol{x}_{11}, \boldsymbol{x}_{12}, \boldsymbol{x}_{15}, \boldsymbol{x}_{18}, \boldsymbol{x}_{19}, \boldsymbol{x}_{20}\};$$
$$C_3 = \{\boldsymbol{x}_3, \boldsymbol{x}_5, \boldsymbol{x}_9, \boldsymbol{x}_{13}, \boldsymbol{x}_{14}, \boldsymbol{x}_{16}, \boldsymbol{x}_{17}, \boldsymbol{x}_{21}\}.$$

---

**Algorithm 13.5** Constrained seed $k$-means clustering

**Input:** Data set $D = \{x_1, x_2, \ldots, x_m\}$;
　　　A small number of labeled samples $S = \bigcup_{j=1}^{k} S_j$;
　　　Number of clusters $k$.

$S \subset D, |S| \ll |D|$.

**Process:**
1: **for** $j = 1, 2, \ldots, k$ **do**
2: 　　$\mu_j = \frac{1}{|S_j|} \sum_{x \in S_j} x$;
3: **end for**
4: **repeat**
5: 　　$C_j = \varnothing (1 \leqslant j \leqslant k)$;
6: 　　**for** $j = 1, 2, \ldots, k$ **do**
7: 　　　　**for all** $x \in S_j$ **do**
8: 　　　　　　$C_j = C_j \cup \{x\}$;
9: 　　　　**end for**
10: 　　**end for**
11: 　　**for all** $x_i \in D \backslash S$ **do**
12: 　　　　Compute the distance between sample $x_i$ and each mean
　　　　　　vector $\mu_j (1 \leqslant j \leqslant k)$: $d_{ij} = \|x_i - \mu_j\|_2$;
13: 　　　　Find the nearest cluster of $x_i$: $r = \arg\min_{j \in \{1,2,\ldots,k\}} d_{ij}$;
14: 　　　　Move $x_i$ to the corresponding cluster: $C_r = C_r \cup \{x_i\}$;
15: 　　**end for**
16: 　　**for** $j = 1, 2, \ldots, k$ **do**
17: 　　　　$\mu_j = \frac{1}{|C_j|} \sum_{x \in C_j} x$;
18: 　　**end for**
19: **until** All mean vectors remain unchanged
**Output:** Clusters $C = \{C_1, C_2, \ldots, C_k\}$.

Initialize the cluster centroids using labeled samples.

Initialize the $k$ clusters using labeled samples.

Update the mean vectors.

## 13.7　Further Reading

It is generally agreed that semi-supervised learning was first studied in Shahshahani and Landgrebe (1994). Due to the vast demand for utilizing unlabeled data in real-world applications, semi-supervised learning became a flourishing area of research in the late twentieth century and the early twenty-first century. Since 2008, the International Conference on Machine Learning (ICML) started to give out "10-Year Best Paper Award" each year, and works on semi-supervised learning have taken the award three times in six years: Blum and Mitchell (1998) won the award in 2008 for the contributions to disagreement-based methods; Joachims (1999) won the award in 2009 for the contributions to semi-supervised SVM; and Zhu et al. (2003) won the award in 2013 for the contributions to graph-based semi-supervised learning. These three works covered three out of the four mainstream semi-supervised learning paradigms.

Generative semi-supervised learning first appeared in the work of Shahshahani and Landgrebe (1994), and is mainly studied in specific application domains since the model assumption relies on sufficiently reliable domain knowledge.

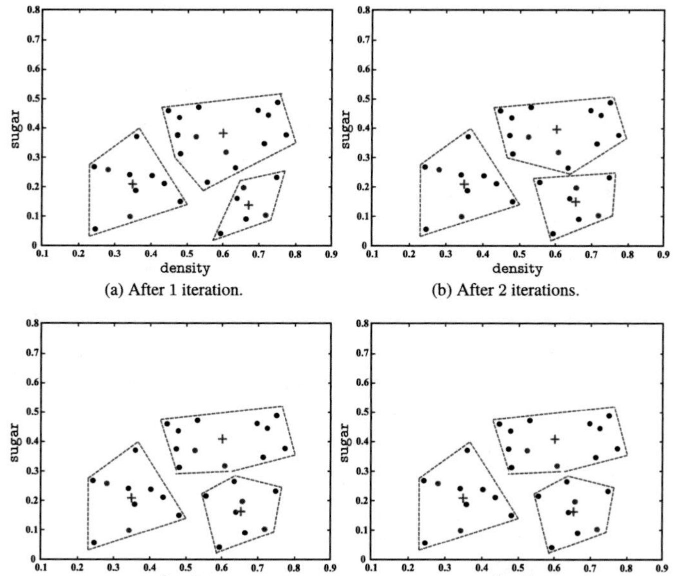

(a) After 1 iteration.

(b) After 2 iterations.

(c) After 3 iterations.

(d) After 4 iterations.

**Fig. 13.5** Results of the constrained seed $k$-means algorithm after different iterations on the watermelon data set 4.0 with $k = 3$. The symbols "•" and "+" represent, respectively, the samples and the mean vectors. The seed samples are shown in red and the red dashed lines show the clusters

The objective function of semi-supervised SVM is non-convex, and efforts have been made to reduce the adverse effects of non-convexity. For example, Chapelle and Zien (2006) employed a continuation method to gradually transform a simple convex objective function to a non-convex S3VM objective function, and Sindhwani et al. (2006) employed a deterministic annealing technique to convert a non-convex problem to a series of convex problems, which are then solved sequentially from the easiest to the most difficult one. Collobert et al. (2006) utilized a CCCP method to optimize the non-convex objective function.

Early graph-based semi-supervised learning methods (Blum and Chawla 2001) took the clustering assumption and regarded the learning problem as finding the *mincut* of the graph. For such methods, the quality of graph is vital. Commonly used graph structures include Gaussian distance graph, $k$-nearest neighbor graph, and $\epsilon$-nearest neighbor graph. Besides, studies on graph construction (Wang and Zhang 2006; Jebara et al. 2009) and graph kernel methods are also related to graph-based semi-supervised learning (Chapelle et al. 2003).

See Sect. 10.5.1 for $k$-nearest neighbor graph and $\epsilon$-nearest neighbor graph.

Disagreement-based methods originated from co-training, which was initially designed to choose one learner for prediction (Blum and Mitchell 1998). *Tri-training* used three learners to generate pseudo-labels by majority voting and ensemble these learners (Zhou and Li 2005b). Subsequent studies found that combining more learners can further improve performance, and this important fact bridges two independently developed areas: ensemble learning and semi-supervised learning (Zhou 2009). Besides, such methods can be easily applied to multi-view data and can be naturally combined with active learning (Zhou and Li 2010).

Many researchers on ensemble learning argue that the performance of weak learners can be boosted to a high level by employing multiple learners, and therefore, there is no need to use unlabeled samples. By contrast, many researchers on semi-supervised learning argue that by employing unlabeled samples, the performance of weak learners can be boosted to a high level, and therefore, there is no need to combine multiple learners. However, both arguments have their limitations.

Belkin et al. (2006) proposed the manifold regularization framework for semi-supervised learning. The framework took a local smoothness assumption to regularize the loss function defined over labeled samples such that the prediction function possessed local smoothness.

Utilizing unlabeled samples through semi-supervised learning does not necessarily guarantee the improvement in generalization performance, and it may even decrease the performance sometimes. For generative methods, it is believed that inappropriate model assumptions cause performance degeneration (Cozman and Cohen 2002), and hence such methods rely on sufficiently reliable domain knowledge. For semi-supervised SVM, the performance degeneration is believed to be caused by the existence of multiple "low-density separations" in the training set such that the learning algorithm may make a poor selection. S4VM (Li and Zhou 2015) improved the "safeness" of semi-supervised SVM by optimizing the worst-case performance by exploiting the multiple low-density separators. A general approach to "safe" semi-supervised learning has yet to be developed.

Here, "safe" means the generalization performance after utilizing unlabeled samples is at least no worse than that of just using labeled samples.

Besides classification and clustering problems, semi-supervised learning has also been widely used in other machine learning problems, such as semi-supervised regression (Zhou and Li 2005a) and dimensionality reduction (Zhang et al. 2007). See Chapelle et al. (2006), Zhu (2006) for more information about semi-supervised learning. Detailed discussions on disagreement-based methods can be found in Zhou and Li (2010). Settles (2009) provided an introduction to active learning.

## Exercises

**13.1** Derive (13.5), (13.6), (13.7), (13.8).

**13.2** Derive a generative semi-supervised learning algorithm based on the naïve Bayes model.

**13.3** Suppose that the data set is generated from a mixture of experts model, that is, the data is generated using the probability density obtained from the mixture of $k$ components:

$$p(\boldsymbol{x} \mid \boldsymbol{\theta}) = \sum_{i=1}^{k} \alpha_i \cdot p(\boldsymbol{x} \mid \boldsymbol{\theta}_i), \tag{13.22}$$

where $\boldsymbol{\theta} = \{\boldsymbol{\theta}_1, \boldsymbol{\theta}_2, \ldots, \boldsymbol{\theta}_k\}$ is the model parameter, $p(\boldsymbol{x} \mid \boldsymbol{\theta}_i)$ is the probability density of the $i$th mixture component, $\alpha_i \geqslant 0$ is the mixture coefficient, and $\sum_{i=1}^{k} \alpha_i = 1$. Assuming each mixture component corresponds to one class, but each class may contain multiple mixture components. Derive the corresponding generative semi-supervised learning algorithm.

**13.4** Download or implement the TSVM algorithm, and apply it to two UCI data sets. Use 30% of the samples as testing samples, 10% of the samples as labeled samples, and 60% of the samples as unlabeled samples. Train a TSVM model that utilizes unlabeled samples and a SVM model that only uses labeled samples, and compare the performance.

The UCI data sets can be found at ▶ http://archive.ics.uci.edu/ml/.

**13.5** For the TSVM algorithm, the class imbalance problem may occur during the label assignment or label adjustment. Design an improved version of the TSVM algorithm that considers the class imbalance problem.

**13.6** * The label assignment and label adjustment in the TSVM algorithm are computationally expensive. Design an improved version of the TSVM algorithm that is more computationally efficient.

**13.7** * Design a graph-based semi-supervised learning method that can classify new coming samples.

**13.8** Self-training is one of the oldest semi-supervised learning methods. It uses the labeled samples to train a classifier for predicting the pseudo-labels of the unlabeled samples. After that, it uses the labeled samples and the unlabeled samples with pseudo-labels to train another classifier for revising the pseudo-labels, and this process repeats. Discuss the pros and cons of this method.

**13.9** * Suppose there are two views in the features of a given data set, but we do not know which features belong to each view. Design an algorithm that can separate the two views.

**13.10** For line 10 of ◘ Algorithm 13.4, write down the violation checking algorithm (for checking if any constraints are violated).

## Break Time

### Short Story: Manifold and Bernhard Riemann

The name *manifold* comes from the original German term *Mannigfaltigkeit* proposed by the great German mathematician Bernhard Riemann (1826–1866), who was born in Breselenz, Germany. Riemann appeared gifted in mathematics when he was a child. In 1946, his father sent him to the University of Göttingen to study Theology. However, after attending Gauss's lectures on the method of least squares, Riemann decided to study mathematics instead. In 1847, Riemann transferred to the University of Berlin for two years and obtained his Ph.D. degree in mathematics under the supervision of Gauss in 1851. During his study, Riemann was influenced by many great mathematicians, including C. Jacobi and P. Dirichlet. In 1854, Riemann was invited by Gauss to give his inaugural lecture "On the Hypotheses which lie at the Bases of Geometry", which created a new research area known as Riemannian geometry. In the lecture, Riemann integral was proposed, and the term *Mannigfaltigkeit* appeared for the first time. After that, Riemann started to teach at the University of Göttingen and was promoted to professor following the death of Dirichlet.

Riemann is the creator of Riemann geometry and the founder of complex analysis. He has made significant contributions to many areas, including calculus, analytic number theory, combinatorial topology, algebraic geometry, and mathematical physics. Riemann's work influenced the development of mathematics for a century, and many outstanding mathematicians have tried to prove the hypotheses proposed by Riemann. In David Hilbert's list of 23 unsolved problems and the 7 Millennium prize problems raised by the Clay Mathematics Institute, there is one problem in common, that is, *Riemann hypothesis*. Based on the Riemann hypothesis, thousands of mathematical statements have been proposed in different branches of mathematics, and all of them will prompt to be theorems once Riemann hypothesis is proved. It is very rare to have a hypothesis connected to that many mathematical branches and statements, and the Riemann hypothesis is regarded as one of the most important unsolved mathematical problems today.

Among the 7 Millennium prize problems, the only solved problem so far is the *Poincaré conjecture*, which directly relates to manifold: every simply connected, closed 3-manifold is homeomorphic to the 3-sphere.

Riemann hypothesis was proposed in 1859 in Riemann's paper "On the Number of Prime Numbers Less Than a Given Quantity", which states that the real part of every non-trivial zero of the Riemann zeta function is $1/2$.

## References


Basu S, Banerjee A, Mooney RJ (2002) Semi-supervised clustering by seeding. In: Proceedings of the 19th international conference on machine learning (ICML). Sydney, Australia, pp 19–26

Belkin M, Niyogi P, Sindhwani V (2006) Manifold regularization: a geometric framework for learning from labeled and unlabeled examples. J Mach Learn Res 7:2399–2434

Blum, A. and Chawla, S. (2001). Learning from labeled and unlabeled data using graph mincuts. In *Proceedings of the 18th International Conference on Machine Learning (ICML)*, pages 19–26. Williamston, MA

Blum A, Mitchell T (1998) Combining labeled and unlabeled data with co-training. In: Proceedings of the 11th annual conference on computational learning theory (COLT). Madison, pp 92–100

Chapelle O, Chi M, Zien A (2006) A continuation method for semi-supervised SVMs. In: Proceedings of the 23rd international conference on machine learning (ICML). Pittsburgh, pp 185–192

Chapelle O, Schölkopf B, Zien A (eds) (2006) Semi-supervised learning. MIT Press, Cambridge

Chapelle O, Weston J, Schölkopf B (2003) Cluster kernels for semi-supervised learning. In: Becker S, Thrun S, Obermayer K (eds) Advances in neural information processing systems 15 (NIPS). MIT Press, Cambridge, pp 585–592

Chapelle O, Zien A (2005) Semi-supervised learning by low density separation. In: Proceedings of the 10th international workshop on artificial intelligence and statistics (AISTATS). Savannah Hotel, Barbados, pp 57–64

Collobert R, Sinz F, Weston J, Bottou L (2006) Trading convexity for scalability. In: Proceedings of the 23rd international conference on machine learning (ICML). Pittsburgh, pp 201–208

Cozman FG, Cohen I (2002) Unlabeled data can degrade classification performance of generative classifiers. In: Proceedings of the 15th international conference of the florida artificial intelligence research society (FLAIRS). Pensacola, pp 327–331

Goldman S, Zhou Y (2000) Enhancing supervised learning with unlabeled data. In: Proceedings of the 17th international conference on machine learning (ICML). San Francisco, pp 327–334

Jebara T, Wang J, Chang SF (2009) Graph construction and b-matching for semi-supervised learning. In: Proceedings of the 26th international conference on machine learning (ICML). Montreal, Canada, pp 441–448

Joachims T (1999) Transductive inference for text classification using support vector machines. In: Proceedings of the 16th international conference on machine learning (ICML). Bled, Slovenia, pp 200–209

Li Y-F, Kwok JT, Zhou Z-H (2009) Semi-supervised learning using label mean. In: Proceedings of the 26th international conference on machine learning (ICML). Montreal, Canada, pp 633–640

Li Y-F, Zhou Z-H (2015) Towards making unlabeled data never hurt. IEEE Trans Pattern Anal Mach Intell 37(1):175–188

Miller DJ, Uyar HS (1997) A mixture of experts classifier with learning based on both labelled and unlabelled data. In: Mozer M, Jordan MI, Petsche T (eds) Advances in neural information processing systems 9 (NIPS). MIT Press, Cambridge, pp 571–577

Nigam K, McCallum A, Thrun S, Mitchell T (2000) Text classification from labeled and unlabeled documents using EM. Mach Learn 39(2–3):103–134

Settles B (2009) Active learning literature survey. Technical Report 1648, Department of Computer Sciences, University of Wisconsin at Madison, Wisconsin. ► http://pages.cs.wisc.edu/~bsettles/pub/settles.activelearning.pdf

Shahshahani B, Landgrebe D (1994) The effect of unlabeled samples in reducing the small sample size problem and mitigating the hughes phenomenon. IEEE Trans Geosci Remote Sens 32(5):1087–1095

Sindhwani V, Keerthi SS, Chapelle O (2006) Deterministic annealing for semi-supervised kernel machines. In: Proceedings of the 23rd international conference on machine learning (ICML). Pittsburgh, pp 123–130

Wagstaff K, Cardie C, Rogers S, Schrödl S (2001) Constrained $k$-means clustering with background knowledge. In: Proceedings of the 18th international conference on machine learning (ICML). Williamstown, pp 577–584

Wang F, Zhang C (2006) Label propagation through linear neighborhoods. In: Proceedings of the 23rd international conference on machine learning (ICML). Pittsburgh, pp 985–992

Zhang D, Zhou Z-H, Chen S (2007) Semi-supervised dimensionality reduction. In: Proceedings of the 7th SIAM international conference on data mining (SDM). Minneapolis, pp 629–634

Zhou D, Bousquet O, Lal TN, Weston J, Schölkopf B (2004) Learning with local and global consistency. In: Thrun S, Saul L, Scholkopf B (eds) Advances in neural information processing systems 16 (NIPS). MIT Press, Cambridge, pp 284–291

Zhou Z-H (2009) When semi-supervised learning meets ensemble learning. In: Proceedings of the 8th international workshop on multiple classifier systems. Reykjavik, Iceland, pp 529–538

Zhou Z-H, Li M (2005a) Semi-supervised regression with co-training. In: Proceedings of the 19th international joint conference on artificial intelligence (IJCAI). Edinburgh, Scotland, pp 908–913

Zhou Z-H, Li M (2005b) Tri-training: exploiting unlabeled data using three classifiers. IEEE Trans Knowl Data Eng 17(11):1529–1541

Zhou Z-H, Li M (2010) Semi-supervised learning by disagreement. Knowl Inf Syst 24(3):415–439

Zhu X, Ghahramani Z, Lafferty J (2003) Semi-supervised learning using gaussian fields and harmonic functions. In: Proceedings of the 20th international conference on machine learning (ICML). Washington, pp 912–919

Zhu XJ (2006) Semi-supervised learning literature survey. Technical Report 1530, Department of Computer Sciences, University of Wisconsin at Madison, Wisconsin. ► http://www.cs.wisc.edu/~jerryzhu/pub/ssl_survey.pdf

# Probabilistic Graphical Models

Table of Contents




Z.-H. Zhou, *Machine Learning*,
https://doi.org/10.1007/978-981-15-1967-3_14

## 14.1 Hidden Markov Model

The most important problem in machine learning is to estimate and infer the value of unknown variables (e.g., class label) based on the observed evidence (e.g., training samples). Probabilistic models provide a framework that considers learning problems as computing the probability distributions of variables. In probabilistic models, the process of inferring the distributions of unknown variables conditioned on known variables is called *inference*. More specifically, let $Y$ denote the set of target variables, $O$ denote the set of observable variables, and $R$ denote the set of other variables. Then, *generative* models consider the joint distribution $P(Y, R, O)$, while *discriminative* models consider the conditional distribution $P(Y, R \mid O)$. Given the values of a set of observed variables, inference aims to obtain the conditional probability $P(Y \mid O)$ from $P(Y, R, O)$ or $P(Y, R \mid O)$.

A learner makes inference when it predicts the ripeness of watermelon based on information such as texture, color, and root. However, the inference is more than just prediction. For example, when we eat a ripe watermelon and try to infer the shape of its root reversely, it is also inference.

It is impractical to eliminate $R$ by probability marginalization since the computational complexity is prohibitive. For example, it costs at least $O(2^{|Y|+|R|})$ operations even if each variable has just two possible values. Besides, the learning process of probabilistic models, that is, estimating the parameters of variable distributions from the data set, is not easy since there are often complex relationships between variables. Hence, we must develop a methodology to concisely represent the relationships between variables for the study of efficient learning and inference.

Probabilistic Graphical Models (PGM) are a family of probabilistic models that represent the relationships between variables with graph structures, in which, each node (also known as vertex) represents one or a set of random variables and each link (also known as edge) between two nodes represents the probabilistic relationship between the variables. Depending on the properties of edges, probabilistic graphical models can be roughly divided into two categories. The first category is called directed graphical models or Bayesian networks, which employ Directed Acyclic Graphs (DAG) to represent the dependence between variables. The second category is called undirected graphical models or Markov networks, which employ undirected graphs to represent the dependence between variables.



Bayesian networks are often used when explicit causal relationships exist between variables. Markov networks are often used when correlations exist between variables while explicit causal relationships are difficult to obtain.

See Sect. 7.5 for static Bayesian network.

The simplest *dynamic Bayesian network* is Hidden Markov Model (HMM), a well-known directed graphical model commonly used for modeling time-series data and has been widely used in speech recognition and natural language processing.

As illustrated in ◘ Figure 14.1, there are two sets of variables in an HMM. The first set of variables are state variables $\{y_1, y_2, \ldots, y_n\}$, where $y_i \in \mathcal{Y}$ represents the system state at

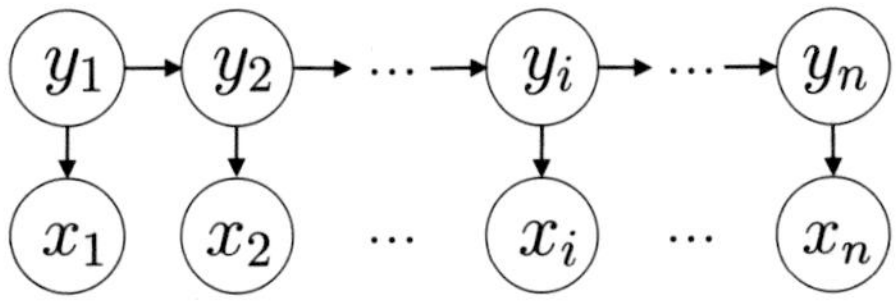


**Fig. 14.1** The graph structure of HMM

the $i$th time point. The state variables are usually assumed to be hidden and unobserved, hence they are also called *hidden variables*. The second set of variables are observed variables $\{x_1, x_2, \ldots, x_n\}$, where $x_i \in \mathcal{X}$ represents the observation at the $i$th time point. In HMM, the system changes between different states $\{s_1, s_2, \ldots, s_N\}$, and hence the state space $\mathcal{Y}$ is usually a discrete space with $N$ possible values. The observed variables $x_i$, however, can be either discrete or continuous. For ease of discussion, we assume that the observed variables are discrete, i.e., $\mathcal{X} = \{o_1, o_2, \ldots, o_M\}$.

The directed links in ◘ Figure 14.1 represent the dependence between variables. For each time point, the value of an observed variable only depends on the state variable, that is, $x_t$ is solely determined by $y_t$. Meanwhile, the state $y_t$ at time point $t$ only depends on the state $y_{t-1}$ at time point $t-1$ and is independent of the previous $t-2$ states. Such a model is known as *Markov chain*, in which the system state at the next time point does not depend on any previous states but the current state. Under this dependence setting, the joint probability of all variables is

"The future depends on what you do today."—Mahatma Gandhi.

$$P(x_1, y_1, \ldots, x_n, y_n) = P(y_1)P(x_1 \mid y_1) \prod_{i=2}^{n} P(y_i \mid y_{i-1})P(x_i \mid y_i). \tag{14.1}$$

In addition to the structure information, an HMM has three more sets of parameters

- State transition probabilities: the probabilities that the model changes between states, usually denoted by matrix $\mathbf{A} = [a_{ij}]_{N \times N}$, where

$$a_{ij} = P(y_{t+1} = s_j \mid y_t = s_i), \quad 1 \leqslant i, j \leqslant N$$

  indicates the probability that the next state is $s_j$ when the current state is $s_i$ at time point $t$.
- Output observation probabilities: the probabilities of observations based on the current state, usually denoted by matrix $\mathbf{B} = [b_{ij}]_{N \times M}$, where

$$b_{ij} = P(x_t = o_j \mid y_t = s_i), \quad 1 \leqslant i \leqslant N, 1 \leqslant j \leqslant M$$

indicates the probability of observing $o_j$ when the current state is $s_i$ at time point $t$.

- Initial state probabilities: the probability of each state that appears at the initial time point, usually denoted by $\boldsymbol{\pi} = (\pi_1, \pi_2, \ldots, \pi_N)$, where

$$\pi_i = P(y_1 = s_i), \quad 1 \leqslant i \leqslant N$$

indicates the probability that the initial state is $s_i$.

The above three sets of parameters together with the state space $\mathcal{Y}$ and the observation space $\mathcal{X}$ determine an HMM, usually denoted by $\lambda = [\mathbf{A}, \mathbf{B}, \boldsymbol{\pi}]$. Given an HMM $\lambda$, it generates the observed sequence $\{x_1, x_2, \ldots, x_n\}$ by the following process:

(1) Set $t = 1$ and select the initial state $y_1$ based on the initial state probability $\boldsymbol{\pi}$;
(2) Select the value of the observed variable $x_t$ based on the state variable $y_t$ and the output observation probability matrix $\mathbf{B}$;
(3) Transition to the next state $y_{t+1}$ based on the current state $y_t$ and the state transition probability matrix $\mathbf{A}$;
(4) If $t < n$, set $t = t+1$ and return to step (2); otherwise, stop.

$y_t \in \{s_1, s_2, \ldots, s_N\}$ and $x_t \in \{o_1, o_2, \ldots, o_M\}$ are the state and observation at time point $t$, respectively.

There are three fundamental problems when applying HMM in practice:

- Given a model $\lambda = [\mathbf{A}, \mathbf{B}, \boldsymbol{\pi}]$, how can we effectively calculate the probability $P(\mathbf{x} \mid \lambda)$ for generating the observed sequence $\mathbf{x} = \{x_1, x_2, \ldots, x_n\}$? In other words, how to evaluate the matching degree between a model and an observed sequence?
- Given a model $\lambda = [\mathbf{A}, \mathbf{B}, \boldsymbol{\pi}]$ and an observed sequence $\mathbf{x} = \{x_1, x_2, \ldots, x_n\}$, how can we find the best state sequence $\mathbf{y} = \{y_1, y_2, \ldots, y_n\}$ that matches $\mathbf{x}$? In other words, how to infer the hidden states from the observed sequence?
- Given an observed sequence $\mathbf{x} = \{x_1, x_2, \ldots, x_n\}$, how can we adjust the model parameter $\lambda = [\mathbf{A}, \mathbf{B}, \boldsymbol{\pi}]$ such that the probability $P(\mathbf{x} \mid \lambda)$ of observing the given sequence is maximized? In other words, how to train the model such that it can better describe the observed data?

The above problems are critical in real-world applications. For example, in many tasks, we need to estimate the most likely value of the current observation $x_n$ based on the previously observed sequence $\{x_1, x_2, \ldots, x_{n-1}\}$; this problem can be solved by finding $P(\mathbf{x} \mid \lambda)$, that is, the first listed problem. In speech recognition problems, the observations are audio signals, the hidden states are spoken texts, and the task is to infer the most likely state sequence (i.e., spoken texts) based on the observed sequence (i.e., audio signals), that is, the second listed problem. In many applications, manually specifying model parameters is becoming impractical, and hence model parameters need to be learned from data, that is, the third listed problem. Fortunately, thanks to the conditional independence given in (14.1), all of the three listed problems can be solved efficiently.

## 14.2 Markov Random Field

Markov Random Field (MRF) is a typical Markov network and a well-known undirected graphical model. In MRF, each node represents one or a set of variables, and the edges between nodes represent the variable dependence. Besides, there is a set of *potential functions*, also known as *factors*, which are non-negative real-valued functions defined over variable subsets mainly for defining probability distribution functions.

A simple MRF is illustrated in ◘ Figure 14.2. A subset of nodes in the graph is called a *clique* if there exists a link between any two nodes. We say a clique is a *maximal clique* if adding any extra node makes it no longer a clique; in other words, a maximal clique is a clique that is not contained in any other cliques. For example, the cliques in ◘ Figure 14.2 are $\{x_1, x_2\}$, $\{x_1, x_3\}$, $\{x_2, x_4\}$, $\{x_2, x_5\}$, $\{x_2, x_6\}$, $\{x_3, x_5\}$, $\{x_5, x_6\}$, and $\{x_2, x_5, x_6\}$, which are also maximal cliques except $\{x_2, x_5\}$, $\{x_2, x_6\}$, and $\{x_5, x_6\}$; $\{x_1, x_2, x_3\}$ is not a clique since there is no link between $x_2$ and $x_3$. We notice that every node appears in at least one maximal clique.

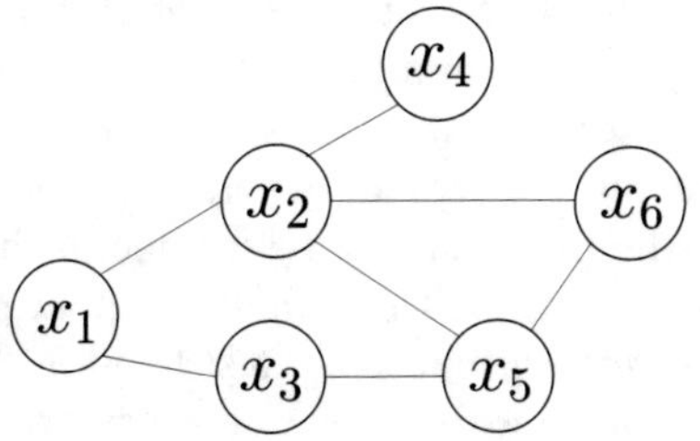


**Fig. 14.2** A simple MRF

In MRF, the joint probability of multiple variables can be decomposed into the product of multiple factors based on the cliques, and each factor corresponds to one clique. To be specific, let $\mathbf{x} = \{x_1, x_2, \ldots, x_n\}$ denote the set of $n$ variables, $\mathcal{C}$ denote the set of all cliques, and $\mathbf{x}_Q$ denote the set of variables in clique $Q \in \mathcal{C}$, then the joint probability $P(\mathbf{x})$ is defined as

$$P(\mathbf{x}) = \frac{1}{Z} \prod_{Q \in \mathcal{C}} \psi_Q(\mathbf{x}_Q), \tag{14.2}$$

where $\psi_Q$ is the potential function that captures the dependence between variables in clique $Q$; $Z = \sum_{\mathbf{x}} \prod_{Q \in \mathcal{C}} \psi_Q(\mathbf{x}_Q)$ is the normalization factor that ensures $P(\mathbf{x})$ is a properly defined probability. In practice, it is often difficult to calculate $Z$ exactly, but we usually do not need the exact value.

When there are many variables, the number of cliques can be quite large. For example, every pair of linked variables forms a clique, and hence there will be many terms multiplied in (14.2), leading to high computational cost. We notice that, if clique $Q$ is not a maximal clique, then it is contained in a maximal clique $Q^*$ (i.e., $\mathbf{x}_Q \subseteq \mathbf{x}_{Q^*}$). Hence, the dependence between variables $\mathbf{x}_Q$ is not only encoded in the potential function $\psi_Q$, but also the potential function $\psi_{Q^*}$. Therefore, it is also possible to define the joint probability $P(\mathbf{x})$ based on the maximal cliques. Let $\mathcal{C}^*$ denote the set of all maximal cliques, we have

$$P(\mathbf{x}) = \frac{1}{Z^*} \prod_{Q \in \mathcal{C}^*} \psi_Q(\mathbf{x}_Q), \tag{14.3}$$

where $Z^* = \sum_{\mathbf{x}} \prod_{Q \in \mathcal{C}^*} \psi_Q(\mathbf{x}_Q)$ is the normalization factor. Taking ◘ Figure 14.2 as an example, the joint probability $P(\mathbf{x})$ can be defined as



$$P(\mathbf{x}) = \frac{1}{Z}\psi_{12}(x_1, x_2)\psi_{13}(x_1, x_3)\psi_{24}(x_2, x_4)\psi_{35}(x_3, x_5)\psi_{256}(x_2, x_5, x_6),$$

where the potential function $\psi_{256}(x_2, x_5, x_6)$ is defined over the maximal clique $\{x_2, x_5, x_6\}$. Since we have $\psi_{256}(x_2, x_5, x_6)$, there is no need to construct the potential functions for the cliques $\{x_2, x_5\}$, $\{x_2, x_6\}$, and $\{x_5, x_6\}$.

How can we obtain *conditional independence* in MRF? We still utilize the concept of *separation*. As illustrated in ◘ Figure 14.3, the path connecting a node in the node set $A$ to a node in node set $B$ passes through the node set $C$, and we say $C$ is a *separating set* that separates $A$ and $B$. For MRF, we have the *global Markov property*: two subsets of variables are conditionally independent given a separating set of these two subsets.

See Sect. 7.5.1.

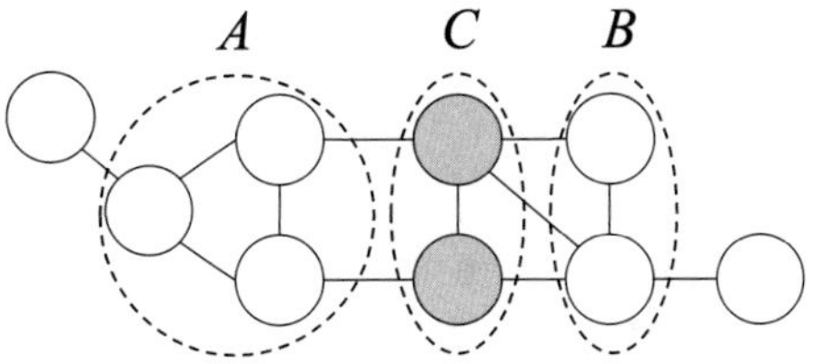


**Fig. 14.3** The node set $C$ separates the node set $A$ and the node set $B$

Taking ◘ Figure 14.3 as an example, let $\mathbf{x}_A$, $\mathbf{x}_B$, and $\mathbf{x}_C$ denote the sets of variables for $A$, $B$, and $C$, respectively. Then, $\mathbf{x}_A$ and $\mathbf{x}_B$ are conditionally independent given $\mathbf{x}_C$, denoted by $\mathbf{x}_A \perp \mathbf{x}_B \mid \mathbf{x}_C$.

Now, let us do a simple verification. For ease of discussion, let $A$, $B$, and $C$ correspond to single variables $x_A$, $x_B$, and $x_C$, respectively. Then, ◘ Figure 14.3 is simplified to ◘ Figure 14.4.

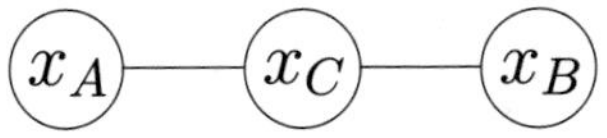


**Fig. 14.4** A simplified version of ◘ Figure 14.3

From (14.2), the joint probability of the variables in ◘ Figure 14.4 is given by

$$P(x_A, x_B, x_C) = \frac{1}{Z}\psi_{AC}(x_A, x_C)\psi_{BC}(x_B, x_C). \tag{14.4}$$

According to the definition of conditional probability, we have

$$\begin{aligned} P(x_A, x_B \mid x_C) &= \frac{P(x_A, x_B, x_C)}{P(x_C)} = \frac{P(x_A, x_B, x_C)}{\sum_{x'_A}\sum_{x'_B} P(x'_A, x'_B, x_C)} \\ &= \frac{\frac{1}{Z}\psi_{AC}(x_A, x_C)\psi_{BC}(x_B, x_C)}{\sum_{x'_A}\sum_{x'_B}\frac{1}{Z}\psi_{AC}(x'_A, x_C)\psi_{BC}(x'_B, x_C)} \\ &= \frac{\psi_{AC}(x_A, x_C)}{\sum_{x'_A}\psi_{AC}(x'_A, x_C)} \cdot \frac{\psi_{BC}(x_B, x_C)}{\sum_{x'_B}\psi_{BC}(x'_B, x_C)}. \end{aligned} \tag{14.5}$$

$$\begin{aligned} P(x_A \mid x_C) &= \frac{P(x_A, x_C)}{P(x_C)} = \frac{\sum_{x'_B} P(x_A, x'_B, x_C)}{\sum_{x'_A}\sum_{x'_B} P(x'_A, x'_B, x_C)} \\ &= \frac{\sum_{x'_B}\frac{1}{Z}\psi_{AC}(x_A, x_C)\psi_{BC}(x'_B, x_C)}{\sum_{x'_A}\sum_{x'_B}\frac{1}{Z}\psi_{AC}(x'_A, x_C)\psi_{BC}(x'_B, x_C)} \\ &= \frac{\psi_{AC}(x_A, x_C)}{\sum_{x'_A}\psi_{AC}(x'_A, x_C)}. \end{aligned} \tag{14.6}$$

From (14.5) and (14.6), we have

$$P(x_A, x_B \mid x_C) = P(x_A \mid x_C)P(x_B \mid x_C), \tag{14.7}$$

that is, $x_A$ and $x_B$ are conditionally independent given $x_C$.

From the global Markov property, we can derive the following two useful corollaries:

- **Local Markov property**: a variable is conditionally independent of other variables given its adjacent variables. Formally, we have $\mathbf{x}_v \perp \mathbf{x}_{V \setminus n^*(v)} \mid \mathbf{x}_{n(v)}$, where $V$ is the set of all nodes in the graph, $n(v)$ are the adjacent nodes of node $v$ in the graph, and $n^*(v) = n(v) \cup \{v\}$.
- **Pairwise Markov property**: two non-adjacent variables are conditionally independent given all other variables. Formally, we have $\mathbf{x}_u \perp \mathbf{x}_v \mid \mathbf{x}_{V \setminus \langle u,v \rangle}$ if $\langle u, v \rangle \notin E$, where $u$ and $v$ are two nodes in the graph, and $V$ and $E$ are, respectively, the set of all nodes and the set of all edges in the graph.

The set of parents, children, and children's parents is called the *Markov blanket* of a variable.

Now, let us take a look at the potential functions in MRF. A potential function $\psi_Q(\mathbf{x}_Q)$ describes the dependence between a set of variables $\mathbf{x}_Q$. It should be a non-negative function that returns a large value when the variables take preferred values. For example, suppose that all variables in ■ Figure 14.4 are binary variables, and the potential functions are

$$\phi_{AC}(x_A, x_C) = \begin{cases} 1.5, & \text{if } x_A = x_C; \\ 0.1, & \text{otherwise,} \end{cases}$$

$$\phi_{BC}(x_B, x_C) = \begin{cases} 0.2, & \text{if } x_B = x_C; \\ 1.3, & \text{otherwise,} \end{cases}$$

then the model is biased towards $x_A = x_C$ and $x_B \neq x_C$, that is, $x_A$ and $x_C$ are positively correlated, while $x_B$ and $x_C$ are negatively correlated. From (14.2), we know that the joint probability would be high when the variable assignments satisfy $x_A = x_C$ and $x_B \neq x_C$.

To satisfy the non-negativity, we often use exponential functions to define potential functions

$$\psi_Q(\mathbf{x}_Q) = e^{-H_Q(\mathbf{x}_Q)}. \tag{14.8}$$

$H_Q(\mathbf{x}_Q)$ is real-valued function defined on variable $\mathbf{x}_Q$, usually in the form of

$$H_Q(\mathbf{x}_Q) = \sum_{u,v \in Q, u \neq v} \alpha_{uv} x_u x_v + \sum_{v \in Q} \beta_v x_v, \tag{14.9}$$

where $a_{uv}$ and $\beta_v$ are parameters. In (14.9), the first term considers pairs of nodes and the second term considers individual nodes.

14

## 14.3 Conditional Random Field

Conditional Random Field (CRF) is a discriminative undirected graphical model. In Sect. 14.1, we mentioned that generative models consider joint distributions, while discriminative models consider conditional distributions. The previously introduced HMM and MRF are examples of generative models, and now we introduce CRF as an example of discriminative models.

We can regard CRF as MRF with observed values, or as an extension of logistic regression. See Sect. 3.3.

CRF aims to model the conditional probability of multiple variables given some observed values. To be specific, CRF constructs a conditional probability model $P(\mathbf{y} \mid \mathbf{x})$, where $\mathbf{x} = \{x_1, x_2, \ldots, x_n\}$ is the observed sequence, and $\mathbf{y} = \{y_1, y_2, \ldots, y_n\}$ is the corresponding label sequence. Note that the label variable $\mathbf{y}$ can be structural, that is, there are some correlations among its components. For example, in part-of-speech tagging problems, the observations are natural language sentences (i.e., sequences of words), and the labels are sequences of part-of-speech tags, as shown in ◘ Figure 14.5(a). In syntactic analysis, the output labels are parse trees, as shown in ◘ Figure 14.5(b).

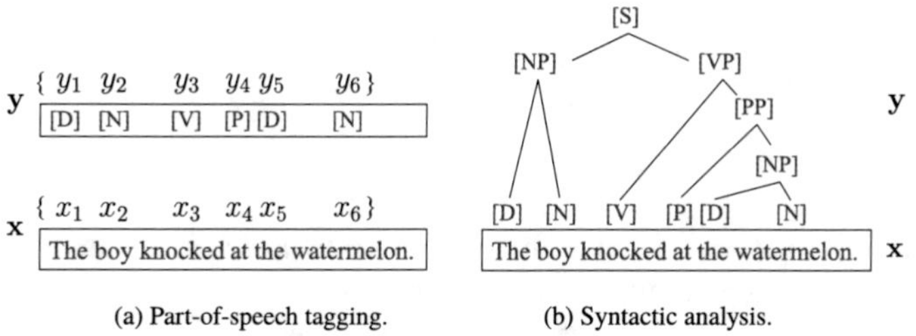


**Fig. 14.5** The part-of-speech tagging problem and the syntactic analysis problem in natural language processing

Let $G = \langle V, E \rangle$ be an undirected graph in which each node corresponds to one component in the label vector $\mathbf{y}$, where $y_v$ is the component corresponding to node $v$, and let $n(v)$ denote the adjacent nodes of node $v$. Then, we say $(\mathbf{y}, \mathbf{x})$ forms a CRF if every label variable $y_v$ in graph $G$ satisfies the Markov property

$$P(y_v \mid \mathbf{x}, \mathbf{y}_{V \setminus \{v\}}) = P(y_v \mid \mathbf{x}, \mathbf{y}_{n(v)}), \tag{14.10}$$

Theoretically, the structure of graph $G$ can be arbitrary as long as it encodes the conditional independence between label variables. In practice, however, especially when modeling label sequences, the most common structure is the chain structure,

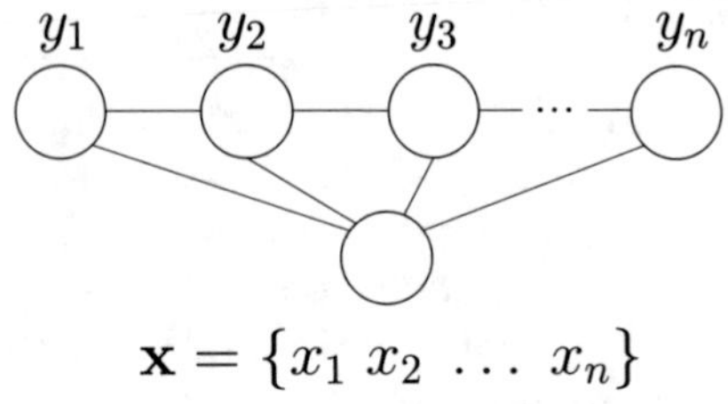

**Fig. 14.6** The graph structure of chain-structured CRF

as illustrated in ◘ Figure 14.6. Such a CRF is called *chain-structured CRF*, which is the focus of the rest of our discussions.

Similar to how joint probability is defined in MRF, CRF defines conditional probability $P(\mathbf{y} \mid \mathbf{x})$ according to the potential functions and cliques in the graph structure. Given an observed sequence $\mathbf{x}$, the chain-structured CRF in ◘ Figure 14.6 mainly contains two types of cliques about label variables, that is, one for single label variable $\{y_i\}$ and the other for adjacent label variables $\{y_{i-1}, y_i\}$. With appropriate potential functions, we can define conditional probability like (14.2). In CRF, by using exponential potential functions and introducing *feature functions*, the conditional probability is defined as

$$P(\mathbf{y} \mid \mathbf{x}) = \frac{1}{Z} \exp\left(\sum_j \sum_{i=1}^{n-1} \lambda_j t_j(y_{i+1}, y_i, \mathbf{x}, i) + \sum_k \sum_{i=1}^{n} \mu_k s_k(y_i, \mathbf{x}, i)\right), \tag{14.11}$$

where $t_j(y_{i+1}, y_i, \mathbf{x}, i)$ is the *transition feature function* defined on two adjacent labels in the observed sequence, describing the relationship between the two adjacent labels as well as measuring the impact of the observed sequence on them; $s_k(y_i, \mathbf{x}, i)$ is the *status feature function* defined on the label index $i$ in the observed sequence, describing the impact of the observed sequence on the label variable; $\lambda_j$ and $\mu_k$ are parameters; $Z$ is the normalization factor that ensures (14.11) to be a properly defined probability.

We also need to define appropriate feature functions, which are usually real-valued functions that describe empirical properties that are likely or expected to be held about the data. Taking the part-of-speech tagging in ◘ Figure 14.5 (a) as an example, we can employ the following transition feature function:

$$t_j(y_{i+1}, y_i, \mathbf{x}, i) = \begin{cases} 1, & \text{if } y_{i+1} = [P],\ y_i = [V] \text{ and } x_i = \text{“}knock\text{”}; \\ 0, & \text{otherwise,} \end{cases}$$

14

which says the labels $y_i$ and $y_{i+1}$ are likely to be $[V]$ and $[P]$ when the $i$th observation $x_i$ is the word "knock". We can also employ the following status feature function:

$$s_k(y_i, \mathbf{x}, i) = \begin{cases} 1, & \text{if } y_i = [V] \text{ and } x_i = \text{"}knock\text{"}; \\ 0, & \text{otherwise,} \end{cases}$$

which says that the label $y_i$ is likely to be $[V]$ if the observation $x_i$ is the word "knock".

By comparing (14.11) and (14.2), we can observe that both CRF and MRF define the probabilities using potential functions on cliques. The difference is that CRF models conditional probabilities, whereas MRF models joint probabilities.

## 14.4 Learning and Inference

Given the joint probability distributions defined on probabilistic graphical models, we can infer the *marginal distribution* or *conditional distribution* of the target variables. We have encountered conditional distributions previously. For example, in HMM, we infer the conditional probability distribution of an observed sequence $\mathbf{x}$ given certain parameter $\lambda$. By contrast, marginal distribution refers to probabilities obtained by summing out or integrating out irrelevant variables. Taking Markov networks as an example, the joint distribution of variables is expressed as the product of maximal cliques' potential functions, and therefore, finding the distribution of variable $x$ given parameter $\Theta$ is equivalent to integrating out irrelevant variables in the joint distribution, known as *marginalization*.

In probabilistic graphical models, we also need to determine the parameters of distributions by parameter estimation (i.e., parameter learning), which is often solved via maximum likelihood estimation or maximum a posteriori estimation. If we consider the parameters as variables to be inferred, then the parameter estimation process is similar to the inference process, that is, it can be absorbed into the inference problem. Hence, we mainly discuss the inference methods for the rest of our discussions.

The Bayesian school thinks that unknown parameters are random variables, just like all other variables. Hence, parameter estimation and variable inference can be performed within the same inference framework. The frequentist school disagrees.

To be specific, suppose that the set of variables $\mathbf{x} = \{x_1, x_2, \ldots, x_N\}$ in a graphical model can be divided into two disjoint variable sets $\mathbf{x}_E$ and $\mathbf{x}_F$, then the inference problem is about finding the marginal probability $P(\mathbf{x}_F)$ or the conditional probability $P(\mathbf{x}_F \mid \mathbf{x}_E)$. From the definition of conditional probability, we have

$$P(\mathbf{x}_F \mid \mathbf{x}_E) = \frac{P(\mathbf{x}_E, \mathbf{x}_F)}{P(\mathbf{x}_E)} = \frac{P(\mathbf{x}_E, \mathbf{x}_F)}{\sum_{\mathbf{x}_F} P(\mathbf{x}_E, \mathbf{x}_F)}, \tag{14.12}$$

where the joint probability $P(\mathbf{x}_E, \mathbf{x}_F)$ can be obtained from the probabilistic graphical model. Hence, the core of the inference problem is how to efficiently compute the marginal distribution, that is

$$P(\mathbf{x}_E) = \sum_{\mathbf{x}_F} P(\mathbf{x}_E, \mathbf{x}_F). \tag{14.13}$$

There are two types of inference methods for probabilistic graphical models: exact inference methods and approximate inference methods. Exact inference methods compute the exact values of marginal distributions or conditional distributions. However, such methods are often impractical since their computational complexity increases exponentially to the number of maximal cliques. By contrast, approximate inference methods find approximate solutions with tractable time complexity and are more practical in real-world applications. The rest of this section introduces two representative exact inference methods, and we will introduce approximate inference methods in the next section.

### 14.4.1 Variable Elimination

Exact inference methods are essentially a kind of dynamic programming methods. Such methods attempt to reduce the cost of computing the target probability by exploiting the conditional independence encoded by the graphical model. Among them, *variable elimination* is the most intuitive one and is the basis of other exact inference methods.

We demonstrate variable elimination with the directed graphical model in ◘ Figure 14.7 (a).

14

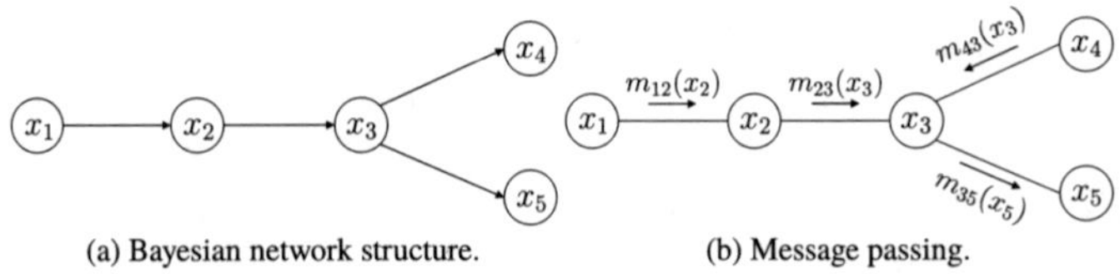


**Fig. 14.7** The process of variable elimination and message passing

Suppose that the inference objective is to compute the marginal probability $P(x_5)$. To compute it, we only need to eliminate the variables $\{x_1, x_2, x_3, x_4\}$ by summation, that is

$$P(x_5) = \sum_{x_4}\sum_{x_3}\sum_{x_2}\sum_{x_1} P(x_1, x_2, x_3, x_4, x_5)$$
$$= \sum_{x_4}\sum_{x_3}\sum_{x_2}\sum_{x_1} P(x_1)P(x_2 \mid x_1)P(x_3 \mid x_2)P(x_4 \mid x_3)P(x_5 \mid x_3). \quad (14.14)$$

Using the conditional independence encoded by the directed graphical model.

By doing the summations in the order of $\{x_1, x_2, x_4, x_3\}$, we have

$$P(x_5) = \sum_{x_3} P(x_5 \mid x_3) \sum_{x_4} P(x_4 \mid x_3) \sum_{x_2} P(x_3 \mid x_2) \sum_{x_1} P(x_1)P(x_2 \mid x_1)$$
$$= \sum_{x_3} P(x_5 \mid x_3) \sum_{x_4} P(x_4 \mid x_3) \sum_{x_2} P(x_3 \mid x_2) m_{12}(x_2), \quad (14.15)$$

where $m_{ij}(x_j)$ is an intermediate result in the summation, the subscript $i$ indicates that the term is the summation result with respect to $x_i$, and the subscript $j$ indicates other variables in the term. We notice that $m_{ij}(x_j)$ is a function of $x_j$. By repeating the process, we have

$$P(x_5) = \sum_{x_3} P(x_5 \mid x_3) \sum_{x_4} P(x_4 \mid x_3) m_{23}(x_3)$$
$$= \sum_{x_3} P(x_5 \mid x_3) m_{23}(x_3) \sum_{x_4} P(x_4 \mid x_3)$$
$$= \sum_{x_3} P(x_5 \mid x_3) m_{23}(x_3) m_{43}(x_3)$$
$$= m_{35}(x_5). \quad (14.16)$$

$m_{35}(x_5)$ is a function of $x_5$ and only depends on the value of $x_5$.

The above method also applies to undirected graphical models. For example, if we ignore the directions of the edges in ◘ Figure 14.7 (a) and consider it as an undirected graphical model, then we have

$$P(x_1, x_2, x_3, x_4, x_5) = \frac{1}{Z}\psi_{12}(x_1, x_2)\psi_{23}(x_2, x_3)\psi_{34}(x_3, x_4)\psi_{35}(x_3, x_5), \quad (14.17)$$

where $Z$ is the normalization factor. The marginal distribution $P(x_5)$ is given by

$$P(x_5) = \frac{1}{Z}\sum_{x_3} \psi_{35}(x_3, x_5) \sum_{x_4} \psi_{34}(x_3, x_4) \sum_{x_2} \psi_{23}(x_2, x_3) \sum_{x_1} \psi_{12}(x_1, x_2)$$
$$= \frac{1}{Z}\sum_{x_3} \psi_{35}(x_3, x_5) \sum_{x_4} \psi_{34}(x_3, x_4) \sum_{x_2} \psi_{23}(x_2, x_3) m_{12}(x_2)$$
$$= \cdots$$
$$= \frac{1}{Z} m_{35}(x_5). \quad (14.18)$$

By using the distributive law of multiplication to addition, variable elimination converts the problem of calculating summations of products of multiple variables to the problem of alternately calculating summations and products of some of the variables. Doing so simplifies the calculations by restricting the summations and products to local regions that involve only some of the variables.

Nevertheless, variable elimination has a clear disadvantage: there is a considerable amount of redundancy in the calculations of multiple marginal distributions. Taking the Bayesian network in ◘ Figure 14.7 (a) as an example, if we compute $P(x_4)$ after computing $P(x_5)$, then the calculations of $m_{12}(x_2)$ and $m_{23}(x_3)$ are repetitive when the summation order is $\{x_1, x_2, x_5, x_3\}$.

### 14.4.2 Belief Propagation

Also known as the *sum-product* algorithm.

The *belief propagation* algorithm avoid repetitive calculations by considering the summation operations in variable elimination as a process of message passing. In variable elimination, a variable $x_i$ is eliminated by the summation operation

$$m_{ij}(x_j) = \sum_{x_i} \psi(x_i, y_j) \prod_{k \in n(i) \setminus j} m_{ki}(x_i), \tag{14.19}$$

where $n(i)$ are the adjacent nodes of $x_i$. In belief propagation, however, the operation is considered as passing the message $m_{ij}(x_j)$ from $x_i$ to $x_j$. By doing so, the variable elimination process in (14.15) and (14.16) becomes a message passing process, as illustrated in ◘ Figure 14.7 (b). We see that each message passing operation involves only $x_i$ and its adjacent nodes, and hence the calculations are restricted to local regions.

14

In belief propagation, a node starts to pass messages after receiving the messages from all other nodes. The marginal distribution of a node is proportional to the product of all received messages, that is

$$P(x_i) \propto \prod_{k \in n(i)} m_{ki}(x_i). \tag{14.20}$$

Taking ◘ Figure 14.7 (b) as an example, $x_3$ must receive the messages from $x_2$ and $x_4$ before it passes the message to $x_5$, and the message $m_{35}(x_5)$ that $x_3$ passes to $x_5$ is exactly $P(x_5)$.

When there is no cycle in the graph, belief propagation can compute marginal distributions of all variables via the following two steps:

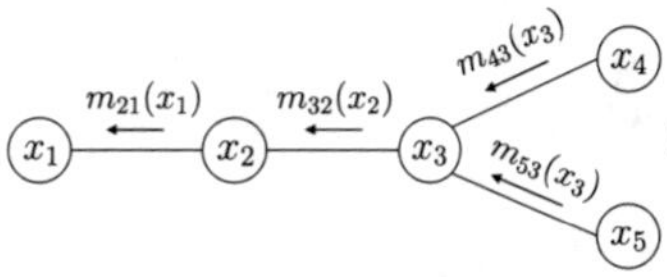


(a) Passing messages to the root node.

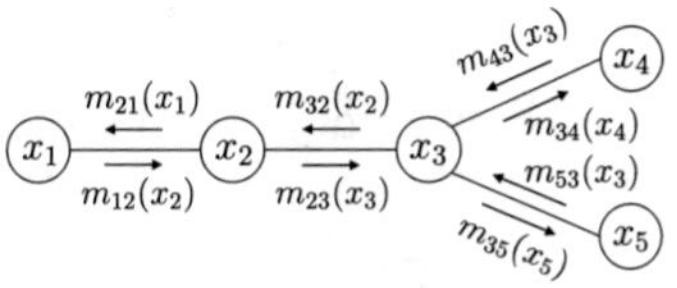


(b) Passing messages from the root node.

**Fig. 14.8** An illustration of the belief propagation algorithm

- Select a root node, and then pass messages from all leaf nodes to the root node until the root node has received messages from all adjacent nodes;
- Pass messages from the root node toward leaf nodes until all leaf nodes have received messages.

Taking ◘ Figure 14.7 (a) as an example, let $x_1$ be the root node, and $x_4$ and $x_5$ be the leaf nodes. The two steps of message passing are illustrated in ◘ Figure 14.8, where each edge has two messages on it with different directions. From the messages and (14.20), we have the marginal probabilities of all variables.

## 14.5 Approximate Inference

Exact inference methods are usually computationally expensive, and hence we often use approximate inference methods in practice. Roughly speaking, there are two types of approximate inference methods, namely sampling, which accomplishes approximation by stochastic methods, and deterministic approximations, represented by variational inference.

### 14.5.1 MCMC Sampling

In many tasks, we are interested in probability distributions just because we need them to calculate some expectations for decision-making. Taking the Bayesian network in ◘ Figure 14.7 (a) as an example, the goal of inference could be finding the expectation of $x_5$. It turns out that, sometimes, it can be more efficient to calculate or approximate the expectations directly without finding the probability distributions first.

The above idea motivates the sampling methods. Suppose our objective is to find the expectation of the function $f(x)$ with respect to the probability density function $p(x)$

Replace integration with summation if $x$ is discrete.

$$\mathbb{E}_p[f] = \int f(x)p(x)dx. \tag{14.21}$$

We can approximate the objective expectation $\mathbb{E}[f]$ by sampling a set of samples $\{x_1, x_2, \ldots, x_N\}$ from $p(x)$ and then compute the mean of $f(x)$ on these samples

Or a distribution related to $p(x)$.

$$\hat{f} = \frac{1}{N}\sum_{i=1}^{N} f(x_i), \tag{14.22}$$

According to the law of large numbers, we can obtain an accurate approximation from the *i.i.d.* samples $\{x_1, x_2, \ldots, x_N\}$ by large-scale sampling. The problem here is how to sample? For example, in probabilistic graphical models, how can we efficiently obtain samples from the probability distribution described by the graphical model?

One of the most commonly used sampling techniques for probabilistic graphical models is the Markov Chain Monte Carlo (MCMC) method. Given the probability density function $p(x)$ of a continuous variable $x \in X$, the probability that $x$ lies in the interval $A$ is

$$P(A) = \int_A p(x)dx. \tag{14.23}$$

If $f : X \mapsto \mathbb{R}$, then the expectation of $f(x)$ is given by

$$p(f) = \mathbb{E}_p[f(X)] = \int_x f(x)p(x)dx. \tag{14.24}$$

However, the integration in (14.24) is not easy to compute when $x$ is not univariate but a high-dimensional multivariate variable $\mathbf{x}$ that follows a complex distribution. Hence, MCMC first constructs some *i.i.d.* samples $\mathbf{x}_1, \mathbf{x}_2, \ldots, \mathbf{x}_N$ that follow the distribution $p$, and then obtains an unbiased estimate of (14.24) as

14

$$\tilde{p}(f) = \frac{1}{N}\sum_{i=1}^{N} f(\mathbf{x}_i). \tag{14.25}$$

Nevertheless, constructing *i.i.d.* samples that follow the distribution $p$ can still be difficult if the probability density function $p(\mathbf{x})$ is complex. The key idea of MCMC is to generate samples by constructing a "Markov chain with stationary distribution $p$". To be specific, by letting the Markov chain run for a sufficiently long time (i.e., converged to a stationary distribution), the generated samples approximately follow the distribution $p$. How do we know if the Markov chain has arrived at a stationary state? We say a Markov chain $T$ has arrived at a stationary state with a stationary distribution $p(\mathbf{x}^t)$ once the following stationary condition is met at time point $t$:

$$p(\mathbf{x}^t)T(\mathbf{x}^{t-1} \mid \mathbf{x}^t) = p(\mathbf{x}^{t-1})T(\mathbf{x}^t \mid \mathbf{x}^{t-1}), \tag{14.26}$$

where $T(\mathbf{x}' \mid \mathbf{x})$ is the state transition probability (i.e., the probability of transitioning from state $\mathbf{x}$ to state $\mathbf{x}'$), and $p(\mathbf{x}^t)$ is the distribution at time point $t$.

In short, MCMC starts by constructing a Markov chain and let it converge to the stationary distribution, which is exactly the posterior distribution of the parameters to be estimated. Then, it uses the Markov chain to generate the desired samples for further estimations. A vital step in this process is constructing the state transition probabilities of the Markov chain, and different construction methods lead to different MCMC algorithms.

The Metropolis–Hastings (MH) algorithm is an important representative of MCMC methods, which approximates the stationary distribution $p$ via *reject sampling*. The MH algorithm is given in ◘ Figure 14.1. In each round, the MH algorithm draws a candidate state sample $\mathbf{x}^*$ based on the sample $\mathbf{x}^{t-1}$ of the last round, where $x^*$ has a certain probability of being "rejected". Once $\mathbf{x}^*$ converged to a stationary state, from (14.26), we have

The Metropolis–Hastings algorithm is named after the original authors Metropolis et al. (1953) and Hastings (1970), who extended the algorithm to a general form afterwards.

$$p(\mathbf{x}^{t-1})Q(\mathbf{x}^* \mid \mathbf{x}^{t-1})A(\mathbf{x}^* \mid \mathbf{x}^{t-1}) = p(\mathbf{x}^*)Q(\mathbf{x}^{t-1} \mid \mathbf{x}^*)A(\mathbf{x}^{t-1} \mid x^*), \tag{14.27}$$

where $Q(\mathbf{x}^* \mid \mathbf{x}^{t-1})$ is the user-specified prior probability, $A(\mathbf{x}^* \mid \mathbf{x}^{t-1})$ is the probability of accepting $\mathbf{x}^*$, and $Q(\mathbf{x}^* \mid \mathbf{x}^{t-1})A(\mathbf{x}^* \mid \mathbf{x}^{t-1})$ is the state transition probability from state $\mathbf{x}^{t-1}$ to state $\mathbf{x}^*$. To arrive at the stationary state, we just need to set the acceptance probability to

$$A(\mathbf{x}^* \mid \mathbf{x}^{t-1}) = \min\left(1, \frac{p(\mathbf{x}^*)Q(\mathbf{x}^{t-1} \mid \mathbf{x}^*)}{p(\mathbf{x}^{t-1})Q(\mathbf{x}^* \mid \mathbf{x}^{t-1})}\right). \tag{14.28}$$

**Algorithm 14.1** Metropolis–Hastings Sampling

**Input:** Prior probability $Q(\mathbf{x}^* \mid \mathbf{x}^{t-1})$.

**Process:**

```
Initialize x^0;
for t = 1, 2, ... do
    Sample the candidate sample x* according to Q(x* | x^{t-1});
    Sample the threshold u from range (0, 1) according to uniform distribution;
    if u ⩽ A(x* | x^{t-1}) then
        x^t = x*;
    else
        x^t = x^{t-1}.
    end if
end for
return x^1, x^2, ...
```

**Output:** A list of sampled samples: $\mathbf{x}^1, \mathbf{x}^2, \ldots$

Repeat enough times to arrive at the stationary distribution.

According to (14.28).

In practice, we often discard the samples in the beginning of the list since we wish to use samples generated from the stationary distribution.

See Sect. 7.5.3 for Gibbs sampling.

Gibbs sampling is sometimes considered as a special case of the MH algorithm, since it also obtains samples using Markov chains with the target sampling distribution $p(\mathbf{x})$ as the stationary distribution. Specifically, let $\mathbf{x} = \{x_1, x_2, \ldots, x_N\}$ be the set of variables, and $p(\mathbf{x})$ be the objective distribution, then the Gibbs sampling algorithm generates samples by repeating the following steps after initializing $\mathbf{x}$:

(1) Select a variable $x_i$ either randomly or according to a certain ordering;

(2) Compute the conditional probability $p(x_i \mid \mathbf{x}_{\bar{i}})$, where $\mathbf{x}_{\bar{i}} = \{x_1, x_2, \ldots, x_{i-1}, x_{i+1}, \ldots, x_N\}$ is the current value of $\mathbf{x}$ excluding $x_i$;

(3) Sample a new value of $x_i$ from $p(x_i \mid \mathbf{x}_{\bar{i}})$ and replace the original value.



### 14.5.2 Variational Inference

*Variational inference* approximates complex distributions with simple and known distributions. It restricts the type of the approximate distribution, such that the approximate posterior distribution is locally optimal with a deterministic solution.

Before introducing the details of variational inference, let us see a concise way of representing graphical models—*plate notation* (Buntine 1994). Figure 14.9 gives an example. ◘ Figure 14.9 (a) shows that there are $N$ variables $\{x_1, x_2, \ldots, x_N\}$ dependent on the variable $\mathbf{z}$. In ◘ Figure 14.9 (b), the plate notation compactly describes the same relationship, where multiple variables independently generated by the same mechanism are placed in the same rectangle (plate), which allows nesting, and there is a label $N$ indicating the number of repeti-

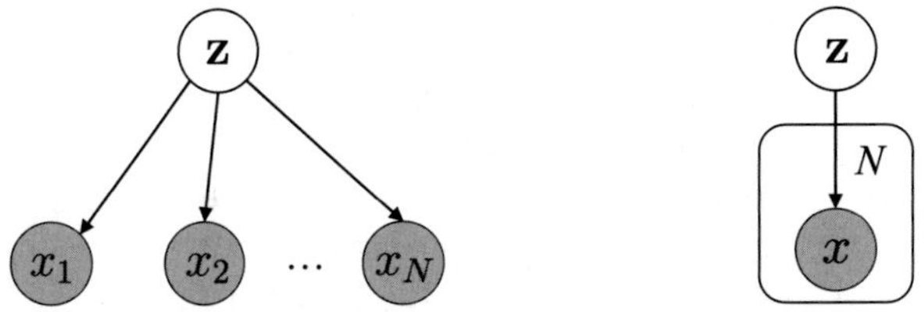


(a) Typical variable dependency graph. (b) Plate notation.

**Fig. 14.9** An example of plate notation

tions. Observable or known variables are usually shaded, e.g., $x$ in ◘ Figure 14.9. The plate notation provides a very concise way of representing variable relationships in various learning problems.

In ◘ Figure 14.9 (b), the probability density function of all observable variables $x$ is

$$p(\mathbf{x} \mid \Theta) = \prod_{i=1}^{N} \sum_{\mathbf{z}} p(x_i, \mathbf{z} \mid \Theta), \tag{14.29}$$

Approximate distributions used in variational inference should have nice mathematical properties. They are usually probability density functions of continuous variables.

which has the corresponding log-likelihood function

$$\ln p(\mathbf{x} \mid \Theta) = \sum_{i=1}^{N} \ln \left\{ \sum_{\mathbf{z}} p(x_i, \mathbf{z} \mid \Theta) \right\}, \tag{14.30}$$

where $\mathbf{x} = \{x_1, x_2, \ldots, x_N\}$, $\Theta$ includes the parameters of the distributions that $\mathbf{x}$ and $\mathbf{z}$ follow.

Generally speaking, the inference and learning task in ◘ Figure 14.9 is mainly estimating the hidden variable $\mathbf{z}$ and the distribution parameter variable $\Theta$, that is, finding $p(\mathbf{z} \mid \mathbf{x}, \Theta)$ and $\Theta$.

The parameters of graphical models are often estimated by maximum likelihood estimation. For (14.30), we can apply the EM algorithm. In the E-step, we infer $p(\mathbf{z} \mid \mathbf{x}, \Theta^t)$ by the parameter variable $\Theta^t$ at time point $t$, and then compute the joint likelihood function $p(\mathbf{x}, \mathbf{z} \mid \Theta)$. In the M-step, we use the current parameter $\Theta^t$ obtained in the E-step to find the parameter $\Theta^{t+1}$ of the next time point by optimizing the function $\mathcal{Q}(\Theta; \Theta^t)$

See Sect. 7.6 for the EM algorithm.

$$\begin{aligned}\Theta^{t+1} &= \arg\max_{\Theta} \mathcal{Q}(\Theta; \Theta^t) \\ &= \arg\max_{\Theta} \sum_{\mathbf{z}} p(\mathbf{z} \mid \mathbf{x}, \Theta^t) \ln p(\mathbf{x}, \mathbf{z} \mid \Theta).\end{aligned} \tag{14.31}$$

where $\mathcal{Q}(\Theta; \Theta^t)$ is actually the expectation of the joint log-likelihood function $\ln p(\mathbf{x}, \mathbf{z} \mid \Theta)$ with respect to the distribution $p(\mathbf{z} \mid \mathbf{x}, \Theta^t)$. It approximates the log-likelihood function when the distribution $p(\mathbf{z} \mid \mathbf{x}, \Theta^t)$ equals to the ground-truth posterior distribution of the hidden variable $\mathbf{z}$. Hence, the EM algorithm estimates not only the parameter $\Theta$ but also the distribution of the hidden variable $\mathbf{z}$.

Note that $p(\mathbf{z} \mid \mathbf{x}, \Theta^t)$ is not necessarily the ground-truth distribution of $\mathbf{z}$ but only an approximate distribution. Let $q(\mathbf{z})$ denote the approximate distribution, we can derive

$$\ln p(\mathbf{x}) = \mathcal{L}(q) + \mathrm{KL}(q \| p), \tag{14.32}$$

where

$$\mathcal{L}(q) = \int q(\mathbf{z}) \ln \left\{ \frac{p(\mathbf{x}, \mathbf{z})}{q(\mathbf{z})} \right\} d\mathbf{z}, \tag{14.33}$$

$$\mathrm{KL}(q \| p) = -\int q(\mathbf{z}) \ln \frac{p(\mathbf{z} \mid \mathbf{x})}{q(\mathbf{z})} d\mathbf{z}. \tag{14.34}$$

See Appendix C.3 for the KL divergence.

In practice, however, it may be difficult to find $p(\mathbf{z} \mid \mathbf{x}, \Theta^t)$ in the E-step due to the intractable multivariate $\mathbf{z}$, and this is when variational inference comes in handy. We typically assume that $\mathbf{z}$ follows the distribution

$$q(\mathbf{z}) = \prod_{i=1}^{M} q_i(\mathbf{z}_i). \tag{14.35}$$

In other words, we assume that we can decompose the complex multivariate variable $\mathbf{z}$ into a series of independent multivariate variables $\mathbf{z}_i$. With this assumption, the distribution $q_i$ can be made simple or has a good structure. For example, suppose $q_i$ is an *exponential family* distribution, then we have

To make the notation uncluttered, we abbreviate $q_i(\mathbf{z}_i)$ as $q_i$.

$$\begin{aligned}
\mathcal{L}(q) &= \int \prod_i q_i \left\{ \ln p(\mathbf{x}, \mathbf{z}) - \sum_i \ln q_i \right\} d\mathbf{z} \\
&= \int q_j \left\{ \int \ln p(\mathbf{x}, \mathbf{z}) \prod_{i \neq j} q_i d\mathbf{z}_i \right\} d\mathbf{z}_j - \int q_j \ln q_j d\mathbf{z}_j + \mathrm{const} \\
&= \int q_j \ln \tilde{p}(\mathbf{x}, \mathbf{z}_j) d\mathbf{z}_j - \int q_j \ln q_j d\mathbf{z}_j + \mathrm{const},
\end{aligned} \tag{14.36}$$

const is a constant.

14

where

$$\ln \tilde{p}(\mathbf{x}, \mathbf{z}_j) = \mathbb{E}_{i \neq j}[\ln p(\mathbf{x}, \mathbf{z})] + \mathrm{const}, \tag{14.37}$$

$$\mathbb{E}_{i \neq j}[\ln p(\mathbf{x}, \mathbf{z})] = \int \ln p(\mathbf{x}, \mathbf{z}) \prod_{i \neq j} q_i d\mathbf{z}_i. \tag{14.38}$$

Since we are interested in $q_j$, we can maximize $\mathcal{L}(q)$ with $q_{i \neq j}$ fixed. We notice that (14.36) equals to $-\mathrm{KL}(q_j \parallel \tilde{p}(\mathbf{x}, \mathbf{z}_j))$, that is, $\mathcal{L}(q)$ is maximized when $q_j = \tilde{p}(\mathbf{x}, \mathbf{z}_j)$. Hence, the optimal distribution $q_j^*$ that the variable subset $\mathbf{z}_j$ follows should satisfy

$$\ln q_j^*(\mathbf{z}_j) = \mathbb{E}_{i \neq j}[\ln p(\mathbf{x}, \mathbf{z})] + \text{const}, \tag{14.39}$$

so we have

$$q_j^*(\mathbf{z}_j) = \frac{\exp\left(\mathbb{E}_{i \neq j}[\ln p(\mathbf{x}, \mathbf{z})]\right)}{\int \exp\left(\mathbb{E}_{i \neq j}[\ln p(\mathbf{x}, \mathbf{z})]\right) d\mathbf{z}_j}. \tag{14.40}$$

In other words, with the assumption (14.35), (14.40) provides the best approximation to the ground-truth distribution of the variable subset $\mathbf{z}_j$.

With the assumption (14.35), we can often find a closed form solution to $\mathbb{E}_{i \neq j}[\ln p(\mathbf{x}, \mathbf{z})]$ by properly partitioning variable subsets $\mathbf{z}_j$ and selecting the distribution that $q_i$ follows; hence the hidden variable $\mathbf{z}$ can be inferred efficiently by (14.40). From (14.38), we observe that the estimation of the distribution $q_j^*$ of $\mathbf{z}_j$ not only considers $\mathbf{z}_j$ but also $\mathbf{z}_{i \neq j}$. Since this is achieved by finding the expectation of the joint log-likelihood function $\ln p(\mathbf{x}, \mathbf{z})$ with respect to $\mathbf{z}_{i \neq j}$, such a method is also called the *mean field* method.

When applying variational inference in practice, the most important thing is to find the proper hidden variable decomposition and the proper distribution hypothesis of each subset of hidden variables. With the hidden variable decomposition and distribution hypotheses, the parameter estimation and inference of probabilistic graphical models can be made by the EM algorithm and the consideration of (14.40). Clearly, the performance of variational inference is subject to the quality of hidden variable decomposition and distribution hypotheses.

"mean" refers to expectation and "field" refers to distribution.

## 14.6 Topic Model

*Topic model* is a family of generative directed graphical models mainly used for modeling discrete data, e.g., text corpus. As represented by Latent Dirichlet Allocation (LDA), topic models have been widely used in information retrieval and natural language processing.

There are three key concepts in topic models, namely *word*, *document*, and *topic*. A *word* is the basic discrete unit in the data, e.g., an English word in text processing. A *document* is a data object, such as a paper or a Web page, containing a set of words without considering the ordering of words. Such a representation is known as *bag-of-words*. Topic models apply

We can describe an image using bag-of-words by considering the small blocks in the image as words, and then topic models are applicable.

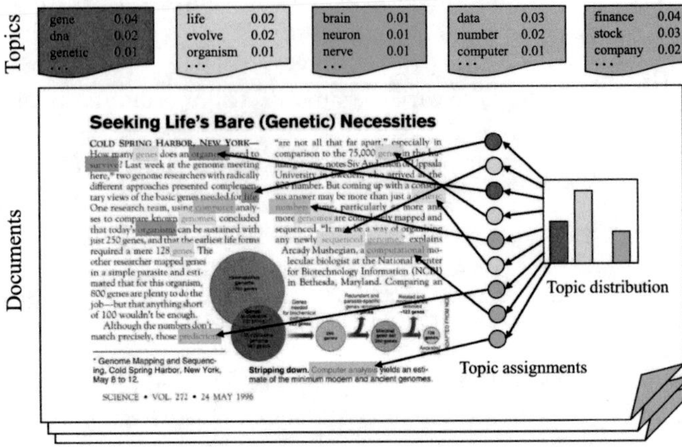


**Fig. 14.10** An illustration of the document generation process of LDA

to any data objects that can be described by bag-of-words. A *topic* describes a concept represented by a series of related words together with the probabilities that they appear in the concept.

Figure 14.10 provides an intuitive example of topic model. A topic is like a box containing those words with high probability to appear under the concept of the topic. Suppose that we have a data set of $T$ documents on $K$ topics, where all words in the documents are from a dictionary of $N$ distinct words. The data set (i.e., collection of documents) is denoted by $T \times N$-dimensional vectors $\mathbf{W} = \{\mathbf{w}_1, \mathbf{w}_2, \ldots, \mathbf{w}_T\}$, where $w_{t,n}$ (i.e., the $n$th component of $\mathbf{w}_t \in \mathbb{R}^N$) is the frequency of the word $n$ appeared in the document $t$. The topics are denoted by $K$ $N$-dimensional vectors $\boldsymbol{\beta}_k$ $(k = 1, 2, \ldots, K)$, where $\beta_{k,n}$ (i.e., the $n$th component of $\boldsymbol{\beta}_k \in \mathbb{R}^N$) is the frequency of the word $n$ in the topic $k$.

In practice, we can obtain the word frequency vectors $\mathbf{w}_i$ $(i = 1, 2, \ldots, T)$ by counting the words in documents, though we do not know which topic is mentioned in which document. LDA assumes that each document contains multiple topics that can be modeled by a generative model. More specifically, let $\Theta_t \in \mathbb{R}^K$ denote the proportion of each topic in document $t$, and $\Theta_{t,k}$ denote the proportion of topic $k$ in document $t$. Then, LDA assumes a document $t$ is "generated" by the following steps:

Some words are usually excluded, such as *stop words*.

(1) Randomly draw a topic distribution $\Theta_t$ from a Dirichlet distribution with parameter $\boldsymbol{\alpha}$;
(2) Generate $N$ words for document $t$ by the following steps:
   (a) Obtain a topic assignment $z_{t,n}$ according to topic distribution $\Theta_t$ for each word $n$ in document $t$;

See Appendix C.1.6 for Dirichlet distribution.

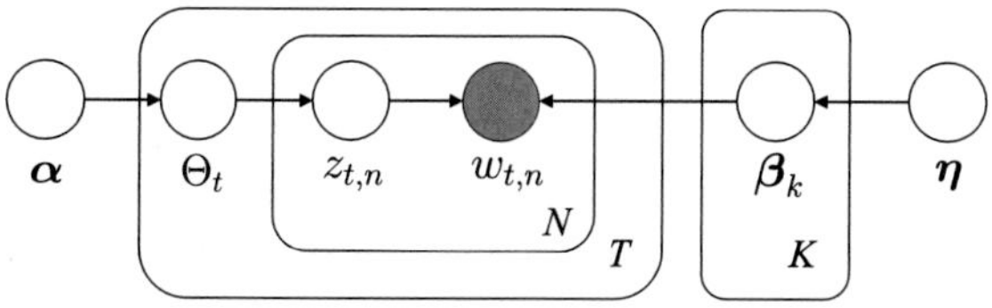


**Fig. 14.11** The plate notation of LDA

(b) Generate a word through random sampling according to the word frequency distribution $\boldsymbol{\beta}_k$ corresponding to topic assignment $z_{t,n}$.

The above document generation process is illustrated in ◘ Figure 14.10. Note that a generated document will have different proportions of topics (step 1), and each word in the document comes from a topic (step 2b) generated according to the topic distribution (step 2a).

The plate notation in ◘ Figure 14.11 describes the relationships between variables, where word frequency $w_{t,n}$ is the only observed variable that depends on topic assignment $z_{t,n}$ and the corresponding word frequency distribution $\boldsymbol{\beta}_k$. The topic assignment $z_{t,n}$ depends on topic distribution $\Theta_t$, which depends on a parameter $\boldsymbol{\alpha}$. The word frequency distribution $\boldsymbol{\beta}_k$ depends on a parameter $\boldsymbol{\eta}$. Then, the probability distribution of LDA is

$$p(\mathbf{W}, \mathbf{z}, \boldsymbol{\beta}, \Theta \mid \boldsymbol{\alpha}, \boldsymbol{\eta}) =$$
$$\prod_{t=1}^{\mathrm{T}} p(\Theta_t \mid \boldsymbol{\alpha}) \prod_{i=1}^{K} p(\boldsymbol{\beta}_k \mid \boldsymbol{\eta}) \left( \prod_{n=1}^{N} P(w_{t,n} \mid z_{t,n}, \boldsymbol{\beta}_k) P(z_{t,n} \mid \Theta_t) \right), \tag{14.41}$$

where $p(\Theta_t \mid \boldsymbol{\alpha})$ is usually set to a $K$-dimensional Dirichlet distribution with a parameter $\boldsymbol{\alpha}$, and $p(\boldsymbol{\beta}_k \mid \boldsymbol{\eta})$ is usually set to an $N$-dimensional Dirichlet distribution with a parameter $\boldsymbol{\eta}$. For example

$$p(\Theta_t \mid \boldsymbol{\alpha}) = \frac{\Gamma(\sum_k \alpha_k)}{\prod_k \Gamma(\alpha_k)} \prod_k \Theta_{t,k}^{\alpha_k - 1}, \tag{14.42}$$

where $\Gamma(\cdot)$ is the Gamma function. Clearly, $\boldsymbol{\alpha}$ and $\boldsymbol{\eta}$ in (14.41) are the model parameters to be determined.

See Appendix C.1.5 for Gamma function.

Given a data set $\mathbf{W} = \{\mathbf{w}_1, \mathbf{w}_2, \ldots, \mathbf{w}_T\}$, the parameters of LDA can be estimated by maximum likelihood estimation, that is, finding $\boldsymbol{\alpha}$ and $\boldsymbol{\eta}$ by maximizing the log-likelihood

The word frequencies in training documents.

$$LL(\alpha, \eta) = \sum_{t=1}^{T} \ln p(\mathbf{w}_t \mid \alpha, \eta). \tag{14.43}$$

However, it is difficult to solve (14.43) directly since $p(\mathbf{w}_t \mid \alpha, \eta)$ is not easy to compute. In practice, we often use variational inference to find an approximate solution.

Once $\alpha$ and $\eta$ are found, we can use word frequency $w_{t,n}$ to infer the topic structure of a document, that is, inferring $\Theta_t$, $\beta_k$, and $z_{t,n}$ by solving

$$p(\mathbf{z}, \boldsymbol{\beta}, \Theta \mid \mathbf{W}, \alpha, \eta) = \frac{p(\mathbf{W}, \mathbf{z}, \boldsymbol{\beta}, \Theta \mid \alpha, \eta)}{p(\mathbf{W} \mid \alpha, \eta)}. \tag{14.44}$$

Similarly, (14.44) is hard to solve since $p(\mathbf{w} \mid \alpha, \eta)$ is not easy to compute. In practice, Gibbs sampling or variational inference is often employed to find an approximate solution.

## 14.7 Further Reading

Koller and Friedman (2009) is a book dedicated to probabilistic graphical models. Pearl (1982) initialized the study on Bayesian networks, and (Pearl 1988) summarized relevant studies in the early days. Geman and Geman (1984) proposed Markov random fields, which are often used together with Bayesian networks in real-world applications. Hidden Markov model and its application to speech recognition can be found in Rabiner (1989). Lafferty et al. (2001) proposed conditional random fields, and more information can be found in Sutton and McCallum (2012).

Pearl (1986) proposed the belief propagation algorithm as an exact inference method, and it was derived into many approximate inference methods. For typical cyclic graphs, the initialization and message passing mechanisms of belief propagation need to be modified, resulting in the Loopy Belief Propagation algorithm (Murphy et al. 1999). Its theoretical properties are still unclear, though some progress can be found in Mooij and Kappen (2007), Weiss (2000). Some cyclic graphs can be described by *factor graphs* (Kschischang et al. 2001), and then converted into factor trees for belief propagation. There are attempts (Lauritzen and Spiegelhalter 1988) to enable belief propagation for arbitrary graph structures. Recent advances in parallel computing motivated studies on parallelized belief propagation. For example, Gonzalez et al. (2009) proposed the concept of $\tau_\epsilon$ approximate inference and designed a multi-core parallelized belief propagation algorithm with a time complexity decreasing linearly to the number of cores.

The modeling and inference techniques, particularly variational inference, for graphical models became mature in the middle 1990s, and (Jordan 1998) summarized the major outcomes in that period. See (Wainwright and Jordan 2008) for more information about variational inference.

An advantage of graphical models is that we can intuitively and quickly define models for specific learning problems. A prominent representative of such methods is LDA (Blei et al. 2003), which has many variants (Blei 2012). One research direction of probabilistic graphic models is to make the model adaptive to the data (i.e., *non-parametric* methods), such as Hierarchical Dirichlet Processes (Teh et al. 2006) and Infinite Latent Feature Model (Ghahramani and Griffiths 2006).

*Non-parametric* means parameters, such as assumptions on data distribution, are not required to be specified, and this is an important advancement of Bayesian learning. See Sect. 7.7 for Bayesian learning.

Not all topic models are Bayesian learning methods. For example, Probabilistic Latent Semantic Analysis (PLSA) (Hofmann 2001) is a probabilistic extension of Latent Semantic Analysis (LSA).

LSA is a variant of SVD for textual data.

Monte Carlo methods are a family of numerical methods developed in the 1940s based on random numbers, and probability and statistical theory. MCMC is the combination of Markov chains and the Monte Carlo method, and (Pearl 1987) introduced it to Bayesian network inference. See Neal (1993) for more information about applying MCMC to probabilistic inference. See Andrieu et al. (2003), Gilks et al. (1996) for more information about MCMC.

See "Break Time" of Chap. 11 for Monte Carlo methods.

## Exercises

**14.1** Use plate notation to represent conditional random field and naïve Bayes classifier.

**14.2** Prove the local Markov property in graphical models: a variable is conditionally independent of other variables given the adjacent variables.

**14.3** Prove the pairwise Markov property in graphical models: two non-adjacent variables are conditionally independent given all other variables.

**14.4** Explain why the potential functions are only needed for the maximal cliques in Markov random field.

**14.5** Discuss the similarities and differences between conditional random field and logistic regression.

**14.6** Prove that the computational complexity of the variable elimination method increases exponentially to the number of maximal cliques in graphical models, but does not necessarily increase exponentially to the number of nodes.

**14.7** Gibbs sampling can be seen as a special case of the Metropolis–Hastings algorithm, but it does not take the reject sampling strategy. Discuss the advantages of doing so.

**14.8** Mean field is an approximate inference method. Considering (14.32), discuss the difference between the approximated problem solved by the mean field method and the original problem, and discuss how to select the prior probability of variables in practice.



**14.9** * Download or implement the LDA algorithm, and apply it to a novel book (e.g., Robinson Crusoe) to see how the topics evolve over chapters.

**14.10** * Design an improved LDA algorithm that does not require the predefined number of topics.

## Break Time

### Short Story: Judea Pearl—Pioneer of Probabilistic Graphical Models

We must mention the Israeli-American computer scientist Judea Pearl (1936–) when talking about graphical probabilistic models. Pearl was born in Tel Aviv. After obtaining his B.S. degree from the Technion in 1960, Pearl emigrated to the United States to continue his study at the Newark College of Engineering and received his Ph.D. degree in Electrical Engineering from the Polytechnic Institute of Brooklyn in 1965. After graduation, he worked at RCA Research Laboratories on superconductive amplifiers and storage devices, and later on, in 1970, he joined the University of California, Los Angeles.

Research on artificial intelligence in the early days focused on symbolism learning and logic reasoning, which can hardly process and represent uncertainties in a quantitative manner. In the 1970s, Pearl introduced probabilistic methods into artificial intelligence and invented a series of techniques, including Bayesian network and Belief propagation, which led to the framework of probabilistic graphical models. By using Bayesian networks as a tool, Pearl created a new research area known as causal inference. Pearl received the ACM/AAAI Allen Newell Award in 2003, and later on, the Turing Award in 2011 for his fundamental contributions to artificial intelligence through the development of a calculus for probabilistic and causal reasoning. ACM commented that "*Pearl's work not only revolutionized the field of artificial intelligence, but also became an important tool for many other branches of engineering and the natural sciences.*" In 2011, Pearl also received the Lakatos Award, which is the most prestigious award in the philosophy of science.

See Sect. 1.5.

The ACM/AAAI Allen Newell Award is presented to people for career contributions that have breadth within computer science, or that bridge computer science and other disciplines. The award is named after Allen Newell (1927–1992), who is a Turing Award winner and a pioneer in the field of artificial intelligence. The second recipient of this award from the field of machine learning is Michael Jordan, who received the award in 2009.

## References


Andrieu C, Freitas ND, Doucet A, Jordan MI (2003) An introduction to MCMC for machine learning. Mach Learn 50(1–2):5–43

Blei DM (2012) Probabilistic topic models. Commun ACM 55(4):77–84

Blei DM, Ng A, Jordan MI (2003) Latent Dirichlet allocation. J Artif Intell Res 3:993–1022

Buntine W (1994) Operations for learning with graphical models. J Artif Intell Res 2:159–225

Geman S, Geman D (1984) Stochastic relaxation, Gibbs distributions, and the Bayesian restoration of images. IEEE Trans Patt Anal Mach Intell 6(6):721–741

Ghahramani Z, Griffiths TL (2006) Infinite latent feature models and the Indian buffet process. In: Weiss Y, Scholkopf B, Platt JC (eds) Advances in Neural Information Processing Systems 18 (NIPS). MIT Press, Cambridge, MA, pp 475–482

Gilks WR, Richardson S, Spiegelhalter DJ (1996) Markov chain monte carlo in practice. Chapman & Hall/CRC, Boca Raton, FL

Gonzalez JE, Low Y, Guestrin C (2009) Residual splash for optimally parallelizing belief propagation. In Proceedings of the 12th International Conference on Artificial Intelligence and Statistics (AISTATS), Clearwater Beach, FL, pp 177–184

Hastings WK (1970) Monte Carlo sampling methods using Markov chains and their applications. Biometrica 57(1):97–109

Hofmann T (2001) Unsupervised learning by probabilistic latent semantic analysis. Mach Learn 42(1):177–196

Jordan MI (ed) (1998) Learning in graphical models. Kluwer, Dordrecht, Netherlands

Koller D, Friedman N (2009) Probabilistic graphical models: principles and techniques. MIT Press, Cambridge, MA

Kschischang FR, Frey BJ, Loeliger H-A (2001) Factor graphs and the sum-product algorithm. IEEE Trans Inf Theory 47(2):498–519

Lafferty JD, McCallum A, Pereira FCN (2001) Conditional random fields: Probabilistic models for segmenting and labeling sequence data. In: Proceedings of the 18th International Conference on Machine Learning (ICML), Williamstown, MA, pp 282–289

Lauritzen SL, Spiegelhalter DJ (1988) Local computations with probabilities on graphical structures and their application to expert systems. J R Stat Soc -Ser B 50(2):157–224

Metropolis N, Rosenbluth AW, Rosenbluth MN, Teller AH, Teller E (1953) Equations of state calculations by fast computing machines. J Chem Phys 21(6):1087–1092

Mooij JM, Kappen HJ (2007) Sufficient conditions for convergence of the sum-product algorithm. IEEE Trans Inf Theory 53(12):4422–4437

Murphy KP, Weiss Y, Jordan MI (1999) Loopy belief propagation for approximate inference: an empirical study. In: Proceedings of the 15th Conference on Uncertainty in Artificial Intelligence (UAI), Stockholm, Sweden, pp 467–475

Neal RM (1993) Probabilistic inference using Markov chain Monte Carlo methods. Technical Report CRG-TR-93-1, Department of Computer Science, University of Toronto

Pearl J (1982) Asymptotic properties of minimax trees and game-searching procedures. In: Proceedings of the 2nd National Conference on Artificial Intelligence (AAAI), Pittsburgh, PA

Pearl J (1986) Fusion, propagation and structuring in belief networks. Artif Intell 29(3):241–288

Pearl J (1987) Evidential reasoning using stochastic simulation of causal models. Artif Intell 32(2):245–258

Pearl J (1988) Probabilistic reasoning in intelligent systems: networks of plausible inference. Morgan Kaufmanns, San Francisco, CA

Rabiner LR (1989) A tutorial on hidden Markov model and selected applications in speech recognition. Proc IEEE 77(2):257–286

Sutton C, McCallum A (2012) An introduction to conditional random fields. Found Trends Mach Learn 4(4):267–373

Teh YW, Jordan MI, Beal MJ, Blei DM (2006) Hierarchical Dirichlet processes. J Am Stat Assoc 101(476):1566–1581

Wainwright MJ, Jordan MI (2008) Graphical models, exponential families, and variational inference. Found Trends Mach Learn 1(1–2):1–305

Weiss Y (2000) Correctness of local probability propagation in graphical models with loops. Neural Comput 12(1):1–41

# Rule Learning

Table of Contents




Z.-H. Zhou, *Machine Learning*,
https://doi.org/10.1007/978-981-15-1967-3_15

## 15.1 Basic Concepts

Broadly speaking, all predictive models can be seen as one or a set of rules. In rule learning, we refer to logic rules with the term "logic" omitted.

In machine learning, *rules* usually refer to logic rules in the form of "if ..., then ..." that can describe regular patterns or domain concepts with clear semantics (Fürnkranz et al. 2012). *Rule learning* is about learning a set of rules from training data for predicting unseen samples.

Formally, a rule is in the form of

$$\oplus \leftarrow \mathbf{f}_1 \wedge \mathbf{f}_2 \wedge \cdots \wedge \mathbf{f}_L, \tag{15.1}$$

In formal logic, a *literal* is an atomic formula (atom) or its negation.

where the right-hand side of implication symbol "$\leftarrow$" is called the *antecedent* or *body* of the rule, and the left-hand side is called the *consequent* or *head* of the rule. The body is a *conjunction* of *literals* $\mathbf{f}_k$, where "$\wedge$" represents "AND". Each literal $\mathbf{f}_k$ is a Boolean expression on a feature, e.g., $(\texttt{color} = \texttt{dark})$ or $\neg(\texttt{root} = \texttt{straight})$. $L$ is the length of rule indicating the number of literals in the rule body. The head $\oplus$ consists of literals representing decisions or concepts, e.g., ripe. Such logic rules are also known as *if-then rules*.

Compared to *black boxes*, such as neural networks and support vector machines, the decision process of rule learning is more intuitive and transparent, leading to better interpretability. Besides, most human knowledge can be concisely described and represented by formal logic. For example, a piece of knowledge like "the father of father is grandfather" is difficult to be expressed by numerical functions, but it can be conveniently expressed by a first-order logic formula "$\texttt{grandfather}(X, Y) \leftarrow \texttt{father}(X, Z) \wedge \texttt{father}(Z, Y)$". Hence, we can easily introduce domain knowledge into rule learning. The abstraction ability of logic rules leads to significant advantages in dealing with some highly complex AI problems. For example, in question answering systems, we may have many or even infinite possible answers, and performing abstraction and reasoning using logic rules will bring apparent convenience.

Suppose we learned the following rule set $\mathcal{R}$ from the watermelon data set:

Rule 1: $\texttt{ripe} \leftarrow (\texttt{root} = \texttt{curly}) \wedge (\texttt{umbilicus} = \texttt{hollow})$

Rule 2: $\neg\texttt{ripe} \leftarrow (\texttt{texture} = \texttt{blurry})$.

The watermelon data set 2.0 is given in Table 4.1.

Rule 1 has a length of 2, and it classifies samples by checking the *valuation* (or truth-value assignment) of two literals. A sample (e.g., sample 1 in the watermelon data set 2.0) that satisfies the rule body is said to be *covered* by the rule. However, not being covered by rule 1 does not imply the watermelon is unripe. We classify a watermelon as unripe if it is covered by rules with the head "$\neg$ripe" (e.g., rule 2).

Each rule in the rule set can be seen as a submodel, and the rule set is an ensemble of submodels. When a sample is covered by multiple rules with different classification outcomes, we say there is a *conflict*, which can be resolved by *conflict resolution*. Common conflict resolution strategies include (weighted) voting, ordering, and meta-rule methods. The voting strategy returns the most agreed prediction. The ordering strategy defines a preference order over the rule set, and the highest-ordered rule is used when there is a conflict; the corresponding rule learning process is called *ordered rule* learning or *priority rule* learning. The meta-rule method defines a set of *meta-rules* based on domain knowledge for resolving conflicts, that is, rules about rules, e.g., "Choosing the rule with the minimum length when there is a conflict."

See Chap. 8 for ensemble learning.

The rules learned from training data may be unable to cover all unseen samples. For example, the rule set $\mathcal{R}$ cannot classify samples with "`root` = `curly`", or samples with "`umbilicus` = `slightly hollow`" and "`texture` = `clear`". Such cases are common when the number of features is large. Hence, rule learning algorithms often set a *default rule* that handles uncovered samples, e.g., a default rule for $\mathcal{R}$ could be "all watermelons not covered by rule 1 and rule 2 are unripe."

A default rule can be seen as a special meta-rule.

Rules can be divided into two classes by the expressive power of formal languages: *propositional rule* and *first-order rule*. A propositional rule is a plain statement consisting of *propositional atoms* and logical connectors such as *conjunction* ($\wedge$), *disjunction* ($\vee$), *negation* ($\neg$), and *implication* ($\leftarrow$). For example, the rule set $\mathcal{R}$ is a propositional rule set, where `root` = `curly` and `umbilicus` = `hollow` are propositional atoms. By contrast, the basic elements of first-order rules are *atomic formulas* that describe the features or relations of objects. For example, the predicate `father`$(X, Y)$ is an atomic formula that describes the father–son relationship. The successor function $\sigma(X) = X + 1$ is also an atomic formula. Let the predicate $N(X)$ denote "$X$ is a natural number", and $\forall X$ denote "holds for all $X$", $\exists Y$ denote "exists $Y$ such that", then we can write the statement "all natural numbers incremented by 1 are natural numbers" as $\forall X \exists Y (N(Y) \leftarrow N(X) \wedge (Y = \sigma(X)))$, or more concisely, $\forall X (N(\sigma(X)) \leftarrow N(X))$. Such a rule is called first-order rule, where $X$ and $Y$ are logic variables, $\forall$ ("for all") and $\exists$ ("exists") are *quantifiers* that specify the quantity of specimens in the domain of discourse that satisfy the rule. First-order rules are also called *relational rules* since they can express complex relations between objects. Taking the watermelon data set as an example, if we simply use the names of features as predicates to define the relations between their values and samples, then propositional rule set $\mathcal{R}$ can be rewritten as the following first-order rule set $\mathcal{R}'$:

Rule 1: $\texttt{ripe}(X) \leftarrow \texttt{root}(X, \texttt{curly}) \wedge \texttt{umbilicus}(X, \texttt{hollow})$;

Rule 2: $\neg\texttt{ripe}(X) \leftarrow \texttt{texture}(X, \texttt{blurry})$.

From the perspective of formal language, propositional rules are special cases of first-order rules, and hence first-order rule learning is more complicated than propositional rule learning.

## 15.2 Sequential Covering

Rule learning aims to find a rule set that can cover as many samples as possible. The most straightforward approach is *sequential covering*, which induces one rule at a time: every time a rule is learned, all samples covered by it are removed from the training set, and the learning process repeats with the remaining samples in the training set. Such a strategy is also called *separate-and-conquer*, since only part of the training set is processed in each round.

Let us take a closer look at sequential covering with propositional rule learning. The body of a propositional rule consists of Boolean functions for testing the feature values (e.g., $\texttt{color} = \texttt{green}$ and $\texttt{sugar} \leqslant 0.2$), and the head of rule is the class label. The core of sequential covering is how to learn individual rules from the training set. Given a rule head $\oplus$, rule learning is a search problem of finding the optimal set of literals for the rule body. Formally, when given a positive data set and a negative data set, the learning task is to induce the optimal rule **r** based on the candidate literal set $\mathcal{F} = \{\mathbf{f}_k\}$. In propositional rule learning, a candidate literal is a Boolean expression in the form of "$R(\text{feature}_i, \text{feature}_{i,j})$", where $\text{feature}_i$ is the $i$th feature, $\text{feature}_{i,j}$ is the $j$th candidate value of $\text{feature}_i$, and $R(x, y)$ is a binary Boolean function that tests whether $x$ and $y$ satisfy relation $R$.

The simplest approach is to start with an empty rule "$\oplus \leftarrow$" with the positive class as the rule head and then iterate over every candidate value of every feature: the rule can be constructed by conjunctively adding them to the rule body as literals. Once the current rule covers only positive samples, we return the rule and remove the covered samples from the training set, and then induce the next rule with the remaining samples.

Taking the watermelon training set 2.0 as an example, we generate the literals ripe and color = green from the sample 1:

The watermelon training set 2.0 is in the first part of ◘ Table 4.2.

$$\texttt{ripe} \leftarrow (\texttt{color} = \texttt{green}).$$

This rule covers the samples 1, 6, 10, and 17, where two of them are positive and the rests are negative. Since the rule does not meet the requirement "covering only positive samples", we replace the proposition with another atomic proposition about feature color (e.g., color = dark). However, the new rule still fails to meet the requirement, hence we fallback to color = green and try to add a propositional atom about other features, e.g., root = curly:

For ease of discussion, we do not consider literals of negative atoms in the rest of this chapter, that is, we only consider candidate literals in the form of **f** but not ¬**f**.

$$\text{ripe} \leftarrow (\text{color} = \text{green}) \wedge (\text{root} = \text{curly}),$$

which still covers negative sample 17. Therefore, we replace the second proposition with another atomic proposition about this feature, e.g., root = slightly curly:

$$\text{ripe} \leftarrow (\text{color} = \text{green}) \wedge (\text{root} = \text{slightly curly}),$$

which finally excludes negative samples and satisfies the requirement of "covering only positive samples", though it covers only one positive sample. We keep this rule and remove the covered sample 6 and then use the rest 9 samples as the updated training set. By repeating this process, we have

Rule 1: ripe ← (color = green) ∧ (root = slightly curly);
Rule 2: ripe ← (color = green) ∧ (sound = muffled);
Rule 3: ripe ← (color = dark) ∧ (root = curly);
Rule 4: ripe ← (color = dark) ∧ (texture = slightly blurry),

which is a rule set covering all positive samples but none of the negative samples, and this is the output of sequential covering.

When there are many features and candidate values, the above exhaustive search becomes infeasible due to the combinatorial explosion. Therefore, in practice, we often take either a *top-down* strategy or a *bottom-up* strategy. The *top-down* strategy, also called the *generate-then-test* method, starts with a general rule and gradually adds more literals to reduce the coverage of samples until the pre-specified conditions are met. Such a process is known as *specialization* of rules. The bottom-up strategy, also called the *data-driven* method, starts with a specialized rule and gradually removes literals to increase the coverage of samples until the pre-specified conditions are met. Such a process is known as *generalization* of rules. The top-down strategy, which searches for rules from high coverage to low coverage, often generates rules with better generalization and noise-tolerance than the rules generated by the bottom-up strategy, though the bottom-up strategy is more suitable with limited training samples. In practice, propositional rule learning often takes the top-down strategy, whereas first-order rule learning, which deals with more complex hypothesis space, often takes the bottom-up strategy.

For example, an empty rule without any features is a general rule that covers all samples.

For example, a rule generated from all the feature values of a sample is a specialized rule that covers only this sample.

The watermelon data set 2.0 is given in the first part of ◘ Table 4.2.

Let us see a demonstration of the top-down strategy using the watermelon training set 2.0. We start with an empty rule "`ripe` ←", and then gradually add each "feature = value" as a propositional atom to the empty rule for consideration. Suppose the quality of a rule is measured by its accuracy on the training set, and let $n/m$ be the accuracy of the new rule obtained by adding a proposition, where $n$ is the number of positive samples in the $m$ samples covered. As illustrated in ◘ Figure 15.1, both `color` = `dark` and `umbilicus` = `hollow` have achieved the highest accuracy 3/4 in the first round.

Adding the first encountered literal `color` = `dark` to the empty rule, gives

`ripe` ← (`color` = `dark`).

Then, the samples covered by the above rule are used as the training set in the second round, in which, we find that adding any of the five literals in ◘ Figure 15.1 can achieve an accuracy of 100%. By choosing the literal `root` = `curly`, which locates first and covers the most samples, we have

`ripe` ← (`color` = `dark`) ∧ (`root` = `curly`).

The rule induction process needs criteria for evaluating the quality of rules. For example, in the above example, we first consider the accuracy of rules. With the same accuracy, we further consider the sample coverage, followed by the order of propositions. Such criteria can be varied according to different learning problems.

The above example is greedy to consider only one "optimal" literal in each round, and such an approach can easily lead to local optimum. To alleviate this problem, we can take a more



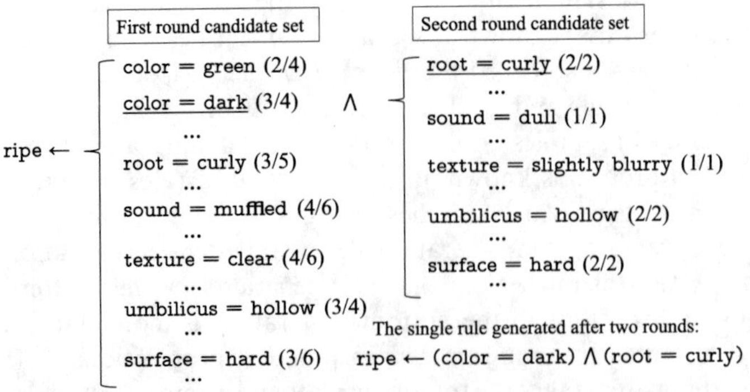


**Fig. 15.1** Taking the top-down strategy to generate single rules from the watermelon data set 2.0

gentle approach such as *beam search*, which adds the best $b$ literals in the current round to the candidate literal set of the next round. Taking Figure 15.1 as an example, if we let $b = 2$, then both literals that have achieved an accuracy of 3/4 in the first round are kept for the second round; after the second round, we obtain the following rule, which not only achieves an accuracy of 100% but also covers 3 positive samples:

$$\texttt{ripe} \leftarrow (\texttt{umbilicus} = \texttt{hollow}) \wedge (\texttt{root} = \texttt{curly}).$$

Due to the simplicity and effectiveness of sequential covering, it is the basis for almost all rule learning algorithms. Sequential covering can be easily extended to multiclass problems by considering each class in turn: when learning rules for class $c$, considering all class $c$ samples as positive and the rest samples as negative.

## 15.3 Pruning Optimization

Rule induction is essentially a greedy search process that needs a mechanism to alleviate the risk of overfitting, and one common approach is *pruning*. Like decision tree learning, pruning can take place during the rule generation (i.e., *pre-pruning*) or after the rules have been generated (i.e., *post-pruning*). Whether pruning is needed or not is usually decided by comparing the performance of the rule or rule set before and after adding/removing its literals and rules, respectively.

See Sect. 4.3 for decision tree pruning.

Pruning can also be decided by statistical significance test. For example, CN2 (Clark and Niblett 1989) performs pre-pruning by assuming that the predictive performance of the rule set must be significantly better than that of simply making predictions according to the posterior probability distribution exhibited by the training set. For ease of calculation, CN2 employs Likelihood Ratio Statistics (LRS). Let $m_+$ and $m_-$ denote, respectively, the number of positive samples and the number of negative samples in the training set; let $\hat{m}_+$ and $\hat{m}_-$ denote, respectively, the number of positive and negative samples covered by the rule set. Then, LRS can be defined as

See Sect. 2.4 for statistical significance test.

$$\mathrm{LRS} = 2 \cdot \left( \hat{m}_+ \log_2 \frac{\left(\frac{\hat{m}_+}{\hat{m}_+ + \hat{m}_-}\right)}{\left(\frac{m_+}{m_+ + m_-}\right)} + \hat{m}_- \log_2 \frac{\left(\frac{\hat{m}_-}{\hat{m}_+ + \hat{m}_-}\right)}{\left(\frac{m_-}{m_+ + m_-}\right)} \right), \tag{15.2}$$

which is essentially a measure of information. To be specific, LRS measures the difference between the distribution of samples covered by the rule (set) and the empirical distribution of all the training samples: a larger LRS implies that the predictions made by the rule (set) are more likely to be different from the predictions made by simply conjecturing based on the ratio of positive/negative samples in the training set; and a smaller LRS implies that the performance of the rule (set) is more likely to be by chance. In real-world applications with a large amount of data, the LRS threshold is often set to a large value (e.g., 0.99) for stopping the growth of rule (set).

A widely used approach of post-pruning is the Reduced Error Pruning (REP) (Brunk and Pazzani 1991), which takes the following procedure: split samples into a training set and a validation set, and then do multiple rounds of pruning on the rule set $\mathcal{R}$ learned from the training set; in each round, use the validation set to find the best rule set by evaluating all possible pruning operations, such as deleting a literal from the rule body, deleting the last one or multiple literals in the rule, and deleting the entire rule. The above procedure repeats until pruning no longer improves the performance on the validation set.

In rule learning, they are often called *growing set* and *pruning set*.

Though REP is often effective (Brunk and Pazzani 1991), it has a complexity of $O(m^4)$ for $m$ training samples. Incremental REP (IREP) (Fürnkranz and Widmer 1994) managed to reduce the complexity to $O(m \log^2 m)$ with the following procedure: before generating a rule, the current data set is split into a training set and a validation set; then, a rule $\mathbf{r}$ is generated from the training set and is immediately pruned by REP on the validation set to get $\mathbf{r}'$; remove all samples covered by $\mathbf{r}'$ and repeat the above process on the remaining samples. IREP is more efficient since it only prunes one rule at a time, whereas REP prunes the entire rule sets.

Better performance can often be achieved by combining pruning with other post-process techniques for rule-set optimization. Taking the well-known rule learning algorithm RIPPER (Cohen 1995) as an example, it can achieve both better generalization performance and faster learning speed than many decision tree algorithms, and the secrete behind it is combing pruning with post-processing optimization.

RIPPER stands for Repeated Incremental Pruning to Produce Error Reduction, which is named JRIP in WEKA.

15

Given the number of repetitions $k$ in ◘ Algorithm 15.1, RIPPER is also called RIPPER$k$, e.g., RIPPER5 means $k = 5$.

The pseudocode of RIPPER is given in ◘ Algorithm 15.1. RIPPER starts by generating rule set $\mathcal{R}$ based on the pruning mechanism of IREP* (Cohen 1995). IREP* improves IREP by replacing the evaluation heuristic of accuracy in IREP with $\frac{\hat{m}_+ + (m_- - \hat{m}_-)}{m_+ + m_-}$, pruning each single rule by deleting multiple literals from the end of the body, and performing a last-time pruning with IREP on the entire rule set that has been learned. The post-processing mechanism in RIPPER aims to further improve the performance after the pruning for every single rule has been done. To this end, RIPPER generates two variants for each rule $\mathbf{r}_i \in \mathcal{R}$

- $\mathbf{r}'_i$: use IREP* to generate a *replacement rule* $\mathbf{r}'_i$ from the samples covered by $\mathbf{r}_i$;
- $\mathbf{r}''_i$: specialize $\mathbf{r}_i$ by adding literals, and then use IREP* to generate a *revised rule* $\mathbf{r}''_i$.

**Algorithm 15.1** RIPPER

**Input:** The data set $D$;
The number of repetitions $k$.

**Process:**

1: $\mathcal{R} = \text{IREP*}(D)$;
2: $i = 0$;
3: **repeat**
4: $\quad \mathcal{R}^* = \text{PostOpt}(\mathcal{R})$;
5: $\quad D_i = \text{NotCovered}(\mathcal{R}^*, D)$;
6: $\quad \mathcal{R}_i = \text{IREP*}(D_i)$;
7: $\quad \mathcal{R} = \mathcal{R}^* \cup \mathcal{R}_i$;
8: $\quad i = i + 1$.
9: **until** $i = k$

**Output:** Rule set $\mathcal{R}$.

Generate rule set based on IREP*.

Post-processing.

Remove covered samples.

After that, $\mathbf{r}'_i$ and $\mathbf{r}''_i$ are added to $\mathcal{R}$ without $\mathbf{r}_i$, giving $\mathcal{R}'$ and $\mathcal{R}''$, respectively. Then, the rule sets $\mathcal{R}$, $\mathcal{R}'$ and $\mathcal{R}''$ are compared and the optimal one is kept as $\mathcal{R}^*$. This process is denoted by PostOpt($\mathcal{R}$) in line 4 of ◘ Algorithm 15.1,

Why would the optimization strategy of RIPPER work? The reason is simple: the rules are generated under a particular order in the initial $\mathcal{R}$ ignoring the subsequently learned rules, and in this way, a greedy approach can easily be stuck at a local optimum. The post-processing optimization of RIPPER alleviates the locality problem of greedy approach by revisiting $\mathcal{R}$ at the end to achieve a better performance (Fürnkranz et al. 2012).

## 15.4 First-Order Rule Learning

Due to the limitation of the expressive power of propositional logic, propositional rule learning can hardly handle more complex *relations* between objects, although such relation information is crucial in many applications. For example, when we pick watermelons in a supermarket, it can be difficult to describe all watermelons with precise feature values: how green is color = green and how dull is sound = dull? A more practical way is to compare watermelons. For example, "watermelon 1 is riper than watermelon 2" since "watermelon 1 has a greener color and curlier root than watermelon 2". However, such an argument is beyond the expressive power of propositional logic, and hence we need to employ the first-order rule learning.

Let us define the following concepts for our watermelon data:

- darkness of color: dark > green > light;
- curliness of root: curly > slightly curly > straight;
- dullness of sound: dull > muffled > crisp;
- clearness of texture: clear > slightly blurry > blurry;
- hollowness of umbilicus: hollow > slightly hollow > flat;
- hardness of surface: hard > soft.

With these concepts, we convert the watermelon data set 2.0 to the watermelon data set 5.0 as shown in Table 15.1. Data in such a format is called *relational data*, which describes the relations between samples. The atomic formulas, such as darker_color and curlier_root, that are converted from the original features are called *background knowledge*. The atomic formulas, such as riper and ¬riper, that are converted from the class labels are called relational data *examples*. From the watermelon data set 5.0, we can learn first-order rules, such as

Such rules are also called first-order logic clause.

$$(\forall X, \forall Y)(\mathrm{riper}(X, Y) \leftarrow \mathrm{curlier\_root}(X, Y) \wedge \mathrm{hollower\_umbilicus}(X, Y)).$$

Though the above first-order rule is still in the form of (15.1), the head and body of the rule are first-order logic expressions. riper(·, ·), curlier_root(·, ·), and hollower_umbilicus(·, ·) are predicates describing relations, and the individual objects "watermelon 1" and "watermelon 2" are replaced with logic variables $X$ and $Y$. The universal quantifier ∀ indicates that the rule holds for all individual objects in the domain. Since all variables in first-order rules are quantified by the universal quantifier, we omit the quantifiers in subsequent discussions when the context is clear.

First-order rules have strong expressive power. For example, recursion can be concisely expressed as

$$\mathrm{riper}(X, Y) \leftarrow \mathrm{riper}(X, Z) \wedge \mathrm{riper}(Z, Y).$$

**Tab. 15.1** The watermelon data set 5.0.

| | | |
|---|---|---|
| darker_color(2, 1) | darker_color(2, 6) | darker_color(2, 10) |
| darker_color(2, 14) | darker_color(2, 16) | darker_color(2, 17) |
| ... | ... | ... |
| darker_color(3, 1) | darker_color(3, 6) | darker_color(15, 16) |
| ... | ... | ... |
| darker_color(15, 17) | darker_color(17, 14) | darker_color(17, 16) |
| curlier_root(1, 6) | curlier_root(1, 7) | curlier_root(1, 10) |
| ... | ... | ... |
| curlier_root(1, 14) | curlier_root(17, 7) | curlier_root(17, 10) |
| curlier_root(17, 14) | curlier_root(17, 15) | duller_sound(2, 1) |
| duller_sound(2, 3) | duller_sound(2, 6) | duller_sound(2, 7) |
| ... | ... | ... |
| duller_sound(17, 7) | duller_sound(17, 10) | duller_sound(17, 15) |
| duller_sound(17, 16) | clearer_texture(1, 7) | clearer_texture(1, 14) |
| clearer_texture(1, 16) | clearer_texture(1, 17) | clearer_texture(15, 14) |
| ... | ... | ... |
| clearer_texture(15, 16) | clearer_texture(15, 17) | clearer_texture(17, 16) |
| hollower_umbilicus(1, 6) | hollower_umbilicus(1, 7) | hollower_umbilicus(1, 10) |
| ... | ... | ... |
| hollower_umbilicus(1, 15) | hollower_umbilicus(15, 10) | hollower_umbilicus(15, 16) |
| hollower_umbilicus(17, 10) | hollower_umbilicus(17, 16) | harder_surface(1, 6) |
| harder_surface(1, 7) | harder_surface(1, 10) | harder_surface(1, 15) |
| ... | ... | ... |
| harder_surface(17, 6) | harder_surface(17, 7) | harder_surface(17, 10) |
| harder_surface(17, 15) | | |
| riper(1, 10) | riper(1, 14) | riper(1, 15) |
| ... | ... | ... |
| riper(1, 16) | riper(7, 14) | riper(7, 15) |
| riper(7, 16) | riper(7, 17) | ¬riper(10, 1) |
| ¬riper(10, 2) | ¬riper(10, 3) | ¬riper(10, 6) |
| ... | ... | ... |
| ¬riper(17, 2) | ¬riper(17, 3) | ¬riper(17, 6) |
| ¬riper(17, 7) | | |

The numbers in parentheses correspond to the ID in ◘ Table 4.2.

The formulas above the divider are background knowledge, and the formulas below the divider are samples.

Another advantage of first-order rules over propositional rules is the ability to incorporate domain knowledge. In propositional rule learning or even in general statistical learning, there are two typical approaches for incorporating domain knowledge: using domain knowledge to construct new features on top of current features, or design a mechanism according to domain knowledge (e.g., regularization) to constrain the hypothesis space. However, in practice, not all domain knowledge can be easily included through reconstructing features or introducing regularization. For example, suppose we obtain a chemical compound $X$ composed of unknown chemical elements, and wish to find how $X$ is reacted with known chemical compound $Y$. To this end, we repeat the reaction multiple times, and each time, we analyze the constituent elements from the outcome. Though we know nothing about the properties of the unknown chemical elements, some general chemical principles are available as domain knowledge, such as metal atoms

Statistical learning are generally based on *attribute-value* representation, whose expressive power is equivalent to propositional logic. Such learning approaches can be collectively called *propositional learning*.

produce ionic bond and hydrogen atoms share covalent bond, as well as some possible reactions between known chemical elements. With such domain knowledge, it is relatively easy to find the reaction formula between $X$ and $Y$, and we can also infer the properties of $X$ or even discover new chemical molecules or elements. Such kind of domain knowledge is common in real-world applications, but it can hardly be utilized in propositional learning.

One famous first-order rule learning algorithm, FOIL (First-Order Inductive Learner) (Quinlan 1990), follows the sequential covering framework and takes a top-down inductive strategy. FOIL is similar to the propositional rule learning discussed in Sect. 15.2, but it needs to consider more variable combinations due to the existence of logic variables in the first-order setting. Taking the watermelon data set 5.0 as an example, we start with an empty rule for the concept riper($X$, $Y$)

riper($X$, $Y$) $\leftarrow$ .

Then, we consider combinations of all possible predicates and variables as candidate literals, which must include at least one variable that have already appeared in the rule; otherwise the literal would be meaningless. In this example, we consider the following candidate literals:

| | |
|---|---|
| darker_color($X$, $Y$), | darker_color($Y$, $X$), |
| darker_color($X$, $Z$), | darker_color($Z$, $X$), |
| darker_color($Y$, $Z$), | darker_color($Z$, $Y$), |
| darker_color($X$, $X$), | darker_color($Y$, $Y$), |
| curlier_root($X$, $Y$), | duller_sound($X$, $Y$), |
| ... | ... |

FOIL selects literals by evaluating the *FOIL gain*

$$\text{F\_Gain} = \hat{m}_+ \times \left(\log_2 \frac{\hat{m}_+}{\hat{m}_+ + \hat{m}_-} - \log_2 \frac{m_+}{m_+ + m_-}\right), \tag{15.3}$$

where $m_+$ and $m_-$ are, respectively, the number of positive samples and negative samples that are covered by the original rule; $\hat{m}_+$ and $\hat{m}_-$ are, respectively, the number of positive samples and negative samples that are covered by the new rule after adding a literal. Unlike information gain used in decision trees, FOIL gain considers only the information of positive samples and uses the coverage of positive samples of the new rule as

See Sect. 4.2.1 for information gain used in decision trees.

15

weight. The reason is that the number of positive samples in relational data is often far fewer than the number of negative samples, and hence we pay more attention to positive samples.

Essentially, this is due to class imbalance. See Sect. 3.6.

For our example in the watermelon data set 5.0, the new rule can cover 16 positive samples and 2 negative samples by adding either darker_color($X$, $Y$) or hollower_umbilicus($X$, $Y$) to the empty rule. The corresponding FOIL gain is $16 \times (\log_2 \frac{16}{18} - \log_2 \frac{25}{50}) = 13.28$. Suppose we choose darker_color($X$, $Y$), then we have

riper($X$, $Y$) $\leftarrow$ darker_color($X$, $Y$).

Because the rule still covers 2 negative samples riper(15, 1) and riper(15, 6), FOIL grows the length of the rule body as propositional rule learning does, and adds the final one to the rule set. At the end, FOIL optimizes the rule set with pruning.

FOIL can produce recursive rules if we allow the target predicate to appear in candidate literals. Besides, the rule set is often more concise if we allow negation literals (i.e., $\neg\mathbf{f}$).

FOIL sits somewhere between the propositional rule learning and inductive logic programming, which will be discussed in the next section. The expressive power of FOIL is still limited since its top-down strategy does not support function symbols and nested expressions; however, FOIL converts propositional rule learning directly to first-order rule learning by operations such as replacing constants representing objects with logic variables, and therefore, FOIL is often more efficient than typical inductive logic programming methods.

## 15.5 Inductive Logic Programming

Inductive Logic Programming (ILP) gives stronger expressive power to machine learning systems by supporting function symbols and allowing nested expressions in first-order rule learning. Besides, ILP can be seen as using machine learning techniques to induce logic programs from background knowledge. The learned *rules* can be directly used in logic programming languages (i.e. PROLOG).

Nevertheless, the nesting of functions and logic expressions also brings great challenges to computing. For example, given unary predicate $P$ and unary function $f$, we can construct infinite number of literals such as $P(X)$, $P(f(X))$ and $P(f(f(X)))$. Then, the number of candidate atomic formulas becomes infinite. As a result, top-down strategies such as FOIL and propositional logic rule learning will fail, since we cannot enumerate all candidate literals when growing the rule. Moreover, the calculation of FOIL gain needs to calculate the coverage of rules, and this becomes infeasible after introducing function symbols and nested expressions.

### 15.5.1 Least General Generalization

ILP takes a bottom-up strategy which directly uses the *grounded facts* of one or more positive samples as the bottom rules, and then gradually generalizes the rules to improve the coverage. The generalization operation could be deleting literals from the rule body or replacing constants with logic variables.

The watermelon data set 5.0 is in Table 15.1.

Taking the watermelon data set 5.0 as an example. For ease of discussion, suppose riper$(X, Y)$ only depends on the relations involving $(X, Y)$. Then, the bottom rules of positive samples riper(1, 10) and riper(1, 15) are, respectively

Here the numbers are the ID of watermelons.

$$\begin{aligned} \text{riper}(1, 10) \leftarrow\ & \text{curlier\_root}(1, 10) \wedge \text{duller\_sound}(1, 10) \\ & \wedge \text{hollower\_umbilicus}(1, 10) \wedge \text{harder\_surface}(1, 10); \\ \text{riper}(1, 15) \leftarrow\ & \text{curlier\_root}(1, 15) \wedge \text{hollower\_umbilicus}(1, 15) \\ & \wedge \text{harder\_surface}(1, 15). \end{aligned}$$

The above two rules have limited generalization abilities since they only describe two specific relational data samples. Hence, we wish to convert such specific rules to rules that are more general. To achieve this goal, Least General Generalization (LGG) (Plotkin 1970) is the most fundamental technique.

Given first-order formulas $\mathbf{r}_1$ and $\mathbf{r}_2$, LGG starts by finding all literals that have the same predicate, and then checks each pair of constants at the same position in the literals: if the two constants are the same, then they remain unchanged, denoted by LGG$(t, t) = t$; otherwise, the constants are replaced by a new variable that also applies to all other places in the formulas. For example, if the two different constants are $s$ and $t$ and the new variable is $V$, then we have LGG$(s, t) = V$, which means whenever $s$ or $t$ appears in the formula, we replace it by $V$. Let us take the two rules in the above as an example and LGG would start by comparing riper(1, 10) and riper(1, 15); since "10"≠"15", both constants are replaced with $Y$, and all pairs of "10" and "15" in $\mathbf{r}_1$ and $\mathbf{r}_2$ are replaced with $Y$. Then, we have

$$\begin{aligned} \text{riper}(1, Y) \leftarrow\ & \text{curlier\_root}(1, Y) \wedge \text{duller\_sound}(1, 10) \\ & \wedge \text{hollower\_umbilicus}(1, Y) \wedge \text{harder\_surface}(1, Y); \\ \text{riper}(1, Y) \leftarrow\ & \text{curlier\_root}(1, Y) \wedge \text{hollower\_umbilicus}(1, Y) \\ & \wedge \text{harder\_surface}(1, Y). \end{aligned}$$

15

After that, LGG ignores all literals with different predicates in both formulas $\mathbf{r}_1$ and $\mathbf{r}_2$. Because the LGG cannot specialize to a formula $\mathbf{r}$ when the LGG contains a predicate that does not present in the formula $\mathbf{r}$. In our example, the literal duller_sound(1, 10) is ignored, gives the LGG

$$\begin{aligned}\mathtt{riper}(1, Y) \leftarrow & \ \mathtt{curlier\_root}(1, Y) \wedge \mathtt{hollower\_umbilicus}(X, Y_2) \\ & \wedge \mathtt{harder\_surface}(1, Y).\end{aligned} \tag{15.4}$$

We see that (15.4) can only decide whether watermelon 1 is better than the others. To improve its generalization ability, suppose we have another bottom rule about watermelon 2

$$\begin{aligned}\mathtt{riper}(2, 10) \leftarrow & \ \mathtt{darker\_color}(2, 10) \wedge \mathtt{curlier\_root}(2, 10) \\ & \wedge \mathtt{duller\_sound}(2, 10) \wedge \mathtt{hollower\_umbilicus}(2, 10) \\ & \wedge \mathtt{harder\_surface}(2, 10),\end{aligned} \tag{15.5}$$

then we can find the LGG of (15.4) and (15.5). We notice that the constant "10" and variable "$Y$" appear in the same position of the literals riper(2, 10) and riper(1, $Y$). Hence, we let LGG(10, $Y$) $= Y_2$ and replace all pairs of "10" and "$Y$" in the formulas with $Y_2$. Finally, by letting LGG(2, 1) $= X$ and deleting all literals that do not have a common predicate in both formulas, we have the following rule without any constant:

$$\begin{aligned}\mathtt{riper}(X, Y_2) \leftarrow & \ \mathtt{curlier\_root}(X, Y_2) \wedge \mathtt{hollower\_umbilicus}(X, Y_2) \\ & \wedge \mathtt{harder\_surface}(X, Y_2).\end{aligned}$$

The above example only considers literals without negation ($\neg$). In fact, LGG is able to perform more complex generalization operations. Besides, we assumed that the bottom rules of riper($X$, $Y$) only includes relations about ($X$, $Y$), but there are often other useful relations in background knowledge as well. Therefore, many ILP systems take different approaches to select bottom rules, where the most commonly used one is Relative Least General Generalization (RLGG) (Plotkin 1971). When computing LGG, RLGG takes background knowledge into consideration by setting the bottom rule of sample $e$ as $e \leftarrow K$, where $K$ is the conjunction of all atoms in background knowledge.

See Chap. 3 of Lavrac and Dzeroski (1993).

It is easy to prove that LGG is the most specialized formula among all first-order formulas that can specialize to $\mathbf{r}_1$ and $\mathbf{r}_2$: there is no first-order formula $\mathbf{r}'$ that can specialize to both $\mathbf{r}_1$ and $\mathbf{r}_2$ while can also generalize to the LGG of $\mathbf{r}_1$ and $\mathbf{r}_2$.

In many ILP systems, after obtaining the LGG, we add it to the rule set just like an individual rule, and then optimize the rule set using the techniques that we introduced in earlier sections, such as post-pruning.

### 15.5.2 Inverse Resolution

In logic, *deduction* and *induction* are fundamental approaches that humans use to understand the world. Roughly speaking, deduction infers specific phenomenons from general rules, whereas induction summarizes general rules from specific observations. The poofs of mathematical theorems are representative examples of applying deduction, whereas machine learning falls into induction category. In 1965, logician J. A. Robinson proposed that the deductive reasoning in first-order logic can be described by a simple rule, that is, the famous *resolution principle* (Robinson 1965). Two decades later, computer scientists S. Muggleton and W. Buntine proposed *inverse resolution* (Muggleton and Buntine 1988) for inductive reasoning, which played an important role in the development of ILP.

Induction was mathematically proved to be the inverse of deduction by W. S. Jevons, who was the British economist and logician in the nineteenth century.

Using the resolution principle, we can link first-order logic rules with background knowledge for simplification; whereas using inverse resolution, we can develop new concepts and relations from background knowledge. Next, we take the simpler propositional reasoning as an example to demonstrate how resolution principle and inverse resolution work.

Suppose logic expressions $C_1$ and $C_2$ hold, containing complementary literals $L_1$ and $L_2$, respectively; without loss of generality, letting $L = L_1 = \neg L_2$, $C_1 = A \vee L$, and $C_2 = B \vee \neg L$. The resolution principle tells that we can obtain *resolvent* $C = A \vee B$ by eliminating $L$ using deductive reasoning. If we define the deletion operation on disjunctive normal form as

$$(A \vee B) - \{B\} = A, \tag{15.6}$$

then the resolution process can be expressed as

15

$$C = (C_1 - \{L\}) \vee (C_2 - \{\neg L\}), \tag{15.7}$$

abbreviated by

$$C = C_1 \cdot C_2. \tag{15.8}$$

■ Figure 15.2 provides an illustrating example of the resolution principle.

Oppositely, inverse resolution is about how to obtain $C_j$ $(i \neq j)$ given $C$ and $C_i$. Suppose that we wish to find $C_2$ given $C$ and $C_1$, then the inverse resolution process is

$$C_2 = (C - (C_1 - \{L\})) \vee \{\neg L\}. \tag{15.9}$$

How can we conduct inverse resolution in practice? Muggleton (1995) defined four complete inverse resolution operations. Let rule $p \leftarrow q$ be an equivalent expression of $p \vee \neg q$, let lowercase letters be atomic literals, and let uppercase letters be logic clauses consisting of conjunctions, then the four inverse resolution operations are

$$\text{Absorption:} \quad \frac{p \leftarrow A \wedge B \qquad q \leftarrow A}{p \leftarrow q \wedge B \qquad q \leftarrow A}. \tag{15.10}$$

$$\text{Identification:} \quad \frac{p \leftarrow A \wedge B \qquad p \leftarrow A \wedge q}{q \leftarrow B \qquad p \leftarrow A \wedge q}. \tag{15.11}$$

$$\text{Intra-construction:} \quad \frac{p \leftarrow A \wedge B \qquad p \leftarrow A \wedge C}{q \leftarrow B \qquad p \leftarrow A \wedge q \qquad q \leftarrow C}. \tag{15.12}$$

$$\text{Inter-construction:} \quad \frac{p \leftarrow A \wedge B \qquad q \leftarrow A \wedge C}{p \leftarrow r \wedge B \qquad r \leftarrow A \qquad q \leftarrow r \wedge C}. \tag{15.13}$$

Here, $\frac{X}{Y}$ represents "$X$ implies $Y$", which is written as $X \vdash Y$ in formal logic. In the above rules, a clause in $X$ is either the resolvent of $Y$ or an equivalent term of a clause in $Y$; the new literals appear in $Y$ can be seen as new propositions learned via induction.

Both resolution and inverse resolution can be easily extended to first-order logic. Unlike propositional logic, the resolution and inverse resolution of first-order logic often need unification and substitution operations.

*Substitution* refers to substituting the variables in logic expressions with other terms. For example, by substituting $C$ = darker_color($X$, $Y$) $\wedge$ duller_sound($X$, $Y$) with $\theta = \{1/X, 2/Y\}$, we have $C' = C\theta =$ darker_color(1, 2) $\wedge$ duller_sound(1, 2), where $\{X, Y\}$ is called the domain of $\theta$. Similar to substitutions in algebra, we also have *composition of substitutions* and *inverse substitution*. For example, we first use $\theta = \{Y/X\}$ to substitute $X$ with $Y$, and then use $\lambda = \{1/Y\}$ to substitute $Y$ with 1. Such a composition of substitution is denoted by $\theta \circ \lambda$, and the inverse substitution is denoted by $\theta^{-1} = \{X/Y\}$.

*Unification* refers to making two or more logic expressions equal by variable substitutions. For example, given two logic expressions $A$ = darker_color(1, $X$) and $B$ = darker_color ($Y$, 2), we can apply $\theta = \{2/X, 1/Y\}$ such that $A\theta = B\theta =$

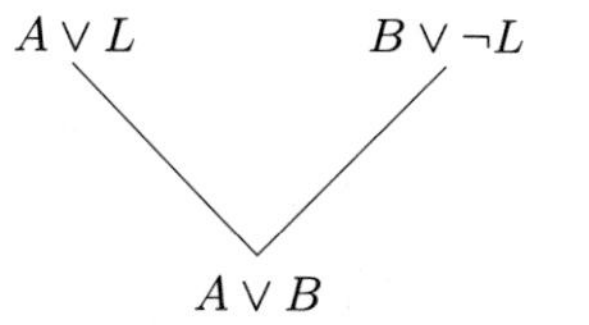


**Fig. 15.2** An example of the resolution principle

darker_color(1, 2); in such case, we say that $A$ and $B$ are *unifiable* and $\theta$ is the *unifier* of $A$ and $B$. Let $\delta$ be the unifier of a set of first-order logic expressions $W$, then we say $\delta$ is the Most General Unifier (MGU) of $W$ if there exists a substitution $\lambda$ for any unifier $\theta$ of $W$ such that $\theta = \delta \circ \lambda$. MGU is also called the *most general substitution*, which is one of the most important concepts of ILP. For example, the logic expressions darker_color(1, $Y$) and darker_color($X$, $Y$) can be unified by both $\theta_1 = \{1/X\}$ and $\theta_2 = \{1/X, 2/Y\}$, where only $\theta_1$ is their MGU.

When resolving first-order logical clauses, we need to use unification to search complementary terms $L_1$ and $L_2$. Given two first-order logic expressions $C_1 = A \vee L_1$ and $C_2 = B \vee L_2$, if there exists an unifier $\theta$ such that $L_1\theta = \neg L_2\theta$, then we can resolve the expressions as

$$C = (C_1 - \{L_1\})\theta \vee (C_2 - \{L_2\})\theta. \tag{15.14}$$

Similarly, we can perform inverse resolution of first-order logic by extending (15.9) with unifier. Based on (15.8), we define $C_1 = C/C_2$ and $C_2 = C/C_1$ as *resolution quotients*, and then the objective of inverse resolution is to find resolution quotient $C_2$ given $C$ and $C_1$. For $L_1 \in C_1$, suppose $\phi_1$ is a substitution such that

For $C = A \vee B$, we have $A \vdash C$ and $\exists B(C = A \vee B)$ equivalent.

$$(C_1 - \{L_1\})\phi_1 \vdash C, \tag{15.15}$$

where the domain of $\phi_1$ is all variables in $C_1$, denoted by vars($C_1$). The purpose of $\phi_1$ is to make the corresponding literals in $C_1 - \{L_1\}$ and $C$ unifiable. Let $\phi_2$ be a substitution with the domain vars($L_1$) − vars($C_1 - \{L_1\}$), $L_2$ be the literal to be eliminated from the resolution quotient $C_2$, $\theta_2$ be a substitution with the domain vars($L_2$), and both $\phi_1$ and $\phi_2$ operate on $L_1$ such that $\neg L_1\phi_1 \circ \phi_2 = L_2\theta_2$. Then, $\phi_1 \circ \phi_2 \circ \theta_2$ is the MGU of $\neg L_1$ and $L_2$. If we denote the composition of substitutions $\phi_1 \circ \phi_2$ as $\theta_1$ an denote $\theta_2^{-1}$ as the inverse substitution of $\theta_2$, then we have $(\neg L_1\theta_1)\theta_2^{-1} = L_2$. Therefore, like (15.9), the first-order inverse resolution can be represented as

$$C_2 = (C - (C_1 - \{L_1\})\theta_1 \vee \{\neg L_1\theta_1\})\theta_2^{-1}. \tag{15.16}$$

In first-order inverse resolution, the choices of $L_1$, $L_2$, $\theta_1$, and $\theta_2$ are usually not unique, and hence other decision criteria are required, such as coverage, accuracy, and information entropy.

Taking the watermelon data set 5.0 as an example, suppose we have already obtained the following rules:

$C_1$ = riper(1, $X$) ← curlier_root(1, $X$) ∧ clearer_texture(1, $X$);
$C_2$ = riper(1, $Y$) ← curlier_root(1, $Y$) ∧ duller_sound(1, $Y$).

15

We recognize that the rules are in the forms of $p \leftarrow A \wedge B$ and $p \leftarrow A \wedge C$, therefore, we can perform inverse resolution via intra-construction in (15.12). Since the predicates in both $C_1$ and $C_2$ are binary, we invent a new binary predicate $q(M, N)$ to represent the generalized information in our new rule. Then, from (15.12), we have

$$C' = \texttt{riper}(1, Z) \leftarrow \texttt{curlier\_root}(1, Z) \wedge q(M, N).$$

The other two terms at the bottom of (15.12) are the resolution quotients of $C_1/C'$ and $C_2/C'$, respectively. For $C_1/C'$, we have the choices $\neg\texttt{curlier\_root}(1, Z)$ and $\neg q(M, N)$ for eliminating $L_1$ from $C'$. $q$ is a newly invented predicate, and sooner or later, we need to learn a new rule $q(M, N) \leftarrow ?$ to define it. According to the Occam's razor principle, the fewer rules are better given the same expressive power, we therefore, let $\neg q(M, N)$ be $L_1$. From (15.16), we have $L_2 = q(1, S)$, $\phi_1 = \{X/Z\}$, $\phi_2 = \{1/M, X/N\}$, and $\theta_2 = \{X/S\}$. By simple calculation, we can obtain that the resolution quotient for $C_1/C'$ is $q(1, S) \leftarrow \texttt{clearer\_texture}(1, S)$. Similarly, we can find the resolution quotient for $C_2/C'$ is $q(1, T) \leftarrow \texttt{duller\_sound}(1, T)$.

See Sect. 1.4 for the Occam's razor principle.

An important ability of inverse resolution is inventing new predicates, which potentially correspond to new knowledge that does not exist in sample features or background knowledge. Such an ability is important for knowledge discovery and refinement. Nevertheless, the actual semantics of automatically invented predicates, e.g., whether $q$ means `fresher`, `sweeter`, or `more sun exposure` can only be decided by users based on further understanding of the task domain.

In the above example, we only showed how to perform inverse resolution based on two rules. In practice, ILP systems often take a bottom-up strategy to generate a set of rules, which are further processed by inverse resolution and the LGG.

## 15.6 Further Reading

Rule learning is the main representative of *symbolism learning* and also one of the earliest machine learning techniques (Michalski 1983). A comprehensive summary of rule learning can be found in Fürnkranz et al. (2012).

The basic framework of rule learning is sequential covering, which was first advocated in Algorithm Quasi-optimal (AQ) in Michalski (1969); later on, AQ has been developed to a family of algorithms, and two famous representatives are AQ15 (Michalski et al. 1986) and AQ17-HCI (Wnek and Michalski 1994). Limited by computing power, AQ in the early days can only randomly pick a pair of positive and negative samples as seeds to start the training, and therefore, the learning performance of AQ is unstable due to the randomness of sampling. The problem was solved by PRISM (Cendrowska 1987), which takes a top-down search strategy. PRISM shows an advantage of rule learning over decision tree learning: decision trees try to split sample space into non-overlapped equivalent classes, whereas rule learning does not impose this constraint, and hence the models learned from rule learning tend to be less complex. The performance of PRISM is weaker than AQ, but it is a milestone of rule learning when we look back today.

Each leaf node in a decision tree corresponds to an equivalent class.

PRISM has been implemented in WEKA.

CN2 (Clark and Niblett 1989) used beam search and is the earliest rule learning algorithm that takes overfitting into consideration. Fürnkranz (1994) showed the advantage of using post-pruning to alleviate overfitting in rule learning. The pinnacle of propositional rule learning is RIPPER (Cohen 1995), which unified different tricks in the area, and for the first time, it made rule learning won the long-term competition against decision trees. An implementation of RIPPER written in the C programming language can be found on the author's homepage.

RIPPER outperforms C4.5 in terms of both efficiency and effectiveness on many tasks.

It is commonly agreed that the study on relational learning started in Winston (1970). Since it was difficult to solve relational learning problems with propositional rule learning, first-order rule learning got developed. FOIL converts propositional rule learning to first-order rule learning with operations such as replacing objects with logic variables, and FOIL is still being used today. For example, the Never-Ending Language Learning (NELL) program launched by Carnegie Mellon University in 2010 employed FOIL to learn semantic relations in natural languages (Carlson et al. 2010). Many literatures simply classified all first-order rule learning methods to ILP, whereas we distinguish them in a more strict sense in this book.

The terminology "Inductive Logic Programming" was proposed by Muggleton (1991). The GOLEM (Muggleton and Feng 1990) algorithm overcame many difficulties in transforming propositional logic to first-order rule learning, and it established the ILP framework that takes the bottom-up search strategy. LGG was first proposed in Plotkin (1970), and GOLEM employs RLGG. PROGOL (Muggleton 1995) improved inverse resolution to *inverse entailment* and achieved better performance. Studies on predicate invention have made some progress in recent years (Muggleton and Lin 2013). Rules learned by ILP can be almost directly used by logic programming interpreters such as PROLOG that is commonly used in expert systems, and therefore, ILP is an important bridge between machine learning and knowledge engineering. PROGOL (Muggleton 1995) and ALEPH (Srinivasan 1999) are widely used ILP systems, and their basic idea has been covered in our discussion of ILP in this chapter. Datalog (Ceri et al. 1989) has made significant influence to the area of database research. For example, it has influenced the SQL 1999 standard and IBM DB2. Classic literature on ILP includes (Muggleton 1992; Lavrac and Dzeroski 1993), and there is also a dedicated International Conference on Inductive Logic Programming (ILP).

See Sect. 1.5 for knowledge engineering and expert systems.

ILP has achieved some successes in biological data mining and natural language processing (Bratko and Muggleton 1995), but its high complexity makes it difficult to handle large-scale problems. For this reason, studies on this topic are somewhat suppressed by the rise of statistical learning. As machine learning techniques have been applied to a wider range of fields in recent years, the importance of ILP has been increasingly recognized in learning problems involving rich structured information and domain knowledge. In light of this fact, some effort has been made to join rule learning and statistical learning together, such as Probabilistic Inductive Logic Programming (PILP) (De Raedt et al. 2008) which incorporates probabilistic models in ILP, and Relational Bayesian Network (RBN) (Jaeger 2002) which assigns logic semantics to nodes in Bayesian network. In fact, joining relational learning and statistical learning is an important trend in the development of machine learning, where PILP is an important representative in this line of research. Other important representatives include Probabilistic Relational Model (PRM) (Friedman et al. 1999), Bayesian Logic Program (BLP) (Kersting et al. 2000), and Markov Logic Network (MLN) (Richardson and Domingos 2006). These efforts are known under the umbrella of *statistical relational learning* (Getoor and Taskar 2007).

## Exercises

**15.1** Given that negation literals are permitted, take the top-down strategy to learn the propositional rule set from the watermelon data set 2.0.

The watermelon data set 2.0 is in ◘ Table 4.1.

**15.2** Given that we can generalize rules by deleting literals or replacing constants with variables during the learning process, take the bottom-up strategy to learn the propositional rule set from the watermelon data set 2.0.

**15.3** Download or implement the RIPPER algorithm, and apply it to the watermelon data set 2.0 to learn the propositional rule set.

**15.4** Rule learning algorithms can also learn from incomplete data sets. Analogously to how missing values are handled by decision tree algorithms, use sequential covering to learn the propositional rule set from the watermelon data set 2.0$\alpha$.

The watermelon data set 2.0$\alpha$ is in ◘ Table 4.4.

**15.5** Download or implement the RIPPER algorithm, and apply it to the watermelon data set 5.0 to learn the first-order rule set with negation literals permitted.

The watermelon data set 5.0 is in ◘ Table 15.1.

**15.6** Use ILP to learn the concept $\texttt{unriper}(X, Y)$ from the watermelon data set 5.0.

**15.7** Prove that there is no first-order formula $\mathbf{r}'$ that can specialize to the first-order formulas $\mathbf{r}_1$ and $\mathbf{r}_2$ while also can generalize to their LGG.

**15.8** Find the LGG set of the watermelon data set 5.0.

**15.9** * First-order atomic formulas are recursively defined in the form of $P(t_1, t_2, \ldots, t_n)$, where $P$ is a predicate or a function, and $t_i$ is called a "term" that can be logical constants, variables, or other atomic formulas. Let $S = \{E_1, E_2, \ldots, E_n\}$ be the set of first-order atomic formulas $E_i$, design an algorithm to find their MGU.

Output "unsolvable" when $S$ cannot be unified.

**15.10** * Sequential covering-based rule learning algorithms will, before learning the next rule, remove training samples that have already been covered by the current rule set. With this greedy approach, the subsequent learning process only looks at the uncovered samples without considering the correlation between the previous rules and the current rules when calculating the coverage of rules. Nevertheless, such an approach will keep reducing the number of samples in the subsequent learning processes. Design a rule learning algorithm that does not remove samples.

## Break Time

### Short Story: Ryszard S. Michalski—Pioneer of Machine Learning

The series of AQ algorithms is an important work in the early days of rule learning research, mainly developed by the Polish-American computer scientist Ryszard S. Michalski (1937–2007), a pioneer in machine learning.

Michalski was born in Kalusz. He earned a Ph.D. degree in computer science from the Silesian University of Technology, Gliwice in 1969. In the same year, he published the AQ technique at the FCIP conference held in Bled, Yugoslavia (present-day Slovenia). In 1970, Michalski started to teach at the University of Illinois Urbana-Champaign and continued his study on AQ technique. In 1980, Michalski, together with J. G. Carbonell and T. Mitchell, organized the first machine learning workshop at Carnegie Mellon University. The workshop was held again in 1983 and 1985, and later on, it became the International Conference on Machine Learning (ICML). In 1983, Michalski, Carbonell, and Mitchell edited a book called *Machine Learning: An Artificial Intelligence Approach*, which is a milestone in the history of machine learning. In 1986, *Machine Learning*, the first journal dedicated to machine learning, was established, and Michalski was one of the three founding editors.

Kalusz was part of different countries in history, including Poland, Russia, Germany, and Ukraine.

See Sect. 1.5.

## References

Bratko I, Muggleton S (1995) Applications of inductive logic programming. Commun ACM 38(11):65–70

Brunk CA, Pazzani MJ (1991) An investigation of noise-tolerant relational concept learning algorithms. In: Proceedings of the 8th international workshop on machine learning (IWML), pp 389–393. Evanston, IL

Carlson A, Betteridge J, Kisiel B, Settles B, Hruschka ER, Mitchell TM (2010) Toward an architecture for never-ending language learning. In: Proceedings of the 24th AAAI conference on artificial intelligence (AAAI), pp 1306–1313. Atlanta, GA

Cendrowska J (1987) PRISM: an algorithm for inducing modular rules. Int J Man Mach Stud 27(4):349–370

Ceri S, Gottlob G, Tanca L (1989) What you always wanted to know about Datalog (and never dared to ask). IEEE Trans Knowl Data Eng 1(1):146–166

Clark P, Niblett T (1989) The CN2 induction algorithm. Mach Learn 3(4):261–283

Cohen WW (1995) Fast effective rule induction. In: Proceedings of the 12th international conference on machine learning (ICML), pp 115–123. Tahoe, CA

De Raedt L, Frasconi P, Kersting K, Muggleton S (eds) (2008) Probabilistic inductive logic programming: theory and applications. Springer, Berlin

Friedman N, Getoor L, Koller D, Pfeffer A (1999) Learning probabilistic relational models. In: Proceedings of the 16th international joint conference on artificial intelligence (IJCAI), pp 1300–1307. Stockholm, Sweden

Fürnkranz, J. (1994). Top-down pruning in relational learning. In: Proceedings of the 11th European conference on artificial intelligence (ECAI), pp 453–457. Amsterdam, The Netherlands

Fürnkranz J, Gamberger D, Lavrac N (2012) Foundations of rule learning. Springer, Berlin

Fürnkranz J, Widmer G (1994) Incremental reduced error pruning. In: Proceedings of the 11th international conference on machine learning (ICML), pp 70–77. New Brunswick, NJ

Getoor L, Taskar B (2007) Introduction to statistical relational learning. MIT Press, Cambridge

Jaeger M (2002) Relational Bayesian networks: a survey. Electron Trans Artif Intell 6, Article 15

Kersting K. De Raedt L, Kramer S (2000) Interpreting Bayesian logic programs. In: Proceedings of the AAAI'2000 workshop on learning statistical models from relational data, pp 29–35. Austin, TX

Lavrac N, Dzeroski S (1993) Inductive logic programming: techniques and applications. Ellis Horwood, New York

Michalski RS (1969). On the quasi-minimal solution of the general covering problem. In: Proceedings of the 5th international symposium on information processing (FCIP), vol A3, pp 125–128. Bled, Yugoslavia

Michalski RS (1983) A theory and methodology of inductive learning. In: Michalski RS, Carbonell J, Mitchell T (eds) Machine learning: an artificial intelligence approach. Tioga, Palo Alto, CA, pp 111–161

Michalski RS, Mozetic I, Hong J, Lavrac N (1986) The multi-purpose incremental learning system AQ15 and its testing application to three medical domains. In: Proceedings of the 5th national conference on artificial intelligence (AAAI), pp 1041–1045. Philadelphia, PA

Muggleton S (1991) Inductive logic programming. New Gener Comput 8(4):295–318

Muggleton S (ed) (1992) Inductive logic programming. Academic, London
Muggleton S (1995) Inverse entailment and Progol. New Gener Comput 13(3–4):245–286
Muggleton S, Buntine W (1988) Machine invention of first order predicates by inverting resolution. In: Proceedings of the 5th International Workshop on Machine Learning (IWML), pp 339–352. Ann Arbor, MI
Muggleton S, Feng C (1990) Efficient induction of logic programs. In: Proceedings of the 1st international workshop on algorithmic learning theory (ALT), pp 368–381. Tokyo, Japan
Muggleton S, Lin D (2013) Meta-interpretive learning of higher-order dyadic datalog: predicate invention revisited. In: Proceedings of the 23rd international joint conference on artificial intelligence (IJCAI), pp 1551–1557. Beijing, China
Plotkin GD (1970) A note on inductive generalization. In: Meltzer B, Mitchie D (eds) Machine intelligence, vol 5. Edinburgh University Press, Edinburgh, pp 153–165
Plotkin GD (1971) A further note on inductive generalization. In: Meltzer B, Mitchie D (eds) Machine intelligence, vol 6. Edinburgh University Press, Edinburgh, pp 107–124
Quinlan JR (1990) Learning logical definitions from relations. Mach Learn 5(3):239–266
Richardson M, Domingos P (2006) Markov logic networks. Mach Learn 62(1–2):107–136
Robinson JA (1965) A machine-oriented logic based on the resolution principle. J ACM 12(1):23–41
Srinivasan A (1999) The Aleph manual. ► http://www.cs.ox.ac.uk/activities/machlearn/Aleph/aleph.html
Winston PH (1970) Learning structural descriptions from examples. PhD thesis, Department of Electrical Engineering, MIT, Cambridge, MA
Wnek J, Michalski RS (1994) Hypothesis-driven constructive induction in AQ17-HCI: a method and experiments. Mach Learn 2(14):139–168

# Reinforcement Learning

Table of Contents




Z.-H. Zhou, *Machine Learning*,
https://doi.org/10.1007/978-981-15-1967-3_16

## 16.1 Task and Reward

Planting watermelon involves many steps, such as seed selection, regular watering, fertilization, weeding, and insect control. We usually do not know the quality of the watermelons until harvesting. If we consider the harvesting of ripe watermelons as a reward for planting watermelons, then we do not receive the final reward immediately after each step of planting, e.g., fertilization. We do not even know the exact impact of the current action on the final reward. Instead, we only receive feedback about the current status, e.g., the watermelon seedling looks healthier. After planting watermelons many times and exploring different planting methods, we may finally come up with a good strategy for planting watermelons. Such a process, when abstracted, is called *reinforcement learning*.

As illustrated in ▪ Figure 16.1, we usually use Markov Decision Process (MDP) to describe reinforcement learning problems: an agent is in an environment $E$ with a state space $X$, where each state $x \in X$ is a description of the environment perceived by the agent, e.g., the growing trend of the watermelon seedling in watermelon planting. The actions that the agent can perform form an action space $A$. For example, in watermelon planting, the actions include watering, using different types of fertilizers, and applying different types of pesticides. When an action $a \in A$ is performed on the current state $x$, the underlying transition function $P$ will transit the environment from the current state to another with a certain probability, e.g., watering a dehydrated seedling may or may not recover it to a healthy state. After the transition from one state to another, the environment sends the agent a reward based on the underlying *reward* function $R$, e.g., $+1$ for healthy seedling, $-10$ for withered seedling, and $+100$ for harvesting a ripe watermelon. In short, reinforcement learning involves a quadruplet $E = \langle X, A, P, R\rangle$, where $P : X \times A \times X \mapsto \mathbb{R}$ gives the state transition probability, and $R : X \times A \times X \mapsto \mathbb{R}$ gives the reward. In some applications, the reward function may only depend on state transitions, that is, $R : X \times X \mapsto \mathbb{R}$.



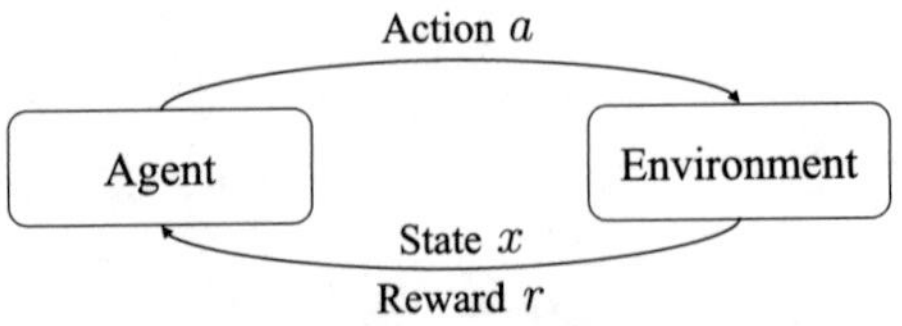


**Fig. 16.1** A diagram of reinforcement learning

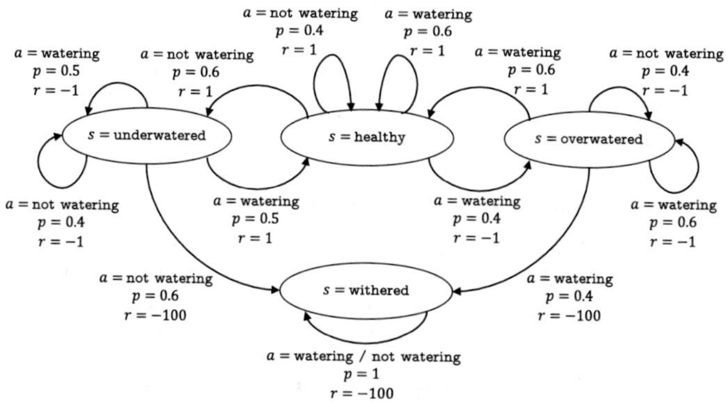


**Fig. 16.2** The MDP of watering watermelons

◘ Figure 16.2 provides an example showing the MDP of watering watermelons. In this example, we have four states (i.e., healthy, underwatered, overwatered and withered) and two actions (i.e., watering, not watering). After each transition, the agent receives a reward of 1 if the state is healthy and a reward of −1 if the state is underwatered or overwatered. The state can be recovered to healthy by watering or not watering. When the seedling is withered, it is unrecoverable, and the agent receives the minimum reward value of −100. The arrows in ◘ Figure 16.2 represent state transitions, where $a$, $p$, and $r$ are, respectively, the action, the state transition probability, and the reward. We can easily figure out that the optimal strategy is to take the action watering in the state healthy, the action not watering in the state overwatered, the action watering in the state underwatered, and any actions in the state withered.

The agent has no direct control of the state transition or reward. It can only influence the environment by taking actions and perceive the environment by observing transited states and returned rewards. For example, in the watermelon planting problem, the environment is the natural world in which watermelons grow; in a chess game, the environment is the chessboard and the opponent; in robot control, the environment is the body of the robot and the physical world.

By interacting with the environment, the agent tries to learn a *policy* $\pi$ that can select the action $a = \pi(x)$ at state $x$. For example, when the agent observes the state underwatered, it takes the action watering. There are two ways of representing policies. The first one is representing policies with functions $\pi : X \mapsto A$, and we often use this representation for deterministic policies. The other one is representing policies with probabilities $\pi : X \times A \mapsto \mathbb{R}$, and we often use this representation for stochastic policies. For the probability representation,

$\pi(x, a)$ is the probability of choosing the action $a$ at state $x$, and we must ensure $\sum_a \pi(x, a) = 1$.

The quality of policy is measured by the cumulative rewards of executing this policy in the long term. For example, a policy that leads to a withered seedling may accumulate only a small amount of reward, whereas a policy that leads to a ripe watermelon can accumulate a large amount of reward. The objective of reinforcement learning is to find a policy that maximizes the long-term cumulative rewards. There are different ways to calculate cumulative rewards, and the commonly used ones are *T-step cumulative rewards* $\mathbb{E}[\frac{1}{T}\sum_{t=1}^{T} r_t]$ and *γ-discounted cumulative rewards* $\mathbb{E}[\sum_{t=0}^{+\infty} \gamma^t r_{t+1}]$, where $r_t$ is the reward at step $t$ and $\mathbb{E}$ is the expectation with respect to all random variables.

Readers may have recognized the differences between reinforcement learning and supervised learning. If we consider "states" and "actions", respectively, as "samples" and "labels" in supervised learning, then "policies" correspond to "classifiers" (for discrete actions) or "regressors" (for continuous actions), and hence the two learning paradigms are somewhat similar. The main difference is that there are no labeled samples in reinforcement learning. In other words, there is no supervised information that tells the agent which action it should take with the given state. Instead, the agent has to wait for the outcome and then "reflects" its previous actions. From this perspective, reinforcement learning can be seen as supervised learning with time-delayed labels.

## 16.2 *K*-Armed Bandit

### 16.2.1 Exploration Versus Exploitation

Unlike common supervised learning, the final amount of rewards in reinforcement learning is only observed after multiple actions. Let us start our discussion with the simplest case: we maximize the reward of each step, that is, consider only one step at a time. Note that reinforcement learning and supervised learning are still quite different even in this simplified scenario since the agent needs to try out different actions to collect the respective outcomes. In other words, there is no training data that tells the agent which actions to take.

To maximize the one-step reward, we need to consider two aspects: find the reward corresponding to each action and take the highest-rewarded action. If the reward of each action is a definite value, then we can find the highest-rewarded action by trying out all actions. In practice, however, the reward of action is usually a random variable sampled from a probability distribution, and hence we cannot accurately determine the mean reward only in one trial.

The above one-step reinforcement learning scenario corresponds to a theoretical model called *K-armed bandit*. As illustrated in ◘ Figure 16.3, there are $K$ arms in the K-armed bandit. After inserting a coin, the player can pull an arm, and the machine will return some coins at a certain probability (of this arm) that is unknown to the player. The player's objective is to develop a policy that maximizes the cumulative reward, that is, more coins.

If we only want to know the expected reward of each arm, then we can employ the *exploration-only* method: equally allocate the pulling opportunities to the arms (i.e., pull each arm in turn), and then calculate the average number of coins returned by each arm as the approximation of expectation. In contrast, if we only want to take the highest-rewarded action, then we can employ the *exploitation-only* method: pull the currently best arm (i.e., the one with the highest average reward), or randomly choose one when there are multiple best arms. By comparing these two methods, we see that the *exploration-only* method can estimate the reward of each arm reasonably well at the cost of losing many opportunities to pull the optimal arm, whereas the *exploitation-only* method is likely to miss the optimal arm since it does not have a good estimation of the expected reward of each arm. Therefore, we are unlikely to maximize the cumulative rewards using either method.

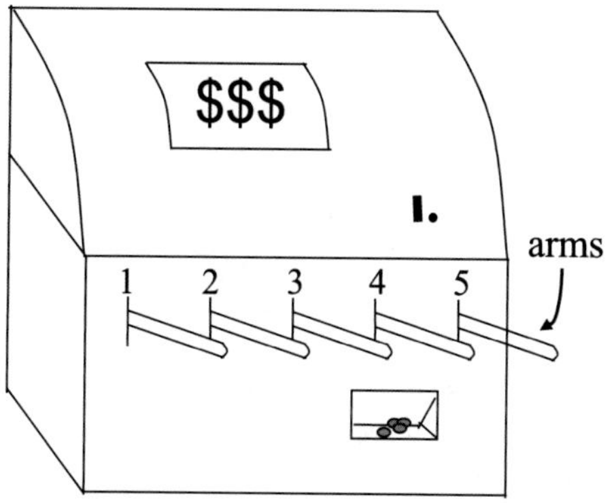


**Fig. 16.3** An example of $K$-armed bandit

Actually, *exploration* (i.e., estimating the quality of each arm) and *exploitation* (i.e., choosing the currently best arm) are conflicted since the total number of trials (i.e., pulling opportunities) is fixed. In other words, improving one aspect will weaken the other, and such a situation is known as the

*Exploration-Exploitation dilemma* in reinforcement learning. To maximize the cumulative rewards, we need to make a good trade-off between exploration and exploitation.

### 16.2.2 $\epsilon$-Greedy

The $\epsilon$-greedy method makes a trade-off between exploration and exploitation based on a probability value. In each trial, it chooses exploration (i.e., randomly select an arm) with a probability of $\epsilon$ or chooses exploitation (i.e., select the arm with the highest reward at the moment, or randomly select from tied arms with the highest reward) with the probability of $1 - \epsilon$.

Formally, let $Q(k)$ denote the average reward of arm $k$. By pulling the arm $k$ for $n$ times, we obtain a series of rewards $v_1, v_2, \ldots, v_n$, and the average reward is given by

$$Q(k) = \frac{1}{n} \sum_{i=1}^{n} v_i. \tag{16.1}$$

Instead of using (16.1) to calculate the average of $n$ rewards, we can also calculate the average incrementally by updating $Q(k)$ after each trial. Let the subscripts denote the number of trials, and we have $Q_0(k) = 0$ at the initial point. For every $n \geqslant 1$, $Q_{n-1}(k)$ is the average reward after $n - 1$ trials, and the average reward, after the $n$th trial with a reward of $v_n$, is updated to

We will use (16.3) in Sect. 16.4.2.

$$Q_n(k) = \frac{1}{n}\left((n-1) \times Q_{n-1}(k) + v_n\right) \tag{16.2}$$

$$= Q_{n-1}(k) + \frac{1}{n}\left(v_n - Q_{n-1}(k)\right). \tag{16.3}$$

In this way, we only need to record two values no matter how many trials we run, that is, the number of completed trials $n-1$ and the most recent average reward $Q_{n-1}(k)$. The $\epsilon$-greedy algorithm is given in ◘ Algorithm 16.1.

When the rewards of arms are highly uncertain (e.g., wide probability distribution), we need more explorations (i.e., larger $\epsilon$). In contrast, when the rewards of arms have low uncertainty (e.g., concentrated probability distribution), we can obtain good approximations with a small number of trials, and hence a small $\epsilon$ is sufficient. In practice, we usually let $\epsilon$ be a small constant, such as 0.1 or 0.01. As the number of trials increases, we get a more accurate estimation of the rewards of all arms and eventually no longer need exploration. In such cases, we can let $\epsilon$ decrease as the number of trials increases, e.g., letting $\epsilon = 1/\sqrt{t}$.

**Algorithm 16.1** $\epsilon$-greedy

**Input:** Number of arms $K$;
Reward function $R$;
Number of trials $T$;
Exploration probability $\epsilon$.

**Process:**

$r = 0$;
$\forall i = 1, 2, \ldots, K : Q(i) = 0, \text{count}(i) = 0$;
**for** $t = 1, 2, \ldots, T$ **do**
  **if** rand() $< \epsilon$ **then**
    $k =$ uniformly and randomly selected from $1, 2, \ldots, K$;
  **else**
    $k = \arg\max_i Q(i)$;
  **end if**
  $v = R(k)$;
  $r = r + v$;
  $Q(k) = \frac{Q(k) \times \text{count}(k) + v}{\text{count}(k) + 1}$;
  $\text{count}(k) = \text{count}(k) + 1$.
**end for**

**Output:** Cumulative rewards $r$.

$Q(i)$ and count$(i)$ are, respectively, the average rewards of arm $i$ and the number of times arm $i$ is chosen.

Generate a random number from $[0, 1]$.

The reward of the current trial.

(16.2) updates the average reward.

### 16.2.3 Softmax

The *Softmax* algorithm makes a trade-off between exploration and exploitation based on the current average rewards. The basic idea is that arms with similar average rewards should have similar probabilities to be chosen, and arms with higher average rewards than the others should have higher probabilities to be chosen.

The Softmax algorithm allocates the probabilities of being chosen based on the Boltzmann distribution

$$P(k) = \frac{e^{\frac{Q(k)}{\tau}}}{\sum_{i=1}^{K} e^{\frac{Q(i)}{\tau}}}, \tag{16.4}$$

where $Q(i)$ is the average reward of the current arm, and $\tau > 0$ is known as *temperature*. A smaller $\tau$ makes the arms with high average rewards more likely to be chosen. Softmax moves towards *exploitation-only* as $\tau$ approaches 0 and moves towards *exploration-only* as $\tau$ approaches the positive infinity. The Softmax algorithm is given in ◘ Algorithm 16.2.

The choice between $\epsilon$-greedy and Softmax mainly depends on the specific applications. To understand their difference, let us take a look at an intuitive example. Considering the following 2-armed bandit: the arm 1 returns a reward of 1 at a probability of 0.4 and returns a reward of 0 at a probability of 0.6, and the arm 2 returns a reward of 1 at a probability of 0.2 and returns a reward of 0 at a probability 0.8. For this 2-armed

**Algorithm 16.2** Softmax

**Input:** Number of arms $K$;
Reward function $R$;
Number of trials $T$;
Temperature parameter $\tau$.

*$\tau$ is a parameter used in line 4.*

**Process:**

*$Q(i)$ and count$(i)$ are, respectively, the average rewards of arm $i$ and the number of times arm $i$ is chosen.*

1: $r = 0$;
2: $\forall i = 1, 2, \ldots, K : Q(i) = 0, \text{count}(i) = 0$;
3: **for** $t = 1, 2, \ldots, T$ **do**
4: $\quad k =$ randomly selected from $1, 2, \ldots, K$ according to (16.4);
5: $\quad v = R(k)$;
6: $\quad r = r + v$;
7: $\quad Q(k) = \frac{Q(k) \times \text{count}(k) + v}{\text{count}(k) + 1}$;
8: $\quad \text{count}(k) = \text{count}(k) + 1$.
9: **end for**

**Output:** Cumulative rewards $r$.

*The reward of the current trial.*

*(16.2) updates the average reward.*

bandit, ◘ Figure 16.4 shows the cumulative rewards of different algorithms under different parameter settings, where each curve is the average result of 1000 repeated experiments. We can observe that the curve of Softmax almost overlaps with the curve of *exploitation-only* when $\tau = 0.01$.

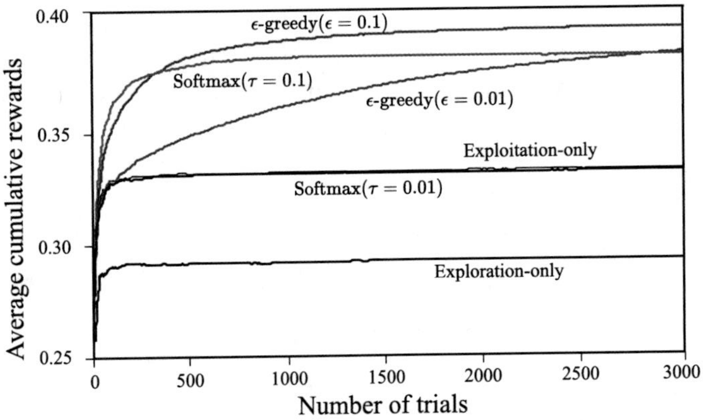


**Fig. 16.4** The performance comparison of different algorithms on the 2-armed bandit

For multi-step reinforcement learning problems with discrete state space and discrete action space, a straightforward approach is to consider the action selection of each state as a $K$-armed bandit problem, where the cumulative rewards in the reinforcement learning problem replace the reward function in the $K$-armed bandit problem, that is, applying bandit algorithms to each state: we keep records of the number of trails and the current cumulative rewards for each state, and decide the action based on the bandit algorithm. However, there are limitations in such an approach since it does not consider the structural information in the MDP of reinforcement learning. As we will see in Sect. 16.3, there are better methods once we take advantage of the properties of MDP.

## 16.3 Model-Based Learning

We start our discussions on multi-step reinforcement learning problems with the scenario of *known model*, that is, the quadruplet $E = \langle X, A, P, R \rangle$ in MDP is known. In other words, the agent has already modeled the environment and can simulate the environment exactly or approximately. We call such a learning task *model-based learning*. Here, the probability of transition from state $x$ to state $x'$ via action $a$ is known, denoted by $P^a_{x \to x'}$, and the reward is also known, denoted by $R^a_{x \to x'}$. For ease of discussion, we assume both the state space $X$ and the action space $A$ be finite. For general model-based reinforcement learning, we need to consider how to learn the environment model and how to learn the policy in the model. Here, we only discuss the latter, given that the environment model is known.

We will discuss *unknown model* in Sect. 16.4.

We will discuss infinite space in Sect. 16.5.

### 16.3.1 Policy Evaluation

With the known model, we can estimate the expected cumulative rewards of using policy $\pi$. Let the function $V^\pi(x)$ denote the cumulative rewards of using policy $\pi$ from the starting state $x$, and the function $Q^\pi(x, a)$ denote the cumulative rewards of using policy $\pi$ after taking action $a$ at state $x$. Here, we call $V(\cdot)$ the *state value function* representing the cumulative rewards for a given state, and $Q(\cdot)$ the *state-action value function* representing the cumulative rewards for a given state-action pair.

From the definition of cumulative rewards, we have the state value functions

$$\begin{cases} V_T^\pi(x) = \mathbb{E}_\pi\left[\frac{1}{T}\sum_{t=1}^{T} r_t \mid x_0 = x\right], & T - \text{step cumulative rewards;} \\ V_\gamma^\pi(x) = \mathbb{E}_\pi\left[\sum_{t=0}^{+\infty}\gamma^t r_{t+1} \mid x_0 = x\right], & \gamma - \text{discounted cumulative rewards.} \end{cases} \tag{16.5}$$

To keep our discussion concise, we omit the type of above two cumulative rewards in subsequent discussions when the context is clear. Let $x_0$ denote the initial state, $a_0$ denote the first action performed on the initial state, and $t$ denote the number of steps for $T$-step cumulative rewards. Then, we have the state-action value functions

$$\begin{cases} Q_T^\pi(x, a) = \mathbb{E}_\pi\left[\frac{1}{T}\sum_{t=1}^{T} r_t \mid x_0 = x, a_0 = a\right]; \\ Q_\gamma^\pi(x, a) = \mathbb{E}_\pi\left[\sum_{t=0}^{+\infty}\gamma^t r_{t+1} \mid x_0 = x, a_0 = a\right]. \end{cases} \tag{16.6}$$

Such recursive equations are known as Bellman equations.

Since MDP possesses the Markov property, that is, the state at the next step only depends on the current state rather than any previous states, and therefore, the value function has a simple form of recursion. Specifically, for $T$-step cumulative rewards, we have

Expanded using the law of total probability.

$$\begin{aligned} V_T^\pi(x) &= \mathbb{E}_\pi\left[\frac{1}{T}\sum_{t=1}^{T} r_t \mid x_0 = x\right] \\ &= \mathbb{E}_\pi\left[\frac{1}{T}r_1 + \frac{T-1}{T}\frac{1}{T-1}\sum_{t=2}^{T} r_t \mid x_0 = x\right] \\ &= \sum_{a\in A}\pi(x, a)\sum_{x'\in X} P_{x\to x'}^a\left(\frac{1}{T}R_{x\to x'}^a \right. \\ &\quad \left. + \frac{T-1}{T}\mathbb{E}_\pi\left[\frac{1}{T-1}\sum_{t=1}^{T-1} r_t \mid x_0 = x'\right]\right) \\ &= \sum_{a\in A}\pi(x, a)\sum_{x'\in X} P_{x\to x'}^a\left(\frac{1}{T}R_{x\to x'}^a + \frac{T-1}{T}V_{T-1}^\pi(x')\right). \end{aligned} \tag{16.7}$$

Similarly, for $\gamma$-discounted cumulative rewards, we have

$$V_\gamma^\pi(x) = \sum_{a\in A}\pi(x, a)\sum_{x'\in X} P_{x\to x'}^a(R_{x\to x'}^a + \gamma V_\gamma^\pi(x')). \tag{16.8}$$

Note that we can expand using the law of total probability because $P$ and $R$ are known.

Readers may have recognized that the above calculations of value functions using recursive equations are essentially a kind of dynamic programming. For $V_T^\pi$, we can imagine that the recursion continues until it reaches the initial starting point. In other words, from the initial value $V_0^\pi$ of the value function, we can calculate the one-step rewards $V_1^\pi$ of every state in one iteration. Then, from the one-step rewards, we can calculate the two-step cumulative rewards $V_2^\pi$, and so on. Such a procedure is shown in ◻ Algorithm 16.3, and it takes only $T$ iterations to precisely calculate the value function of $T$-step cumulative rewards.

**Algorithm 16.3** $T$-step cumulative rewards-based policy evaluation

**Input:** MDP quadruplet $E = \langle X, A, P, R \rangle$;
Evaluating policy $\pi$;
Cumulative rewards parameter $T$.

**Process:**

1: $\forall x \in X : V(x) = 0$;
2: **for** $t = 1, 2, \ldots$ **do**
3: $\quad \forall x \in X : V'(x) = \sum_{a \in A} \pi(x, a) \sum_{x' \in X} P^a_{x \to x'} \left( \frac{1}{t} R^a_{x \to x'} + \frac{t-1}{t} V(x') \right)$;
4: $\quad$ **if** $t = T + 1$ **then**
5: $\quad\quad$ **break**;
6: $\quad$ **else**
7: $\quad\quad$ $V = V'$.
8: $\quad$ **end if**
9: **end for**

**Output:** State value functions $V$.

$V(x)$ is the cumulative rewards of $x$.

(16.7) updates the value function.

We write in this format so that the $T$-step cumulative rewards and the $\gamma$-discounted cumulative rewards can be considered under the same algorithmic framework.

There is also a similar algorithm for $V_\gamma^\pi$ since $\gamma^t$ approaches 0 when $t$ is large. To this end, we just need to modify line 3 of ◻ Algorithm 16.3 based on (16.8). Besides, the algorithm may iterate for quite many rounds, and therefore, we need to set a stopping criterion, e.g., set a threshold $\theta$ and stop the algorithm when the change of value function after one iteration is smaller than $\theta$. Accordingly, $t = T + 1$ in line 4 of ◻ Algorithm 16.3 is replaced by

See Exercise 16.2.

$$\max_{x \in X} \left| V(x) - V'(x) \right| < \theta. \tag{16.9}$$

With the state value functions $V$, we can calculate the state-action value functions

$$\begin{cases} Q_T^\pi(x, a) = \sum_{x' \in X} P^a_{x \to x'} (\frac{1}{T} R^a_{x \to x'} + \frac{T-1}{T} V_{T-1}^\pi(x')); \\ Q_\gamma^\pi(x, a) = \sum_{x' \in X} P^a_{x \to x'} (R^a_{x \to x'} + \gamma V_\gamma^\pi(x')). \end{cases} \tag{16.10}$$

### 16.3.2 Policy Improvement

After evaluating the cumulative rewards of a policy, we naturally wish to improve it if it is not the optimal one. An ideal policy should maximize the cumulative rewards

$$\pi^* = \arg\max_\pi \sum_{x\in X} V^\pi(x). \tag{16.11}$$

A reinforcement learning problem could have more than one optimal policy, and the value function $V^*$ corresponding to the optimal policies is called the optimal value function

$$\forall x \in X : V^*(x) = V^{\pi^*}(x). \tag{16.12}$$

Note that $V^*$ in (16.12) is the value function of the optimal policies only if it imposes no constraint on the policy space. For example, for discrete state space and discrete action space, the policy space is the combination of all actions over all states, containing $|A|^{|X|}$ different policies. However, once there are constraints on the policy space, then the policies that break the constraints are illegitimate policies whose value functions are not the optimal value function even if they have the highest cumulative rewards.

Since the cumulative rewards of the optimal value function are maximized, we can make some small adjustments to the Bellman equations (16.7) and (16.8), that is, change the summation over actions to selecting the best action

$$\begin{cases} V_T^*(x) = \max_{a\in A}\sum_{x'\in X} P^a_{x\to x'}\left(\frac{1}{T}R^a_{x\to x'} + \frac{T-1}{T}V^*_{t-1}(x')\right); \\ V_\gamma^*(x) = \max_{a\in A}\sum_{x'\in X} P^a_{x\to x'}\left(R^a_{x\to x'} + \gamma V^*_\gamma(x')\right). \end{cases} \tag{16.13}$$

In other words,

$$V^*(x) = \max_{a\in A} Q^{\pi^*}(x, a). \tag{16.14}$$

16

Substituting it into (16.10), we have the optimal state-action value function

$$\begin{cases} Q_T^*(x,a) = \sum_{x'\in X} P^a_{x\to x'}\left(\frac{1}{T}R^a_{x\to x'} + \frac{T-1}{T}\max_{a'\in A} Q^*_{T-1}(x', a')\right); \\ Q_\gamma^*(x,a) = \sum_{x'\in X} P^a_{x\to x'}\left(R^a_{x\to x'} + \gamma \max_{a'\in A} Q^*_\gamma(x', a')\right). \end{cases} \tag{16.15}$$

The above equations about the optimal value functions are called the Bellman optimality equations, whose unique solutions are the optimal value functions.

The Bellman optimality equations suggest a method for improving non-optimal policies, that is, changing the action chosen by the policy to the currently best action. Clearly, doing so will improve the quality of policies. Let $\pi'$ denote the policy with changed action, and $Q^{\pi}(x, \pi'(x)) \geqslant V^{\pi}(x)$ denote the condition of action change. Taking the $\gamma$-discounted cumulative rewards as an example, from (16.10), we have the recursive inequality

$$\begin{aligned} V^{\pi}(x) &\leqslant Q^{\pi}(x, \pi'(x)) \\ &= \sum_{x' \in X} P_{x \to x'}^{\pi'(x)} (R_{x \to x'}^{\pi'(x)} + \gamma V^{\pi}(x')) \\ &\leqslant \sum_{x' \in X} P_{x \to x'}^{\pi'(x)} (R_{x \to x'}^{\pi'(x)} + \gamma Q^{\pi}(x', \pi'(x'))) \\ &= \ldots \\ &= V^{\pi'}(x). \end{aligned} \tag{16.16}$$

Since the value function is monotonically increasing with respect to every improvement made on the policy, we can safely improve the current policy $\pi$ to

$$\pi'(x) = \arg\max_{a \in A} Q^{\pi}(x, a) \tag{16.17}$$

until $\pi'$ and $\pi$ remain the same, and hence the Bellman optimality equation is achieved, that is, the optimal policy is found.

### 16.3.3 Policy Iteration and Value Iteration

The previous two subsections discussed how to evaluate the value function of a policy, as well as how to improve a policy to the optimum. Joining them gives the method for finding the optimal solution: start with an initial policy (usually a random policy), and then alternately iterate through policy evaluation and policy improvement until the policy converges with no more changes. Such a method is known as *policy iteration*.

***

**Algorithm 16.4** $T$-step cumulative rewards-based policy iteration

**Input:** MDP quadruplet $E = \langle X, A, P, R \rangle$;
      Cumulative rewards parameter $T$.

**Process:**

1: $\forall x \in X : V(x) = 0, \pi(x, a) = \frac{1}{|A(x)|}$;
2: **loop**
3:     **for** $t = 1, 2, \ldots$ **do**
4:         $\forall x \in X : V'(x) = \sum_{a \in A} \pi(x, a) \sum_{x' \in X} P^a_{x \to x'} \left( \frac{1}{t} R^a_{x \to x'} + \frac{t-1}{t} V(x') \right)$;
5:         **if** $t = T + 1$ **then**
6:             **break**;
7:         **else**
8:             $V = V'$;
9:         **end if**
10:     **end for**
11:     $\forall x \in X : \pi'(x) = \arg\max_{a \in A} Q(x, a)$;
12:     **if** $\forall x : \pi'(x) = \pi(x)$ **then**
13:         **break**;
14:     **else**
15:         $\pi = \pi'$.
16:     **end if**
17: **end loop**

**Output:** Optimal policy $\pi$.

***

$|A(x)|$ is the number of all possible actions under state $x$.

(16.7) updates the value function.

(16.10) calculates $Q$.

See Exercise 16.3.

◻ Algorithm 16.4 shows the pseudocode of policy iteration, which improves ◻ Algorithm 16.3 by adding policy improvement. Similarly, we can derive the policy iteration algorithm based on $\gamma$-discounted cumulative rewards. Policy iteration algorithms are often time-consuming since they need to re-evaluate policies after each policy improvement.

From (16.16), we see that the policy improvement and the value function improvement are equivalent. Therefore, we can improve policy via value function improvement, that is, from (16.13), we have

$$
\begin{cases}
V_T(x) = \max_{a \in A} \sum_{x' \in X} P^a_{x \to x'} \left( \frac{1}{T} R^a_{x \to x'} + \frac{T-1}{T} V_{T-1}(x') \right); \\
V_\gamma(x) = \max_{a \in A} \sum_{x' \in X} P^a_{x \to x'} \left( R^a_{x \to x'} + \gamma V_\gamma(x') \right).
\end{cases}
$$
$$(16.18)$$

This gives the value iteration algorithm, as shown in ◻ Algorithm 16.5.

**16**

**Algorithm 16.5** $T$-step cumulative rewards-based value iteration

**Input:** MDP quadruplet $E = \langle X, A, P, R \rangle$;
Cumulative rewards parameter $T$;
Convergence threshold $\theta$.

**Process:**

1: $\forall x \in X : V(x) = 0$;
2: **for** $t = 1, 2, \ldots$ **do**
3: $\quad \forall x \in X : V'(x) = \max_{a \in A} \sum_{x' \in X} P^a_{x \to x'} \left( \frac{1}{t} R^a_{x \to x'} + \frac{t-1}{t} V(x') \right)$;
4: $\quad$ **if** $\max_{x \in X} \left| V(x) - V'(x) \right| < \theta$ **then**
5: $\quad\quad$ **break**;
6: $\quad$ **else**
7: $\quad\quad V = V'$.
8: $\quad$ **end if**
9: **end for**

**Output:** Policy $\pi(x) = \arg\max_{a \in A} Q(x, a)$.

(16.18) updates the value function.

(16.10) calculates $Q$.

When $\gamma$-discounted cumulative rewards are used, we just need to replace line 3 of ◘ Algorithm 16.5 with

$$\forall x \in X : V'(x) = \max_{a \in A} \sum_{x' \in X} P^a_{x \to x'} \left( R^a_{x \to x'} + \gamma V(x') \right). \tag{16.19}$$

From the above algorithms, we see that reinforcement learning problems with known models can be regarded as dynamic programming-based optimization. Unlike supervised learning, there is no generalization considered but just finding the best action for each state.

## 16.4 Model-Free Learning

In real-world reinforcement learning problems, it is often difficult to obtain state transition probabilities and reward functions of the environment, and it can even be difficult to know the number of possible states. When the environment model is unknown, we call the learning task *model-free learning*, which is much more difficult than learning with known models.

### 16.4.1 Monte Carlo Reinforcement Learning

In model-free learning, the first problem faced by policy iteration algorithms is that policies become unevaluable since we cannot apply the law of total probability without knowledge about the model. As a result, the agent has to try actions and observe state transitions and rewards. Inspired by $K$-armed bandit, a straightforward replacement of policy evaluation is to approximate the expected cumulative rewards by averaging over the cumulative rewards of multiple samplings, and such an approach is called Monte Carlo Reinforcement Learning (MCRL). Since the number of samplings must be finite, the method is more suitable for reinforcement learning problems with $T$-step cumulative rewards.

See Sect. 14.7 for Monte Carlo methods. We have seen the MCMC method in Sect. 14.5.1.

The other difficulty is that policy iteration algorithms only estimate the value function $V$, but the final policy is obtained from the state-action value function $Q$. Converting $V$ to $Q$ is easy when the model is known, but it can be difficult when the model is unknown. Therefore, the target of our estimation is no longer $V$ but $Q$, that is, estimating the value function for every state-action pair.

Besides, when the model is unknown, the agent can only start from the initial state (or initial state set) to explore the environment, and hence policy iteration algorithms are not applicable since they need to estimate every individual state. For example, the exploration of watermelon planting can only start from sowing but not other states. As a result, we can only gradually discover different states during the exploration and estimate the value functions of state-action pairs.

Putting them all together, when the model is unknown, we start from the initial state and take a policy for sampling. That is, by executing the policy for $T$ steps, we obtain a trajectory

$$\langle x_0, a_0, r_1, x_1, a_1, r_2, \ldots, x_{T-1}, a_{T-1}, r_T, x_T \rangle.$$

Then, for each state-action pair in the trajectory, we sum up its subsequent rewards as a sample of its cumulative rewards. Once we have multiple trajectories, we take the average cumulative rewards for each state-action pair as an estimate of the state-action value function.

To obtain a reliable estimation of value functions, we need many different trajectories. However, if our policy is deterministic, then we will end up with identical trajectories since the policy will always select the same action for the same state. We recognize that this is similar to the problem in the *exploitation-only* method of $K$-armed bandit, and therefore, we can learn from the idea of exploration-exploitation trade-off. Taking $\epsilon$-greedy method as an example, it selects an action at random with probability $\epsilon$ and the current best action with probability $1 - \epsilon$. We call the deterministic policy $\pi$ as *original policy*, and the policy of applying the $\epsilon$-greedy method on the original policy is denoted by

16

$$\pi^{\epsilon}(x) = \begin{cases} \pi(x), & \text{with probability } 1-\epsilon; \\ \text{Uniformly and randomly} & \\ \text{select an action from } A & \text{with probability } \epsilon. \end{cases} \tag{16.20}$$

For the original policy $\pi = \arg\max_a Q(x, a)$ that maximizes the value function, the current best action has a probability of $1 - \epsilon + \frac{\epsilon}{|A|}$ to be selected in the $\epsilon$-greedy policy $\pi^{\epsilon}$, and each non-optimal action has a probability of $\frac{\epsilon}{|A|}$ to be selected. In this way, there is a chance for every action to be selected, and hence we can obtain different trajectories.

Assuming there is only one best action.

After Monte Carlo-based policy evaluation, we also need policy improvement as we had in policy iteration algorithms. Recall that we exploited the monotonicity revealed by (16.16) to improve policy using the current best action. For any original policy $\pi$, its $\epsilon$-greedy policy $\pi^{\epsilon}$ equally allocates the probability $\epsilon$ to all actions, and therefore, we have $Q^{\pi}(x, \pi'(x)) \geqslant V^{\pi}(x)$ for the original policy $\pi'$ that maximizes the value function. Hence, (16.16) still holds, and we can use the same policy improvement method.

The pseudocode of the above procedure is given in ◘ Algorithm 16.6. Since the evaluation and improvement are on the same policy, the algorithm is called the *on-policy* Monte Carlo reinforcement learning algorithm. The algorithm computes the average reward incrementally by updating the value function using all state-action pairs in the trajectory once sampled.

---

**Algorithm 16.6** On-policy Monte Carlo reinforcement learning

---

**Input:** Environment $E$;
Action space $A$;
Initial state $x_0$;
Number policy steps $T$.

**Process:**

$Q(x, a) = 0$, $\text{count}(x, a) = 0$, $\pi(x, a) = \frac{1}{|A(x)|}$; — By default, the action is chosen with uniform probability.

**for** $s = 1, 2, \ldots$ **do** — Sample the $s$th trajectory.

  Obtain a trajectory by executing the policy $\pi$ in $E$:
  $\langle x_0, a_0, r_1, x_1, a_1, r_2, \ldots, x_{T-1}, a_{T-1}, r_T, x_T \rangle$;

  **for** $t = 0, 1, \ldots, T-1$ **do**

    $R = \frac{1}{T-t} \sum_{i=t+1}^{T} r_i$; — For each state-action pair, compute the cumulative rewards of the trajectory.

    $Q(x_t, a_t) = \frac{Q(x_t, a_t) \times \text{count}(x_t, a_t) + R}{\text{count}(x_t, a_t) + 1}$; — (16.2) updates the average reward.

    $\text{count}(x_t, a_t) = \text{count}(x_t, a_t) + 1$;

  **end for**

  For each seen state $x$: — Obtain the policy from the value function.

$$\pi(x) = \begin{cases} \arg\max_{a'} Q(x, a'), & \text{with probability } 1 - \epsilon; \\ \text{Uniformly and randomly select an action from } A, & \text{with probability } \epsilon. \end{cases}$$

**end for**

**Output:** Policy $\pi$.

---

The on-policy Monte Carlo reinforcement learning algorithm produces an $\epsilon$-greedy policy. However, introducing the $\epsilon$-greedy policy is just for the convenience of policy evaluation rather than using it as the final policy. The one we wish to improve is indeed the original (non $\epsilon$-greedy) policy. Then, is it possible to use $\epsilon$-greedy only in policy evaluation and improve the original policy in policy improvement? The answer is "yes".

Suppose we sample trajectories using two different policies $\pi$ and $\pi'$, where the difference is that the same state-action pair has different probabilities of being sampled. In general, the expectation of function $f$ over probability distribution $p$ is given by

$$\mathbb{E}[f] = \int_x p(x)f(x)dx, \tag{16.21}$$

which can be approximated using the samples $\{x_1, x_2, \ldots, x_m\}$ obtained from the probability distribution $p$, that is

$$\hat{\mathbb{E}}[f] = \frac{1}{m}\sum_{i=1}^{m} f(x_i). \tag{16.22}$$

Suppose we introduce another distribution $q$, then the expectation of function $f$ over the probability distribution $p$ can be equivalently written as

$$\mathbb{E}[f] = \int_x q(x)\frac{p(x)}{q(x)}f(x)dx, \tag{16.23}$$

which can be seen as the expectation of $\frac{p(x)}{q(x)}f(x)$ over the distribution $q$, and hence can be estimated using the samples $\{x'_1, x'_2, \ldots, x'_m\}$ drawn from $q$

The method of estimating the expectation of distribution using the samples of another distribution is known as *importance sampling*.

$$\hat{\mathbb{E}}[f] = \frac{1}{m}\sum_{i=1}^{m} \frac{p(x'_i)}{q(x'_i)}f(x'_i). \tag{16.24}$$

Back to our original problem, using the sampled trajectory of the policy $\pi$ to evaluate $\pi$ is actually estimating the expectation of the cumulative rewards

$$Q(x, a) = \frac{1}{m}\sum_{i=1}^{m} R_i, \tag{16.25}$$

where $R_i$ is the cumulative rewards from the state $x$ to the end of the $i$th trajectory. If we evaluate the policy $\pi$ using the trajectories sampled from the policy $\pi'$, then we just need to add weights to the cumulative rewards, that is

$$Q(x, a) = \frac{1}{m}\sum_{i=1}^{m} \frac{P_i^{\pi}}{P_i^{\pi'}} R_i, \tag{16.26}$$

where $P_i^{\pi}$ and $P_i^{\pi'}$ are the two probabilities of producing the $i$th trajectory by the two policies. Given a trajectory $\langle x_0, a_0, r_1, \ldots, x_{T-1}, a_{T-1}, r_T, x_T\rangle$, its probability of being produced by the policy $\pi$ is given by

$$P^{\pi} = \prod_{i=0}^{T-1} \pi(x_i, a_i) P_{x_i \to x_{i+1}}^{a_i}. \tag{16.27}$$

Though the state transition probability $P_{x_i \to x_{i+1}}^{a_i}$ appears in (16.27), (16.24) only needs the ratio of the two probabilities

$$\frac{P^{\pi}}{P^{\pi'}} = \prod_{i=0}^{T-1} \frac{\pi(x_i, a_i)}{\pi'(x_i, a_i)}. \tag{16.28}$$

If $\pi$ is a deterministic policy and $\pi'$ is its $\epsilon$-greedy policy, then $\pi(x_i, a_i)$ is always 1 for $a_i = \pi(x_i)$ and $\pi'(x_i, a_i)$ is either $\frac{\epsilon}{|A|}$ or $1 - \epsilon + \frac{\epsilon}{|A|}$, and hence we can evaluate the policy $\pi$. The *off-policy* Monte Carlo reinforcement learning algorithm is given in ◘ Algorithm 16.7.

---

**Algorithm 16.7** Off-policy Monte Carlo reinforcement learning

---

**Input:** Environment $E$;
Action space $A$;
Initial state $x_0$;
Number policy steps $T$.

**Process:**

$Q(x, a) = 0$, $\text{count}(x, a) = 0$, $\pi(x, a) = \frac{1}{|A(x)|}$;
**for** $s = 1, 2, \ldots$ **do**
  Obtain a trajectory by executing the $\epsilon$-greedy policy $\pi$ in $E$: $\langle x_0, a_0, r_1, x_1, a_1, r_2, \ldots, x_{T-1}, a_{T-1}, r_T, x_T\rangle$;
  $p_i = \begin{cases} 1 - \epsilon + \epsilon / |A|, & a_i = \pi(x_i); \\ \epsilon / |A|, & a_i \neq \pi(x_i); \end{cases}$
  **for** $t = 0, 1, \ldots, T-1$ **do**
    $R = \frac{1}{T-t}\left(\sum_{i=t+1}^{T} r_i\right)\prod_{i=t+1}^{T-1} \frac{\mathbb{I}(a_i = \pi(x_i))}{p_i}$;
    $Q(x_t, a_t) = \frac{Q(x_t, a_t) \times \text{count}(x_t, a_t) + R}{\text{count}(x_t, a_t) + 1}$;
    $\text{count}(x_t, a_t) = \text{count}(x_t, a_t) + 1$;
  **end for**
  $\pi(x) = \arg\max_{a'} Q(x, a')$.
**end for**

**Output:** Policy $\pi$.

---

By default, the action is chosen with uniform probability.
Sample the $s$th trajectory.

Compute the adjusted cumulative rewards. The terms in the product with the subscription greater than the superscription take value 1.

(16.2) updates the average reward.
Obtain the policy from the value function.

### 16.4.2 Temporal Difference Learning

When the model is unknown, Monte Carlo reinforcement learning algorithms use trajectory sampling to overcome the difficulty in policy evaluation. Such algorithms update value functions after each trajectory sampling. In contrast, the dynamic programming-based policy iteration and value iteration algorithms update value functions after every step of policy execution. Comparing these two approaches, we see that Monte Carlo reinforcement learning algorithms are far less efficient, mainly because they do not take advantage of the MDP structure. We now introduce Temporal Difference (TD) learning, which enables efficient model-free learning by joining the ideas of dynamic programming and Monte Carlo methods.

Essentially, Monte Carlo reinforcement learning algorithms approximate the expected cumulative rewards by taking the average across different trials. The averaging operation is in batch mode, which means state-action pairs are updated together after sampling an entire trajectory. To improve efficiency, we can make this updating process incremental. For state-action pair $(x, a)$, suppose we have estimated the value function $Q_t^{\pi}(x, a) = \frac{1}{t}\sum_{i=1}^{T} r_i$ based on the $t$ state-action samples, then, similar to (16.3), after we obtained the $(t+1)$-th sample $r_{t+1}$, we have

$$Q_{t+1}^{\pi}(x, a) = Q_t^{\pi}(x, a) + \frac{1}{t+1}(r_{t+1} - Q_t^{\pi}(x, a)), \quad (16.29)$$

which increments $Q_t^{\pi}(x, a)$ by $\frac{1}{t+1}(r_{t+1} - Q_t^{\pi}(x, a))$. More generally, by replacing $\frac{1}{t+1}$ with coefficient $\alpha_{t+1}$, we can write the increment as $\alpha_{t+1}(r_{t+1} - Q_t^{\pi}(x, a))$. In practice, we often set $\alpha_t$ to a small positive value $\alpha$. If we expand $Q_t^{\pi}(x, a)$ to the sum of step-wise cumulative rewards, then the sum of the coefficients is 1, that is, letting $\alpha_t = \alpha$ does not change the fact that $Q_t$ is the sum of cumulative rewards. The larger the step-size $\alpha$ is, the more important the later cumulative rewards are.

Taking $\gamma$-discounted cumulative rewards as an example, suppose we use dynamic programming and state-action functions, which are convenient when the model is unknown. Then, from (16.10), we have



$$\begin{aligned} Q^{\pi}(x, a) &= \sum_{x' \in X} P_{x \to x'}^{a}(R_{x \to x'}^{a} + \gamma V^{\pi}(x')) \\ &= \sum_{x' \in X} P_{x \to x'}^{a}(R_{x \to x'}^{a} + \gamma \sum_{a' \in A} \pi(x', a') Q^{\pi}(x', a')). \end{aligned} \quad (16.30)$$

With cumulative sum, we have

$$Q^{\pi}_{t+1}(x,a) = Q^{\pi}_{t}(x,a)+\alpha(R^{a}_{x\to x'}+\gamma Q^{\pi}_{t}(x',a')-Q^{\pi}_{t}(x,a)), \tag{16.31}$$

where $x'$ is the state transitioned from the state $x$ after executing the action $a$, and $a'$ is the action for $x'$ selected by the policy $\pi$.

With (16.31), value functions are updated after each action, as shown in ◘ Algorithm 16.8. Each update on value functions requires to know the previous **s**tate, the previous **a**ction, the **r**eward, the current **s**tate, and the **a**ction to be executed. Putting the initials together gives the algorithm name *Sarsa* (Rummery and Niranjan 1994). Sarsa is an on-policy algorithm, in which both the evaluation (line 6) and the execution (line 5) use the $\epsilon$-greedy policy.

By modifying Sarsa to an off-policy algorithm, we have the Q-Learning algorithm (Watkins and Dayan 1992), as shown in ◘ Algorithm 16.9, in which the evaluation (line 6) uses the original policy and the execution (line 4) uses the $\epsilon$-greedy policy.

**Algorithm 16.8** Sarsa

**Input:** Environment $E$;
Action space $A$;
Initial state $x_0$;
Reward discount $\gamma$;
Step-size $\alpha$.

**Process:**

1: $Q(x,a)=0$, $\pi(x,a)=\frac{1}{|A(x)|}$; — By default, the action is chosen with uniform probability.
2: $x=x_0$, $a=\pi(x)$;
3: **for** $t=1,2,\ldots$ **do**
4: $r, x'$ = the rewards and transitioned state through executing the action $a$ in $E$; — One-step execution approach.
5: $a'=\pi^{\epsilon}(x')$; — The $\epsilon$-greedy policy of the original policy.
6: $Q(x,a)=Q(x,a)+\alpha\left(r+\gamma Q(x',a')-Q(x,a)\right)$; — (16.31) updates the value function.
7: $\pi(x)=\arg\max_{a''} Q(x,a'')$;
8: $x=x'$, $a=a'$.
9: **end for**

**Output:** Policy $\pi$.

## 16.5 Value Function Approximation

Our previous discussions assumed finite state space with each state assigned an index. The value function is a *tabular value function* of the finite state space, that is, it can be represented as an array with the $i$th element in the array as the output value for the input $i$, and modifying the value of a state does not affect the values of other states. In practical reinforcement learning problems, however, the state space is often continuous with an

**Algorithm 16.9** Q-learning

**Input:** Environment $E$;
Action space $A$;
Initial state $x_0$;
Reward discount $\gamma$;
Step-size $\alpha$.

**Process:**

$Q(x, a) = 0, \pi(x, a) = \frac{1}{|A(x)|}$;

By default, the action is chosen with uniform probability.

$x = x_0$;

**for** $t = 1, 2, \ldots$ **do**

  $r, x'$ = the rewards and transitioned state through executing the action $a = \pi^{\epsilon}(x)$ in $E$;

One-step execution approach.

  $a' = \pi(x')$;

The original policy.

  $Q(x, a) = Q(x, a) + \alpha \left(r + \gamma Q(x', a') - Q(x, a)\right)$;

(16.31) updates the value function.

  $\pi(x) = \arg\max_{a''} Q(x, a'')$;

  $x = x'$.

**end for**

**Output:** Policy $\pi$.

infinite number of states. To solve this problem, what should we do?

Intuitively, we can discretize the continuous state space into a finite state space and then apply the previously introduced techniques. Unfortunately, it is difficult to discretize the state space effectively, especially before we have explored the state space.

As a workaround, we can try to model the value function in continuous state space directly. Suppose the state space is an $n$-dimensional real-valued space $X = \mathbb{R}^n$, whose state values cannot be recorded with a tabular value function. Let us start with a simple case in which the value function is represented as a linear function of states (Busoniu et al. 2010):

$$V_{\boldsymbol{\theta}}(\boldsymbol{x}) = \boldsymbol{\theta}^{\top} \boldsymbol{x}, \tag{16.32}$$

where $\boldsymbol{x}$ is the state vector, and $\boldsymbol{\theta}$ is the parameter vector. Such a value function cannot precisely record the value of every state as we did for the finite state space, and hence solving such a value function is called *value function approximation*.

We wish that the value function learned from (16.32) can approximate the ground-truth value function $V^{\pi}$ as precise as possible, and we often measure the difference using the least squares error

$$E_{\boldsymbol{\theta}} = \mathbb{E}_{\boldsymbol{x} \sim \pi} \left[ (V^{\pi}(\boldsymbol{x}) - V_{\boldsymbol{\theta}}(\boldsymbol{x}))^2 \right], \tag{16.33}$$

where $\mathbb{E}_{\boldsymbol{x} \sim \pi}$ is the expectation over the states sampled from the policy $\pi$.

We can use a gradient descent method to minimize the error. Finding the negative derivative of the error, we have

$$\begin{aligned} -\frac{\partial E_{\theta}}{\partial \theta} &= \mathbb{E}_{x \sim \pi}\left[2(V^{\pi}(x) - V_{\theta}(x))\frac{\partial V_{\theta}(x)}{\partial \theta}\right] \\ &= \mathbb{E}_{x \sim \pi}\left[2(V^{\pi}(x) - V_{\theta}(x))x\right], \end{aligned} \tag{16.34}$$

from which we have the update rule of individual samples

$$\theta = \theta + \alpha(V^{\pi}(x) - V_{\theta}(x))x. \tag{16.35}$$

Though we do not know the ground-truth value function $V^{\pi}$ of the policy, we can learn from temporal difference learning: based on $V^{\pi}(x) = r + \gamma V^{\pi}(x')$, we substitute the current estimated value function for the ground-truth value function, that is

$$\begin{aligned} \theta &= \theta + \alpha(r + \gamma V_{\theta}(x') - V_{\theta}(x))x \\ &= \theta + \alpha(r + \gamma \theta^{\top} x' - \theta^{\top} x)x, \end{aligned} \tag{16.36}$$

where $x'$ is the state at the next moment.

Temporal difference learning requires the state-action value function to obtain the policy. A simple approach is to apply $\theta$ to the joint vector of states and actions. For example, we can add one more dimension to the state vector for storing the action indices, that is, replacing $x$ in (16.32) with $(x; a)$. Another approach is to encode actions with 0/1 to get the vector $a = (0; \ldots; 1; \ldots; 0)$, where "1" indicates the selected action. Then, we combine $a$ with the state vector into $(x; a)$, and use it to replace $x$ in (16.32). By doing so, the linear approximation is done on the state-action value function.

By replacing the value function in the Sarsa algorithm with linear value function, we have the Sarsa algorithm with linear value function approximation, as shown in ◘ Algorithm 16.10. Similarly, we can derive the Q-Learning algorithm with linear value function approximation. We can also easily replace the linear learner in (16.32) with other learning methods, e.g., kernel methods for non-linear value function approximation.

See Chap. 6 for kernel methods.

## 16.6 Imitation Learning

In classic reinforcement learning settings, the only feedback information received by the agent is the cumulative rewards after making multiple decisions. In real-world applications, however, we can often obtain some examples of how human experts make decisions. For example, we may ask experienced

**Algorithm 16.10** Sarsa with linear value function approximation

**Input:** Environment $E$;
Action space $A$;
Initial state $x_0$;
Reward discount $\gamma$;
Step-size $\alpha$.

**Process:**

$\boldsymbol{\theta} = \mathbf{0}$;
$\boldsymbol{x} = \boldsymbol{x}_0, a = \pi(\boldsymbol{x}) = \arg\max_{a''} \boldsymbol{\theta}^{\top}(\boldsymbol{x}; a'')$;
**for** $t = 1, 2, \ldots$ **do**
  $r, \boldsymbol{x}'$ = the rewards and transitioned state through executing the action $a$ in $E$;
  $a' = \pi^{\epsilon}(\boldsymbol{x}')$;
  $\boldsymbol{\theta} = \boldsymbol{\theta} + \alpha(r + \gamma\boldsymbol{\theta}^{\top}(\boldsymbol{x}'; a') - \boldsymbol{\theta}^{\top}(\boldsymbol{x}; a))(\boldsymbol{x}; a)$;
  $\pi(\boldsymbol{x}) = \arg\max_{a''} \boldsymbol{\theta}^{\top}(\boldsymbol{x}; a'')$;
  $\boldsymbol{x} = \boldsymbol{x}', a = a'$.
**end for**

**Output:** Policy $\pi$.

The $\epsilon$-greedy policy of the original policy.

(16.36) updates the parameters.

farmers to demonstrate how to plant watermelons. Learning from such demonstrations is called *imitation learning*.

Also known as *apprenticeship learning*, *learning from demonstration*, and *learning by watching*. This learning paradigm has a connection to *learning from instruction*. See Sect. 1.5.

### 16.6.1 Direct Imitation Learning

The multi-step decision process in reinforcement learning faces a huge search space, and hence it is not easy to use the cumulative rewards to learn the appropriate decisions made many steps ago. Such a difficulty, however, can be well alleviated by directly imitating the state-action pairs provided by human experts, and we call it *direct imitation learning* (a.k.a. behavior cloning).

Suppose that human experts have supplied us a set of trajectories $\{\tau_1, \tau_2, \ldots, \tau_m\}$, where each trajectory includes a sequence of states and actions

$$\tau_i = \langle s_1^i, a_1^i, s_2^i, a_2^i, \ldots, s_{n_i+1}^i \rangle,$$

where $n_i$ is the number of transitions in the $i$th trajectory. Such information tells the agent what to do under each state, and therefore, we can use supervised learning to learn the policy that matches the trajectory data obtained from human experts.

We can extract the state-action pairs from all trajectories to make a new data set

$$D = \{(s_1, a_1), (s_2, a_2), \ldots, (s_{\sum_{i=1}^m n_i}, a_{\sum_{i=1}^m n_i})\},$$

16

which encodes states as features and actions as labels. Then, from data set $D$, we can learn a policy model using either classification algorithms for discrete actions or regression algorithms for continuous actions. The agent can use the learned policy model to set the initial policy, which is then further improved by reinforcement learning techniques.

### 16.6.2 Inverse Reinforcement Learning

In many applications, it is often challenging to design the reward function, but we may inversely derive the reward function from the examples provided by human experts, and we call it *inverse reinforcement learning* (Abbeel and Ng 2004).

In inverse reinforcement learning, the state space $X$ and the action space $A$ are known, and we also have a trajectory data set $\{\tau_1, \tau_2, \ldots, \tau_m\}$ just like we have in direct imitation learning. The basic idea of inverse reinforcement learning is as follows: letting the agent take the actions that are consistent with the provided examples is equivalent to finding the optimal policy in the environment of a reward function, where the optimal policy produces the same trajectories as the provided examples. In other words, we look for a reward function such that the provided samples are optimal and then use this reward function to train the policy in reinforcement learning.

Suppose that the reward function can be represented as a linear function of states, that is, $R(\boldsymbol{x}) = \mathbf{w}^{\top}\boldsymbol{x}$. Then, the cumulative rewards of policy $\pi$ can be written as

$$\begin{aligned}\rho^{\pi} &= \mathbb{E}\left[\sum_{t=0}^{+\infty} \gamma^t R(\boldsymbol{x}_t) \mid \pi\right] = \mathbb{E}\left[\sum_{t=0}^{+\infty} \gamma^t \mathbf{w}^{\top}\boldsymbol{x}_t \mid \pi\right]\\ &= \mathbf{w}^{\top}\mathbb{E}\left[\sum_{t=0}^{+\infty} \gamma^t \boldsymbol{x}_t \mid \pi\right],\end{aligned} \tag{16.37}$$

which is the inner product of the coefficients $\mathbf{w}$ and the expectation of the weighted sum of state vectors.

Let $\bar{\boldsymbol{x}}^{\pi}$ denote the expectation of state vectors $\mathbb{E}[\sum_{t=0}^{+\infty} \gamma^t \boldsymbol{x}_t \mid \pi]$. The expectation $\bar{\boldsymbol{x}}^{\pi}$ can be approximated using the Monte Carlo method: the example trajectories can be seen as a sampling of the optimal policy, and hence we can calculate the weighted sum of states in each example trajectory and then take the average, denoted by $\bar{\boldsymbol{x}}^{*}$. Then, for the optimal reward function $R(\boldsymbol{x}) = \mathbf{w}^{*\top}\boldsymbol{x}$ and the expectation $\bar{\boldsymbol{x}}^{\pi}$ produced by any other policies, we have

$$\mathbf{w}^{*\top}\bar{\boldsymbol{x}}^{*} - \mathbf{w}^{*\top}\bar{\boldsymbol{x}}^{\pi} = \mathbf{w}^{*\top}(\bar{\boldsymbol{x}}^{*} - \bar{\boldsymbol{x}}^{\pi}) \geqslant 0. \tag{16.38}$$

If we can calculate $(\bar{x}^* - \bar{x}^\pi)$ for every policy, then we can solve

$$\mathbf{w}^* = \arg\max_{\mathbf{w}} \min_{\pi} \mathbf{w}^\top(\bar{x}^* - \bar{x}^\pi) \quad \text{s.t.} \quad \|\mathbf{w}\| \leqslant 1. \tag{16.39}$$

Since it is difficult to obtain all policies, an alternative approach is to start with a random policy and use it to find a better reward function iteratively. The reward function is then used to find a better policy. This process continues until we have the reward function and policy that fit the example trajectories. The pseudocode of the above procedure is given in ◘ Algorithm 16.11. In the step of finding a better reward function, the minimization in (16.39) should only include the previously learned policies instead of all policies.

**Algorithm 16.11** Iterative inverse reinforcement learning

**Input:** Environment $E$;
State space $X$;
State space $A$;
Trajectory data set $D = \{\tau_1, \tau_2, \ldots, \tau_m\}$.

**Process:**
1: $\bar{x}^*$ = the mean vector of the weighted sum of the state vectors in the example trajectories;
2: $\pi$ = random policy;
3: **for** $t = 1, 2, \ldots$ **do**
4: $\bar{x}_t^\pi$ = the mean vector of the weighted sum of the state vectors in the trajectories sampled from $\pi$;
5: Solve $\mathbf{w}^* = \arg\max_{\mathbf{w}} \min_{i=1}^{t} \mathbf{w}^\top(\bar{x}^* - \bar{x}_i^\pi)$ s.t. $\|\mathbf{w}\| \leqslant 1$;
6: $\pi$ = the optimal policy in the environment $\langle X, A, R(\boldsymbol{x}) = \mathbf{w}^{*\top}\boldsymbol{x}\rangle$.
7: **end for**

**Output:** Reward function $R(\boldsymbol{x}) = \mathbf{w}^{*\top}\boldsymbol{x}$ and policy $\pi$.

## 16.7 Further Reading

16

Sutton and Barto (1998) wrote a famous book dedicated to reinforcement learning. Gosavi (2003) discussed reinforcement learning from the optimization perspective, and Whiteson (2010) discussed reinforcement learning with an emphasis on evolutionary search algorithms. Mausam and Kolobov (2012) introduced reinforcement learning from the Markov decision process point of view. Sigaud and Buffet (2010) has broad coverage, including Partially Observable MDP (POMDF) and policy gradient methods. See Busoniu et al. (2010) for more information about reinforcement learning with value function optimization.

The European Workshop on Reinforcement Learning (EWRL) is a series of workshops dedicated to reinforcement learning. The Multi-disciplinary Conference on Reinforcement Learning and Decision Making (RLDM) is a new conference that started in 2013.

Kaelbling et al. (1996) wrote an early survey on reinforcement learning. Deisenroth et al. (2013), Kober et al. (2013) surveyed the applications of reinforcement learning in the field of robotics.

Vermorel and Mohri (2005) introduced and compared several $K$-armed bandit algorithms. Multi-armed bandit models have been extensively studied Berry and Fristedt (1985) in the field of statistics, which have been widely used, for problems such as *online learning* and *adversarial learning*, in recent years. Bubeck and Cesa-Bianchi (2012) provided an overview on *regret bound* analysis of multi-armed bandit problems.

"regret" refers to the difference between the rewards of the decision made under uncertain conditions and the decision made under certain conditions.

Temporal decision (TD) learning was first proposed by A. Samuel in his famous checker project. Sutton (1988) proposed the TD($\lambda$) algorithm, from which (Tesauro 1995) developed the TD-Gammon program that popularized TD learning by reaching the level of top human backgammon players at that time. Watkins and Dayan (1992) proposed the Q-Learning algorithm, which was then improved to Sarsa by Rummery and Niranjan (1994). TD learning is still being improved and extended in recent years, such as generalized TD learning (Ueno et al. 2011) and TD learning with eligibility traces (Geist and Scherrer 2014). Dann et al. (2014) compared different policy evaluation methods for TD learning.

See Break Time of Chap. 1 for Samuel's checker project.

Imitation learning is an important approach for speeding up reinforcement learning (Lin 1992, Price and Boutilier 2003) and has been widely used in the field of robotics (Argall et al. 2009). Abbeel and Ng (2004), Langford and Zadrozny (2005) proposed inverse reinforcement learning methods.

Reinforcement learning is known as *approximate dynamic programming* in operations research and cybernetics. See Bertsekas (2012) for more information.

## Exercises

**16.1** In the $K$-armed bandit problem, the Upper Confidence Bound (UCB) method selects arm that has the largest value of $Q(k) + UC(k)$ in each trial, where $Q(k)$ is the current average reward of the arm $k$ and $UC(k)$ is the confidence interval. For example

$$Q(k) + \sqrt{\frac{2\ln n}{n_k}},$$

where $n$ is the total number of pulls, and $n_k$ is the total number of pulls of the arm $k$. Discuss the difference and commonality between the UCB method, the $\epsilon$-greedy method, and the Softmax method.

**16.2** Taking ◘ Algorithm 16.3 as a reference, write down the policy evaluation algorithm based on the $\gamma$-discounted reward function.

**16.3** Taking ◘ Algorithm 16.4 as a reference, write down the policy iteration algorithm based on the $\gamma$-discounted reward function.

**16.4** When the MDP is unknown, we can learn the MDP first and then apply model-based reinforcement learning methods. For example, we sample from random policies and estimate the transition function and the reward function. Discuss the pros and cons of such a method compared to model-free reinforcement learning methods.

**16.5** Derive the update rule (16.31) of the Sarsa algorithm.

**16.6** Taking ◘ Algorithm 16.10 as a reference, write down the Q-Learning algorithm with linear value function approximation.

16

**16.7** In practice, linear value function approximation has relatively large errors. Extend the Sarsa algorithm with linear value function approximation to the Sarsa algorithm with neural network approximation.

**16.8** Extend the Sarsa algorithm with linear value function approximation to the Sarsa algorithm with non-linear value function approximation that uses kernel functions.

**16.9** For goal-directed reinforcement learning, the task is to arrive at a specific state, e.g., to park a car to the designated location. Design a reward function for such problems, and discuss the utility of different reward functions, e.g., setting the rewards to 0, $-1$, or 1 for each step that does not meet the goal.

**16.10** * Unlike traditional supervised learning, direct imitation learning deals with data that may come from different data distributions at different times. Design a direct imitation learning algorithm that considers the changes in data distributions over time.

## Break Time

### Short Story: Markov Decision Process and Andrey Andreyevich Markov

Andrey Andreyevich Markov (1856–1922) was a renowned Russian mathematician and best known for his work on stochastic processes. Markov made significant contributions to probability theory, number theory, approximation theory, differential calculus, etc.

Markov was born in Ryazan, Russia. When he was 17 years old, Markov independently developed a solution to linear ordinary differential equations, and this drew the attention of several mathematicians from St. Petersburg University. In 1874, Markov was admitted to St. Petersburg University, and obtained a lecturer position there in 1878. Supervised by the prominent Russian mathematician P. Chebyshev, Markov earned a Ph.D. degree in 1884 at St. Petersburg University, where he remained for the rest of his career. In the early days of his research, Markov followed Chebyshev's research direction and worked on the law of large numbers and the central limit theorem. However, Markov is best known for his contribution to research on stochastic processes. From 1906 to 1912, Markov proposed the Markov chains and Markov processes. These methods can describe a wide range of real-world processes, ranging from Brownian motion on the microscopic level to the spread of large-scale infectious disease. In his classical textbook *The Calculus of Probabilities*, Markov illustrated Markov chains using the distribution of vowels and consonants in A. S. Pushkin's poem "Eugeny Onegin". In the 1950s, more than 30 years later after Markov's death, the Markov decision process was developed by combining Markov processes and dynamic programming.

Chebyshev has many notable students, including A. A. Markov, A. Lyapunov, A. Korkin, and D. Grave. He has also influenced many mathematicians outside of St. Petersburg University.

The Markov brothers' inequality is named after Markov and his younger brother Vladimir Andreevich Markov (1871–1897). Markov's son, whose name was also Andrey Andreyevich Markov (1903–1979), was also a prominent mathematician. The Turing-complete Markov algorithm, Markov's principle, and Markov's rule are named after the junior Markov.

## References

Abbeel P, Ng AY (2004) Apprenticeship learning via inverse reinforcement learning. In: Proceedings of the 21st international conference on machine learning (ICML). Banff, Canada

Argall BD, Chernova S, Veloso M, Browning B (2009) A survey of robot learning from demonstration. Robot Autonom Syst 57(5):469–483

Berry D, Fristedt B (1985) Bandit problems. Chapman & Hall/CRC, London

Bertsekas DP (2012) Dynamic programming and optimal control: approximate dynamic programming, 4th edn. Athena Scientific, Nashua

Bubeck S, Cesa-Bianchi N (2012) Regret analysis of stochastic and nonstochastic multi-armed bandit problems. Found Trends Mach Learn 5(1):1–122

Busoniu L, Babuska R, Schutter BD, Ernst D (2010) Reinforcement learning and dynamic programming using function approximators. Chapman & Hall/CRC Press, Boca Raton

Dann C, Neumann G, Peters J (2014) Policy evaluation with temporal differences: a survey and comparison. J Mach Learn Res 15:809–883

Deisenroth MP, Neumann G, Peters J (2013) A survey on policy search for robotics. Found Trends Robot 2(1–2):1–142

Geist M, Scherrer B (2014) Off-policy learning with eligibility traces: a survey. J Mach Learn Res 15:289–333

Gosavi A (2003) Simulation-based optimization: parametric optimization techniques and reinforcement learning. Kluwer, Norwell

Kaelbling LP, Littman ML, Moore AW (1996) Reinforcement learning: a survey. J Artif Intell Res 4:237–285

Kober J, Bagnell JA, Peters J (2013) Reinforcement learning in robotics: a survey. Int J Robot Res 32(11):1238–1274

Langford J, Zadrozny B (2005) Relating reinforcement learning performance to classification performance. In: Proceedings of the 22nd international conference on machine learning (ICML). Bonn, Germany, pp 473–480

Lin L-J (1992) Self-improving reactive agents based on reinforcement learning, planning and teaching. Mach Learn 8(3–4):293–321

Mausam, Kolobov A (2012) Planning with Markov decision processes: an AI perspective. Morgan & Claypool, San Rafael

Price B, Boutilier C (2003) Accelerating reinforcement learning through implicit imitation. J Artif Intell Res 19:569–629

Rummery GA, Niranjan M (1994) On-line Q-learning using connectionist systems. Technical report CUED/F-INFENG/TR 166, Engineering Department, Cambridge University, Cambridge

Sigaud O, Buffet O (2010) Markov decision processes in artificial intelligence. Wiley, Hoboken

Sutton RS (1988) Learning to predict by the methods of temporal differences. Mach Learn 3(1):9–44

Sutton RS, Barto A (1998) Reinforcement learning: an introduction. MIT Press, Cambridge

Tesauro G (1995) Temporal difference learning and TD-Gammon. Commun ACM 38(3):58–68

Ueno T, Maeda S, Kawanabe M, Ishii S (2011) Generalized TD learning. J Mach Learn Res 12:1977–2020

Vermorel J, Mohri M (2005) Multi-armed bandit algorithms and empirical evaluation. In: Proceedings of the 16th European conference on machine learning (ECML). Porto, Portugal, pp 437–448
Watkins CJCH, Dayan P (1992) Q-learning. Mach Learn 8(3–4):279–292
Whiteson S (2010) Adaptive representations for reinforcement learning. Springer, Berlin

Z.-H. Zhou, *Machine Learning*,
https://doi.org/10.1007/978-981-15-1967-3

# Appendix A
# Matrix

## A.1 Basic Operations

Let $(\mathbf{A})_{ij} = A_{ij}$ denote the element in the $i$th row and $j$th column of real matrix $\mathbf{A} \in \mathbb{R}^{m \times n}$, and $\mathbf{A}^\top$ denote the transpose of $\mathbf{A}$, where $(\mathbf{A}^\top)_{ij} = A_{ji}$. Then, we have

$$(\mathbf{A} + \mathbf{B})^\top = \mathbf{A}^\top + \mathbf{B}^\top, \tag{A.1}$$

$$(\mathbf{AB})^\top = \mathbf{B}^\top \mathbf{A}^\top. \tag{A.2}$$

When $m = n$, the matrix $\mathbf{A} \in \mathbb{R}^{m \times n}$ is called a square matrix of order $n$. Let $\mathbf{I}_n$ denote the identity matrix of order $n$, then the inverse of $\mathbf{A}$, denoted by $\mathbf{A}^{-1}$, satisfies $\mathbf{AA}^{-1} = \mathbf{A}^{-1}\mathbf{A} = \mathbf{I}$. Then, we have

$\mathbf{I}_n$ is the identity matrix of order $n$, or simply denoted by $\mathbf{I}$ when the context is clear.

$$(\mathbf{A}^\top)^{-1} = (\mathbf{A}^{-1})^\top, \tag{A.3}$$

$$(\mathbf{AB})^{-1} = \mathbf{B}^{-1}\mathbf{A}^{-1}. \tag{A.4}$$

For an $n$-by-$n$ square matrix $\mathbf{A}$, its trace is the sum of the elements on its diagonal, that is, $\mathrm{tr}(\mathbf{A}) = \sum_{i=1}^{n} A_{ii}$. For trace, we have the following properties:

$$\mathrm{tr}(\mathbf{A}^\top) = \mathrm{tr}(\mathbf{A}), \tag{A.5}$$

$$\mathrm{tr}(\mathbf{A} + \mathbf{B}) = \mathrm{tr}(\mathbf{A}) + \mathrm{tr}(\mathbf{B}), \tag{A.6}$$

$$\mathrm{tr}(\mathbf{AB}) = \mathrm{tr}(\mathbf{BA}), \tag{A.7}$$

$$\mathrm{tr}(\mathbf{ABC}) = \mathrm{tr}(\mathbf{BCA}) = \mathrm{tr}(\mathbf{CAB}). \tag{A.8}$$

The *determinant* of an $n$-by-$n$ square matrix $\mathbf{A}$ is given by

$$\det(\mathbf{A}) = \sum_{\boldsymbol{\sigma} \in S_n} \mathrm{par}(\boldsymbol{\sigma}) A_{1\sigma_1} A_{2\sigma_2} \ldots A_{n\sigma_n}, \tag{A.9}$$

where $S_n$ is the set of *permutations* of order $n$. The value of $\mathrm{par}(\boldsymbol{\sigma})$ can be either $-1$ or $+1$ depending on whether the permutation $\boldsymbol{\sigma} = (\sigma_1, \sigma_2, \ldots, \sigma_n)$ is odd or even, that is, whether the number of descending value pairs is odd or even. For example, the number of descending value pairs in $(1, 3, 2)$ is 1, and 2 for $(3, 1, 2)$. For the identify matrix, we have $\det(\mathbf{I}) = 1$. For a 2-by-2 $\mathbf{A}$, we have

$$\det(\mathbf{A}) = \det \begin{pmatrix} A_{11} & A_{12} \\ A_{21} & A_{22} \end{pmatrix} = A_{11}A_{22} - A_{12}A_{21}.$$

The determinant of an $n$-by-$n$ square matrix $\mathbf{A}$ has the following properties:

$$\det(c\mathbf{A}) = c^n \det(\mathbf{A}), \quad (A.10)$$

$$\det(\mathbf{A}^\top) = \det(\mathbf{A}), \quad (A.11)$$

$$\det(\mathbf{AB}) = \det(\mathbf{A})\det(\mathbf{B}), \quad (A.12)$$

$$\det(\mathbf{A}^{-1}) = \det(\mathbf{A})^{-1}, \quad (A.13)$$

$$\det(\mathbf{A}^n) = \det(\mathbf{A})^n. \quad (A.14)$$

The Frobenius norm of the matrix $\mathbf{A} \in \mathbb{R}^{m\times n}$ is given by

$$\|\mathbf{A}\|_F = (\mathrm{tr}(\mathbf{A}^\top\mathbf{A}))^{1/2} = \left(\sum_{i=1}^{m}\sum_{j=1}^{n} A_{ij}^2\right)^{1/2}. \quad (A.15)$$

We can see that the Frobenius norm of a matrix is the $L_2$ norm of its spanned vectors.

## A.2 Derivative

The derivative of a vector $\boldsymbol{a}$ with respect to a scalar $x$ and the derivative of $x$ with respect to $\boldsymbol{a}$ are both vectors, where the $i$th components of them are, respectively

$$\left(\frac{\partial \boldsymbol{a}}{\partial x}\right)_i = \frac{\partial a_i}{\partial x}, \quad (A.16)$$

$$\left(\frac{\partial x}{\partial \boldsymbol{a}}\right)_i = \frac{\partial x}{\partial a_i}. \quad (A.17)$$

Similarly, the derivative of a matrix $\mathbf{A}$ with respect to a scalar $x$ and the derivative of $x$ with respect to $\mathbf{A}$ are both matrices, where the components on the $i$th row and $j$th column are, respectively

$$\left(\frac{\partial \mathbf{A}}{\partial x}\right)_{ij} = \frac{\partial A_{ij}}{\partial x}, \quad (A.18)$$

$$\left(\frac{\partial x}{\partial \mathbf{A}}\right)_{ij} = \frac{\partial x}{\partial A_{ij}}. \quad (A.19)$$

For a function $f(\boldsymbol{x})$, if it is differentiable with respect to the components of $\boldsymbol{x}$, then the first-order derivative of $f(\boldsymbol{x})$ with respect to $\boldsymbol{x}$ is a vector, whose $i$th component is given by

$$(\nabla f(\boldsymbol{x}))_i = \frac{\partial f(\boldsymbol{x})}{\partial x_i}. \quad (A.20)$$

The second-order derivative of $f(\boldsymbol{x})$ with respect to $\boldsymbol{x}$ is a square matrix, known as Hessian matrix, whose element in the $i$th row and $j$th column is given by

$$\left(\nabla^2 f(\boldsymbol{x})\right)_{ij} = \frac{\partial^2 f(\boldsymbol{x})}{\partial x_i \partial x_j}. \tag{A.21}$$

The derivative of vectors and matrices obeys the product rule

$\boldsymbol{a}$ is a constant vector with respect to $\boldsymbol{x}$.

$$\frac{\partial \boldsymbol{x}^\top \boldsymbol{a}}{\partial \boldsymbol{x}} = \frac{\partial \boldsymbol{a}^\top \boldsymbol{x}}{\partial \boldsymbol{x}} = \boldsymbol{a}, \tag{A.22}$$

$$\frac{\partial \mathbf{AB}}{\partial \boldsymbol{x}} = \frac{\partial \mathbf{A}}{\partial \boldsymbol{x}}\mathbf{B} + \mathbf{A}\frac{\partial \mathbf{B}}{\partial \boldsymbol{x}}. \tag{A.23}$$

From $\mathbf{A}^{-1}\mathbf{A} = \mathbf{I}$ and (A.23), the derivative of the inverse matrix can be written as

$$\frac{\partial \mathbf{A}^{-1}}{\partial x} = -\mathbf{A}^{-1}\frac{\partial \mathbf{A}}{\partial x}\mathbf{A}^{-1}. \tag{A.24}$$

If the derivative of matrix $\mathbf{A}$ is taken with respect to an element of it, then we have

$$\frac{\partial \operatorname{tr}(\mathbf{AB})}{\partial A_{ij}} = B_{ji}, \tag{A.25}$$

$$\frac{\partial \operatorname{tr}(\mathbf{AB})}{\partial \mathbf{A}} = \mathbf{B}^\top, \tag{A.26}$$

and subsequently, we have

$$\frac{\partial \operatorname{tr}(\mathbf{A}^\top \mathbf{B})}{\partial \mathbf{A}} = \mathbf{B}, \tag{A.27}$$

$$\frac{\partial \operatorname{tr}(\mathbf{A})}{\partial \mathbf{A}} = \mathbf{I}, \tag{A.28}$$

$$\frac{\partial \operatorname{tr}(\mathbf{ABA}^\top)}{\partial \mathbf{A}} = \mathbf{A}(\mathbf{B} + \mathbf{B}^\top). \tag{A.29}$$

From (A.15) and (A.29), we have

$$\frac{\partial \|\mathbf{A}\|_F^2}{\partial \mathbf{A}} = \frac{\partial \operatorname{tr}(\mathbf{AA}^\top)}{\partial \mathbf{A}} = 2\mathbf{A}. \tag{A.30}$$

The chain rule is an important technique for calculating the derivatives of composite functions. In simple words, if a function $f$ is the composition of $g$ and $h$, that is, $f(x) = g(h(x))$, then we have

$$\frac{\partial f(x)}{\partial x} = \frac{\partial g(h(x))}{\partial h(x)} \cdot \frac{\partial h(x)}{\partial x}. \tag{A.31}$$

For example, $\mathbf{A}\boldsymbol{x} - \boldsymbol{b}$ can be seen as a whole to simplify the calculation of the following equation

$$\frac{\partial}{\partial \boldsymbol{x}}(\mathbf{A}\boldsymbol{x} - \boldsymbol{b})^{\top}\mathbf{W}(\mathbf{A}\boldsymbol{x} - \boldsymbol{b}) = \frac{\partial(\mathbf{A}\boldsymbol{x} - \boldsymbol{b})}{\partial \boldsymbol{x}} \cdot 2\mathbf{W}(\mathbf{A}\boldsymbol{x} - \boldsymbol{b}) = 2\mathbf{A}^{\top}\mathbf{W}(\mathbf{A}\boldsymbol{x} - \boldsymbol{b}). \quad \text{(A.32)}$$

In machine learning, $\mathbf{W}$ is usually a symmetric matrix.

## A.3 Singular Value Decomposition

Any real matrix $\mathbf{A} \in \mathbb{R}^{m \times n}$ can be decomposed as

$$\mathbf{A} = \mathbf{U}\boldsymbol{\Sigma}\mathbf{V}^{\top}, \quad \text{(A.33)}$$

where $\mathbf{U} \in \mathbb{R}^{m \times m}$ is a *unitary matrix* of order $m$ satisfying $\mathbf{U}^{\top}\mathbf{U} = \mathbf{I}$, $\mathbf{V} \in \mathbb{R}^{n \times n}$ is a unitary matrix of order $n$ satisfying $\mathbf{V}^{\top}\mathbf{V} = \mathbf{I}$, and $\sigma \in \mathbb{R}^{m \times n}$ is a $m \times n$ matrix with all of its elements taking the value 0 except $(\sigma)_{ii} = \sigma_i$, where $\sigma_i$ are non-negative real numbers and $\sigma_1 \geqslant \sigma_2 \geqslant \ldots \geqslant 0$.

The decomposition in (A.33) is known as Singular Value Decomposition (SVD), where the column vectors $\boldsymbol{u}_i \in \mathbb{R}^m$ of $\mathbf{U}$ are called the left-singular vectors, the column vectors $\mathbf{v}_i \in \mathbb{R}^n$ of $\mathbf{V}$ are called the right-singular vectors, and $\sigma_i$ are called the singular values. The number of non-zero singular values is called the rank of $\mathbf{A}$.

The singular values are typically arranged in descending order to ensure **6** is unique.

The results of SVD and eigendecomposition are the same when $\mathbf{A}$ is a symmetric positive definite matrix.

SVD has a wide range of applications. Taking the low-rank matrix approximation problem as an example, suppose we want to approximate a $r$-rank matrix $\mathbf{A}$ with a $k$-rank matrix $\widetilde{\mathbf{A}}$, where $k \leqslant r$, then, the problem can be formulated as

$$\begin{aligned} \min_{\widetilde{\mathbf{A}} \in \mathbb{R}^{m \times n}} \quad & \|\mathbf{A} - \widetilde{\mathbf{A}}\|_F \\ \text{s.t.} \quad & \operatorname{rank}(\widetilde{\mathbf{A}}) = k. \end{aligned} \quad \text{(A.34)}$$

SVD provides an analytical solution to the above problem: after performing SVD on $\mathbf{A}$, we obtain a matrix $\sigma_k$ by setting the $r - k$ smallest singular values in $\sigma$ to zero, that is, we keep only the $k$ largest singular values. Then, the optimal solution of (A.34) is given by

$$\mathbf{A}_k = \mathbf{U}_k\boldsymbol{\Sigma}_k\mathbf{V}_k^{\top}, \quad \text{(A.35)}$$

where $\mathbf{U}_k$ and $\mathbf{V}_k$ are, respectively, the first $k$ columns of $\mathbf{U}$ and $\mathbf{V}$ in (A.33). The result is referred to as the Eckart–Young–Mirsky theorem.

# Appendix B Optimization

## B.1 Lagrange Multiplier Method

The method of Lagrange multipliers is a technique for finding the extrema of a multivariate function subject to a set of constraints. By introducing Lagrange multipliers, an optimization problem with $d$ variables and $k$ constraints can be converted into an unconstrained optimization problem with $d + k$ variables.

Let us consider an optimization problem with one equality constraint. Suppose $\boldsymbol{x}$ is a $d$-dimensional vector, and we want to find a certain value $\boldsymbol{x}^*$ that minimizes the function $f(\boldsymbol{x})$ subject to the constraint $g(\boldsymbol{x}) = 0$. Geometrically, the task is to find a point that minimizes the objective function $f(x)$ on the $(d-1)$-dimensional space determined by $g(x) = 0$. It is not difficult to obtain the following conclusions:

The point is the tangent point of the constraint surface and the contour line of $f(\boldsymbol{x})$.

- The gradient $\nabla g(\boldsymbol{x})$ at any point $\boldsymbol{x}$ on the constraint surface is orthogonal to the constraint surface;
- The gradient $\nabla f(\boldsymbol{x}^*)$ at the optimal point $\boldsymbol{x}^*$ is orthogonal to the constraint surface.

Proof by contradiction: if the gradient $\nabla f(\boldsymbol{x}^*)$ is not orthogonal to the constraint surface, then we can shift $\boldsymbol{x}^*$ on the constraint surface to further reduce the objective value.

From the above conclusions, we know that, at the optimal point $\boldsymbol{x}^*$ shown in Figure B.1, the gradients $\nabla g(\boldsymbol{x})$ and $\nabla f(\boldsymbol{x})$ are either in the same direction or the opposite directions. That is, there exists a Lagrange multiplier $\lambda \neq 0$ such that

For equality constraints, $\lambda$ could be positive or negative.

$$\nabla f(\boldsymbol{x}^*) + \lambda \nabla g(\boldsymbol{x}^*) = 0. \tag{B.1}$$

We can define the Lagrange function as

$$L(\boldsymbol{x}, \lambda) = f(\boldsymbol{x}) + \lambda g(\boldsymbol{x}), \tag{B.2}$$

from which we can get (B.1) by setting its partial derivative with respect to $\boldsymbol{x}$, i.e., $\nabla_{\boldsymbol{x}} L(\boldsymbol{x}, \lambda)$ or partial derivative $\nabla_{\boldsymbol{x}} L(\boldsymbol{x}, \lambda)$ to zero. Also, we can obtain the constraint $g(\boldsymbol{x}) = 0$ by setting the partial derivative with respect to $\boldsymbol{x}$, i.e., $\nabla_{\boldsymbol{x}} L(\boldsymbol{x}, \lambda)$ or partial derivative $\nabla_{\boldsymbol{x}} L(\boldsymbol{x}, \lambda)$ to zero. Hence, the original constrained optimization problem can be converted into an unconstrained optimization problem of the Lagrange function $L(\boldsymbol{x}, \lambda)$.

Now let us take a look at the inequality constraint $g(\boldsymbol{x}) \leqslant 0$. As shown in Figure B.1b, the optimal point $\boldsymbol{x}^*$ lies on the boundary corresponding to $g(\boldsymbol{x}) = 0$ or falls into the area corresponding to $g(\boldsymbol{x}) < 0$. In the case of $g(\boldsymbol{x}) < 0$, the constraint $g(\boldsymbol{x}) \leqslant 0$ has no affect, and the optimal point can be

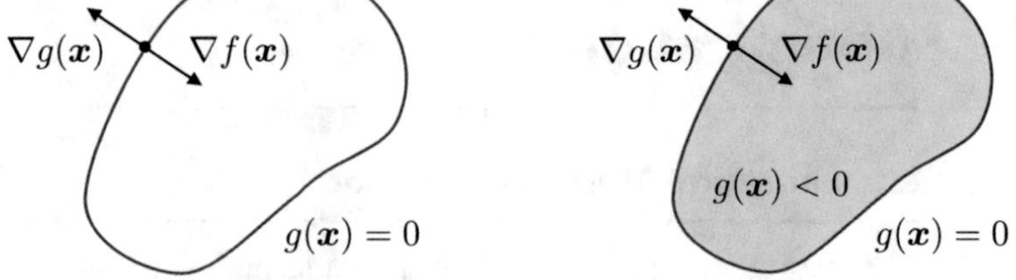


(a) Equality constraint. (b) Inequality constraint.

**Fig. B.1** The geometric interpretation of Lagrange multipliers. Minimizing the objective function $f(\boldsymbol{x})$ subject to (a) the equality constraint $g(\boldsymbol{x}) = 0$ or (b) the inequality constraint $g(\boldsymbol{x}) \leqslant 0$. The red surface corresponds to $g(\boldsymbol{x}) = 0$ and the shaded area corresponds to $g(\boldsymbol{x}) < 0$

obtained using the condition $\nabla f(\boldsymbol{x}) = 0$, which is equivalent to finding the optimal point by letting $\lambda = 0$ and then setting $\nabla_{\boldsymbol{x}} L(\boldsymbol{x}, \lambda)$ as zero. In the case of $g(\boldsymbol{x}) = 0$, the optimization works similarly as our previous discussions on equality constraints except that the directions of $\nabla f(\boldsymbol{x}^*)$ and $\nabla g(\boldsymbol{x}^*)$ are always opposite in this case, that is, there exists $\lambda > 0$ such that $\nabla f(\boldsymbol{x}^*) + \lambda \nabla g(\boldsymbol{x}^*) = 0$. Considering both cases, $\lambda g(\boldsymbol{x}) = 0$ is always satisfied. Hence, minimizing $f(\boldsymbol{x})$ subject to $g(\boldsymbol{x}) \leqslant 0$ is equivalent to minimizing the Lagrange function of (B.2) subject to the following constraints:

$$\begin{cases} g(\boldsymbol{x}) \leqslant 0, \\ \lambda \geqslant 0, \\ \lambda g(\boldsymbol{x}) = 0, \end{cases} \tag{B.3}$$

which are known as the Karush–Kuhn–Tucker (KKT) conditions.

The above method can be generalized to optimization problems with multiple constraints. Considering the following optimization problem with a nonempty feasible region $\mathbb{D} \subset \mathbb{R}^d$ and subject to $m$ equality constraints and $n$ inequality constraints:

$$\begin{aligned} &\min_{\boldsymbol{x}} \ f(\boldsymbol{x}) \\ &\text{s.t. } h_i(\boldsymbol{x}) = 0 \ \ (i = 1, \ldots, m), \\ &\qquad g_j(\boldsymbol{x}) \leqslant 0 \ \ (j = 1, \ldots, n). \end{aligned} \tag{B.4}$$

By introducing the Lagrange multipliers $\boldsymbol{\lambda} = (\lambda_1, \lambda_2, \ldots, \lambda_m)^\top$ and $\boldsymbol{\mu} = (\mu_1, \mu_2, \ldots, \mu_n)^\top$, we have the corresponding Lagrange function

$$L(\boldsymbol{x}, \boldsymbol{\lambda}, \boldsymbol{\mu}) = f(\boldsymbol{x}) + \sum_{i=1}^{m} \lambda_i h_i(\boldsymbol{x}) + \sum_{j=1}^{n} \mu_j g_j(\boldsymbol{x}), \tag{B.5}$$

and the KKT conditions ($j = 1, 2, \ldots, n$) introduced by the inequality constraints are

$$\begin{cases} g_j(\boldsymbol{x}) \leqslant 0, \\ \mu_j \geqslant 0, \\ \mu_j g_j(\boldsymbol{x}) = 0. \end{cases} \tag{B.6}$$

We can examine an optimization problem from two perspectives, that is, the *primal problem* and the *dual problem*. From (B.5), the Lagrange *dual function* $\Gamma : \mathbb{R}^m \times \mathbb{R}^n \mapsto \mathbb{R}$ of the primal problem (B.4) is defined as

$$\begin{aligned} \Gamma(\boldsymbol{\lambda}, \boldsymbol{\mu}) &= \inf_{\boldsymbol{x} \in \mathbb{D}} L(\boldsymbol{x}, \boldsymbol{\lambda}, \boldsymbol{\mu}) \\ &= \inf_{\boldsymbol{x} \in \mathbb{D}} \left( f(\boldsymbol{x}) + \sum_{i=1}^{m} \lambda_i h_i(\boldsymbol{x}) + \sum_{j=1}^{n} \mu_j g_j(\boldsymbol{x}) \right). \end{aligned} \tag{B.7}$$

In order to obtain the dual problem, we often set the derivative of the Lagrange function $L(\boldsymbol{x}, \boldsymbol{\lambda}, \boldsymbol{\mu})$ with respect to $\boldsymbol{x}$ as zero.

Let $\tilde{\boldsymbol{x}} \in \mathbb{D}$ denote a point in the feasible region of the primal problem (B.4), then we have

$$\sum_{i=1}^{m} \lambda_i h_i(\tilde{\boldsymbol{x}}) + \sum_{j=1}^{n} \mu_j g_j(\tilde{\boldsymbol{x}}) \leqslant 0 \tag{B.8}$$

for any $\boldsymbol{\mu} \succeq 0$ and $\boldsymbol{\lambda}$. Thereby, we have

$$\Gamma(\boldsymbol{\lambda}, \boldsymbol{\mu}) = \inf_{\boldsymbol{x} \in \mathbb{D}} L(\boldsymbol{x}, \boldsymbol{\lambda}, \boldsymbol{\mu}) \leqslant L(\tilde{\boldsymbol{x}}, \boldsymbol{\lambda}, \boldsymbol{\mu}) \leqslant f(\tilde{\boldsymbol{x}}). \tag{B.9}$$

$\boldsymbol{\mu} \succeq 0$ indicates that all components of $\boldsymbol{\mu}$ are non-negative.

Let $p^*$ denote the optimal value of the primal problem (B.4), then we have

$$\Gamma(\boldsymbol{\lambda}, \boldsymbol{\mu}) \leqslant p^* \tag{B.10}$$

for any $\boldsymbol{\mu} \succeq 0$ and $\boldsymbol{\lambda}$, that is, the dual function can provide a lower bound of the optimal value of the primal problem. Apparently, the lower bound depends on the value of $\boldsymbol{\mu}$ and $\boldsymbol{\lambda}$. Hence, a question naturally arises: what is the optimal lower bound that we can obtain from the dual function? This question leads to the optimization problem

$$\max_{\boldsymbol{\lambda}, \boldsymbol{\mu}} \ \Gamma(\boldsymbol{\lambda}, \boldsymbol{\mu}) \ \text{ s.t. } \ \boldsymbol{\mu} \succeq 0, \tag{B.11}$$

which is the dual problem of the primal problem (B.4), where $\boldsymbol{\mu}$ and $\boldsymbol{\lambda}$ are called *dual variables*. The dual problem (B.11) is always a convex optimization problem regardless of the convexity of the primal problem (B.4).

For the optimal value $d^*$ of (B.11), we have $d^* \leqslant p^*$, which is called *weak duality*. When $d^* = p^*$, we say *strong duality* holds, and in such a case, the dual problem can provide the optimal lower bound of the primal problem. Strong duality usually does not hold in general optimization problems. However, the strong duality holds when the primal problem is a convex optimization problem (e.g., in (B.4), both $f(\boldsymbol{x})$ and $g_j(\boldsymbol{x})$ are convex and $h_i(\boldsymbol{x})$ is affine) and there is at least one feasible point in the feasible region such that the inequality constraints are satisfied with strict inequalities. It is worth mentioning that, when strong duality holds, we can obtain the numerical relationship between the primal variables and the dual variables by finding the derivatives of the Lagrange function with respect to the primal variables and the dual variables, respectively, and then setting both derivatives as zero. Hence, once the dual problem is solved, the primal problem is solved.

This is known as the Slater's condition.

## B.2 Quadratic Programming

Quadratic Programming (QP) is a typical optimization problem that includes both convex and non-convex cases. In a QP problem, the objective function is a quadratic function of the variables and the constraints are linear inequalities of variables.

Suppose there are $d$ variables and $m$ constraints, then the standard QP problem is formulated as

Non-standard QP problems can include equality constraints. An equality constraint can be replaced with two inequality constraints. An inequality constraint can be converted into an equality constraint by introducing slack variables.

$$\begin{aligned} \min_{\boldsymbol{x}} \ & \frac{1}{2}\boldsymbol{x}^\top \mathbf{Q}\boldsymbol{x} + \boldsymbol{c}^\top \boldsymbol{x} \\ \text{s.t.} \ & \mathbf{A}\boldsymbol{x} \leqslant \boldsymbol{b}, \end{aligned} \tag{B.12}$$

where $\boldsymbol{x}$ is a $d$-dimensional vector, $\mathbf{Q} \in \mathbb{R}^{d\times d}$ is a real symmetric matrix, $\mathbf{A} \in \mathbb{R}^{m\times d}$ is a real matrix, $\boldsymbol{b} \in \mathbb{R}^m$ and $\boldsymbol{c} \in \mathbb{R}^d$ are real vectors, and every row in $\mathbf{A}\boldsymbol{x} \leqslant \boldsymbol{b}$ corresponds to a constraint.

Suppose $\mathbf{Q}$ is positive semidefinite, then the objective function (B.12) is convex, and the corresponding QP problem is a convex optimization problem; meanwhile, a global minimum of the problem exists if the feasible region defined by the constraints $\mathbf{A}\boldsymbol{x} \leqslant \boldsymbol{b}$ is nonempty and the objective function has a lower bound in this feasible region. When $\mathbf{Q}$ is positive definite, the problem has a unique global minimum. When $\mathbf{Q}$ is not positive definite, (B.12) is an NP-hard problem with multiple stationary points and local minima.

Common techniques for solving QP problems include the ellipsoid method, the interior-point method, the augmented Lagrangian method, and the gradient projection method. When $\mathbf{Q}$ is positive definite, the corresponding QP problem can be solved by the ellipsoid method in polynomial time.

## B.3 Semidefinite Programming

Semidefinite Programming (SDP) is a class of convex optimization problems in which the variables can be organized into a symmetric positive semidefinite matrix, and both the objective function and the constraints are linear functions of the variables.

Let $\mathbf{X}$ and $\mathbf{C}$ be two $d \times d$ symmetric matrices, and letting

$$\mathbf{C} \cdot \mathbf{X} = \sum_{i=1}^{d} \sum_{j=1}^{d} C_{ij} X_{ij}. \tag{B.13}$$

The SDP problem can be formulated as

$$\begin{aligned} \min_{\mathbf{X}} \ & \mathbf{C} \cdot \mathbf{X} \\ \text{s.t.} \ & \mathbf{A}_i \cdot \mathbf{X} = b_i, \quad i = 1, 2, \ldots, m \\ & \mathbf{X} \succeq 0. \end{aligned} \tag{B.14}$$

where $\mathbf{A}_i (i = 1, 2, \ldots, m)$ are $d \times d$ symmetric matrices, and $b_i (i =, 1, 2, \ldots, m)$ are $m$ real numbers.

$\mathbf{X} \succeq 0$ indicates that $\mathbf{X}$ is positive semidefinite.

Although both SDP and linear programming (LP) involve linear objective functions and constraints, the constraint $\mathbf{X} \succeq 0$ in SDP is non-linear and non-smooth. In optimization theory, SDP unifies several standard optimization problems, such as LP and QP.

The interior-point method, which is commonly used in LP problems, can solve SDP problems after some modifications. However, it is often difficult to solve large-scale SDP problems due to computational complexity.

## B.4 Gradient Descent Method

Gradient descent is a first-order optimization method, and it is one of the most classic and simplest methods for solving unconstrained optimization problems.

First-order optimization methods use only the first-order derivative rather than any higher-order derivatives.

Considering the unconstrained optimization problem $\min_{\boldsymbol{x}} f(\boldsymbol{x})$ for an continuously differentiable function $f(\boldsymbol{x})$. Suppose we can construct a sequence $\boldsymbol{x}^0, \boldsymbol{x}^1, \boldsymbol{x}^2, \ldots$ satisfying

$$f(\boldsymbol{x}^{t+1}) < f(\boldsymbol{x}^t), \quad t = 0, 1, 2, \ldots \tag{B.15}$$

then the objective function converges to a local minimum by repeating (B.15). From Taylor expansion, we have

$$f(\boldsymbol{x} + \Delta\boldsymbol{x}) \simeq f(\boldsymbol{x}) + \Delta\boldsymbol{x}^{\top} \nabla f(\boldsymbol{x}). \tag{B.16}$$

Hence, (B.15) is equivalent to $f(\boldsymbol{x} + \Delta\boldsymbol{x}) < f(\boldsymbol{x})$, and we can choose

$$\Delta\boldsymbol{x} = -\gamma\nabla f(\boldsymbol{x}), \tag{B.17}$$

where the step size $\gamma$ is a small constant. This method is known as the gradient descent method.

It is possible to have different $\gamma_t$ at different steps.

By choosing the step size(s) approximately the objective function $f(\boldsymbol{x})$ is guaranteed to converge to a local minimum if $f(\boldsymbol{x})$ satisfies certain conditions. For example, if $f(\boldsymbol{x})$ satisfies the $L$-Lipschitz condition, then $f(\boldsymbol{x})$ is guaranteed to converge to a local minimum by setting the step size as $1/(2L)$. When the objective function is convex, the local minimum corresponds to the global minimum, and it can be found by the gradient descent method.

The $L$-Lipschitz condition says that, for any $\mathbf{x}$, there exists a constant $L$ such that $\|f(\mathbf{x}) - f(\mathbf{y})\| \leq L\|\mathbf{x} - \mathbf{y}\|$ holds.

When the objective function $f(\boldsymbol{x})$ is twice continuously differentiable, we can replace (B.16) with the more accurate twice Taylor expansion, resulting in Newton's method, which is a typical second-order optimization method. The number of iterations in Newton's method is far less than that of the gradient descent method. However, since Newton's method involves the second-order derivative $\nabla^2 f(\boldsymbol{x})$, the computational complexity of each iteration is high due to computing the inverse of the Hessian matrix (A.21). We can reduce the computational cost significantly by finding an approximate inverse Hessian matrix that is relatively easier to compute, and this is known as the quasi-Newton method.

## B.5 Coordinate Descent Method

Coordinate descent is a gradient-free optimization method. It searches along a coordinate direction in each iteration and finds the local minima by cycling different coordinate directions.

Also known as coordinate ascent when solving maximization problems.

Suppose we have a minimization problem of the function $f(\boldsymbol{x})$, where $\boldsymbol{x} = (x_1, x_2, \ldots, x_d)^\top \in \mathbb{R}^d$ is a $d$-dimensional vector. Starting from the initial point $\boldsymbol{x}^0$, the coordinate descent method solves the problem by iteratively constructing the sequence $\boldsymbol{x}^0, \boldsymbol{x}^1, \boldsymbol{x}^2, \ldots$, where the $i$th component of $\boldsymbol{x}^{t+1}$ is constructed by

$$x_i^{t+1} = \underset{y \in \mathbb{R}}{\arg\min} f(x_1^{t+1}, \ldots, x_{i-1}^{t+1}, y, x_{i+1}^t, \ldots, x_d^t). \tag{B.18}$$

Following the above process, we have

$$f(\boldsymbol{x}^0) \geqslant f(\boldsymbol{x}^1) \geqslant f(\boldsymbol{x}^2) \geqslant \ldots \tag{B.19}$$

Similar to the gradient descent method, the objective function converges to a local minimum by constructing the sequence $\boldsymbol{x}^0, \boldsymbol{x}^1, \boldsymbol{x}^2, \ldots$ via the above process.

The coordinate descent method does not compute the gradient of the objective function and involves only one-dimensional search in each iteration, hence it is applicable to computing heavy problems. However, the coordinate descent method may get stuck at a non-stationary point if the objective function is non-smooth.

# Appendix C
# Probability Distributions

## C.1 Common Probability Distributions

This section briefly discusses several common probability distributions. For each distribution, we provide its probability density (or mass) function together with a few useful statistics, such as expectation $\mathbb{E}[\cdot]$, variance $\mathrm{var}[\cdot]$, and covariance $\mathrm{cov}[\cdot, \cdot]$.

### C.1.1 Uniform Distribution

The uniform distribution is a simple probability distribution over a continuous variable defined in $[a, b](a < b)$. Its probability density function, as shown in Figure C.1, is

Here, we only discuss continuous uniform distribution.

$$p(x \mid a, b) = \mathrm{U}(x \mid a, b) = \frac{1}{b-a}, \tag{C.1}$$

and its expectation and variance are, respectively

$$\mathbb{E}[x] = \frac{a+b}{2}, \tag{C.2}$$

$$\mathrm{var}[x] = \frac{(b-a)^2}{12}. \tag{C.3}$$

We can observe that $a + (b-a)x$ follows the uniform distribution $\mathrm{U}(x \mid a, b)$ if the variable $x$ follows the uniform distribution $\mathrm{U}(x \mid 0, 1)$ and $a < b$.

### C.1.2 Bernoulli Distribution

The Bernoulli distribution is a probability distribution over a binary variable $x \in \{0, 1\}$. The distribution is governed by a continuous parameter $\mu \in [0, 1]$ representing the probability of $x = 1$. Its probability mass function, expectation, and variance are, respectively

The Bernoulli distribution is named after the Swiss mathematician J. Bernoulli (1654–1705).

$$P(x \mid \mu) = \mathrm{Bern}(x \mid \mu) = \mu^x(1-\mu)^{1-x}, \tag{C.4}$$

$$\mathbb{E}[x] = \mu, \tag{C.5}$$

$$\mathrm{var}[x] = \mu(1-\mu). \tag{C.6}$$

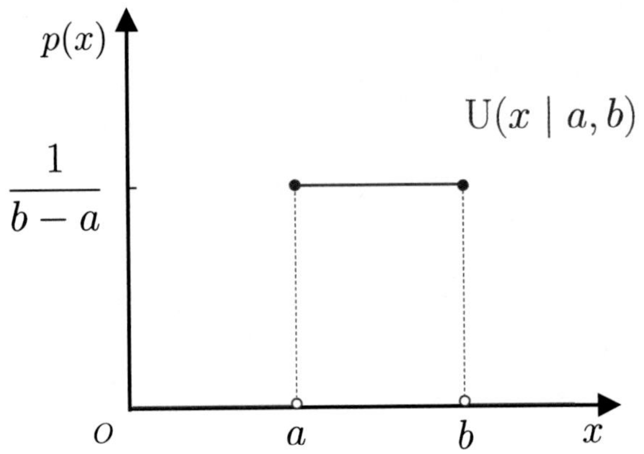


**Fig. C.1** The probability density function of the uniform distribution

### C.1.3 Binomial Distribution

The binomial distribution describes the probability of observing $m$ successes (i.e., $x = 1$) in $N$ independent Bernoulli trials, where the probability of success in each Bernoulli trial is $\mu \in [0, 1]$. Its probability mass function, expectation, and variance are, respectively

$$P(m \mid N, \mu) = \mathrm{Bin}(m \mid N, \mu) = \binom{N}{m} \mu^m (1-\mu)^{N-m}, \tag{C.7}$$

$$\mathbb{E}[x] = N\mu, \tag{C.8}$$

$$\mathrm{var}[x] = N\mu(1-\mu). \tag{C.9}$$

When $N = 1$, a binomial distribution reduces to a Bernoulli distribution.

For $\mu$, the conjugate prior distribution of the binomial distribution is the Beta distribution. See Appendix C.2 for conjugate distributions.

### C.1.4 Multinomial Distribution

We can generalize the binary variable of the Bernoulli distribution to a $d$-dimensional vector $\boldsymbol{x}$, where $x_i \in \{0, 1\}$ and $\sum_{i=1}^{d} x_i = 1$, resulting in the following discrete distribution:

$$P(\boldsymbol{x} \mid \boldsymbol{\mu}) = \prod_{i=1}^{d} \mu_i^{x_i}, \tag{C.10}$$

$$\mathbb{E}[x_i] = \mu_i, \tag{C.11}$$

$$\mathrm{var}[x_i] = \mu_i(1-\mu_i), \tag{C.12}$$

$$\mathrm{cov}[x_j, x_i] = -\mu_i \mu_j, i \neq j, \tag{C.13}$$

where $\mu_i \in [0, 1]$ is the probability of $x_i = 1$ and $\sum_{i=1}^{d} \mu_i = 1$. From the above discrete distribution, we can generalize the binomial distribution to the multinomial distribution that describes the probability of observing $m_i$ occurrences of $x_i = 1$ in $N$ independent trials. For multinomial distribution, the probability mass function, expectation, variance, and covariance are, respectively

For $\boldsymbol{\mu}$, the conjugate prior distribution of the multinomial distribution is the Dirichlet distribution. See Appendix C.2 for conjugate distributions.

$$P(m_1, m_2, \ldots, m_d \mid N, \boldsymbol{\mu}) = \text{Mult}(m_1, m_2, \ldots, m_d \mid N, \boldsymbol{\mu}) = \frac{N!}{m_1!m_2!\ldots m_d!} \prod_{i=1}^{d} \mu_i^{m_i}, \quad \text{(C.14)}$$

$$\mathbb{E}[m_i] = N\mu_i, \quad \text{(C.15)}$$

$$\text{var}[m_i] = N\mu_i(1 - \mu_i), \quad \text{(C.16)}$$

$$\text{cov}[m_j, m_i] = -N\mu_j\mu_i. \quad \text{(C.17)}$$

### C.1.5 Beta Distribution

The Beta distribution is a probability distribution over a continuous variable $\mu \in [0, 1]$ governed by the parameters $a > 0$ and $b > 0$. Its probability density function, as shown in Figure C.2, is

$$p(\mu \mid a, b) = \text{Beta}(\mu \mid a, b) = \frac{\Gamma(a+b)}{\Gamma(a)\Gamma(b)} \mu^{a-1}(1-\mu)^{b-1} = \frac{1}{B(a, b)} \mu^{a-1}(1-\mu)^{b-1}, \quad \text{(C.18)}$$

and its expectation and variance are, respectively

$$\mathbb{E}[\mu] = \frac{a}{a+b}, \quad \text{(C.19)}$$

$$\text{var}[\mu] = \frac{ab}{(a+b)^2(a+b+1)}, \quad \text{(C.20)}$$

where $\Gamma(a)$ is the Gamma function

$$\Gamma(a) = \int_0^{+\infty} t^{a-1} e^{-t} \text{dt}, \quad \text{(C.21)}$$

and $B(a, b)$ is the Beta function

$$B(a, b) = \frac{\Gamma(a)\Gamma(b)}{\Gamma(a+b)}. \quad \text{(C.22)}$$

When $a = b = 1$, the Beta distribution reduces to the uniform distribution.

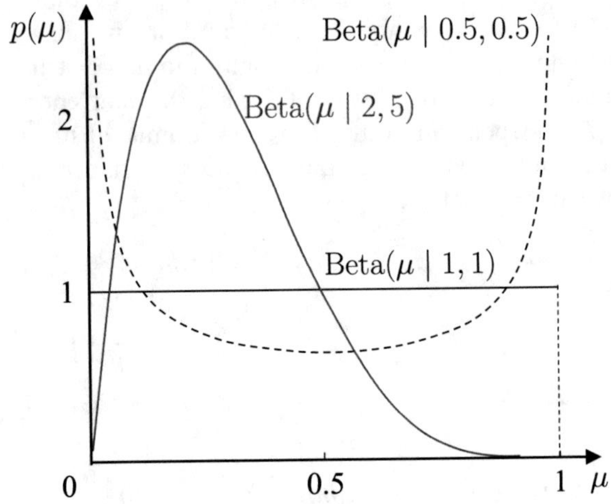


**Fig. C.2** The probability density function of the Beta distribution

### C.1.6 Dirichlet Distribution

The Dirichlet distribution is named after the German mathematician P. Dirichlet (1805–1859).

The Dirichlet distribution is a multivariate probability distribution over a set of $d$ continuous variables $\boldsymbol{\mu} = (\mu_1; \mu_2; \ldots; \mu_d)$, where $\mu_i \in [0, 1]$ and $\sum_{i=1}^{d} \mu_i = 1$. The Dirichlet distribution has the parameters $\boldsymbol{\alpha} = (\alpha_1; \alpha_2; \ldots; \alpha_d)$, where $\alpha_i > 0$ and $\hat{\alpha} = \sum_{i=1}^{d} \alpha_i$. Its probability density function, expectation, variance, and covariance are, respectively

$$p(\boldsymbol{\mu} \mid \boldsymbol{\alpha}) = \mathrm{Dir}(\boldsymbol{\mu} \mid \boldsymbol{\alpha}) = \frac{\Gamma(\hat{\alpha})}{\Gamma(\alpha_1)\ldots\Gamma(\alpha_i)} \prod_{i=1}^{d} \mu_i^{\alpha_i - 1}, \tag{C.23}$$

$$\mathbb{E}[\mu_i] = \frac{\alpha_i}{\hat{\alpha}}, \tag{C.24}$$

$$\mathrm{var}[\mu_i] = \frac{\alpha_i(\hat{\alpha} - \alpha_i)}{\hat{\alpha}^2(\hat{\alpha} + 1)}, \tag{C.25}$$

$$\mathrm{cov}[\mu_j, \mu_i] = \frac{\alpha_j \alpha_i}{\hat{\alpha}^2(\hat{\alpha} + 1)}. \tag{C.26}$$

When $d = 2$, the Dirichlet distribution reduces to the Beta distribution.

### C.1.7 Gaussian Distribution

The Gaussian distribution, also known as the normal distribution, is the most widely used continuous probability distribution.

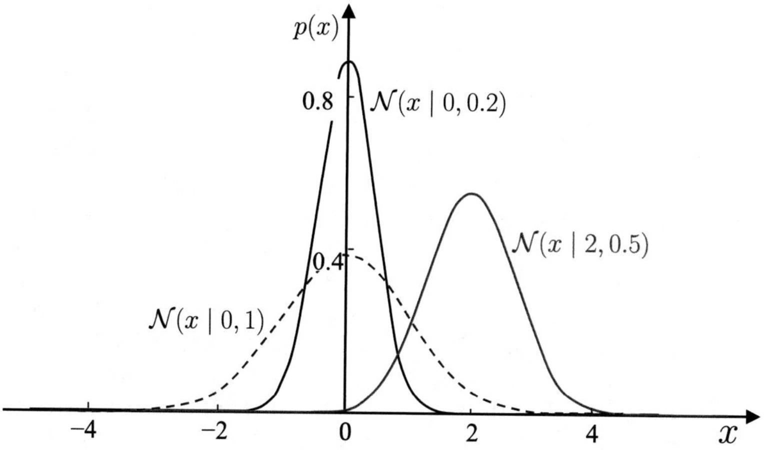


**Fig. C.3** The probability density function of the uniform distribution

For a single variable $x \in (-\infty, \infty)$, the Gaussian distribution is governed by the mean $\mu \in (-\infty, \infty)$ and the variance $\sigma^2 > 0$. Figure C.3 shows the probability density functions of the Gaussian distribution governed by different parameter values. For the univariate Gaussian distribution, its probability density function, expectation, and variance are, respectively

$\sigma$ is the standard deviation.

$$p(x \mid \mu, \sigma^2) = \mathcal{N}(x \mid \mu, \sigma^2) = \frac{1}{\sqrt{2\pi\sigma^2}} \exp\left\{-\frac{(x-\mu)^2}{2\sigma^2}\right\}, \tag{C.27}$$

$$\mathbb{E}[x] = \mu, \tag{C.28}$$

$$\mathrm{var}[x] = \sigma^2. \tag{C.29}$$

In the case of a $d$-dimensional vector $\boldsymbol{x}$, the multivariate Gaussian distribution is governed by a $d$-dimensional mean vector $\boldsymbol{\mu}$ and a $d \times d$ symmetric positive definite covariance matrix $\sigma$. For the multivariate Gaussian distribution, its probability density function, expectation, and covariance are, respectively

$$\begin{aligned} p(\boldsymbol{x} \mid \boldsymbol{\mu}, \boldsymbol{\Sigma}) &= \mathcal{N}(\boldsymbol{x} \mid \boldsymbol{\mu}, \boldsymbol{\Sigma}) \\ &= \frac{1}{\sqrt{(2\pi)^d \det(\boldsymbol{\Sigma})}} \exp\left\{-\frac{1}{2}(\boldsymbol{x}-\boldsymbol{\mu})^\top \boldsymbol{\Sigma}^{-1}(\boldsymbol{x}-\boldsymbol{\mu})\right\}, \end{aligned} \tag{C.30}$$

$$\mathbb{E}[\boldsymbol{x}] = \boldsymbol{\mu}, \tag{C.31}$$

$$\mathrm{cov}[\boldsymbol{x}] = \boldsymbol{\Sigma}. \tag{C.32}$$

## C.2 Conjugate Distribution

When the posterior distribution is in the same probability distribution family as the prior probability distribution, they are called conjugate distributions. Specifically, suppose the variable $x$ follows a distribution $P(x \mid \Theta)$ governed by the parameter $\Theta$. Let $X = \{x_1, x_2, \ldots, x_m\}$ denote the observed samples of the variable $x$, and assuming the parameter $\Theta$ follows a prior distribution $\Pi(\Theta)$. The prior distribution $\Pi(\Theta)$ is said to be the conjugate distribution of the distribution $P(x \mid \Theta)$ or $P(X \mid \Theta)$ if the prior distribution $\Pi(\Theta)$ and the posterior distribution $F(\Theta \mid X)$, which is determined by the prior distribution $\Pi(\Theta)$ and the sample distribution $P(X \mid \Theta)$, belong to the same probability distribution family.

For example, suppose that $x \sim \text{Bern}(x \mid \mu)$, $X = \{x_1, x_2, \ldots, x_m\}$ are the observed samples, $\bar{x}$ is the mean of the observed samples, and $\mu \sim \text{Beta}(\mu \mid a, b)$, where $a$ and $b$ are the known parameters. Then, the posterior distribution

$$\begin{aligned}
F(\mu \mid X) &\propto \text{Beta}(\mu \mid a, b)P(X \mid \mu)\\
&= \frac{\mu^{a-1}(1-\mu)^{b-1}}{B(a, b)}\mu^{m\bar{x}}(1-\mu)^{m-m\bar{x}}\\
&= \frac{1}{B(a+m\bar{x}, b+m-m\bar{x})}\mu^{a+m\bar{x}-1}(1-\mu)^{b+m-m\bar{x}-1}\\
&= \text{Beta}(\mu \mid a', b')
\end{aligned} \tag{C.33}$$

is also a Beta distribution, where $a' = a + m\bar{x}$ and $b' = b+m-m\bar{x}$. This means the Beta distribution and the Bernoulli distribution are conjugate. Similarly, we can find that the conjugate distribution of the multinomial distribution is the Dirichlet distribution, and the conjugate distribution of the Gaussian distribution is itself.

Here, we only consider the cases that the Gaussian distribution has a known variance and its mean follows the prior distribution.

The prior distribution reflects some kind of prior information, whereas the posterior distribution reflects both the information provided by the prior distribution and the information provided by the samples. When the prior distribution and the sample distribution are conjugate, the posterior and prior distributions belong to the same probability distribution family, which means the prior information and the information of samples are the same class of information. Hence, if we use a posterior distribution as the prior distribution for further sampling, then the new posterior distribution is still in the same class. With such a property, the conjugate distribution can simplify many problems. Taking (C.33) as an example, if we apply the Beta distribution to the samples $X$ of the Bernoulli distribution, then the parameters $a$ and $b$ of the Beta distribution can be regarded as an estimate of the ground-truth (success or failure) of the Bernoulli distribution. After receiving some "evidence" (samples), the parameters of the Beta distribution change from $a$ and $b$ to $a + m\bar{x}$ and $b + m - m\bar{x}$ respectively, and $a/(a + b)$ approaches the ground-truth value of the parameter $\mu$ of the Bernoulli distribution as $m$ increases. By using the conjugate prior, the model can be conveniently updated by updating the estimated parameters $a$ and $b$.

## C.3 Kullback–Leibler Divergence

The Kullback–Leibler (KL) divergence, also known as the relative entropy or the information divergence, measures the difference between two probability distributions. Given two probability distributions $P$ and $Q$, their KL divergence is defined as

$$\mathrm{KL}(P \parallel Q) = \int_{-\infty}^{\infty} p(x) \log \frac{p(x)}{q(x)} dx, \tag{C.34}$$

where $p(x)$ and $q(x)$ are the probability density functions of $P$ and $Q$, respectively.

Here, we assume the two distributions are continuous probability distributions. For discrete probability distributions, we only need to change the integration to the summation over the discrete values.

The KL divergence is non-negative, that is

$$\mathrm{KL}(P \parallel Q) \geqslant 0, \tag{C.35}$$

and $\mathrm{KL}(P \parallel Q) = 0$ if and only if $P = Q$. However, the KL divergence is not symmetric, that is

$$\mathrm{KL}(P \parallel Q) \neq \mathrm{KL}(Q \parallel P), \tag{C.36}$$

and hence the KL divergence is not a metric.

By expanding the definition of KL divergence (C.34), we obtain

See Sect. 9.3 for the four basic properties of a valid metric.

$$\begin{aligned}\mathrm{KL}(P \parallel Q) &= \int_{-\infty}^{\infty} p(x) \log p(x) dx - \int_{-\infty}^{\infty} p(x) \log q(x) dx \\ &= -H(P) + H(P, Q),\end{aligned} \tag{C.37}$$

where $H(P)$ is the entropy of $P$ and $H(P, Q)$ is the cross entropy of $P$ and $Q$. In information theory, the entropy $H(P)$ represents the minimum average number of bits needed to encode a variable from the distribution $P$. The cross entropy $H(P, Q)$ represents the minimum average number of bits needed to encode a variable from $P$ if using an encoding scheme based on $Q$. Therefore, the KL divergence can be regarded as the number of the "extra" bits for encoding a variable from $P$ using an encoding scheme based on $Q$. Obviously, the number of the extra bits is non-negative, and it is zero if and only if $P = Q$.

# Index

## D



## E



## F



## G



## H

## I



## J



## K



## L



## M

## N



## O



## P

## Q



## R



## S



## T

## U



## V



## W